Communications in Computer and Information Science 1973

Rationale

The CCIS series is devoted to the publication of proceedings of computer science conferences. Its aim is to efficiently disseminate original research results in informatics in printed and electronic form. While the focus is on publication of peer-reviewed full papers presenting mature work, inclusion of reviewed short papers reporting on work in progress is welcome, too. Besides globally relevant meetings with internationally representative program committees guaranteeing a strict peer-reviewing and paper selection process, conferences run by societies or of high regional or national relevance are also considered for publication.

Topics

The topical scope of CCIS spans the entire spectrum of informatics ranging from foundational topics in the theory of computing to information and communications science and technology and a broad variety of interdisciplinary application fields.

Information for Volume Editors and Authors

Publication in CCIS is free of charge. No royalties are paid, however, we offer registered conference participants temporary free access to the online version of the conference proceedings on SpringerLink (http://link.springer.com) by means of an http referrer from the conference website and/or a number of complimentary printed copies, as specified in the official acceptance email of the event.

CCIS proceedings can be published in time for distribution at conferences or as post-proceedings, and delivered in the form of printed books and/or electronically as USBs and/or e-content licenses for accessing proceedings at SpringerLink. Furthermore, CCIS proceedings are included in the CCIS electronic book series hosted in the SpringerLink digital library at http://link.springer.com/bookseries/7899. Conferences publishing in CCIS are allowed to use Online Conference Service (OCS) for managing the whole proceedings lifecycle (from submission and reviewing to preparing for publication) free of charge.

Publication process

The language of publication is exclusively English. Authors publishing in CCIS have to sign the Springer CCIS copyright transfer form, however, they are free to use their material published in CCIS for substantially changed, more elaborate subsequent publications elsewhere. For the preparation of the camera-ready papers/files, authors have to strictly adhere to the Springer CCIS Authors' Instructions and are strongly encouraged to use the CCIS LaTeX style files or templates.

Abstracting/Indexing

CCIS is abstracted/indexed in DBLP, Google Scholar, EI-Compendex, Mathematical Reviews, SCImago, Scopus. CCIS volumes are also submitted for the inclusion in ISI Proceedings.

How to start

To start the evaluation of your proposal for inclusion in the CCIS series, please send an e-mail to ccis@springer.com.

Sagaya Aurelia · Chandra J. · Ashok Immanuel ·
Joseph Mani · Vijaya Padmanabha
Editors

Computational Sciences and Sustainable Technologies

First International Conference, ICCSST 2023
Bangalore, India, May 8–9, 2023
Revised Selected Papers

Editors
Sagaya Aurelia
CHRIST University
Bangalore, India

Chandra J.
CHRIST University
Bangalore, India

Ashok Immanuel
CHRIST University
Bangalore, India

Joseph Mani
Modern College of Business and Science
Muscat, Oman

Vijaya Padmanabha
Modern College of Business and Science
Muscat, Oman

ISSN 1865-0929 ISSN 1865-0937 (electronic)
Communications in Computer and Information Science
ISBN 978-3-031-50992-6 ISBN 978-3-031-50993-3 (eBook)
https://doi.org/10.1007/978-3-031-50993-3

This Springer imprint is published by the registered company Springer Nature Switzerland AG
The registered company address is: Gewerbestrasse 11, 6330 Cham, Switzerland

Preface

"Research is to see what everybody else has seen, and to think what nobody else has thought" - Albert Szent-Györgyi. What is said is not as important as what we feel. That is why we make an effort to experience conferences in a setting where the still small voice of the spirit can be clearly heard, felt and understood.

The International Conference on Computational Sciences and Sustainable Technologies (ICSSST 2023) was formally inaugurated on 8th May, 2023 by the Vice Chancellor of Christ University, Fr. Dr. Jose C.C., and Mr. Hari Saravanan, IIM-C Bengaluru and Optum Global Solutions, Bengaluru. Chief guest Mr. Saravanan spoke about the modern-day gadgets that use AI as a backend, the knowledge which overburdens the current generation and With the availability of high-end technology, the present generation should focus on finding solutions that have social impact and sustainability.

The conference covered topics such as Artificial Intelligence, Blockchain Technology, Cloud Computing, Cyber Security, Data Science, E-Commerce, Computer Architecture, Image and Video Processing, Pandemic Preparedness and Digital Technology, and Pattern Recognition and Classification.

This book is a compilation of selected papers presented during ICCSST 2023. The primary goal was to bring together academics, researchers, engineers, entrepreneurs, industry experts, and budding research scholars worldwide to exchange and share their research findings and discuss the challenges faced and the solutions adopted.

We are grateful to all the authors who submitted their research work, the reviewers who dedicated their time and made valuable suggestions for the authors to improve their work, the members of the committees and the participants who made this event a dynamic and enriching experience. Without your commitment and dedication, this conference would not have been possible.

The papers in this book span a wide range of ICCSST 2023 themes, reflecting the diversity and depth of the research presented during the conference. We believe that the content in this volume will be of great value to academicians, professionals and researchers.

There are no secrets to success. It is the result of preparation, hard work and learning from failure. The International Conference on Computational Sciences and Sustainable Technologies (ICSSST 2023) has been one such event.

Sagaya Aurelia
Chandra J.
Ashok Immanuel
Joseph Mani
Vijaya Padmanabha

Acknowledgements

You need to be aware of what others are doing,
applaud their efforts,
acknowledge their success,
and encourage them in their pursuits.
When we all help and acknowledge one another, everybody wins

Jim Stovall

On behalf of the International Conference on Computational Sciences and Sustainable Technologies (ICCSST 2023) committee and the entire fraternity of the institution, I, first of all, extend my most sincere thanks to the Almighty God for giving us strength, knowledge, and good health for conducting this international conference successfully.

I am happy to acknowledge Dr. Fr. Jose, our respected Vice-Chancellor, Dr. Joseph Varghese, then Director, Center for Research, beloved registrar Dr. Anil Joseph Pinto, and all the management authorities for giving us permission and logistic support to conduct this international conference in a hybrid mode which made this event a historical benchmark in the history of CHRIST (Deemed to be University) and the Modern College of Business and Science, Sultanate of Oman. I also extend my sincere gratitude to our beloved Dean of Sciences Dr. George Thomas and Associate Dean of Sciences Dr. T.V. Joseph for their kind support throughout the conference.

I offer my wholehearted thanks to Dr. Ashok Immanuel, General Chair and Head, Department of Computer Science, for being the backbone of this entire conference. I will fail in my duty if I don't remember the staff of Springer CCIS, without whom this publication would not have been possible.

I sincerely recognize Mr. Hari Saravanan of IIM-C Bengaluru and Optum Global Solutions, Bengaluru, for delivering a thought-provoking talk during the inaugural ceremony of the conference.

I humbly acknowledge all the eminent keynote speakers: Dr. Pravin Hungund, Chief Technology Officer at MetaZ Digital, Bengaluru, Karnataka, India; Dr. Zahid Akhtarm, State University of New York Polytechnic Institute; and Mr. Bharani P. Jagan Mohan, Manager, Seagen.

Special recognition goes to internal session chairs and external session chairs Dr. Anita, Dr. Kavitha, Dr. Vaidhehi, Dr. Ramesh Chandra, Dr. Senthilnathan, Dr. Deepa V. Jose, Dr. Thirunavukarasu, Dr. Nizar Banu, Dr. Rohini, Dr. Sandeep, Dr. Poornima, Dr. Suresh, Dr. Nismon Rio, Dr. Venkata Subramanian, Dr. Joseph Mani, MCBS, Muscat, Dr. Vijaya Padmanaban, MCBS, Muscat, Dr. Basant Kumar, MCBS, Muscat, Dr. A. Jayanthiladevi, Srinivas University, Ms. Mritula C., Siemens R & D, Dr. Savitha Choudary, Sir MVIT, Dr. J. Thilagavathi, Arjun College of Technology, Dr. Sriram Krish Vasudevan, Intel, and Dr. Prasanna M., VIT.

I acknowledge Dr. Tomy K. Kallarakal, Dean, School of Commerce, Finance and Accountancy, for accepting our invitation and gracing the occasion with his enlightened speech for the valedictory function.

I also acknowledge all the external and internal reviewers for providing their helpful comments and improving the quality of papers presented in the forum.

My sincere gratitude goes to all IT support staff members, student volunteers and non-teaching staff for their support throughout the conference; they worked day and night to make this event a grand success. I thank all the advisory committee members, coordinators, committee heads and members, faculties of Computer Science, CHRIST University, India, and Modern College of Business and Science, Sultanate of Oman. Their names are listed below.

I acknowledge all the authors who submitted papers and congratulate those whose papers were accepted with a deep sense of appreciation.

Sagaya Aurelia
Chandra J.
Ashok Immanuel
Department of Computer Science
CHRIST (Deemed to be University)
Bangalore, Karnataka India

Organization

Organizing Committee

Abraham V. M.	CHRIST (Deemed to be University), India
Ahmed Al Naamany	Modern College of Business and Science, Oman
Joseph C. C.	CHRIST (Deemed to be University), India
Anil Joseph Pinto	CHRIST (Deemed to be University), India
George Thomas C.	CHRIST (Deemed to be University), India
Joseph T. V.	CHRIST (Deemed to be University), India
Moosa Al Kindi	Modern College of Business and Science, Oman
Khalfan Al Asmi	Modern College of Business and Science, Oman
Hothefa Shaker	Modern College of Business and Science, Oman
Ashok Immanuel	CHRIST (Deemed to be University), India
Vinay	CHRIST (Deemed to be University), India

Conference Chairs

Chandra J.	CHRIST (Deemed to be University), India
Mani Joseph	Modern College of Business and Science, Oman

Conveners

Sagaya Aurelia	CHRIST (Deemed to be University), India
Arokia Paul Rajan R.	CHRIST (Deemed to be University), India
Basant Kumar	Modern College of Business and Science, Oman

Organizing Secretaries

Kirubanand V. B.	CHRIST (Deemed to be University), India
Vijaya Padmanabha	Modern College of Business and Science, Oman

Advisory/Review Committee

Ajay K. Sharma	IKG Punjab Technical University, India
Arockiasamy Soosaimanickam	University of Nizwa, Oman
S. Karthikeyan	College of Applied Sciences, Oman
Rajkumar Buyya	University of Melbourne, Australia
Inder Vir Malhan	Central University of Himachal Pradesh, India
Jose Orlando Gomes	Federal University, Brazil
Richmond Adebiaye	University of South Carolina Upstate, USA
Ahmad Sobri Bin Hashim	University Teknologi Petronas, Malaysia
John Digby Haynes	University of Sydney Business School, Australia
Subhash Chandra Yadav	Central University of Jharkhand, India
Rajeev Srivastava	Indian Institute of Technology, Varanasi, India
Inder Vir Malhan	Central University of Himachal Pradesh, India
Abhijit Das	Indian Institute of Technology Kharagpur, India
Muralidhara B. L	Jnanabharathi Campus, Bangalore University, India
K. K. Shukla	Indian Institute of Technology (BHU), Varanasi, India
Pabitra Mitra	Indian Institute of Technology Kharagpur, India
Hanumanthappa M.	Jnanabharathi Campus, Bangalore University, India
Subhrabrata Choudhury	National Institute of Technology Durgapur, India
P. Santhi Thilagam	National Institute of Technology Karnataka, India
Tandra Pal	National Institute of Technology Durgapur, India
Dilip Kumar Yadav	National Institute of Technology Jamshedpur, India
P. Santhi Thilagam	National Institute of Technology Karnataka, India
Annappa	National Institute of Technology, Karnataka, India
R. Thamaraiselvi	Bishop Heber College, India
P. Mukilan	Bule Hora University, Ethiopia
Gnanaprakasam	Gayatri Vidya Parishad College of Engineering (Autonomous), India
Tanmay De	National Institute of Technology Duragpur, India
Saravanan Chandran	National Institute of Technology Duragpur, India
Rupa G. Mehta	Sardar Vallabhbhai National Institute of Technology, Surat, India
Bibhudatta Sahoo	National Institute of Technology Rourkela, India
Baisakhi Chakraborty	National Institute of Technology Durgapur, India
Rajiv Misra	Indian Institute of Technology, Patna, India
Sherimon P.	Arab Open University, Oman
Nebojsa Bacanin	Singidunum University, Serbia

Website, Brochures, Banners and Posters

Tirunavukkarasu (Chair)	CHRIST University, India
Cecil Donald	CHRIST University, India
Santhosh Nair	Modern College of Business and Science, Oman

Communication and Publicity

Rohini (Chair)	CHRIST University, India
Kavitha	CHRIST University, India
Gobi	CHRIST University, India
Lokanayaki	CHRIST University, India
Poornima	CHRIST University, India
Chia Zargeh	Modern College of Business and Science, Oman
Rajeev Rajendran	Modern College of Business and Science, Oman
Muhammad Naeem	Modern College of Business and Science, Oman
Nassor Suleiman	Modern College of Business and Science, Oman

Finance Committee and Sponsorship

Rajesh Kanna (Chair)	CHRIST University, India
Shoney Sebastian	CHRIST University, India
Deepa V. Jose	CHRIST University, India
Sandeep	CHRIST University, India
Nismon Rio	CHRIST University, India
Joseph Ajin	Modern College of Business and Science, Oman

Registration, Mementos, Conference Kits and Certificates

Kavitha (Chair)	CHRIST University, India
Vaidhehi	CHRIST University, India
K. N. Saravanan	CHRIST University, India
Saravanakumar	CHRIST University, India
Ramamurthy	CHRIST University, India
Loganayaki	CHRIST University, India
Reshmi	CHRIST University, India

Session Management, Execution and Technical Support

Senthilnathan (Chair)	CHRIST University, India
Shoney Sebastian	CHRIST University, India
Beulah	CHRIST University, India
Nismon	CHRIST University, India
Gobi	CHRIST University, India
Mohanapriya	CHRIST University, India
Rajesh Kanna	CHRIST University, India
Poornima	CHRIST University, India
Suresh	CHRIST University, India
Somnath	CHRIST University, India

Publication

Nizar Banu (Chair)	CHRIST University, India
Said Nassor (Co-chair)	Modern College of Business and Science, Oman
Anita	CHRIST University, India
Cecil Donald	CHRIST University, India
Deepa V. Jose	CHRIST University, India
Prabu P.	CHRIST University, India
Poornima	CHRIST University, India
Ramamurthy	CHRIST University, India
Ramesh Poonima	CHRIST University, India
Tirunavukkarasu	CHRIST University, India
Manjunatha	CHRIST University, India
Nismon Rio	CHRIST University, India
Mohanapriya	CHRIST University, India
Somnath	CHRIST University, India
Abdelkawy A. Abdelaziz	Modern College of Business and Science, Oman
Ahmed Mounir	Modern College of Business and Science, Oman
Maryam Al Washahi	Modern College of Business and Science, Oman

Chief Guests, Keynote and Session Resource Persons

Manjunatha (Chair)	CHRIST University, India
Anita	CHRIST University, India
Nizar Banu	CHRIST University, India
Ramesh Poonima	CHRIST University, India
Rohini	CHRIST University, India

Hospitality and Report

Smitha Vinod (Chair)	CHRIST University, India
Peter Augustine	CHRIST University, India
Ruchi	CHRIST University, India

Lunch, Refreshments and Accommodation

Sandeep (Chair)	CHRIST University, India
Saravanakumar	CHRIST University, India
K. N. Saravanan	CHRIST University, India
Beaulah	CHRIST University, India
Loganayaki	CHRIST University, India
Peter Augistine	CHRIST University, India
Prabu P.	CHRIST University, India
Suresh	CHRIST University, India

Event Management

Vaishnavi	CHRIST University, India
Ruchi	CHRIST University, India

Contents

Performance Evaluation of Metaheuristics-Tuned Deep Neural Networks for HealthCare 4.0

Luka Jovanovic, Sanja Golubovic, Nebojsa Bacanin(✉), Goran Kunjadic, Milos Antonijevic, and Miodrag Zivkovic

Singidunum University, Danijelova 32, 11000 Belgrade, Serbia
{luka.jovanovic.191,sanja.golubovic.17}@singimail.rs,
nbacanin@signidunum.ac.rs,
{gkunjadic,mantonijevic,mzivkovic}@singidunum.ac.rs

Abstract. The emergence of novel technologies that power advanced networking, coupled with decreasing sizes and lower power demands of chips has given birth to the internet of things. This emerging technology has resulted in a revolution across many fields. A notably interesting application is healthcare where this combination has resulted in Healthcare 4.0. This has enabled better patient monitoring and resulted in more acquired patient data. Novel techniques are needed, capable of evaluating the gathered information and potentially aiding doctors in providing better outcomes. Artificial intelligence provides a promising solution. Methods such as deep neural networks (DDNs) have been used to address similarly difficult tasks with favorable results. However, like many modern algorithms DNNs present a set of control values that require tuning to ensure proper functioning. A popular approach for selecting optimal values is the use of metaheuristic algorithms. This work proposes a novel metaheuristic based on the sine cosine algorithm, that builds on the excellent performance of the original. The introduced approach is then tasked with tuning hyperparameter values of a DNN handling medical diagnostics. This novel approach has been compared to several state-of-the-art algorithms and attained excellent performance applied to three datasets consisting of real-world medical data.

Keywords: Optimization · Sine Cosine Algorithm · Healthcare 4.0 · Artificial Neural Network · Hyperparameter Tuning

1 Introduction

Health 4.0, it alludes to several opportunities for utilizing Industry 4.0 technologies to enhance healthcare, as it puts forth a fresh and cutting-edge perspective for the industry. The goal is to improve the efficacy and efficiency of the healthcare sector while also offering people better, more valuable, and more affordable healthcare services. One of the industries where the 4.0 revolution is expected to produce outstanding outcomes in healthcare.

S. Aurelia et al. (Eds.): ICCSST 2023, CCIS 1973, pp. 1–14, 2024.
https://doi.org/10.1007/978-3-031-50993-3_1

Artificial intelligence (AI) is used in the process of machine learning to give computers the capacity to learn automatically and get better with practice. As a result, the foundation of machine learning is usually the creation of software that can evaluate data and draw conclusions from it. Artificial neural networks (ANNs), referred to as deep learning, are a branch of machine learning that focuses on algorithms that are motivated by the operation and structure of the human brain. A number of interconnected processing nodes (neurons) make up ANNs, which convert a set of inputs into a set of outputs.

Neural networks are faced with two major obstacles. Finding the right network structure is one, and network training is another. The neural networks are commonly trained using typical optimizers. According to a review of the literature, optimizers can be replaced by metaheuristics in some situations. Finding the best network topology for the given job, sometimes referred to as the hyperparameter optimization process, is the second problem. Both issues are thought of as naturally NP-hard problems.

In this work, an improved version of the popular and extensively used sine cosine algorithm (SCA) swarm intelligence metaheuristics is developed in order to overcome the shortcomings of the original method. The introduced techniques are then applied to selecting optimal control parameters including the architecture of a DNN tackling real-world medical data, including liver disorders, hepatitis, and dermatology data.

The rest of this paper is organized as follows: Sect. 2 presents the background and related work, Sect. 3 provides an overview of the introduced algorithm Sect. 4 show experiments and comparative analysis and Sect. 6 concludes this work.

2 Background and Related Works

Artificial neural networks may be used to resolve challenging issues from a variety of domains. When tackling supervised or unsupervised machine learning problems, ANNs can produce respectable outcomes. According to [2], some of these jobs include machine perception issues, where it is not possible to individually understand the set of accessible primary features. As a result, ANNs have been heavily used to implement pattern recognition, classification, clustering, and predicting problems. For instance, many ANN types have been used in the medical field for diagnosis [2]. Healthcare 4.0 remains a promising field however, many challenges persist [4].

Depending on the learning approach that was used for network training, any ANN's capabilities can be significantly increased. There are two primary methods among supervised training techniques: gradient-based and stochastic methods [11]. The gradient-descent strategy that is now most often used is back-propagation. The increased complexity of DNN over traditional approaches allows for the handling of increasingly complex tasks. However, to maintain flexibility, networks present a set of parameters that require appropriate tuning to ensure proper functionality. This process is known as hyperparameter tuning. By formulating this task as an optimization problem, metaheuristic algorithms can be applied to improve performance.

Metaheuristic algorithms have, in recent years, exploded in popularity for tackling complex optimizations. A notable subgroup that models collaborative groups observed in nature is swarm intelligence algorithms that excel when tackling optimizations. This class of algorithms is even capable of addressing NP-hard problems, something considered impossible to do in practice using traditional techniques. Some notable algorithm examples that have emerged in recent years include the bat algorithm (BA) [17], moth algorithm (MA) [15] Harris hawks optimizer (HHO) [3], as well as the particle swarm optimizer (PSO) [14]. These algorithms have seen success when applied to problems in security [5,12], finance [6,7] as well as medicine [8,13].

3 Methods

3.1 Original Sine Cosine Algorithm (SCA)

Despite the variations in stochastic population-based optimization techniques, both stages of the optimization process-exploration and exploitation-are equally important. In the first stage, an optimization algorithm suddenly merges the random solutions in the set of solutions with a high rate of unpredictability to identify the search space's most promising areas. However, there are progressive modifications in the random solutions throughout the exploitation phase, and the random fluctuations are far smaller than they were during the exploration phase.

The position update equations shown below are suggested in this study for both phases:

$$X_i^{t+1} = X_i^t + r1 \times \sin(r2) \times |r3P_i^t - X_i^t| \tag{1}$$

$$X_i^{t+1} = X_i^t + r1 \times \cos(r2) \times |r3P_i^t - X_{i\,i}^t| \tag{2}$$

where X_i^t is the position of the current solution in *i-th* dimension at *t-th* iteration, *r1*/*r2*/*r3* are random values, P_i is the position of the destination point in *i-th* dimension, and || indicates the absolute value.

The following is how these two equations are combined:

$$X_i^{t+1} = \begin{cases} X_i^t + r1 \times \sin(r2) \times |r3P_i^t - X_i^t|, r4 < 0.5 \\ X_i^t + r1 \times \cos(r2) \times |r3P_i^t - X_{i\,i}^t|, r4 \geq 0.5 \end{cases} \tag{3}$$

where $r4$ is a random number in [0, 1].

The four primary SCA parameters are *r1*, *r2*, *r3*, and $r4$, as shown by the equations above. The movement direction is determined by parameter $r1$. How far the movement should be is determined by the parameter $r2$. In order to stochastically emphasize ($r3 > 1$) or deemphasize ($r3 < 1$) the influence of desalination in defining the distance, the parameter $r3$ generates random weights for the destination. Parameter $r4$ alternates evenly between the sine and cosine parts.

3.2 SCA Bat Search Algorithm (SCA-BS)

The original SCA [9] demonstrates amazing performance when tackling optimization. While the SCA strikes a good balance between exploration and exploitation due to the use of both the sine and cosine functions. However, as these functions cannot fully cover the entire search space, there is a high potential for further improvement.

To overcome this, this work employs low-level metaheuristic hybridization. As the BA [17], which models the hunting behaviors of bats, is known to strike a good balance between exploration. In addition, the BA hosts a powerful exploitation mechanism. During the BA search, the location of each agent denoted as X, with a given frequency F_i and velocity V_i is updated according to the following:

$$V_i(t+1) = V_i(t) + (X_i(t) - Gbest) \times F_i \tag{4}$$

$$X_i(t+1) = X_i(t) + V_i(t+1) \tag{5}$$

where $Gest$ represents the best solution. Agent frequencies are updated in every iteration as per:

$$F_i = F_{min} + (F_{max} - F_{min}) \times \beta \tag{6}$$

where β represents an arbitrary value form uniform distribution $[0,1]$. Additionally, a random walk is applied to improve exploitation as described in the following:

$$x_{new} = x_{old} + \epsilon A_t \tag{7}$$

in which ϵ denotes an arbitrary value in $[-1,1]$ range, and A represents the loudness of the sound emitted by the given agent. In each iteration, the amplitude and pulse emission is updated as per:

$$A_i(t+1) = \alpha A_i(t) \tag{8}$$

$$r_i(t+1) = r_i(0)(1 - e^{-\gamma \times t}) \tag{9}$$

where α and γ are constant values in range $[0,1]$ applied to loudness rate A_i as well as pulse rate (r_i).

The exceptionally powerful search mechanism of the BA [17] algorithm hybridized favorably with the original SCA algorithm resulting in better overall performance. However, to allow both algorithms to contribute to the optimization an additional parameter α is introduced with the accompanying DC operator to each individual agent. The DC operator is selected randomly from a uniform distribution in the range $[0,1]$, and its role is to determine if the original SCA search mechanism is used, or the introduced BA mechanism. The parameter α determines the threshold for the DC value and is empirically determined as $alpha = 0.5$. Should the value of $DC \leq \alpha$ the BA search will be used, otherwise the original SCA search is used for that solution. This mechanism has the added benefit of improving population diversity.

The introduced algorithm is dubbed the SCA bat search (SCA-BS) algorithm. The pseudocode for which is shown in Algorithm 1.

Algorithm 1. SCA-BS algorithm pseudocode

Initialization. Chaotically initialized solutions $X_i(1, 2, 3, \cdots, n)$.
Initialize the maximum number of iterations T.
Set α value to 0.5
while $t < T$ **do**
  for Every X in the generated candidates **do**
    Generate random value for DC
    if $DC \geq \alpha$ **then**
      Perfrom SCA search for this solution as outlined in Eq. 2
    else
      Perfrom BA search for this solution as outlined in Eq. 5
    end if
  end for
end while
return P as the optimum solution.

4 Experiments and Comparative Analysis

This work utilized the same datasets as the referenced work, with the exception of it using various ML models. Since this work uses a neural network, in addition to SCA-BA, other various metaheuristics were included to address the same issues. Four datasets from the UCI machine learning repository [1] were used during testing including liver disorder, dermatology, and hepatitis.

4.1 Liver Disorder Dataset Details

The first five factors in the liver disorder dataset [1] are all blood tests, and they are all believed to be sensitive to liver diseases that may result from heavy alcohol intake. Each line in the collection represents a single male person's record.

The available data has been processed and after removing all samples with missing values, the dataset consists of 414 samples in total. A total of 165 patients in the dataset are considered healthy, and 249 are affected by a liver disorder. The dataset is therefore moderately disbalanced.

4.2 Dermatology Dataset Details

The dermatology dataset [1] presents a serious challenge with the differential diagnosis of erythematous-squamous disorders. Typically, a biopsy is required for the diagnosis, but sadly, these illnesses also share a lot of histological characteristics. Another challenge in making a differential diagnosis is that a disease may first resemble another disease while later developing its own distinctive symptoms.

The family history feature in the dataset created for this domain has a value of 1 if any of these diseases have been seen in the family, and 0 otherwise. The age feature only displays the patient's age. Each additional characteristic (clinical and histopathological) received a grade between 0 and 3. Here, 0 denotes the absence of the characteristic, 3 denotes the maximum quantity that might be used, and 1, and 2 denote the relative middle values. It is important to note that several samples are missing in the dermatology dataset. This is typical for medical datasets. However, there are still more than enough samples to train and test the proposed approach.

4.3 Hepatitis Dataset Details

The hepatitis dataset [1] covers the outcomes of 155 infected patients. The predictive features cover 19 clinical and descriptive attributes, describing a total of 155 individuals. The dataset consists of two classes, the first class representing patients with a terminal outcome, while the second represents survivors. The data is, however, imbalanced with 123 samples representing survivors, and 32 terminal cases.

The data consists of qualitative features, coupled with several discreet and continuous quantitative features. Furthermore, some features have several missing values. This data structure, of relatively small sample size, imbalanced structure, and data incompleteness are common when tackling medical classification problems.

4.4 Experimental Setup

The introduced metaheuristic has been tasked with selecting the optimal control parameters for a DNN for each specific problem dataset. The optimization potential of the introduced algorithm has been compared to several metaheuristics tackling the same problem including the original SCA [9] as well as several other algorithms such as the BA [17], GA [10], PSO [14], HHO [3], MS [15].

In each experiment the available data has been divided into a training dataset, consisting of 70% of the available data, validation made up of the following 10%, and finally testing dataset made up of the final 20% of the accessible data.

During testing on the hepatitis and dermatology datasets, each metaheuristic has been assigned a population of five individuals and given 5 iterations to improve. The experiment has also been repeated through a total of 5 independent runs to compensate for the inherent randomness of metaheuristic algorithms. The hyperparameters chosen for optimization include the learning rate from range $[0.0001, 0.005]$, dropout rate $[0.01, 0.5]$, layer count $[1, 2]$, and training epochs count $[100, 300]$. Finally, the number of neurons has also been optimized depending on the number of features available in the dataset, with the maximum number of neurons per layer being the twice number of features, while the minimum is half the number of available features.

Due to increased problem complexity, the liver-distorted dataset has been optimized with a larger population of 8 agents and given a total of 10 iterations for optimizations. The experiments are likewise repeated in 5 executions to account for randomness. Accordingly, the maximum number of layers a DNN could have has been increased to three.

All metaheuristics, as well as the DNN models have been independently implemented for the purposes of this research. The Python programming language has been used alongside supporting libraries. These include: TensorFlow, Sklearn, Pandas and Numpy.

4.5 Evaluation Metrics

To best assess the performance of each optimized model, several evaluation metrics were utilized. These metrics include:
Accuracy: The classifier's total performance is measured and rated as:

$$Accuracy = \left(\frac{TP + TN}{TP + FP + TN + FN}\right) \times 100 \tag{10}$$

Sensitivity is defined as the ratio of true positive cases to all cases who have been exposed to the illness. Precision is another name for sensitivity. The sensitivity is assessed to be:

$$SensitivityPrecision = \left(\frac{TP}{TP + FN}\right) \times 100 \tag{11}$$

Specificity is the proportion of true negative cases to all cases with the condition. Recall is another name for it. The specificity is assessed as:

$$SensitivityPrecision = \left(\frac{TN}{TN + FP}\right) \times 100 \tag{12}$$

The healthcare model's true positive and true negative predictions are denoted by TP and TN, respectively. The healthcare model's false positive and false negative predictions are denoted by FP and FN, respectively.

Finally, due to the imbalance observed in each of the used datasets, an additional metric capable of better evaluating performance under such conditions is introduced. The Cohen's kappa coefficient κ is considered a more fair metric than accuracy when evaluating imbalanced data. Cohen's kappa coefficient can be determined as:

$$\kappa = \frac{p_o - p_e}{1 - p_e} = 1 - \frac{1 - p_o}{1 - p_e} \tag{13}$$

where p_o represents the observed, and p_e the expected values.

With several available metrics, one needed to be selected as an objective function to guide the optimization process. The selected metric for all three problems is the error rate metric.

5 Results and Discussion

The following section presents the results attained during testing for each utilized dataset. The results are presented in both table and graphic forms and discussed in detail.

5.1 Liver Disorder Dataset Results

The liver dataset presented the hardest challenge. It required increased computational resources as well as a larger potential network. Additionally, a larger population for optimization metaheuristics was required. Overall results for Cohen's kappa score for each class are demonstrated in Table 1.

Table 1. Liver disorder overall Cohen's kappa results over 5 independent runs for each class

Method	Best	Worst	Mean	Median	Std	Var	Best1	Worst1	Mean1	Median1	Std1	Var1
DNN-SCA-BS	0.278846	0.298077	**0.282692**	0.288462	0.004711	0.000022	0.423547	**0.390625**	**0.403787**	**0.401587**	**0.010754**	**0.000116**
DNN-SCA	0.288462	0.307692	0.298077	0.298077	0.006081	0.000037	0.398148	0.341772	0.378222	0.383792	0.019649	0.000386
DNN-BA	0.278846	0.326923	0.300000	0.298077	0.016543	0.000274	0.423547	0.273026	0.351357	0.347896	0.049894	0.002489
DNN-GA	0.288462	0.317308	0.301923	0.298077	0.009806	0.000096	0.398148	0.331776	0.359979	0.347896	0.023908	0.000572
DNN-PSO	0.288462	0.298077	0.290385	0.288462	**0.003846**	**0.000015**	0.378981	0.380000	0.391294	0.382911	0.013173	0.000174
DNN-HHO	0.278846	**0.288462**	0.284615	0.288462	0.004711	0.000022	0.401587	0.366883	0.395411	0.394410	0.018262	0.000334
DNN-MS	0.278846	0.317308	0.301923	0.307692	0.013043	0.000170	0.423547	0.344037	0.365570	0.350000	0.029851	0.000891

As demonstrated in Table 1 all tested metaheuristics performed favorably. Indicating that metaheuristic-optimized DNN demonstrates excellent potential for detecting liver disorders. However, it is also important to note that the novel-introduced metaheuristic demonstrated the best performance on average when classifying negative outcomes as well as significantly outperforming all other metaheuristics across all metrics on positive classifications demonstrating increased stability as well.

To further emphasize the comparison, detailed metrics for each metaheuristic best run are demonstrated in Table 2

Table 2. Liver disorder detailed metrics

	DNN-SCA-BS	DNN-SCA	DNN-BA	DNN-GA	DNN-PSO	DNN-HHO	DNN-MS
Accuracy (%)	72.1154	71.1538	72.1154	71.1538	71.1538	72.1154	72.1154
Precision 0	0.741935	0.684211	0.682927	0.684211	0.750000	**0.758621**	0.682927
Precision 1	0.712329	0.727273	0.746032	0.727273	0.697368	0.706667	0.746032
W.Avg. Precision	0.724855	0.709054	0.719334	0.709054	0.719636	**0.728647**	0.719334
Recall 0	0.522727	0.590909	0.636364	0.590909	0.477273	0.500000	0.636364
Recall 1	0.866667	0.800000	0.783333	0.800000	0.883333	0.883333	0.783333
W.Avg. Recall	0.721154	0.711538	0.721154	0.711538	0.711538	0.721154	0.721154
F1 score 0	0.613333	0.634146	0.658824	0.634146	0.583333	0.602740	0.658824
F1 score 1	0.781955	0.761905	0.764228	0.761905	0.779412	**0.785185**	0.764228
W.Avg. F1 score	0.710615	0.707853	0.719634	0.707853	0.696456	0.707997	0.719634

Detailed metrics shown in Table 2 solidify that metaheuristic-optimized DNNs have a high potential for tackling liver disorder detection. While all metaheuristics performed well, the introduced approach outperformed the original SCA algorithm, signifying an improvement. The proposed approach was only slightly outperformed by the HHO. This is to be expected as per the no free lunch theorem (NFL) [16] no approach works best for all applications.

The parameters selected by each respective metaheuristic in the best overall runs are shown in Table 3.

Table 3. Liver disorder optimal parameters

Method	Learning Rate	Dropout	Layers	Neurons	Epochs
DNN-SCA-BS	0.005000	0.500000	3	12	100
DNN-SCA	0.005000	0.238034	3	12	100
DNN-BA	0.005000	0.500000	3	12	100
DNN-GA	0.004972	0.042064	3	10	100
DNN-PSO	0.002045	2.589258	3	12	100
DNN-HHO	0.005000	0.326793	3	11	100
DNN-MS	0.005000	0.182041	3	11	100

Finally, the improvements made can be seen in the convergence rate increases and diversity plot improvements shown in Fig. 1.

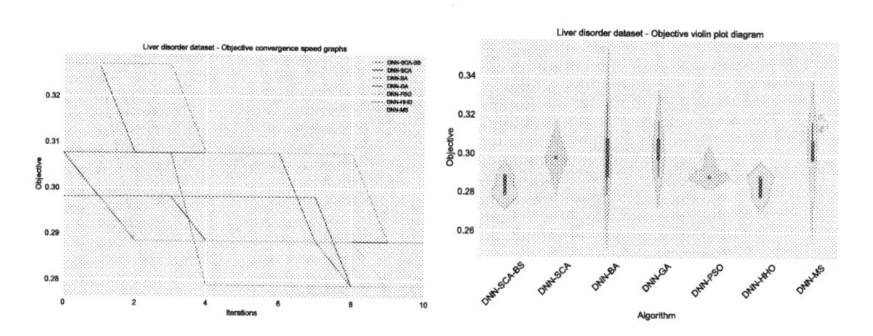

Fig. 1. Liver disorder dataset metaheuristics objective function convergence and distribution violin plots

5.2 Dermatology Dataset Results

The dermatology dataset presents a challenging multiclass classification problem. Overall Cohen's kappa scores for each evaluated metaheuristic are shown in Table 4.

Table 4. Dermatology dataset overall Cohen's kappa scores for each metaheuristic

Method	Best	Worst	Mean	Median	Std	Var	Best1	Worst1	Mean1	Median1	Std1	Var1
DNN-SCA-BS	0.000000	**0.000000**	**0.000000**	**0.000000**	0.000000	0.000000	1.000000	**1.000000**	**1.000000**	1.000000	0.000000	0.000000
DNN-SCA	0.000000	0.009259	0.005556	0.009259	0.005072	0.000021	1.000000	0.988365	0.995346	1.000000	0.005700	0.000033
DNN-BA	0.009259	0.009259	0.009259	0.009259	0.000000	0.000000	0.988365	0.988365	0.988365	0.988365	0.000000	0.000000
DNN-GA	0.000000	0.009259	0.007407	0.009259	0.003704	0.000014	1.000000	0.988365	0.990692	0.988365	0.004654	0.000022
DNN-PSO	0.009259	0.009259	0.009259	0.009259	0.000000	0.000000	0.988365	0.988365	0.988365	0.988365	0.000000	0.000000
DNN-HHO	0.009259	0.009259	0.009259	0.009259	0.000000	0.000000	0.988365	0.988365	0.988367	0.988365	0.000004	0.000000
DNN-MS	0.000000	0.009259	0.007407	0.009259	0.003704	0.000014	1.000000	0.988365	0.990694	0.988365	0.004653	0.000022

As demonstrated in Table 4, while all metaheuristics attained favorable results, the introduced SCA-BS metaheuristic outperformed all competing metaheuristics. These results are further emphasized in Table 5.

Table 5. Dermatology dataset detailed metrics of every metaheuristic for each class

	DNN-SCA-BS	DNN-SCA	DNN-BA	DNN-GA	DNN-PSO	DNN-HHO	DNN-MS
Accuracy (%)	100	100	99.0741	100	99.0741	99.0741	100
Precision 0	1.000000	1.000000	1.000000	1.000000	1.000000	1.000000	1.000000
Precision 1	1.000000	1.000000	0.947368	1.000000	0.947368	0.947368	1.000000
Precision 2	1.000000	1.000000	1.000000	1.000000	1.000000	1.000000	1.000000
Precision 3	1.000000	1.000000	1.000000	1.000000	1.000000	1.000000	1.000000
Precision 4	1.000000	1.000000	1.000000	1.000000	1.000000	1.000000	1.000000
Precision 5	1.000000	1.000000	1.000000	1.000000	1.000000	1.000000	1.000000
W.Avg. Precision	1.000000	1.000000	0.991228	1.000000	0.991228	0.991228	1.000000
Recall 0	1.000000	1.000000	1.000000	1.000000	1.000000	1.000000	1.000000
Recall 1	1.000000	1.000000	1.000000	1.000000	1.000000	1.000000	1.000000
Recall 2	1.000000	1.000000	1.000000	1.000000	1.000000	1.000000	1.000000
Recall 3	1.000000	1.000000	0.928571	1.000000	0.928571	0.928571	1.000000
Recall 4	1.000000	1.000000	1.000000	1.000000	1.000000	1.000000	1.000000
Recall 5	1.000000	1.000000	1.000000	1.000000	1.000000	1.000000	1.000000
W.Avg. Recall	1.000000	1.000000	0.990741	1.000000	0.990741	0.990741	1.000000
F1 score 0	1.000000	1.000000	1.000000	1.000000	1.000000	1.000000	1.000000
F1 score 1	1.000000	1.000000	0.972973	1.000000	0.972973	0.972973	1.000000
F1 score 2	1.000000	1.000000	1.000000	1.000000	1.000000	1.000000	1.000000
F1 score 3	1.000000	1.000000	0.962963	1.000000	0.962963	0.962963	1.000000
F1 score 4	1.000000	1.000000	1.000000	1.000000	1.000000	1.000000	1.000000
F1 score 5	1.000000	1.000000	1.000000	1.000000	1.000000	1.000000	1.000000
W.Avg. F1 score	1.000000	1.000000	0.990694	1.000000	0.990694	0.990694	1.000000

As can be deduced from Table 5 the tested metaheuristic performed excellently across all classification categories. The optimal selected parameters for each metaheuristic can be viewed in Table 6.

Table 6. Parameter selections for best performing runs of each metaherustic

Method	Learning Rate	Dropout	Layers	Neurons	Epochs
DNN-SCA-BS	0.005000	0.400000	2	20.0	100
DNN-SCA	0.000935	0.160560	1	30.0	100
DNN-BA	0.002514	0.358530	1	19.0	100
DNN-GA	0.001007	0.500000	1	34.0	100
DNN-PSO	0.003151	0.328741	1	17.0	100
DNN-HHO	0.000867	0.226853	2	23.0	100
DNN-MS	0.000755	0.256204	1	34.0	100

Finally, the improvements introduced through algorithm hybridization in the proposed metaheuristics can be seen in the convergence and distribution improvements shown in Fig. 2.

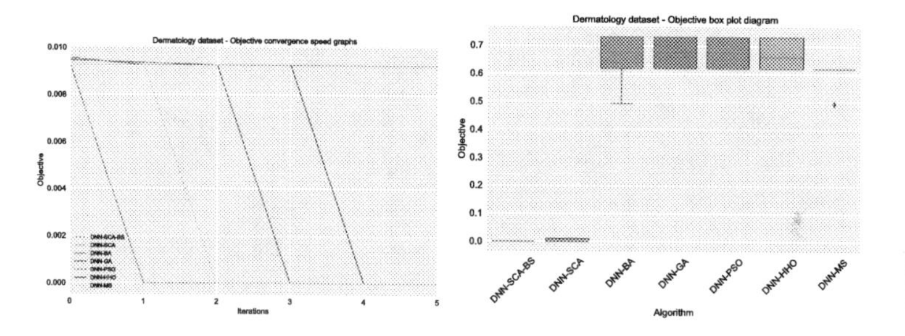

Fig. 2. Dermatology dataset metaheuristics objective function convergence and distribution box plots

5.3 Hepatitis Dataset Results

The hepatitis dataset proved challenging for the metaheuristic algorithms as demonstrated by the overall Cohen's kappa scores over 5 runs shown in Table 7.

Table 7. Overall Cohen's kappa scores for hepatitis dataset

Method	Best	Worst	Mean	Median	Std	Var	Best1	Worst1	Mean1	Median1	Std1	Var1
DNN-SCA-BS	0.069767	**0.069767**	**0.069767**	0.069767	**0.000000**	**0.000000**	0.730689	**0.730689**	**0.730689**	0.730689	**0.000000**	**0.000000**
DNN-SCA	0.069767	0.093023	0.074419	0.069767	0.009302	0.000087	0.730689	0.619469	0.708445	0.730689	0.044488	0.001979
DNN-BA	0.069767	0.116279	0.088372	0.093023	0.017403	0.000303	0.730689	0.494118	0.638887	0.619469	0.087827	0.007713
DNN-GA	0.069767	0.093023	0.079070	0.069767	0.011393	0.000130	0.730689	0.619469	0.686201	0.730689	0.054486	0.002969
DNN-PSO	0.069767	0.093023	0.079070	0.069767	0.011393	0.000130	0.730689	0.619469	0.686201	0.730689	0.054486	0.002969
DNN-HHO	0.069767	0.093023	0.083721	0.093023	0.011393	0.000130	0.730689	0.619469	0.672079	0.660079	0.050100	0.002510
DNN-MS	0.093023	0.116279	0.097674	0.093023	0.009302	0.000087	0.619469	0.494118	0.594399	0.619469	0.050141	0.002514

Nevertheless, the introduced metaheuristic outperformed all the competing algorithms and demonstrated impressive stability. These results indicate a significant potential of the proposed approach for tackling hepatitis classification. The potential of metaheuristic optimized DNNs is further solidified by the detail metrics presented in Table 8.

Table 8. Hepatitis detailed metrics for the best-performing run of each metaherustic

	DNN-SCA-BS	DNN-SCA	DNN-BA	DNN-GA	DNN-PSO	DNN-HHO	DNN-MS
Accuracy (%)	93.0233	93.0233	93.0233	93.0233	93.0233	18.6047	90.6977
Precision 0	1.000000	1.000000	1.000000	1.000000	1.000000	0.153846	1.000000
Precision 1	0.921053	0.921053	0.921053	0.921053	0.921053	0.500000	0.897436
W.Avg. Precision	0.935741	0.935741	0.935741	0.935741	0.935741	0.435599	0.916518
Recall 0	0.625000	0.625000	0.625000	0.625000	0.625000	**0.750000**	0.500000
Recall 1	1.000000	1.000000	1.000000	1.000000	1.000000	0.057143	1.000000
W.Avg. Recall	0.930233	0.930233	0.930233	0.930233	0.930233	0.186047	0.906977
F1 score 0	0.769231	0.769231	0.769231	0.769231	0.769231	0.255319	0.666667
F1 score 1	0.958904	0.958904	0.958904	0.958904	0.958904	0.102564	0.945946
W.Avg. F1 score	0.923616	0.923616	0.923616	0.923616	0.923616	0.130984	0.893987

The parameters selected by each metaheuristic algorithm in the best-performing run of diabetes classification are shown in Table 9.

Table 9. Metrics for best performing DNN optimized by each tested approach for hepatitis classification

Method	Learning Rate	Dropout	Layers	Neurons	Epochs
DNN-SCA-BS	0.005000	0.500000	2	11.0	100
DNN-SCA	0.002396	0.157042	2	19.0	100
DNN-BA	0.005000	0.500000	2	19.0	100
DNN-GA	0.005000	0.010820	1	9.5	100
DNN-PSO	0.001953	0.010000	1	9.5	100
DNN-HHO	0.003296	0.301749	1	14.0	100
DNN-MS	0.001379	0.010000	2	9.5	100

Finally, the improvements introduces though low level hybridization of the SCA with the BA can be seen in the convergence and distribution plots shown in Fig. 3.

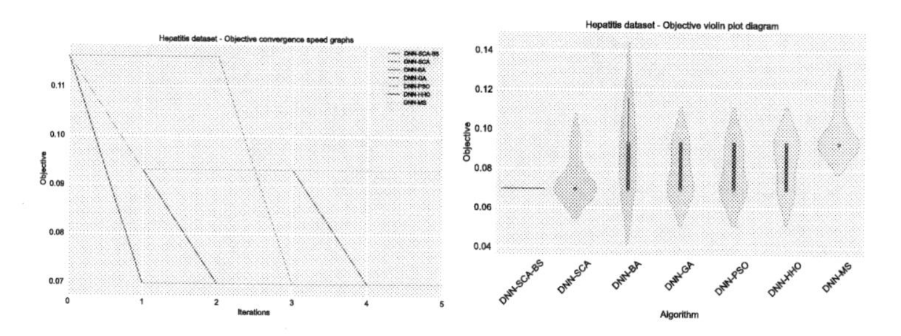

Fig. 3. Hepatitis dataset metaheuristics objective function convergence and distribution box plots

6 Conclusion

This researcher explores the potential of DNNs applied to medical diagnosis within Healthcare 4.0. and demonstrates the importance of optimization. The work presents a novel swarm intelligence algorithm and uses it for tuning hyperparameters of a DNN with the goal of improving performance. The introduced algorithm improves on the performance of the original SCA algorithm, as is dubbed SCA-BA. The improvement is accomplished through the introduction of the powerful search mechanism of the BA algorithm into the basic SCA completed with an additional control parameter the value of which value is empirically determined. The introduced approach has been tested on three medical datasets covering liver disorders, dermatology, and hepatitis. The novel metaheuristic-optimized DNN performed admirably, outperforming competing approaches.

Future work will focus on exploring the potential of DNN for handling emerging problems associated with developing Industry 4.0. Furthermore, the potential of the novel-introduced metaheuristic will be explored for optimizing other approaches.

References

1. Asuncion, A., Newman, D.: UCI machine learning repository (2007)
2. Esteva, A., et al.: Dermatologist-level classification of skin cancer with deep neural networks. Nature **542**(7639), 115–118 (2017)
3. Heidari, A.A., Mirjalili, S., Faris, H., Aljarah, I., Mafarja, M., Chen, H.: Harris hawks optimization: algorithm and applications. Futur. Gener. Comput. Syst. **97**, 849–872 (2019)
4. Jayaraman, P.P., Forkan, A.R.M., Morshed, A., Haghighi, P.D., Kang, Y.B.: Healthcare 4.0: a review of frontiers in digital health. Wiley Interdisc. Rev.: Data Min. Knowl. Discov. **10**(2), e1350 (2020)
5. Jovanovic, D., Antonijevic, M., Stankovic, M., Zivkovic, M., Tanaskovic, M., Bacanin, N.: Tuning machine learning models using a group search firefly algorithm for credit card fraud detection. Mathematics **10**(13) (2022). https://www.mdpi.com/2227-7390/10/13/2272

6. Jovanovic, L., et al.: Multi-step crude oil price prediction based on LSTM approach tuned by salp swarm algorithm with disputation operator. Sustainability **14**(21) (2022). https://www.mdpi.com/2071-1050/14/21/14616
7. Jovanovic, L., et al.: Multi-step crude oil price prediction based on LSTM approach tuned by salp swarm algorithm with disputation operator. Sustainability **14**(21), 14616 (2022)
8. Jovanovic, L., Zivkovic, M., Antonijevic, M., Jovanovic, D., Ivanovic, M., Jassim, H.S.: An emperor penguin optimizer application for medical diagnostics. In: 2022 IEEE Zooming Innovation in Consumer Technologies Conference (ZINC), pp. 191–196. IEEE (2022)
9. Mirjalili, S.: SCA: a sine cosine algorithm for solving optimization problems. Knowl.-Based Syst. **96**, 120–133 (2016)
10. Mirjalili, S.: Genetic algorithm. In: Mirjalili, S. (ed.) Evolutionary Algorithms and Neural Networks: Theory and Applications. Studies in Computational Intelligence, vol. 780, pp. 43–55. Springer, Cham (2019). https://doi.org/10.1007/978-3-319-93025-1_4
11. Ojha, V.K., Abraham, A., Snášel, V.: Metaheuristic design of feedforward neural networks: a review of two decades of research. Eng. Appl. Artif. Intell. **60**, 97–116 (2017)
12. Petrovic, A., Antonijevic, M., Strumberger, I., Jovanovic, L., Savanovic, N., Janicijevic, S.: The XGBoost approach tuned by TLB metaheuristics for fraud detection. In: Proceedings of the 1st International Conference on Innovation in Information Technology and Business (ICIITB 2022), vol. 104, p. 219. Springer Nature (2023)
13. Prakash, S., Kumar, M.V., Ram, S.R., Zivkovic, M., Bacanin, N., Antonijevic, M.: Hybrid GLFIL enhancement and encoder animal migration classification for breast cancer detection. Comput. Syst. Sci. Eng. **41**(2), 735–749 (2022)
14. Wang, D., Tan, D., Liu, L.: Particle swarm optimization algorithm: an overview. Soft. Comput. **22**, 387–408 (2018)
15. Wang, G.G.: Moth search algorithm: a bio-inspired metaheuristic algorithm for global optimization problems. Memetic Comput. **10**(2), 151–164 (2018)
16. Wolpert, D.H., Macready, W.G.: No free lunch theorems for optimization. IEEE Trans. Evol. Comput. **1**(1), 67–82 (1997)
17. Yang, X.S., Hossein Gandomi, A.: Bat algorithm: a novel approach for global engineering optimization. Eng. Comput. **29**(5), 464–483 (2012)

Early Prediction of At-Risk Students in Higher Education Institutions Using Adaptive Dwarf Mongoose Optimization Enabled Deep Learning

P. Vijaya(✉), Rajeev Rajendran, Basant Kumar, and Joseph Mani

Modern College of Business and Science, Bowshar, Oman
Vijaya.Padmanabha@mcbs.edu.om

Abstract. The biggest problem with online learning nowadays is that students aren't motivated to finish their coursework and other assignments. As a result, their performance suffers, which raises the dropout rate, necessitating the need for proactive measures to manage the dropout. Predictions of student performance assist in selecting the best programmers and designing efficient study schedules that are suited to their needs. Additionally, it aids in the development of observation and support tactics for students who require assistance in order to finish the course work by teachers and educational institutions. This paper proposed an efficient method using Adaptive Dwarf Mongoose Optimization (ADMOA)-based Deep Neuro Fuzzy Network (DNFN) for prediction of at-risk students in higher education institutions. Here, DNFN is working to **forecast** at-risk kids and prediction is carried out based on the most pertinent features collected utilizing the created ADMOA algorithm. Additionally, the effectiveness of the proposed ADMOA_DNFN is examined in light of a number of characteristics, including Root MSE (RMSE), Mean Square Error (MSE), Mean Absolute Error (MAE) and Mean Absoulte Percentage Error (MAPE), it attains best values of 0.049, 0.045, 0.212, and 0.022 respectively.

Keywords: predict at-risk students · Dwarf mongoose optimization · Deep Neuro Fuzzy Network · Deep learning · Machine Learning

1 Introduction

Students who have received warnings and have been instructed to leave the classroom often do so today. One of the causes can be that students are unable to accurately assess and forecast their capacity to choose the right courses. Student performance is a crucial duty for higher education institutions since it serves as a criterion for admission to top colleges, which rely their selection on students' stellar academic records. The term "student performance" has numerous definitions. The learning evaluation and curriculum can be used to determine student achievement, claims. However, the majority of research claimed that students' progress was measured by graduation [3, 5, 6]. In recent decades, educational materials have undergone significant change due to the exponential advancement of science and technology. Massive Open Online Courses (MOOCs), a type of

S. Aurelia et al. (Eds.): ICCSST 2023, CCIS 1973, pp. 15–29, 2024.
https://doi.org/10.1007/978-3-031-50993-3_2

virtual learning environment (VLE) that offers lecture videos, discussion forums, online quizzes, and even live video talks via the Internet [7], have been popular, particularly during the COVID-19 pandemic. The rising popularity of online learning is attributed to two of its advantages. First off, by removing time and distance restrictions, VLEs make it convenient for participants to enroll in courses. Additionally, online learning platforms based on the Internet have the capacity to save a sort of data known as trace data [8] that, after proper analysis, significantly aids in the provision of tailored educational services, incorporating data from a user's VLEs and other learning systems.

Online education does, however, have a high dropout rate and a high risk of academic failure [2]. First, the capability to accurately anticipate a student's ultimate presentation and identify that at-risk student early enough for an interference is the most crucial quality of an early warning device. The recall rate is further significant than the total correctness in the evaluation of at-risk prediction. Second, without comparing students' relative engagement levels, most studies use students' absolute behavior frequencies to build prediction models [13]. For instance, 100 accesses per week to course material can be regarded as strong engagement in one course but as poor involvement in another. Third, several research did not attain enough accuracy and recall rates in terms of predictive potential. Numerous investigations showed that the results of predictions were nearly identical to random guesses [14]. Over-simplifying a student's learning behaviors can be a significant factor. Early warning research typically used machine learning modelling techniques, according to the literature. According to the findings of the research, argue that since CNN model can extract features at the micro-level, use it to discover behavioral variances that are influenced by both the traits of individual students and the course's design [4].

In this work, a method for predicting the at-risk students in higher education institutions is developed using the ADMOA_DNFN. The subsequent sections are utilized to carry out prediction procedure: data augmentation, data acquisition, data transformation, feature selection, and prediction of at-risk students. In this case, the input data is first gathered, then converted, and to find most important features in the data in feature selection phase. Following the application of data augmentation to chosen features, an estimation of the at-risk students is made using the DNFN, with the discrimination and parameters of the DNFN being modified by using ADMOA technique.

The major influence of investigation is discussed as follows:

- ***Established ADMOA algorithm for training DNFN to prediction of at-risk students:*** Here, prediction of at-risk students in higher education institutions is carried out using ADMOA_DNFN. Moreover, the introduced ADMOA is devised by the amalgamation of Adaptive concept and Dwarf Mongoose Optimization Algorithm (DMOA). By using Ruzicka's similarity distance as the fitness, the newly developed ADMO algorithm selects the salient characteristics in the best possible way. By improving the weight parameters of the DNFN, the ADMOA is also utilized to improve prediction accuracy.

The association of rest of work is given as follows: In Sect. 2 discusses earlier researches; in Sect. 3 explains established ADMOA_DNFN; predicted results and discussion of developed method is explained in Sect. 4, and in Sect. 5, conclusion of established technique is discussed.

2 Motivation

Numerous research has worked hard to incorporate DL and ML methodologies into learning analytics throughout the years in order to address difficulty of prediction of at-risk students. However, in terms of understandability and efficiency for real - world applications, these methods have shown inconsistent efficacy. As these methods do not take into account the connectivity between the data, it is typically connected to built-in procedures utilized to develop these systems. The several methods for predicting students who are likely to struggle academically or perform poorly are summarized in this section. The methods are considered with their rewards and disadvantages which encouraged the formation of the developed ADMOA_DNFN structure.

2.1 Literature Survey

A handful of the numerous studies that have concentrated on prediction of at-risk students are summarized here. Azcona, D et al., [1], created a brand-new research methodology that classified students and identified students at risk by combining various data sources from the higher education sector. Furthermore, to increase scalability of prediction model so that it could be used to other blended classrooms and to other courses, the approach integrates static and dynamic student data elements. Furthermore, there was a chance of inaccuracy in modelling, which was a severe negative. He, Y et al., [2], presented recurrent neural network (RNN)-gated recurrent unit (GRU) was employed to fit both sequential and static data and the data completion mechanisms were also used to fill in the gaps in the stream of data. The method was failed to develop unified time varying DNN model which eliminates integration of two different methods. The DL technique was developed by Dien, T.T., et al. [3] to predict student performance. Here, a prediction strategy was devised using Convolutional Neural Networks (CNN) and Long Short-Term Memory (LSTM) on one-dimensional (1D) data (CN1D) in order to estimate performance in the upcoming semester while taking the outcomes of the previous one into consideration. The method effectively improved prediction performance, particularly when using data modification. However, it was unable to gather enough data from exercise set to identify features of the test set. Yang, Z. et al., [4], established CNN model for identifying at-risk students. In this instance, student's internet activity was taken into account while defining the behavioural characteristics that were then aggregated to create a picture. Using the features that were collected from the image, CNN was then used to predict whether student would pass. Although taking into account other component arrangement did not improve the scheme's effectiveness, this method achieved a low classification error and good sensitivity.

2.2 Major Challenges

The following is a list of some of the problems that the conventional methods ran into:

- The ML algorithm designed in [1] for defining at-risk students did not consider the participation of other methods, like physical access to the institution, laboratory usage, or video recordings for boosting the student performance by reliably modelling their behavior and helping them in learning.
- For predicting student performance, the RNN-GRU model was put forth in [2]. It achieved good accuracy and recall, and it could identify at-risk students as soon as feasible. Nevertheless, it failed to contemplate possibility of substituting integrated models with a combined time varying DNN.
- DL approaches were considered in [3], wherein LSTM and CNN on one-dimensional (1D) data (CN1D) were developed to create a prediction scheme for estimating the performance. The main issue with this strategy was that there was not enough information in training data to identify features of test data.
- To visualize student course participation for early warning predictive analysis, 1-channel learning image recognition (1-CLIR) and 3-channel learning image recognition (3-CLIR) were devised in [4]. More complicated target factors, such as trends that are increasing or declining or changes in prediction probabilities over the course of the semester, were not considered.

3 Proposed ADMOA_DNFN for Prediction At-Risk Students

The primary goal of this work is to establish an approach for performance prediction of at-risk students in higher education institutions. Initially, the input student behavior data is acquired from the dataset [16], which is later forwarded to data transformation. In this instance, data transformation is performed using Yeo-Johnson Transformation (YJT) [15] is employed to transform the data as desired form [1]. After being modified, the data is supplied into the feature selection step, where the introduced ADMO algorithm selects the salient features in best way possible based on Ruzicka's similarity distance as the fitness. Once, feature selection is finalized, then data augmentation is carried out based on oversampling model, which increases the prediction performance. Finally, performance prediction is performed to find at-risk students using Deep Neuro Fuzzy Network (DNFN) [12]. Moreover, DNFN is trained using proposed Adaptive Dwarf Mongoose Algorithm (ADMOA), which is developed by the integration of Adaptive concept in Dwarf Mongoose Optimization Algorithm (DMOA) [10]. Figure 1 displays the schematic view of proposed prediction of at-risk students utilizing optimized DL method.

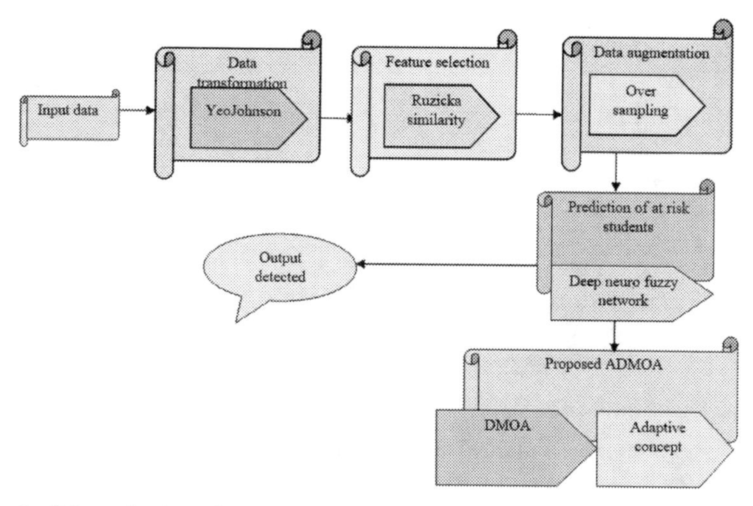

Fig. 1. Schematic view of proposed ADMOA_DNFN for prediction of at-risk students

3.1 Data Acquisition

In this work, determination of which students are at risk is done by taking into account a database that contains information on numerous student characteristics, and is represented by,

$$L = \{L_1, L_2, ..., L_e, ..., L_n\} \tag{1}$$

In this instance, each data has a dimension of $l \times m$, and L_e represents e^{th} data in specified dataset that is consigned to data transformation. In addition, L represents the input student performance database contains n-number of data. The output L_e of the data acquisition process, is forwarded to the data transformation process.

3.2 Data Transformation

The data obtained from dataset L_e is subjected to a data transformation operation, where input data is pre-processed utilizing YJT. The transformation of unstructured data into a structured format makes data more understandable. Regression models that use Box-Cox transformation (BCT) are able to achieve normalcy in random errors, reduce nonlinearity, and reduce data anomalies. It is a non-parametric transformation that obtained a one-parameter transformation family by combining the alterations for negative and positive explanations using a smoothness requirement. This transformation technique is highly flexible and is beneficial in evaluating specific datasets [11]. YJT gives the following expression.

$$\lambda(L_e) = \begin{cases} \frac{(L_e+1)^{\omega}-1}{\omega}, & \omega \neq 0\, and\, L_e \geq 0 \\ Ln(L_e+1), & \omega = 0\, and\, L_e \geq 0 \\ \frac{-((-L_e+1)^{2-\omega}-1)}{2-\omega}, & \omega \neq 2\, and\, L_e < 0 \\ Ln(-L_e+1), & \omega = 2\, and\, L_e < 0 \end{cases} \tag{2}$$

In this instance, index parameter is represented as ω, and $\lambda(L_e)$ gives YJT of data L_e and has a dimension of $l \times m$ is then processed to data transformation, which gives the transformed data of $l \times m$.

3.3 Feature Selection

The most significant features from input data are chosen using the data $\lambda(L_e)$ hence decreasing the number of features. Here, dimension of input data $l \times m$ is minimized to a dimension of $l \times s$, where $m > s$. The Ruzicka similarity is used for feature selection. Below is a description of the feature selection procedure:

3.3.1 Ruzicka Similarity

In Ruzicka similarity, the similarity between two points have been matched using various techniques. The expression of Ruzicka similarity is represented below:

$$Z_{RUZ} = \frac{\sum_{k=1}^{v} \min(M_k, N_k)}{\sum_{k=1}^{v} \max(M_k, N_k)} \tag{3}$$

Here, M_k signifies the feature of candidate

N_k denotes the aim

After calculating the Ruzicka similarity for every feature, it selects top l ranked feature with large score. As a result, ADMOA method successfully chooses best features from input data, and features so selected are indicated as D and have dimension of $m \times s$. The data augmentation phase is where the features are then used.

3.4 Data Augmentation (Oversampling)

The chosen characteristic D is then put through a data augmentation process to increase the sample size. The process of "data augmentation" in data analysis is adding slightly modified copies of either already existing data or entirely new synthetic data that is derived from existing data in order to increase amount of data. In process of training a machine learning model, it acts as a regularizer and helps to decrease overfitting. The data augmentation is done based on the oversampling model, which increases the prediction performance.

Let, $m \times s$ represents data size after feature selection. This is to be enhances to $u \times s$ such that $l < u$. Furthermore, l is enhanced to size u, by oversampling data ie.,) in each column is minimum and maximum values are resolute and lasting samples are randomly generated within that interval.

For example: consider 10×5 is the input selected features. After the data augmentation process, the features are incremented by $10, 00, 000 \times 5$. By generating the 99990×5 samples based on oversampling technique. The data augmentation thus generated is specified by A.

3.5 Performance Prediction to Determine at Risk Students

The augmented data A is taken as input of DNFN which gathers data from both fuzzy and neural representations. The flexibility of fuzzy logic representations allows for the flexible construction of fuzzy rules, which reduces the uncertainties in raw data as compared to traditional deterministic representations. Students who are at risk of failing next exams and who require more learning support are found using DNFN. Therefore, the DNFN employed here is trained using the ADMOA technique. The following gives an explanation of the DNFN structure and training algorithm.

3.5.1 Deep Neuro Fuzzy Network

Fuzzy logic representation: Multiple membership functions connect each node in the input layer, assigning linguist tags to each input variable. In this case, input variable is one of dimensions of input vector. The fuzzy association purpose determines how much a given input node is a member of a given fuzzy collection. On the surface, the cth fuzzy neuron $x_c(\bullet): I \rightarrow [0, 1]$ maps t^{th} input as the fuzzy degree

$$w_c^{(l)} = x_c(b_t^{(l)}) \tag{4}$$

$$w_c^{(l)} = \frac{a^{-(b_t^{(l)} - \eta_c)^2}}{\eta_c^2}, \forall_c \tag{5}$$

The "AND" fuzzy logic operation is carried out via the fuzzy rule layer. i.e., $w_c^{(l)} = \alpha_h o_h^{l-1}, \forall \in \tau_c$, where τ_c indicates node sets on the $(l-1)$ th layer that attach to c. Here, $(l-1)$ is taken as input layer. The fuzzy degree is the output of the section.

NR (blue): This section makes use of the idea of neural learning to turn input into a few high-level representations. Here, $(l)^{th}$ layer is contact with all nodes of $(l-1)^{th}$ layer with parameters $\gamma^{(l)} = \{f^{(l)}, g^{(l)}\}$,

$$\gamma_c^{(l)} = \frac{1}{1 + a^{-b_c^{(l)}}}, b_c^{(l)} \tag{6}$$

$$\gamma_c^{(l)} = f_c^{(l)} w^{(l-1)} + g_c^{(l-1)} \tag{7}$$

Here, $f_c^{(l)}$ and $g_c^{(l)}$ indicates weights and bias linked to node c on l^{th} layer.

Fusion part: The merged concept employed here is primarily motivated by current achievements in multimodal learning [9]. Feature extraction from a single view alone, according to multimodal learning, is insufficient to represent the intricate structure of high-content material. In light of this, these methods consistently produce several structures from several features and combine them into a high level depiction for categorization. Utilizing neural and fuzzy components of the FDNN, better representations were sought by lowering the hesitation and input data noise. Readers should think of the fuzzy part's output rather than its fuzzy underpinning in order to better appreciate our model design. Additionally, the neural network is set up with both the FL and neural learning components. The relation is follows:

$$w_c^{(l)} = \frac{1}{1 + a^{-b_c^l}} \tag{8}$$

$$b_c^{(l)} = (f_z)_c^{(l)} (w_z)^{(l-1)} + (f_d)_c^{(l)} (w_d)^{(l-1)} + g_c^{(l)} \tag{9}$$

In (6), weights f_z and f_d are utilized to combine the deep representation of output component (referred to w_z) and fuzzy logic depiction fragment (referred to w_d). After the fusion layer, other all-connected layers are added to further manipulate the combined information, like in (7). NRs and fuzzy degrees are combined in the outputs, that are no more fuzzy degrees.

First, FL offers a practical method to lower input data uncertainty. Other learning systems are unlikely to be able to replace fuzzy systems' critical quest of ambiguity reduction. Second, the FL naturally results in fuzzy representations in range of (0, 1). Additionally, every measurement in the neuronal outputs falls within range of (0, 1), from Eq. (7). Third, FL component enables parameter learning that is task-driven. Here, sophisticated data-driven learning via backpropagation can take the role of the taxing hand-craft parameter adjustment procedures [12]. The output W_j^* obtained while DNFN is used to performance prediction of at risk students.

3.5.2 Proposed Adaptive Dwarf Mongoose-Training Algorithm

This section, ADMOA is devised for performance prediction of at-risk students. The classifier may alter its learning rate as it is being trained, which increases efficiency, thanks to the use of the adaptive concept. The DMOA is a swarm intelligence-based approach that draws inspiration from animal behavior to find the best answers to world challenges. It imitates the behavioral responses of dwarf mongooses. The DMO is integrated with Adaptive concept to get the developed ADMOA for ascertain the main causes of the inability to pinpoint the ideal remedy in every situation. The proposed ADMOA mimics the compensatory behavioral response of the dwarf mongoose, which is modelled as given below.

Step i) Initialization: The initialization of the search space is initial step in the ADMOA. Consider a search agents population can be expressed as below,

$$V = \{V_1, V_2, ..., V_J, ..., V_K\} \tag{10}$$

Here, K indicates total number of search space, and V_J specified radius.

Step ii) Fitness function:

MSE, which is provided by, is the fitness parameter that is taken into consideration here because the problem is presented as a reduction issue.

$$MSE = \frac{1}{\sigma} \sum_{o=1}^{\sigma} (W_j^o - W_j^{o*})^2 \tag{11}$$

Here, W_j and W_j^* represents target output and battered output of DNFN. Based on the quality of the produced solution, the representatives are ranked.

Step iii) Alpha Group:

The population has initiated, and the effectiveness of each explanation is taken into account. The probability value is determined by Eq. (12), and the δ female is considered likelihood.

$$\delta = \frac{fit_n}{\sum_{n=1}^{i} fit_n} \tag{12}$$

Here, i indicates the number of mongooses in the δ. Where, ψ indicates vocalization of dominant female that develops the family on track [13]. Solution mechanism is specified as follows:

$$S_{n+1} = S_n + T_i * \psi \tag{13}$$

From the above equation, T_i represents the distributed random number and T_i is a uniformly distributed random number [0, 1]. The sleeping mound is given in (14).

$$\varepsilon = \frac{fit_{n+1} - fit_n}{\max\left[|fit_{n+1} - fit_n|\right]} \tag{14}$$

Furthermore, Eq. (15) gives the typical quantity of sleeping mounds.

$$\phi = \sum_{n=1}^{i} \varepsilon \tag{15}$$

As soon as the need for a babysitting exchange is met, an algorithm advances to the scouting stage, when next food supply or sleeping mound is taken into consideration.

Step iv) Scout Group:

The scout mongoose is provided in (16).

$$S_{n+1} = \begin{cases} S_n - GH * T_i * r_1[S_n - \overrightarrow{Q}], & if\ \phi_{n+1} > \phi_n \\ S_n + GH * T_i * r_1[S_n - \overrightarrow{Q}], & otherwise \end{cases} \tag{16}$$

Here, the following condition $\phi_{n+1} > \phi_n$ only. Now, the scout mongoose equation is given by,

$$S_{n+1} = S_n - GH * T_i * r_1[S_n - \overrightarrow{Q}] \tag{17}$$

where, r_1 indicates the random value in range [0, 1]. Moreover, the r_1 is made adaptive. G represents the parameter constant (2). The equation of GH (18) and $\overrightarrow{Q}$ (19) and r_1 (20) is given below:

$$GH = (1 - \frac{\rho}{\max_\rho})^{(2*\frac{\rho}{Max_\rho})} \tag{18}$$

$$\overrightarrow{Q} = \sum_{n=1}^{i} \frac{S_n * \pi_n}{S_n} \tag{19}$$

$$r_1 = G - (1 - \frac{\rho}{\max_\rho}) \tag{20}$$

The DMOA model is detailed in Algorithm 1.

Step v) Re-evaluate fitness:

After the agents' locations have been upgraded, Eq. (10) is utilized to calculate their fitness, and operator with lowest fitness is selected as best option.

Step vi) Termination

Aforementioned method is repeated until the extreme number of repetitions has been reached. Algorithm 1 is used to demonstrate pseudocode of the DMOA algorithm.

Algorithm 1. Pseudocode of DMOA

1	Input: **Set parameters and answers for algorithm.**
2	Set algorithm's inputs and outputs to their initial values.
3	**while** $(\rho < Max_{\rho})$ **do**
4	**for** $(n = 1\, to\, Solutions)$ **do**
5	The Mongoose Fitness Function (FF) should be calculated.
6	Fix the timer (C).
7	Equation (9) is used to get the alpha value.
8	Utilize Equation (10) to carry out a solution.
9	Equation (11) is used to evaluate the sleeping mound.
10	Equation (12) is used to calculate the sleeping mound's average.
11	Equation (16) is used to determine the movement vector.
12	Equation (13) is used to pretend the scout Mongoose for the following answer.
13	**end for**
14	$\rho = \rho + 1$
15	**end while**
16	Output: Return the finest result (x).

4 Results and Discussion

Experimental findings of the ADMOA- DNFN for identifying at risk students are covered in this part. Additionally, a detailed presentation of the investigational setup, algorithm evaluation, dataset, technique, and evaluation metrics are made.

4.1 Experimental Results

Python tool is utilized in the experimentation of the suggested at-risk prediction.

4.2 Dataset Description

The investigation of the introduced ADMOA_DNFN is gifted using the Higher Education Students Performance Evaluation Dataset [16]. Students in the engineering and educational sciences faculties provided the data in 2019. Utilizing ML approaches, the aim is to forecast students' end-of-term performance. The attribute information of student ID is gathered from the dataset is given below: student age, graduated high school type, scholarship type etc.

4.3 Evaluation Metrics

Experimental analysis is used to evaluate the efficiency of ADMOA_DNFN for prediction of at-risk student. Metrics, like RMSE, MAE, MSE, and MAPE are considered. Below is a brief description of the evaluation metrics.

MSE: Using equation, MSE calculates square of error in between original and goal outputs (8). MSE measures how closely the anticipated and actual outputs compare.

RMSE: The root square of MSE is used to calculate RMSE, which is expressed in the following way.

$$RMSE = \sqrt{MSE} = \sqrt{\frac{\sum_{o=1}^{\sigma} (W_j^o - W_j^{o*})^2}{\sigma}} \tag{21}$$

MAE: The MAE, which is determined and utilized to calculate error value among projected and desired production.

$$MAE = \sum_{o=1}^{\sigma} \left| W_j - W_j^* \right| \tag{22}$$

MAPE: It calculates the prediction accuracy of the methods. The formula for computing MAPE is shown in Eq. (23).

$$MAPE = \frac{1}{o} \sum\nolimits_{n=1}^{o} \left| \frac{W_j^o - W_j^{o*}}{W_j^o} \right| \tag{23}$$

4.4 Comparative Techniques

In order to assess effectiveness of ADMOA_DNFN prediction of at risk student considering several measures; an empirical study is conducted. Moreover, to test the effectiveness of the created method, the traditional approaches to predicting student performance, such as ML [1], RNN-GRU [2], DL [3], and CNN [4], SLnS are taken into consideration.

i) ***Assessment using training data***

The developed ADMOA_DNFN is estimated by varying training data, and this is demonstrated in Fig. 2. Figure 2a) displays the assessment of the established ADMOA_DNFN based on MSE. With training data of 60%, the MSE value attained by the at-risk student's prediction schemes, like ML, RNN-GRU, DL, CNN, SLnS-DQN, and the ADMOA_DNFN is 0.900, 0.102, 0.089, 0.082, 0.065, and 0.055, respectively. In Fig. 2b), RMSE concerned with valuation of the established ADMOA_DNFN is illustrated. The various prediction models computed RMSE of 0.948 for ML, 0.320 for RNN-GRU, 0.299 for DL, 0.286 for CNN, 0.255 for CNN, and 0.234 for the current work of ADMOA_DNFN with 60% training data. Figure 2c) represents the analysis of the devised work considered MAE. The value of MAE achieved is 0.269, 0.085, 0.083, 0.082, 0.071, and 0.050, conforming to

ML, DL, RNN-GoRU, CNN, SLnS-DQN, and the presented ADMOA_DNFN, with 60% training data. Figure 2d) shows the MAPE values of proposed technique and prevailing approaches. For 60% of training data, the MAPE values attained by the comparative methods, like ML, DL, RNN-GRU, CNN, SLnS-DQN, and proposed ADMOA_DNFN are 0.052, 0.029, 0.029, 0.029, 0.027, and 0.022, respectively. On seeing Fig. 2d), ADMOA_DNFN has the minimum MAPE value than prevailing approaches, which means that proposed method predicts the at-risk students precisely.

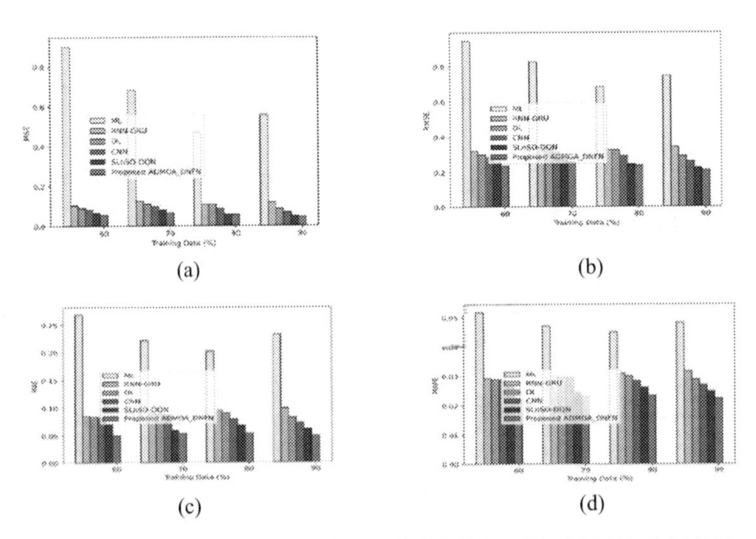

Fig. 2. Valuation using training data a) MSE, b) RMSE, c) MAE, d) MAPE

ii) ***Evaluation using population size***

Figure 3 displays the examination of the established ADMOA_DNFN considering population size. Figure 3a) demonstrates the MSE- oriented assessment of the work. MSE value attained by the techniques like AO+DNFN, JAYA+DNFN, SLnS+DNFN, ROA+DNFN and devised ADMOA_DNFN attained MSE of 0.269, 0.226, 0.177, 0.137, and 0.101 with population 5. In Fig. 3b), the valuation of the established estimate scheme concerning RMSE is offered. The systems like AO+DNFN, JAYA+DNFN, SLnS+DNFN, ROA+DNFN and devised ADMOA_DNFN measured RMSE of 0.519, 0.475, 0.420, 0.370, and 0.317 with population size of 5. The MAE focused on valuation of the work by varying the population size is displayed in Fig. 3c). MAE value enumerated is 0.226 for AO+DNFN, 0.198 for JAYA+DNFN, 0.165 for SLnS+DNFN, 0.147 for ROA+DNFN, and 0.127 for the devised work for population size of 5. Figure 3d) shows the MAPE values of the comparative methods by varying the population size. For the population size 5, the MAPE values of the methods, such as AO+DNFN, JAYA+DNFN, SLnS+DNFN, ROA+DNFN, and devised ADMOA_DNFN are 0.048, 0.044, 0.041, 0.038, and 0.036, respectively. The MAPE values gradually decreases with the increase in the population size.

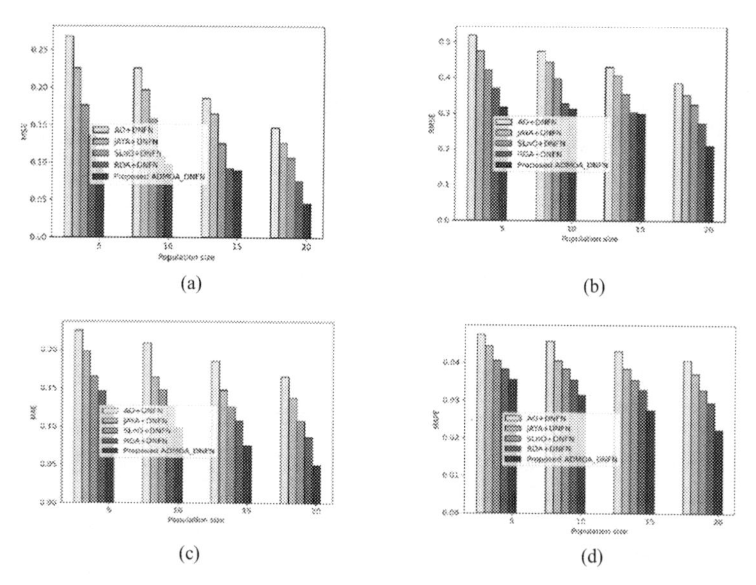

Fig. 3. Analysis using population set a) MSE, b) RMSE, c) MAE, d) MAPE

4.5 Comparative Discussion

This section describes comparative discussion of current work, in which performance of current work is evaluated using several metrics, includes MSE, RMSE, MAE, and MAPE in comparison to prediction at-risk students. Table 1 shows the comparison of ADMOA_DNFN at-risk student prediction.

Table 1. Comparative discussion of the devised ADMOA_DNFN prediction of at risk students

Variations	*Metrics*	*ML*	*RNN_GRU*	*DL*	*CNN*	*SLnS_DQN*	*Proposed ADMOA_DNFN*
Training data	*MSE*	0.557	0.117	0.084	0.068	0.050	0.045
	RMSE	0.746	0.342	0.291	0.260	0.224	0.212
	MAE	0.232	0.099	0.082	0.072	0.061	0.049
	MAPE	0.048	0.031	0.029	0.027	0.025	0.022

Table 2 shows the evaluation of the created ADMOA DNFN. The algorithmic comparisons are performed while taking into account numerous parameters, including MSE, RMSE, MAE, and MAPE. The data in the table represent a population of 20, and it is evident from the table that the newly introduced ADMOA DNFN method produced a reduced MSE of 0.045, MAE of 0.049, RMSE of 0.212, and MAPE of 0.022.

Table 2. Qualified analysis of the established ADMOA_DNFN

Metrics	*AO+DNFN*	*JAYA+DNFN*	*SLnS+DNFN*	*ROA+DNFN*	*Proposed ADMOA_DNFN*
MSE	0.147	0.127	0.107	0.076	0.045
RMSE	0.383	0.356	0.326	0.275	0.212
MAE	0.165	0.138	0.108	0.087	0.049
MAPE	0.041	0.037	0.033	0.029	0.022

5 Conclusion

The ADMOA_DNFN methodology is used in this paper to provide a useful method for identifying at-risk students. Numerous factors, including demographic, socioeconomic, and educational characteristics as well as student grade, are considered when identifying the at-risk students. The proposed ADMOA optimization algorithm is used to train the DNFN which in turn enhances performance of prediction. Additionally, ADMOA trained with newly created DNFN algorithm is used to forecast the at-risk students. The performance of the proposed ADMOA_DNFN for prediction of at-risk students is examined in consideration of parameters, includes MSE, RMSE, MAE and MAPE. The investigational outcomes demonstrate that a least value of MSE at 0.045, RMSE at 0.212, MAE at 0.049 and MAPE at 0.022 is gained, revealing improved performance. In the future, the primary finding of the work will be the linkages and characteristics between multi-dimensional features which are not adequately captured by the data augmentation.

References

1. Azcona, D., Hsiao, I.H., Smeaton, A.F.: Detecting students-at-risk in computer programming classes with learning analytics from students' digital footprints. User Model. User-Adap. Inter. **29**(4), 759–788 (2019)
2. He, Y., et al.: Online at-risk student identification using RNN-GRU joint neural networks. Information **11**(10), 474 (2020)
3. Dien, T.T., Luu, S.H., Thanh-Hai, N., Thai-Nghe, N.: Deep learning with data transformation and factor analysis for student performance prediction. Int. J. Adv. Comput. Sci. Appl. **11**(8) (2020)
4. Yang, Z., Yang, J., Rice, K., Hung, J.L., Du, X.: Using convolutional neural network to recognize learning images for early warning of at-risk students. IEEE Trans. Learn. Technol. **13**(3), 617–630 (2020)
5. Rastrollo-Guerrero, J.L., Gómez-Pulido, J.A., Durán-Domínguez, A.: Analyzing and predicting students' performance by means of machine learning: a review. Appl. Sci. **10**(3), 1042 (2020)
6. Shahiri, A.M., Husain, W.: A review on predicting student's performance using data mining techniques. Procedia Comput. Sci. **72**, 414–422 (2015)
7. Karimi, H., Huang, J., Derr, T.: A deep model for predicting online course performance. CseMsu Educ. **192**, 302 (2014)

8. Silveira, P.D.N., Cury, D., Menezes, C., dos Santos, O.L.: Analysis of classifiers in a predictive model of academic success or failure for institutional and trace data. In: 2019 IEEE Frontiers in Education Conference, FIE, pp. 1–8 (2019)
9. Ngiam, J., Khosla, A., Kim, M., Nam, J., Lee, H., Ng, A.Y.: Multimodal deep learning. In: ICML 689–696 (2011)
10. Aldosari, F., Abualigah, L., Almotairi, K.H.: A normal distributed dwarf mongoose optimization algorithm for global optimization and data clustering applications. Symmetry **14**(5), 1021 (2022)
11. Othman, S.A., Ali, H.T.M.: Improvement of the nonparametric estimation of functional stationary time series using yeo-johnson transformation with application to temperature curves. Adv. Math. Phys. 1–6 (2021)
12. Deng, Y., Ren, Z., Kong, Y., Bao, F., Dai, Q.: A hierarchical fused fuzzy deep neural network for data classification. IEEE Trans. Fuzzy Syst. **25**(4), 1006–1012 (2016)
13. Hung, J.L., Wang, M.C., Wang, S., Abdelrasoul, M., Li, Y., He, W.: Identifying at-risk students for early interventions—a time-series clustering approach. IEEE Trans. Emerg. Top. Comput. **5**(1), 45–55 (2015)
14. Baker, R.S., Lindrum, D., Lindrum, M.J., Perkowski, D.: Analyzing early at-risk factors in higher education E-learning courses. Int. Educ. Data Min. Soc. (2015)
15. Tsiotas, G.: On the use of non-linear transformations in Stochastic Volatility models. Stat. Methods Appl. **18**(4), 555–583 (2009)
16. The dataset. https://archive.ics.uci.edu/ml/datasets/Higher+Education+Students+Performance+Evaluation+Dataset. Assessed November 2022

Decomposition Aided Bidirectional Long-Short-Term Memory Optimized by Hybrid Metaheuristic Applied for Wind Power Forecasting

Luka Jovanovic[1], Katarina Kumpf[1], Nebojsa Bacanin[1(✉)], Milos Antonijevic[1], Joseph Mani[2], Hothefa Shaker[2], and Miodrag Zivkovic[1]

[1] Singidunum University, Danijelova 32, 11000 Belgrade, Serbia
{luka.jovanovic.191,katarina.kumpf.22}@singimail.rs,
{nbacanin,mantonijevic,mzivkovic}@singidunum.ac.rs
[2] Modern College of Business and Science, Muscat, Sultanate of Oman
{drjosephmani,Hothefa.shaker}@mcbs.edu.om

Abstract. Increasing global energy demands and environmental concerns have in recent times lead to a shift in energy production towards green and renewable sources. While renewable energy has many advantages, it also highlights certain challenges in storage and reliability. Since many renewable sources heavily rely on weather forecasting the amount of produced energy with a degree of accuracy becomes crucial. Energy production reliant on wind farms requires accurate forecasts in order to make the most of the generated electricity. Artificial intelligence (AI) has previously been used to make tackle many complex tasks. By formulating wind-farm energy production as a time-series forecasting task novel AI techniques may be applied to address this challenge. This work explores the potential of bidirectional long-short-term (BiLSTM) neural networks for wind power production time-series forecasting. Due to the many complexities affecting wind power production data, a signal decomposition technique, variational mode decomposition (VMD), is applied to help BiLSTM networks accommodate data. Furthermore, to optimize the performance of the network an improved version of the reptile search algorithm, which builds on the admirable capabilities of the original, is introduced to optimize hyperparameter selection. The introduced method has been compared to several state-of-the-art technique forecasting wind energy production on real-world data and has demonstrated great potential, outperforming competing approaches.

Keywords: metaheuristics · BiLSTM · variational mode decomposition · RSA multivariate forecasting

S. Aurelia et al. (Eds.): ICCSST 2023, CCIS 1973, pp. 30–42, 2024.
https://doi.org/10.1007/978-3-031-50993-3_3

1 Introduction

In recent decades, gradually increasing amounts of energy have been consumed due to population growth, and economic and technological development. Energy sources can be classified into renewable and non-renewable energy sources. A problem associated with renewable energy sources is their weather dependence and the fluctuation in the amount of provided power. With the shift towards renewable energy, the analysis and forecasting of the future, electricity generation plays an important strategic role. Forecasting energy production is meaningful for a number of reasons.

The use of artificial neural networks (ANN) models can successfully solve nonlinear problems, but it requires the input of training data and can be time-consuming. It has been found that due to the unknown internal architecture, the use of trial and error, ANN, models has some disadvantages, and even with very large amounts of data, these models become unstable [7,16]. To achieve the best performance, it is often necessary to combine the advantages of traditional and artificial intelligence (AI) methods.

Simple ANNs lack the ability to handle data formulated as a time series. Recurrent neural networks (RNNs) have been developed to address this. However, while improving the ability of basic ANN, RNNs come with their own set of challenges. Issues such as vanishing gradients are a pressing issue associated with RNN. Modern techniques have been developed to tackle these issues such as long-short-term LSTM neural networks, and further improvements have been made by bidirectional-LSTM (BiLSTM) networks that allow for data flow in both directions. It is important to note that like many machine learning (ML) methods, these present a set of adjustable control parameters that require adequate adjustment to attain desirable performance when tackling a problem.

By formulating hyperparameter tuning as an optimization problem metaheuristics can be applied to optimize performance. This work explores the potential of the novel Reptile search algorithm (RSA) [2] for tuning (BiLSTM) hyperparameters. Additionally, an approach improving on the excellent performance of the original is introduced.

The remainder of this work is structured according to the following: Sect. 2 presents preceding works that helped make this work possible. The proposed methods are presented in Sect. 3. The experimental setup is presented in Sect. 4 accompanied by their discussion. Finally, Sect. 6 concludes the work and proposed future work in the field.

2 Related Works

Accurate prediction of power consumption is very difficult due to its fluctuations and non-linearity but is necessary for proper operation and maintenance of the electrical power system [19]. In order to predict the consumption of energy produced by wind farms, an accurate and reliable forecast of wind power is needed. The physical approach utilizes numerical weather prediction results, with the help of which it calculates the wind speed in a certain area using the characteristics of that area and using the turbine power curve to convert the speed into wind power. Models based on single-valued numerical weather forecasts often cannot reliably predict wind speed and power due to inevitable inaccuracies from model initialization and model imperfections [27]. This implies limited technical skills, instrument calibration, and uncertainty can also be caused by inaccurate representation of physical and dynamic processes [27].

Zhao et al. [27] proposed a CSFC-Apriori-WRF method to provide one-day ahead wind speed and power forecasting, which is of great importance for risk reduction when a large amount of wind energy is injected into the grid system. In addition to CSFC-Apriori-WRF, single-valued NWP was tested, which showed large inconsistencies that appear as the degree of dispersion of the prediction error. Statistical models use historical data to predict future values and they are based on statistics, probability theory, and random processes. Jenkins models [25], Bayesian multiple kernel regression [23], and copula theory [22] are some other statistical models that can be used in wind energy forecasting.

Puri et al. [18] proposed an ANN algorithm that provides successful predictions and uses 30-day wind speed, temperature, and air density data for training and wind energy prediction purposes. ML models have excellent prediction performance [1]. Wang et al. [21] proposed a hybrid model that combines the (ICEEMDAN) method for obtaining short-term wind speed forecasts and the autoregressive integrated moving average (ARIMA) method used to determine the input variables. This decomposition is considered to be powerful in non-linear processing. A paper [26] combining full ensemble empirical mode decomposition adaptive noise (CEEMDAN), chaotic local search flower pollination algorithm (CLSFPA), five neural networks, and no-negative constraint theory (NNCT) - is proposed for short-term high-precision wind speed prediction.

2.1 Variational Mode Decomposition VMD

Variational mode decomposition (VMD) [6] is a signal decomposition method that extracts a set of mode functions with different center frequencies and bandwidths from a given signal. The method is based on the concept of a variational principle, which minimizes the difference between the original signal and its decomposed components.

It works by iteratively solving a set of constrained optimization problems. Starting with a signal $x(t)$ that we wish to decompose into K mode functions $u_k(t)$, each with a center frequency ω_k and a bandwidth $\Delta\omega_k$. The VMD method seeks to minimize the following cost function:

$$J = \sum_{k=1}^{K} \left[||u_k||_2^2 + \lambda \Big(||H(u_k) - \omega_k u_k||_2^2 - \mu^2 \Big) \right] \tag{1}$$

Here, H denotes the Hilbert transform, which transforms a real-valued signal to its analytic representation, and λ and μ are regularization parameters that control the smoothness and frequency resolution of the mode functions.

The optimization problem can be solved using an iterative algorithm, where at each iteration step m, we solve the following problem:

$$u_k^{m+1} = arg\ \underset{u_k}{min}\Big[||x - \sum_{j=1}^{K} u_j^m||_2^2 + \lambda(||H(u_k) - \omega_k u_k||_2^2 - \mu^2)\Big] \tag{2}$$

This can be done efficiently using a Fourier-based algorithm that alternates between solving for the mode functions and updating the frequency parameters ω_k.

After obtaining the mode functions $u_k(t)$, we can reconstruct the original signal $x(t)$ as follows:

$$x(t) = \sum_{k=1}^{K} u_k(t) \tag{3}$$

The VMD method has been shown to be effective in a wide range of applications, including crude oil price forecasting [12]. Its advantages over other decomposition methods include its ability to extract modes with arbitrary center frequencies and bandwidths, and its ability to handle signals with nonstationary and nonlinear characteristics.

2.2 Bidirectional Long Short-Term Memory (BiLSTM)

A Bidirectional Long Short-Term Memory (BiLSTM) neural network is built of two LSTM units. To increase the accuracy of prediction the BiLSTM network can use the past and future states. This recurrent neural network, with forward and backward propagation, overcomes the insufficient data information of the LSTM neural network. The BiLSTM reduces the problem of vanishing gradients and exploding gradients.

The same data are input to the forward and the backward LSTM layer. In the forward LSTM layer, the forward calculation is performed from time 1 to time t. In the backward LSTM layer, the backward calculation is performed from time t to time 1.

In the forward direction, the internal state stores information from the past time series values in $H_{f(t)}$, shown in Eq. (4).

$$H_{f(t)} = \psi(W_{fh}X_t + W_{fhh}H_{f(t-1)} + b_{fb}) \tag{4}$$

In the backward direction, information from the future sequence values is stored in $H_{b(t)}$, Eq. (5).

$$H_{b(t)} = \psi(W_{bh}X_t + W_{bhh}H_{b(t+1)} + b_b) \tag{5}$$

where W_{fh}, W_{fhh}, W_{bh} and W_{bhh} are weight matrices b_{fb} and b_b represent bias signals in forward and backward directions, respectively.ψ is the recurrent layer activation function.

The output vector of the BiLSTM Y_t is represented by Eq. (6).

$$Y_t = \sigma(W_{fhy}H_{f(t)} + W_{bhy}H_{b(t)} + b_y) \tag{6}$$

where W_{fhy} and W_{bhy} are the forward and backward weights from the internal unit to the output, respectively, σ is the activation function of the output layer, which can be set to sigmoid or linear functions and b_y denotes the bias vector of the output layer.

2.3 Metaheuristics Optimization

Metaheuristic optimization algorithms are inspired by processes in nature. They work by imitating physical and chemical processes as well as mathematical processes. Also, by imitating the social behavior of different various animals and collective intelligence [8,17] swarm intelligence is one of the most widely used algorithm groups in almost all areas of science, technology, and industry. An additional advantage of swarm intelligence algorithms is their ability to tackle NP-hard problems, something considered impossible via traditional deterministic approaches.

Due to increased computational resources becoming available, a large number of algorithms based on simulating the behavior of different animals have been proposed. Some notably interesting examples the artificial bee colony (ABC) [14] algorithm, Harris hawk optimizer (HHO) [9] and the novel chimp optimization algorithm (ChOA) [15]. However, mathematics have also served as an inspiration for algorithms with a notable examples being the sine cosine algorithm (SCA) [3].

Swarm intelligence algorithms have seen a great range of applications across several fields. Some notable examples include tackling forecasting in finance [12,20]. These algorithms excel at optimizations and have found numerous applications [4,5,11]. An additionally interesting approach is the use of swarm intelligence algorithms for selecting and optimizing hyperparameter values [10,13].

3 Methods

3.1 Original Reptile Search Algorithm (RSA)

At its core of the RSA [2] is a swarm-based algorithm, that draws inspiration from hunting and social behaviors observed in crocodiles. Several phases are at the core of the algorithm. The first stage involves initializing a population according to Eq. (7)

$$X_i = LB + rand(UB - LB) \tag{7}$$

with X_i representing the i-th agent, LB denoting the lower and UB the upper bounds of the search space.

The algorithm performs an optimization through two distinct phases, exploration, and exploitation. Agents simulate crocodile sprawl walks in the former stage. Depending on the current iteration t, in cases when $t \leq 0.25T$, where T is the maximum number of iterations, a high walk will be simulated. In cases where $t > 0.25T$ and $t \leq 0.5T$, a sprawling walk is simulated. These simulations can be defined according to the following:

$$X_i^{t+1} = \begin{cases} X_{best}^t - \eta \times \beta - R_i^t \times rand, & t \leq \frac{T}{4}, \\ X_{best}^t \times X_{rand}^t \times ES \times rand, & t \leq \frac{T}{2} \text{ and } t > \frac{T}{4}, \end{cases} \tag{8}$$

$$\eta_i = X_{best}^t \times P_i, \tag{9}$$

$$R_i = \frac{X_{best}^t - X_i^t}{X_{best}^t + \epsilon}, \tag{10}$$

$$ES = 2 \times r_1 \times (1 - \frac{1}{T}), \tag{11}$$

$$P_i = \alpha + \frac{X_i^t - M(X_i^t)}{X_{best}^t \times (UB - LB) + \epsilon}, \tag{12}$$

in which X_{best}^t denotes the best obtained solution thus far, β represents a constant that defines exploration speed set to 0.1 and X_{rand}^t represent an arbitrary agent. An arbitrary descending value in range $[-2, 2]$ is denoted with ES, and a minimum value ϵ is also introduced to prevent division by zero. A random number in range $[-1, 1]$ is described as r_1, α is a constant with a value of 0.1, and finally $rand$ represents a random value from interval $[0, 1]$.

The exploitation phase simulates hunting behaviors. Coordination is practiced when $t < 0.75T$ and $t \geq 0.5T$. Cooperation is applied when $t < T$ and $t \geq 0.75T$. These actions are mathematically modeled according to the following:

$$X_i^{t+1} = \begin{cases} X_{best}^t \times P_i \times rand, & \frac{3t}{4} \text{ and } t > \frac{T}{2} \\ X_{best}^t - \eta_i \times \epsilon - R_i^t, & t \leq T and\, t > \frac{3T}{4} \end{cases} \tag{13}$$

3.2 Hybrid RSA (HRSA)

Extensive testing has shown that the novel RSA possesses a very powerful exploitative mechanism and performs admirably in most applications. However, later iterations can suffer from a lack of exploitation power. This can lead to less-than-optimal convergence rates and result in reduced overall performance.

To bolster the exploitation mechanism of the original this work proposes a low-level hybrid approach. By introducing search elements from the firefly algorithm (FA), that's known for its distinctly capable exploitation mechanism, performance can be boosted. The elements of the introduced FA are described in Eq. (14).

$$x_i = x_i + \zeta_0 e_i^{-\gamma r_{ij}^2} j(x_j - x_i) + \nu \epsilon_t \tag{14}$$

where x_i represents the position i-th agent, distance to between two agents denoted as $r_{i,j}$ and ν_0 being the attraction of fireflies at distance $r = 0$. The media absorption coefficient is γ, and *rand* is a random value in range $[0, 1]$. Finally, ζ is a randomization factor that is gradually reduced through iterations. The proposed algorithm implements the FA search mechanisms during even other iterations to improve exploration power.

The introduced mechanism is used to enhance exploration in the later stages of the optimization. To accomplish this an additional parameter ψ is introduced that defines the iteration threshold after which the FA mechanism is activated. Empirically, the threshold value was determined to be $\frac{T}{3}$ Additionally, each agent is assigned a parameter ω. One $t > \psi$ each agent selects a value for ω from a uniform distribution in range $[0, 1]$. If $\omega < 0.5$ the agent uses the standard RSA search, otherwise FA search is used.

The described algorithm is named the hybrid RSA due to the utilized hybridization techniques used to enhance it. The pseudocode is described in Alg 1.

Algorithm 1. Pseudocode of the introduced hybrid RSA (HRSA) algorithm

```
Set RSA, FA, and introduced parameter values
Initialize randomized population
while t < T do
  Evaluate agent fitness
  Update best solution
  Update ES parameter
  for all Agents in populations do
    if ψ < T/3 then
      Update agent position using RSA search
    else
      Select random value in range [0, 1] for agent ω
      if ω < 0.5 then
        Update agent position using RSA search
      else
        Update agent position using FA search
      end if
    end if
  end for
  Increment t
end while
```

4 Experimental Setup

The following section describes the utilized dataset, and evaluation metrics used during experimentation. Following this, the detail of the experimental procedure is provided.

4.1 Dataset

The GEFCom2012 challenge dataset was originally created as a competition dataset and was available on Kaggle. However, it has since been repurposed for use in wind energy generation forecasting. The dataset contains weather and generation data for 7 anonymized wind farms in mainland China. The power generation data has been normalized to ensure anonymity. The dataset's aim is to improve forecasting practices across industries and to promote analytics in power engineering education.

For experimental purposes, the meteorological data has been trimmed into 12-h predictions in every forecast, combined with the available normalized real-world wind generation data for each respective wind farm on an hourly basis. The original dataset covered four and a half years' worth of hourly resolution data, with the final half year's worth of data reserved for testing. Due to a large amount of data and missing values, a reduced portion of the available data was used for experimentation. The experimental dataset used in simulations covers 2 years' worth of data (from January 1, 2009 - December 31, 2010) for a single anonymized wind farm, containing a total of 13176 instances for wind farm 3. The initial 70% of the dataset is used for training, followed by 10% for validation. Finally, the last 20% was used for testing. The dataset split for the target feature can be seen in Fig. 1.

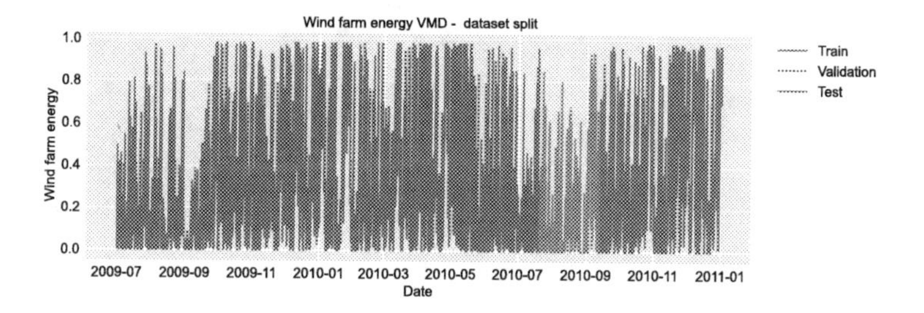

Fig. 1. Target feature dataset split plot

4.2 Metrics

The experimental procedure includes applying various cutting-edge metaheuristic algorithms to determine the best BiLSTM parameters. The performance of the algorithms is measured using several metrics, namely MAE, MSE, $RMSE$, and R^2, which are defined in equations (15), (16), (17), and (18), respectively.

$$MAE = \frac{1}{n}\sum_{i=1}^{n} |y_i - \hat{y_i}| \tag{15}$$

$$MSE = \frac{1}{n}\sum_{i=1}^{n} \frac{(y_i - \hat{y_i})^2}{y_i} \tag{16}$$

$$RMSE = \sqrt{\frac{1}{n}\sum_{i=1}^{n}(y_i - \hat{y_i})^2} \tag{17}$$

$$R^2 = 1 - \frac{\sum_{i=1}^{n}(y_i - \hat{y_i})^2}{\sum_{i=1}^{n}(y_i - \bar{y})^2}, \tag{18}$$

4.3 Setup

For experimentation, each of the features which include the Zonal Wind Component, Meridional Wind Component, Wind Speed, Wind Direction, and Wind Power was subjected to VMD decomposition. Decomposition was conducted with a K value of three. Residuals were also considered giving a total of 4 signals per input component. The decomposition results are shown in Fig. 2.

The BiLSTM models were tasked with forecasting wind power generation three steps ahead, based on 6 steps of input. Each step represents an hour's worth of data. Metaheuristics were tasked with selecting optimal parameters of the BiLSTM models in order to attain the best possible performance. The parameters and their respective ranges are as follows: the number on network layers range $[1, 2]$, and neuron count in each of those layers were independently selected from range $[100, 200]$. Additionally, learning rate was tuned within a range of $[0.001, 0.01]$, dropout from range $[0.05, 0.2]$, and the number of training epochs selected form a range $[300, 600]$.

The evaluated metaheuristics include the introduced HRSA, as well as the original RSA. Additionally, the well-known GA, ABC, HHO, and novel ChOA were included in the assessment. Each assessed metaheuristic was given a VMD-BiLSTM prefix. Each metaheuristic algorithm was assigned a population of five agents and allowed 8 iterations to handle the optimization. Finally, to account for the inherent randomness in these algorithms, the optimizations were carried out in 30 independent.

5 Results and Discussion

The performance of each metaheuristic has been evaluated using the described metrics. Overall objective function scores for each algorithm over 30 independent runs are shown in Table 1.

As the results in Table 1 indicate, the proposed metaheuristic demonstrated the best performance in the best, works and mean cases. Only slightly being outdone by the ABC algorithm in the median case. Further detailed metrics are provided in Table 2.

The detailed metrics shown in Table 2 further emphasize the improvements made by the introduced algorithm, as it attained the best scores in one step ahead of forecasting, as well as overall forecasting accuracy. Two-step ahead and three-step ahead results indicated that the best-performing algorithm for these specific cases was in the GA. This is of course to be expected as no single approach works best for all test cases as per the no free lunch theorem [24].

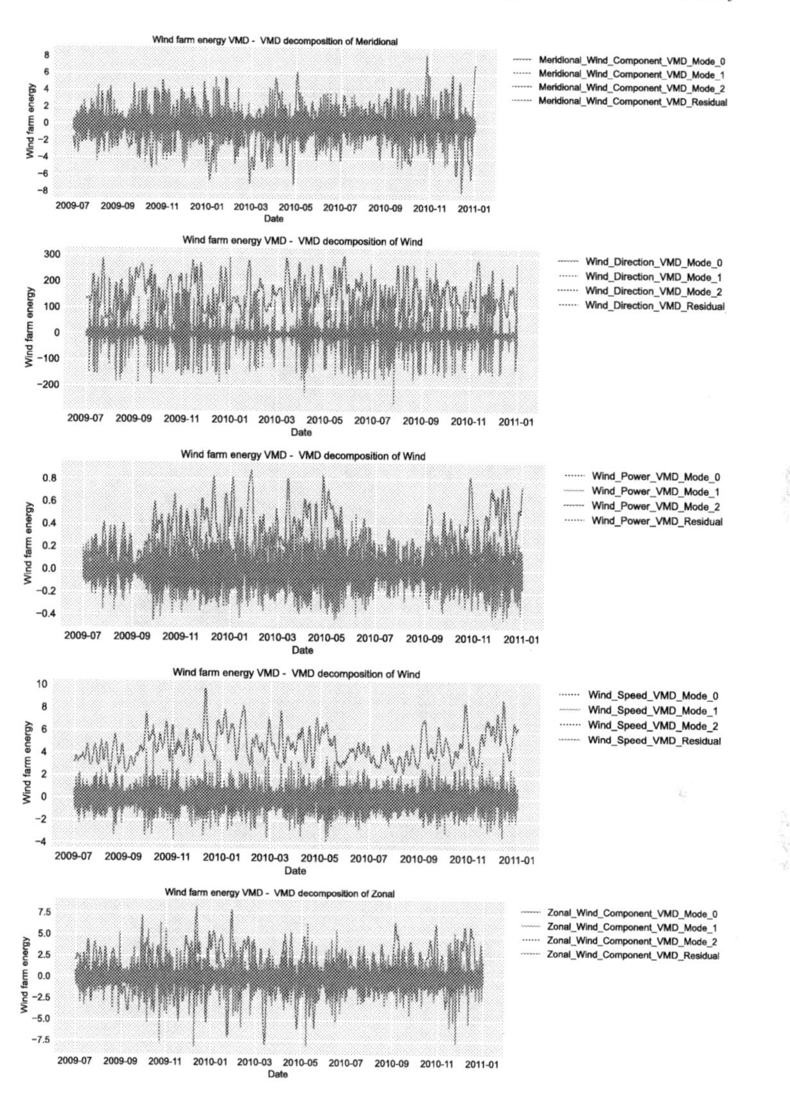

Fig. 2. Target feature decomposition components

Table 1. Overall objective function results of all evaluated approaches over 30 independent runs

Method	Best	Worst	Mean	Median	Std	Var
VMD-BiLSTM-HRSA	**0.009815**	**0.010144**	**0.009997**	0.010035	1.25E−04	1.57E−08
VMD-BiLSTM-RSA	0.010023	0.010189	0.010101	0.010067	**6.48E−05**	**4.20E−09**
VMD-BiLSTM-GA	0.009989	0.010197	0.010072	0.010068	7.54E−05	5.69E−09
VMD-BiLSTM-ABC	0.009874	0.010218	0.010007	**0.009944**	1.51E−04	2.28E−08
VMD-BiLSTM-HHO	0.009910	0.010123	0.010010	0.010024	7.85E−05	6.17E−09
VMD-BiLSTM-ChOA	0.009951	0.010204	0.010095	0.010110	1.07E−04	1.14E−08

Table 2. Detail metrics for each forecasting step of the best-performing model of every evaluated approach

	Error indicator	VMD-BiLSTM-HRSA	VMD-BiLSTM-RSA	VMD-BiLSTM-GA	VMD-BiLSTM-ABC	VMD-BiLSTM-HHO	VMD-BiLSTM-ChOA
One-step ahead	R^2	**0.879415**	0.866626	0.858819	0.871772	0.876294	0.867287
	MAE	0.082132	0.085042	0.086451	0.083652	**0.081983**	0.084678
	MSE	**0.013763**	0.015222	0.016113	0.014635	0.014119	0.015147
	RMSE	**0.117315**	0.123379	0.126939	0.120975	0.118823	0.123073
Two-step ahead	R^2	0.923686	0.927520	**0.932068**	0.927649	0.924854	0.928427
	MAE	0.065635	0.063413	**0.061008**	0.063612	0.065508	0.063160
	MSE	0.008710	0.008272	**0.007753**	0.008258	0.008577	0.008169
	RMSE	0.093327	0.090953	**0.088053**	0.090871	0.092610	0.090381
Three-step ahead	R^2	0.938918	0.942388	**0.946541**	0.941040	0.938358	0.942718
	MAE	0.060081	0.058215	**0.055777**	0.058770	0.060518	0.058145
	MSE	0.006971	0.006575	**0.006101**	0.006729	0.007035	0.006538
	RMSE	0.083495	0.081089	**0.078111**	0.082032	0.083877	0.080856
Overall Results	R^2	**0.914006**	0.912178	0.912476	0.913487	0.913168	0.912811
	MAE	0.069282	0.068890	**0.067745**	0.068678	0.069336	0.068661
	MSE	**0.009815**	0.010023	0.009989	0.009874	0.009910	0.009951
	RMSE	**0.099069**	0.100117	0.099947	0.099368	0.099551	0.099755

Network control selected by each metaheuristic for the best-performing models is shown in Table 3.

Table 3. Parameters selected for the best performing models of each evaluated approach

Method	Layer 1 neurons	Learning rate	Epochs	Dropout	Layer count	Layer 2 neurons
VMD-BiLSTM-HRSA	200	0.010000	569	0.112238	2	100
VMD-BiLSTM-RSA	124	0.009831	464	0.200000	1	104
VMD-BiLSTM-GA	100	0.010000	600	0.200000	2	100
VMD-BiLSTM-ABC	161	0.009894	469	0.197079	1	116
VMD-BiLSTM-HHO	129	0.010000	320	0.058345	1	152
VMD-BiLSTM-ChOA	100	0.009715	520	0.200000	1	100

Finally, forecasts made by the best-performing model compared to actual values are shown in Fig. 3.

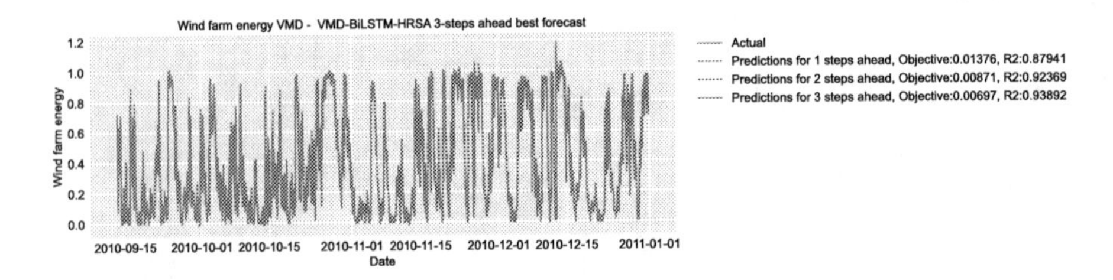

Fig. 3. Best performing model wind energy forecasts

6 Conclusion

This research tackles an important emerging issue with the transition towards renewable energy sources. The pressing challenge of accurately forecasting power generation is an important step in reducing humanity's reliance on limited sources of energy such as fossil fuels. Novel techniques capable of haling non-linear relations between data are needed. In this work, a novel approach for forecasting wind farm power generation is proposed based on time series BiLSTM forecasting. Additionally, to better handle the complexities associated with the use of VMD is explored. A novel metaheuristic, the HRSA, is introduced and tasked with selecting hyperparameter values of BiLSTM networks. The introduced metaheuristic outperformed competing algorithms, emphasizing the improvements made. The proposed model demonstrated admirable performance in tackling this challenging task. Computational demands limit the extent of testing. Future work will focus on further refining the forecasting methodology. Additional decomposition methods and their combinations will be explored for tackling similar forecasting issues. Finally, a further application for the introduced metaheuristic will be explored.

References

1. Abdoos, A.A.: A new intelligent method based on combination of VMD and elm for short term wind power forecasting. Neurocomputing **203**, 111–120 (2016)
2. Abualigah, L., Abd Elaziz, M., Sumari, P., Geem, Z.W., Gandomi, A.H.: Reptile search algorithm (RSA): a nature-inspired meta-heuristic optimizer. Expert Syst. Appl. **191**, 116158 (2022)
3. Abualigah, L., Diabat, A.: Advances in sine cosine algorithm: a comprehensive survey. Artif. Intell. Rev. **54**(4), 2567–2608 (2021)
4. AlHosni, N., et al.: The XGBoost model for network intrusion detection boosted by enhanced sine cosine algorithm. In: Chen, J.I.Z., Tavares, J.M.R.S., Shi, F. (eds.) ICIPCN 2022. LNNS, vol. 514, pp. 213–228. Springer, Cham (2022). https://doi.org/10.1007/978-3-031-12413-6_17
5. Bacanin, N., Antonijevic, M., Bezdan, T., Zivkovic, M., Rashid, T.A.: Wireless sensor networks localization by improved whale optimization algorithm. In: Mathur, G., Bundele, M., Lalwani, M., Paprzycki, M. (eds.) Proceedings of 2nd International Conference on Artificial Intelligence: Advances and Applications. Algorithms for Intelligent Systems, pp. 769–783. Springer, Singapore (2022). https://doi.org/10.1007/978-981-16-6332-1_62
6. Dragomiretskiy, K., Zosso, D.: Variational mode decomposition. IEEE Trans. Signal Process. **62**(3), 531–544 (2013)
7. Foucquier, A., Robert, S., Suard, F., Stéphan, L., Jay, A.: State of the art in building modelling and energy performances prediction: a review. Renew. Sustain. Energy Rev. **23**, 272–288 (2013)
8. Gandomi, A.H., Yang, X.S., Talatahari, S., Alavi, A.H.: Metaheuristic algorithms in modeling and optimization. Metaheuristic Appl. Struct. Infrastruct. **1**, 1–24 (2013)
9. Heidari, A.A., Mirjalili, S., Faris, H., Aljarah, I., Mafarja, M., Chen, H.: Harris hawks optimization: algorithm and applications. Futur. Gener. Comput. Syst. **97**, 849–872 (2019)
10. Jovanovic, D., Antonijevic, M., Stankovic, M., Zivkovic, M., Tanaskovic, M., Bacanin, N.: Tuning machine learning models using a group search firefly algorithm for credit card fraud detection. Mathematics **10**(13) (2022). https://doi.org/10.3390/math10132272

11. Jovanovic, L., et al.: Machine learning tuning by diversity oriented firefly metaheuristics for industry 4.0. Expert Syst. e13293 (2023). https://doi.org/10.1111/exsy.13293
12. Jovanovic, L., et al.: Multi-step crude oil price prediction based on LSTM approach tuned by salp swarm algorithm with disputation operator. Sustainability **14**(21), 14616 (2022)
13. Jovanovic, L., Zivkovic, M., Antonijevic, M., Jovanovic, D., Ivanovic, M., Jassim, H.S.: An emperor penguin optimizer application for medical diagnostics. In: 2022 IEEE Zooming Innovation in Consumer Technologies Conference (ZINC), pp. 191–196. IEEE (2022)
14. Karaboga, D., Basturk, B.: On the performance of artificial bee colony (ABC) algorithm. Appl. Soft Comput. **8**(1), 687–697 (2008)
15. Khishe, M., Mosavi, M.R.: Chimp optimization algorithm. Expert Syst. Appl. **149**, 113338 (2020)
16. Kisi, O., Ozkan, C., Akay, B.: Modeling discharge-sediment relationship using neural networks with artificial bee colony algorithm. J. Hydrol. **428**, 94–103 (2012)
17. Kumar, A.: Application of nature-inspired computing paradigms in optimal design of structural engineering problems-a review. Nat.-Inspired Comput. Paradigms Syst. 63–74 (2021)
18. Puri, V., Kumar, N.: Wind energy forecasting using artificial neural network in Himalayan region. Model. Earth Syst. Environ. **8**(1), 59–68 (2022)
19. Solyali, D.: A comparative analysis of machine learning approaches for short-/long-term electricity load forecasting in Cyprus. Sustainability **12**(9), 3612 (2020)
20. Stankovic, M., Jovanovic, L., Bacanin, N., Zivkovic, M., Antonijevic, M., Bisevac, P.: Tuned long short-term memory model for ethereum price forecasting through an arithmetic optimization algorithm. In: Abraham, A., Bajaj, A., Gandhi, N., Madureira, A.M., Kahraman, C. (eds.) IBICA 2022. LNNS, vol. 649, pp. 327–337. Springer, Cham (2023)
21. Wang, L., Li, X., Bai, Y.: Short-term wind speed prediction using an extreme learning machine model with error correction. Energy Convers. Manage. **162**, 239–250 (2018)
22. Wang, Y., et al.: A new method for wind speed forecasting based on copula theory. Environ. Res. **160**, 365–371 (2018)
23. Wang, Y., Hu, Q., Meng, D., Zhu, P.: Deterministic and probabilistic wind power forecasting using a variational Bayesian-based adaptive robust multi-kernel regression model. Appl. Energy **208**, 1097–1112 (2017)
24. Wolpert, D.H., Macready, W.G.: No free lunch theorems for optimization. IEEE Trans. Evol. Comput. **1**(1), 67–82 (1997)
25. Yatiyana, E., Rajakaruna, S., Ghosh, A.: Wind speed and direction forecasting for wind power generation using ARIMA model. In: 2017 Australasian Universities Power Engineering Conference (AUPEC), pp. 1–6. IEEE (2017)
26. Zhang, W., Qu, Z., Zhang, K., Mao, W., Ma, Y., Fan, X.: A combined model based on CEEMDAN and modified flower pollination algorithm for wind speed forecasting. Energy Convers. Manage. **136**, 439–451 (2017)
27. Zhao, J., Guo, Y., Xiao, X., Wang, J., Chi, D., Guo, Z.: Multi-step wind speed and power forecasts based on a WRF simulation and an optimized association method. Appl. Energy **197**, 183–202 (2017)

Interpretable Drug Resistance Prediction for Patients on Anti-Retroviral Therapies (ART)

Jacob Muhire, Ssenoga Badru, Joyce Nakatumba-Nabende, and Ggaliwango Marvin(✉)

College of Computing and Information Sciences, Department of Computer Science, Makerere University, P.O Box 7062, Kampala, Uganda
ggaliwango.marvin@mak.ac.ug

Abstract. The challenge of eliminating HIV transmission is a critical and complex under taking, particularly in Africa, where countries like Uganda are grappling with a staggering 1.6 million people living with the disease. The virus's fast pace of mutation is one of the main challenges in this battle, which often leads to the development of drug resistance and makes it difficult to provide effective treatment through AntiRetroviral Therapies (ART). By leveraging the latest innovations in Smart Technologies and Systems, such as Machine Learning, Artificial Intelligence, and Deep Learning, we can create novel approaches to tackle this issue. We presented a model that predicts which HIV patients are likely to develop drug resistance using viral load laboratory test data and machine learning algorithms. On the remaining 30% of the data, we tested our algorithms after painstakingly training and validating them on the previous 70%. Our findings were remarkable: the Decision Tree algorithm outperformed four other comparative algorithms with an f1 scoring mean of 0.9949, greatly improving our ability to identify drug resistance in HIV patients. Our research highlights the potential of combining data from viral load tests with machine learning techniques to identify patients who are likely to develop treatment resistance. These findings are a significant step forward in our ongoing fight against HIV, and we are confident that they will pave the way for new, innovative solutions to address this global health crisis.

Keywords: Explainable Artificial Intelligence (XAI) · Machine Learning · Viral Load · Drug Resistance · Anti-Retroviral Therapies (ART)

1 Introduction

One of the world's fastest-growing infections is HIV, which results in AIDS (Acquired Immuno Deficiency Syndrome), an untreatable illness that severely impairs the immune system.

Antiretroviral therapies (ARVs) have been instrumental in the fight against the spread of HIV/AIDS, whose benefits include control of morbidity and mortality of patients. However, for the treatment options to be effective, adherence is extremely important to prevent the development of ART drug resistance. Also, routine testing is essential to monitor treatment and to determine suitable interventions or detect drug resistance to the

S. Aurelia et al. (Eds.): ICCSST 2023, CCIS 1973, pp. 43–53, 2024.
https://doi.org/10.1007/978-3-031-50993-3_4

regimen the patient is on [1]. It has been reported that most patients on ART treatment are switched from one treatment line to another within 2 to 3 years of initiation of ART treatment [2]. With this information, there is an urgent need to predict patients that are likely to switch from the first treatment line to the second to improve chances of a better quality of life, control the mutations of the virus, lower the cost burden of treatment by the public health care system and most of all reduce the spread of the HIV.

2 Background and Motivation

HIV (Human Immunodeficiency Virus) is a grave concern. Sadly, there were 650,000 deaths resulting from Acquired Immune Deficiency Syndrome (AIDS) related causes in the same year [3]. In Uganda, the number of individuals living with HIV ranges from 1.3 to 1.6 million [4]. The first known cases of HIV prevalence in Uganda were recorded in the mid-1980s among expectant mothers in Kampala, with a prevalence rate of 11%. Shockingly, by 1992, over 3.3 million Ugandans were infected, representing 18% of the general population [5]. However, Uganda has made commendable progress in reducing HIV prevalence, which currently stands at 5.8%. The assessment also defined adults aged between 15 and 64 years [6]. One of the most significant challenges in managing HIV is the development of drug resistance due to mutations in the virus's genomic composition. Such resistance can significantly impact the effectiveness of Anti Retroviral Therapies (ART) in treating HIV [7]. Machine learning (ML) is a powerful tool that can help us analyze and interpret patterns and structures in data to facilitate learning, synthesis, and decision- making [8, 9]. In recent years, machine learning has demonstrated its potential [10]. Notably, Douglas et al. employed ML to develop and validate. [11]. In light of these developments, the objective of this paper is to evaluate ML models that tell the likelihood of drug failure among HIV patients undergoing ART treatment using viral load test data. By leveraging machine learning approaches, we aim to improve our ability to identify individuals at high risk of developing drug resistance, thereby facilitating more effective treatment and management of HIV.

3 Literature Review

To accelerate the battle against HIV and AIDS globally, the resolution seeks to end the AIDS pandemic by 2030 [12]. However, realizing this ambitious goal requires innovative approaches, particularly in resource-constrained environments like Uganda.

However, realizing this ambitious goal requires innovative approaches, particularly in resource-constrained environments like Uganda which have limited access to resources like testing equipment, computers, electricity, trained health workers among others. Fortunately, the advent of Artificial Intelligence (AI), and Machine Learning (ML), offers a promising solution to the challenges such as traceability of patients' treatment longitudinally. A comprehensive literature review has identified several areas where ML has the potential to significantly advance the fight against HIV/AIDS. For instance, ML can facilitate the early identification of potential candidates for Pre Exposure Prophylaxis (PrEP), a critical approach in controlling the spread of HIV [13]. Additionally, ML can be used to estimate a patient's HIV risk, informing decisions around diagnostics and the use of PrEP [14]. ML techniques have also been effective in identifying HIV predictors [15]. There have been successful applications of Machine Learning which include the prediction of HIV status and drug resistance among same sex partners [16], as a risk-prediction tool for sexually transmitted infections over a year [17]. The significant impact of Machine Learning in the fight against HIV/AIDS underscores the importance of further application in areas such as the prediction of drug resistance in patients on ART treatment. With the right tools, equipped with information from previously unseen patterns, we can take the right and important steps towards achieving the 2030 goals to ultimately end the AIDS epidemic.

3.1 Research Gaps

While the application of machine learning algorithms in HIV treatment and prevention shows immense potential, several limitations must be addressed to optimize their effectiveness. Specifically, one significant gap in the reviewed literature is that many of the algorithms utilized linear models that may not adequately capture the complex, non-linear relationships between patient characteristics and HIV treatment outcomes. Additionally, a key issue is that many of the algorithms were trained on small datasets, with an average of less than 100,000 samples. This can lead to poor predictive results and undermine the accuracy of these tools.

Another critical issue is the lack of interpretability of many machine learning algorithms. Inadequate clarity may reveal potential biases in the data or limit their effectiveness in clinical settings. To fully realize the potential of machine learning in the fight against HIV, it is crucial to address these limitations and develop more robust and interpretable models that capture the nuances of this complex disease.

By expanding the scope of machine learning research to include more complex and diverse data sets, we can improve the predictive power of these algorithms and achieve more accurate results. Additionally, developing transparent and interpretable machine learning models can help to build trust in these tools and foster greater acceptance in clinical settings. Ultimately, by addressing these limitations and unlocking the full potential of machine learning in HIV prevention and treatment, we can take a significant step forward in the global fight against this devastating disease.

3.2 Paper Contributions

Our implementation is designed to effectively tackle the challenges associated with small data sets. To ensure good generalizability, we utilized a robust data set consisting of 998,295 records. Furthermore, we adopted the Decision Tree algorithm, a proven non-linear method that yielded exceptional results with an F1-score of 0.99715. This approach not only enhances the reliability of our findings, but also underscores the importance of leveraging cutting-edge techniques to overcome limitations inherent in working with small data sets.

4 Data Analysis and Methods

4.1 Dataset Description

We used Viral Load laboratory test data from the Central Public Health Laboratories, a department in the Ministry of Health of Uganda. Results of HIV samples collected through the national sample transportation network that consists of 100 hubs, hub riders, drivers and hub coordinators. We performed the prediction on data set for 2019, comprising 349,284 male and 686,272 female respondents. With 91.7%, 4.2% and 0.5% ARV adherence for good, fair and poor categories respectively.

4.2 Data Preparation and Exploratory Data Analysis

The data set contained 31 features which were reduced to 12 relevant features for this paper as shown in Table 1. The features removed were indeterminate and non informative such as the sample id, date of sample collection, date of sample reception, dispatch date among others. We did data cleaning by deletion and data transformation on the various features because some samples were missing or had aberrant values like age, treatment duration, ARV adherence and gender. We considered adherence to treatment as good $\geq 95\%, fair\ \ 85-94\%\ and\ poor\ \leq 85\%$

We eliminated 37,690 records that did not have any record for ARV adherence. Also we eliminated 13,019 gender records that did not have a clear indicator of either male or female or were left blank. The characteristics of the data are reflected in Table 1. This resulted in 328,704 male and 669,591 female in the final data set. Using the label encode method properly on the aspects of greatest interest, we appropriately encoded both the nominal and ordinal variables which were treatment duration, gender, indication for viral load testing, ARV adherence, suppressed, and result alphanumeric.

4.3 Methodology

We explored five machine learning algorithms for predicting patients on ART treatment likely to develop drug resistance which were Logistic Regression (LR), Random Forest (RF),Gaussian Naive Bayes (GNB), Decision Tree (DT), and k-nearest neigh bors (KNN). Definitions for LR [18], DT [19], RF [20], KNN [21] and GNB [22] are highlighted here.

4.4 Model Evaluation

We adopted the standard indicators evaluate the selected models' performance which were accuracy, precision, recall, and F1-score.

On completion of data processing and feature selection, the 2019 data was split (see Fig. 1 flow chart) into three groups ARV adherence of good, fair and bad adherence. All the data analyses were performed with Python version 3.9.9.

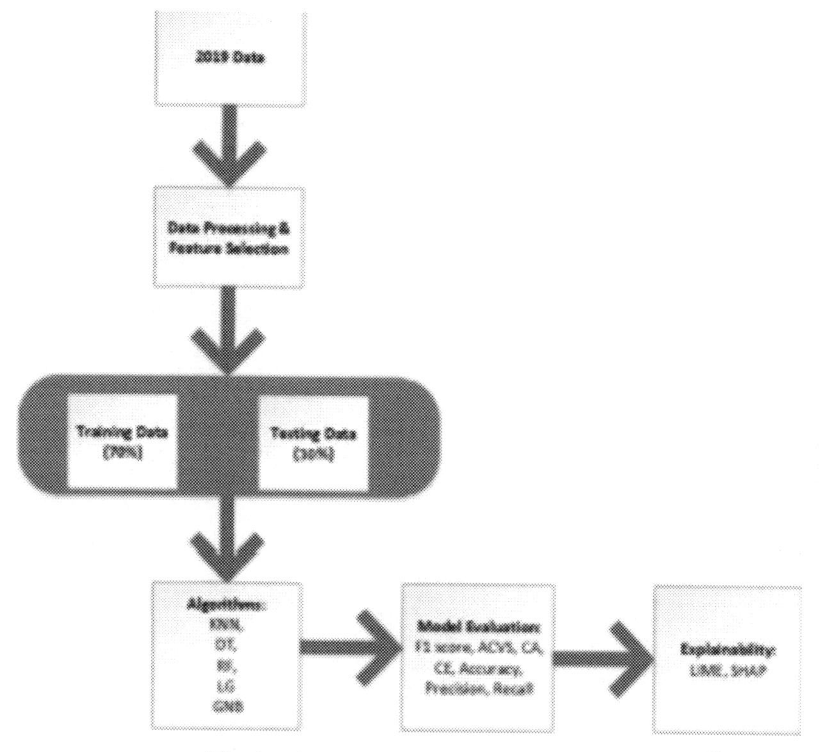

Fig. 1. Flow Chart of models development

Table 1. Summary of Viral Load Laboratory Dataset

No	Column	Data Type
0	ART NUMBER	object
1	Gender	object
2	DOB	object
3	Age	object
4	Treatment initiation date	object
5	Treatment duration	object
6	Indication for VL Testing	object
7	ARV adherence	object
8	Result alphanumeric	object
9	Suppressed	int64
10	Current WHO stage	object

4.5 Feature Importance

The intricate model is first trained by the explanation. The value of the characteristic are displayed. This demonstrates the significance of importance in features in order of importance such as result alphanumeric, current regimen, and Indication for viral load Testing in feature prediction shown in Fig. 2.

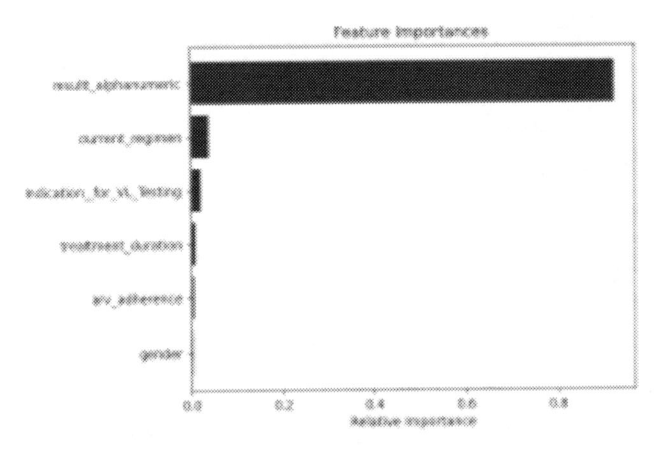

Fig. 2. Feature Importance

5 Results and Discussion

Two common visualizations were chosen to report the results of the our exploratory analysis which are the correlation heatmap and box plot. The heatmap (Fig. 3) shows correlation between variables in a range of -1 to $+1$. We see that as suppression increases the result alphanumeric (patient viral load count) reduces, shown by the correlation value 0.7. The diagonals are all yellow indicating that the squares are positively correlating to themselves. In Fig. 4, a box plot representing patient regimen against the suppression in three categories 1, 2 and 3 which are good, fair and poor suppression respectively. The box plots of the three categories show that the overall patients have high agreement in their respective categories with more patients in category 2 having a similar regimen.

In Fig. 4, a box plot representing patient regimen against the suppression in two categories namely suppressed and non-suppressed. The box plots of the two categories show that the overall patients have high agreement in their respective categories with more patients non-suppressing.

5.1 ML Model Selection and Optimization

Our work demonstrates that a good portion of high accuracy can be achieved by applying machine learning concepts. We contrasted DT, LR, GNB, RF, and KNN. The following measures for each model were presented; accuracy, precision, recall,

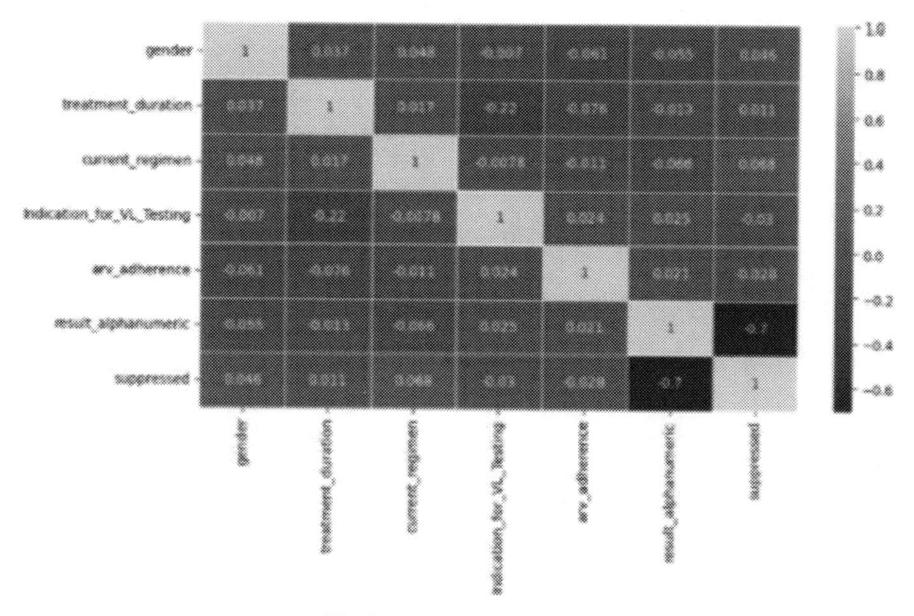

Fig. 3. Correlation Heatmap

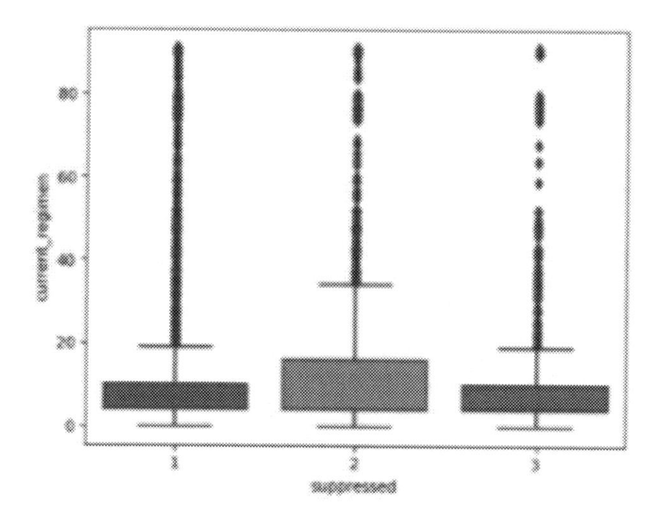

Fig. 4. Suppression Box Plot

F-measure, specificity, classification accuracy, classification error and average cross validation score values.

A confusion matrix (Fig. 5) shows the results of Random Forest indicating 263,809 correctly predicted true positive values, 27878 true negatives however in Table 2, we observed that Decision Tree classification model performed best for prediction with class labels.

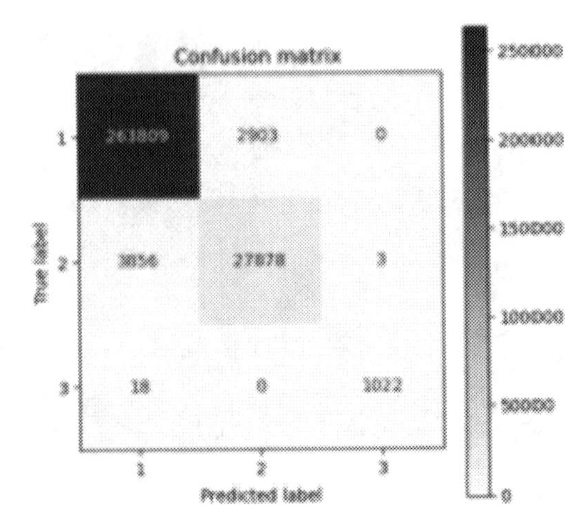

Fig. 5. Random Forest confusion matrix

5.2 ML Model Selection Accountability

Users must be able to comprehend models for AI systems to be trusted by people. LIME (Local Interpretable Model-Agnostic Explanations) offers a broad frame work for understanding "black boxes" and clarifies the "why" of predictions or recommendations provided by AI. [23] and also explainability according to [24].

The observation that needs to be explained and the model prediction that needs to be interpreted are passed to the LIME explainer. The LIME method creates new samples in the vicinity of the observation to be explained and uses the model to anticipate what will happen in those new samples. We just utilize num features to describe the quantity of features that will be visible. The results from LIME show in Fig. 6 and Fig. 7, that the variables affecting the overall outcome for a patient to be HIV positive and likely to develop drug resistance are poor arv adherence, viral load results, indication for viral load testing, gender, current regimen and the treatment duration not being a variable to rely on as a predictor for drug resistance.

The SHAP plot (Fig. 8), results show that variables which are the result alphanumeric (viral load result), current regimen and indication for viral load testing as the best predictors for drug resistance while ARV adherence was of low value as a predictor for drug resistance.

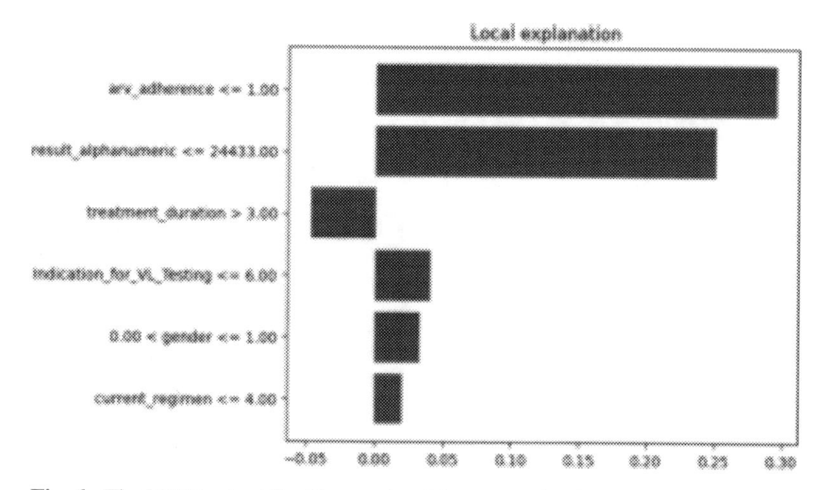

Fig. 6. The LIME output for Observation 1 in x test using Random forest classifier

Table 2. Statistical metrics for the five models

Models	RF	GNB	KNN	DT	LR
F1	0.9874	0.9684	0.9939	0.9949	0.9634
Model accuracy	0.9772	0.9439	0.9889	0.9949	0.9297
Precision	0.9891	0.9654	0.9982	0.9982	0.9759
Recall	0.9856	0.9714	0.9895	0.9895	0.9504
Specificity	0.9057	0.7236	0.9841	0.9841	0.9800
Average cross validation score	0.9769	0.9440	0.9888	0.9888	0.9310
Classification accuracy	0.9774	0.9437	0.9890	0.9949	0.9330
Classification error	0.0226	0.0563	0.0110	0.0051	0.0670
True Positives(TP)	263809	257484	266244	266061	260293
True Negatives(TN)	27878	24161	28903	30874	18151
False Positives(FP)	2903	9228	468	651	6419
False Negatives(FN)	3856	7576	2829	863	13586

6 Conclusion and Future Works

In this work, we have demonstrated the promising potential of utilizing viral load test data, in conjunction with the Decision Tree algorithm. Our analysis was conducted using secondary laboratory test data, limited by constraints of time and finances. However, our results suggest that with sufficient resources, further prospective studies incorporating clinical data could be conducted to enhance the accuracy and robustness of our predictive model. Such studies could capture additional predictors, including behavioral characteristics, and provide a more comprehensive understanding of the factors driving drug resistance development in this patient population. These findings have important implications for clinical decision-making and may ultimately con tribute to the development of more personalized and effective treatment strategies for HIV-positive individuals.

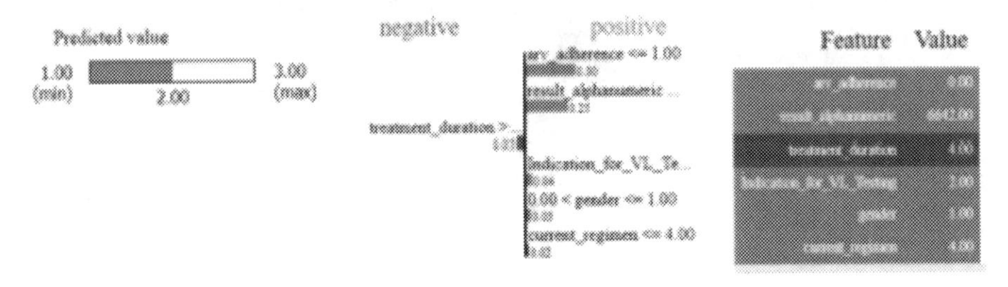

Fig. 7. LIME outcome for Observation 1 using Random forest classifier

Fig. 8. SHAP values for Random Forest Classifier

Acknowledgment. We thank Dr. Isaac Ssewanyana at the Central Public Health Laboratories for his support in granting access to the data and providing guidance during the paper development process.

References

1. Watera, C., et al.: HIV drug resistance among adults initiating antiretroviral therapy in Uganda. J. Antimicrob. Chemother. **76**(9), 2407–2414 (2021). https://doi.org/10.1093/jac/dkab159. Accessed 16 Feb 2023
2. Collins, I.J., et al.: Incidence of switching to second-line antiretroviral therapy and associated factors in children with HIV: an international cohort collaboration. Lancet HIV **6**(2), e105-e115 (2019). https://doi.org/10.1016/s2352-3018(18)30319-9. Accessed 16 Feb 2023
3. Yebra, G.: Analysis of the history and spread of HIV-1 in Uganda using phylo dynamics. J. Gen. Virol. **96**(7), 1890–1898 (2015). https://doi.org/10.1099/vir.0.000107
4. Estimated number of people (all ages) living with HIV. World Health Organization (WHO). https://www.who.int/data/gho/data/indicators/indicatordetails/GHO/estimated-number-of-people-living-with-hiv. Accessed 10 Feb 2023
5. Population, total - Uganda—data. World Bank Open Data—Data. https://data.worldbank.org/indicator/SP.POP.TOTL?locations=UG. Accessed 16 Feb 2023
6. Uganda summary sheet 2020-2021 - PHIA project. PHIA Project. https://phia.icap.columbia.edu/uganda-summary-sheet-2020-2021/. Accessed 16 Feb 2023
7. HIV drug resistance. World Health Organization (WHO). https://www.who.int/teams/global-hiv-hepatitis-and-stisprogrammes/hiv/treatment/hiv-drug-resistance. Accessed 16 Feb 2023

8. What is machine learning - ML - and why is it important?—netapp. Data Man agement Solutions for the Cloud—NetApp. https://www.netapp.com/artificialintelligence/what-is-machine-learning. Accessed 16 Feb 2023
9. Marvin, G., Alam, M.G.R.: A machine learning approach for predicting therapeutic adherence to osteoporosis treatment. In: 2021 IEEE Asia-Pacific Conference on Computer Science and Data Engineering (CSDE), Brisbane, Australia, pp. 1–6 (2021)
10. Mutai, C.K., McSharry, P.E., Ngaruye, I., Musabanganji, E.: Use of machine learning techniques to identify HIV predictors for screening in sub-Saharan Africa. BMC Med. Res. Methodol. **21**(1) (2021). https://doi.org/10.1186/s12874-021-01346-2. Accessed 16 Feb 2023
11. Krakower, D.S., et al.: Development and validation of an automated HIV prediction algorithm to identify candidates for pre-exposure prophylaxis: a mod elling study. Lancet HIV **6**(10), e696–e704 (2019). https://doi.org/10.1016/s2352-3018(19)30139-0
12. Countries Commit to Action to End AIDS by 2030. United Nations. https://www.un.org/en/academic-impact/countries-commit-action-end-aids2030. Accessed 16 Feb 2023
13. Marcus, J.L., Sewell, W.C., Balzer, L.B., Krakower, D.S.: Artificial intelligence and machine learning for HIV prevention: emerging approaches to ending the epidemic. Curr. HIV/AIDS Rep. **17**(3), 171–179 (2020). https://doi.org/10.1007/s11904-020-00490-6
14. Haas, O., Maier, A., Rothgang, E.: Machine learning-based HIV risk estimation using incidence rate ratios. Front. Reprod. Health **3** (2021). https://doi.org/10.3389/frph.2021.756405
15. Chingombe, I., et al.: Predicting HIV status among men who have sex with men in Bulawayo Harare, Zimbabwe using bio-behavioural data, recurrent neural networks, and machine learning techniques. Trop. Med. Infect. Dis. **7**(9), 231 (2022). https://doi.org/10.3390/tropicalmed7090231
16. He, J., et al.: Application of machine learning algorithms in predicting HIV infection among men who have sex with men: Model development and validation. Front. Public Health **10** (2022). https://doi.org/10.3389/fpubh.2022.967681. Accessed 16 Feb 2023
17. Xu, X., et al.: A Machine-Learning-Based Risk-Prediction Tool for HIV and Sexually Transmitted Infections Acquisition over the Next 12 Months. J. Clin. Med. **11**(7), 1818 (2022). https://doi.org/10.3390/jcm11071818
18. IBM United States (2015). https://www.ibm.com/topics/logistic-. Accessed 16 Feb 2023
19. IBM. What is a Decision Tree—IBM. www.ibm.com, https://www.ibm.com/topics/decision-trees
20. Simplilearn. Random forest algorithm. https://www.simplilearn.com/tutorials/machine-learning-tutorial/randomforest-algorithm. Accessed 16 Feb 2023
21. F1 score in machine learning: Intro and calculation. V7 - AI Data Platform for Computer Vision. https://www.v7labs.com/blog/f1-score-guide. Accessed 16 Feb 2023
22. Gaussian naive bayes: What you need to know? upGrad blog. https://www.upgrad.com/blog/gaussian-naive-bayes/. Accessed 16 Feb 2023
23. Why should I trust you?: Explaining the predictions of any classifier. https://arxiv.org/abs/1602.04938. Accessed 16 Feb 2023
24. Marvin, G., Alam, M.G.R.: Explainable feature learning for predicting neonatal intensive care unit (NICU) admissions. In: 2021 IEEE International Conference on Biomedical Engineering, Computer and Information Technology for Health (BECITHCON), Dhaka, Bangladesh, pp. 69–74 (2021)

Development of a Blockchain-Based Vehicle History System

Mani Joseph[(✉)], Hothefa Shaker, and Nadheera Al Hosni

Modern College of Business and Science, Bowshar, Muscat, Oman
drjosephmani@mcbs.edu.om

Abstract. A complete vehicle history can be obtained by using the recording, storing, retrieving, and tracking of pertinent information re garding a vehicle throughout its entire lifecycle, beginning with its initial sale by the manufacturer/dealer to its ultimate disposal as scrap. All par ties in the Registration Chain, including the vehicle supplier, owner, various government agencies, and the general public, can have confidence, trust, and authentication of records kept by the agency by incorporating a tamper-proof approach like blockchain to capture this crucial information in a distributed ledger register. A decentralized, secure, digital ledger that documents such transactions, avoiding or minimizing the limitations of conventional methods, can be advantageous to all parties involved in the Vehicle Registry. Traditional systems also lack the ability to con firm whether any data has been altered or falsified, a problem that can be addressed with blockchain's trusted data. Decentralizing the Registrar process, enhancing data accessibility, and boosting security are all possible with the implementation of a vehicle registration ledger system based on blockchain technology. This paper explores the development of blockchain-based vehicle history system in Oman.

Keywords: Blockchain Technology · Vehicle Registration · Vehicle History Authentication · Distributed Ledger · Smart Contract

1 Introduction

Due to rapid growth of urbanization and economic prosperity, there is an increased demand for vehicles and associated infrastructure. The vehicle registration procedure, which can be time-consuming and labor-intensive, is one of the biggest problems with vehicle tracking today. Filling out a registration form, giving shipping or purchasing documentation, customs cards, original certificates of cancellation, and an official deed for auction purchases are all required in the Sultanate of Oman in order to register a vehicle. A letter form from the mortgage business may occasionally be needed as well. The owner or authorized representative must also be present, and the car must be insured. Manually managing enormous quantities of data from various agencies presents a number of difficulties, including the risk of manipulation and duplication, the potential loss of information, and the possibility of fake ownership deeds being created by criminals. Gathering data from a variety of sources, including the population, and government agencies.

S. Aurelia et al. (Eds.): ICCSST 2023, CCIS 1973, pp. 54–65, 2024.
https://doi.org/10.1007/978-3-031-50993-3_5

As technology develops, it becomes more challenging to manage the massive quantity of data involved in vehicle registration using simple database systems. Countries like India, China, and Bangladesh have adopted registration tracking systems using blockchain technology to collect and organize data from various sources securely in order to address these issues. Although big data technology was initially utilized for this purpose, its restrictions on the management of personal information prompted the creation of blockchain technology, a decentralized and distributed system that securely stores information from multiple sources in the form of blocks. In addition to reducing counterfeiting, monitoring failures, identifying fake goods, ensuring data security, and being immutable, transparent, and a distributed system for information exchange, blockchain technology has many advantages.

This paper discusses the development of a blockchain-based vehicle history platform that enhances the existing registration and tracking processes of a vehicle and it will ease the Registration and Renewal of the registration process by ROP in Oman.

2 Literature Review

Due to its capacity to offer a safe, distributed, and transparent framework for data exchange and transaction processing, blockchain technology has garnered significant attention during the last few years. The applications of blockchain technology are diverse and growing rapidly across various sectors. It has been extensively incorporated into financial services for smart contracts, cross-border trade, and digital payments. According to research, blockchain technology can boost security, lower expenses, and increase efficiency in financial transactions [1].

Supply networks can benefit from increased transparency, traceability, and accountability thanks to blockchain technology. Blockchain-based systems can aid in the tracking of goods from the manufacturer to the final consumer, as well as in stopping fraud and decreasing counterfeiting [2]. By enabling precision medicine, enhancing the drug supply chain, and enabling secure and quick access to medical data, blockchain technology has the capacity to completely revolutionize the healthcare industry [3]. Secure voting, identity management, and the maintenance of public records can all be provided by government agencies using blockchain technology [4].

Energy management can be improved using blockchain by enabling peer-to peer energy trading, tracking renewable energy certificates, and creating more efficient energy systems [5]. Intellectual property rights can be protected by creating immutable and transparent records of ownership, transfer, and licensing using blockchain [6]. In education, secure and verifiable academic credentials can be created using blockchain, enabling lifelong learning and skill development [7].

In the Middle East, several industries have started exploring the use of blockchain technology to enhance their operations. Abu-Samra and Hammouri explored the capacity of blockchain-based technology in the healthcare industry to improve the quality of healthcare services in Jordan. The study found that blockchain technology could enhance patient data security, facilitate the sharing of medical records, and reduce medical errors [8].

The application of blockchain technology in the Sultanate of Oman was began a few years back. Dhofar, a private sector bank, was among the first in Oman to adopt blockchain methodology in its services. This technology ensures the accuracy of financial records and resources by storing them in a decentralized ledger system. For example, the money transfer process in banks using blockchain involves sending money from user A to B, which goes to a block of transactions, gets broadcast to all parties, is approved to be added to the blockchain, and then securely transferred to user B using an immutable and permanent method. Despite the benefits, implementing blockchain methods in the banking sector in the Sultanate of Oman faces several challenges. These include a lack of adequate tools and knowledge among banking staff, economic factors such as employee training, deployment and implementation, and resource and infrastructure costs, technical issues including security, scalability, self-mining, and privacy leakage, and legal and regulatory challenges such as uncertainty about regulations, standards, and agreements, and sharing sensitive data [9, 10].

3 Methodology

The vehicle registration process requires the participation of several stakeholders, which can make data collection, transfer, security, storage, and processing complex and vulnerable to manipulation, theft, duplication, and other errors. Additionally, data tampering and fraudulent use of data pose significant risks, while data loss can render information untraceable. Blockchain technology is essential and highly beneficial given the high value of the assets involved and the multiple stakeholders' roles. As a single data source, blockchain enables stake holders to share data with each other as a smart contract when needed, eliminating many data risks. Only authorized personnel with the private key can update data, providing a complete lifecycle view of the vehicle for various purposes. This section examines the different stakeholders in the vehicle registration system's blockchain implementation, and Fig. 1 represents vehicle registration system stakeholders. The key stakeholders are:

- Manufacturer: The production and delivery of the vehicle to dealerships across the country is the manufacturer's duty, and this establishes the vehicle's initial inclusion in the blockchain. Important details or data, including the make, model, engine, and chassis number, as well as the dates of manufacturing, order, and delivery, are provided by the manufacturer.
- Dealer: After the sale of a vehicle, the dealer provides a temporary registration number to a vehicle that could be used till they get the permanent registration number from the registration authority.
- Pollution Control Board: It provides a carbon emissions report of the vehicle, so the utility of vehicle from the environmental conservation and preservation point of view data can be collected and stored into the blockchain.
- Insurance Agency: This agency gathers the customer details, bank details, credit record and the vehicle data from sales record of dealer and assemble them to provide the right insurance policy to the customer, the insurance policy, premium details, and the interest rates data come into blockchain from Insurance agency.
- Bank: Provide loans based upon the customer easy monthly instalments plan selection. So, banks would provide the loans, interest rates, pendency of loan payments data into blockchain.

- Registration Authority: ROP will gather data from the customer like citizen cards, sales invoice, vehicle insurance details, driving license, loan status, engine, and chassis details of the vehicle plus the temporary registration details. ROP is in charge of providing registration plates. ROP can act as centralized block in the decentralized blockchain since it needs data from all the stakeholders and create the final data point which is the registration number at the time of new purchase, transfers, resales and issuing no objection certificates (NOC) for temporary use of vehicle and interstate travel. The registration numbers and NOCs are the data which comes into the blockchain from ROP.
- Police: Traffic offenses are reported by the police.
- Service Centre: Provides services as per schedule through Job cards, take preventive and reactive services such as part cleaning, oiling, and replacements. The service centers bring the service history data into the blockchain.
- Customer/Vehicle owners: Shares and approves personally identifiable in formation like citizen card, tax id and driving license.

The purpose of using the blockchain in the vehicle registration system is that the blockchain can provide the updated status of the vehicle at any stage of its lifecycle since it is a high-value product and causes a significant impact on the environment and human lives. So, it is important to have their accurate status which can be relied upon by various stakeholders. Blockchain can provide the status of the vehicle Fig. 2 as followed:

1. Active state: whether it is operational and currently in use by its owner or lessee.
2. Inactive state: vehicle not operational and considered sold, unused but not yet destroyed or recycled.
3. Destruction state: the vehicle which is about to be destroyed because it doesnt fit the use in terms of safety of human lives, ageing and threat to the environment.
4. Suspended: this is the stage before destruction when the vehicle is declared unfit to use yet to be destroyed.
5. Stolen: a vehicle that has been reported by its owner to regulatory authorities as stolen and the authorities also filed the vehicle as stolen in their records.

The major transactions which take place among various stakeholders are

- Sales between manufacturer and dealer
- Sales between dealer and consumer
- Transfer of vehicle to be use between owner and the lessee
- Re-sale of vehicle from current owner to new owner
- Service center maintenance details
- Pollution control board data
- Regulatory authorities Data

4 Design

4.1 Manufacturer-Dealer Workflow

The new vehicles are added by the manufacturer into the blockchain network as a smart contract and a general overview of the manufacturer-dealer workflow in a vehicle transaction involves the following steps:

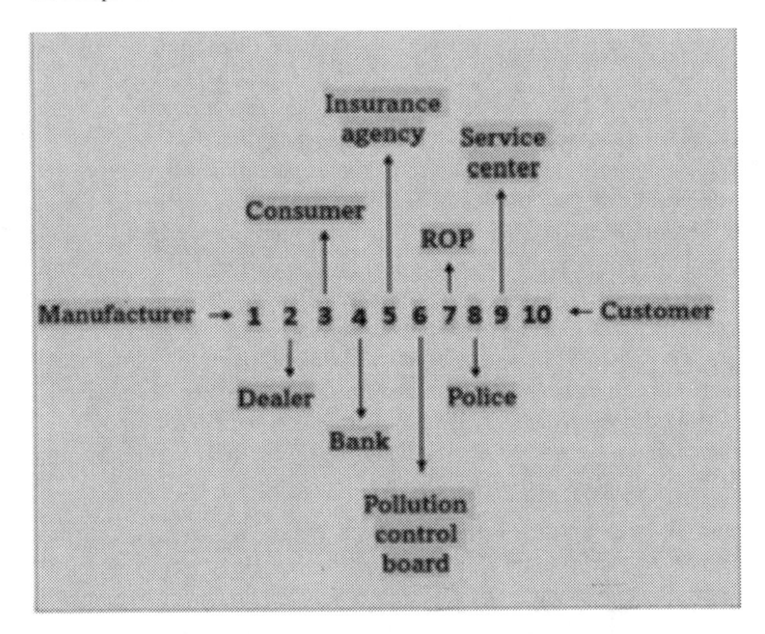

Fig. 1. Vehicle registration system stakeholders

- Basic information about a vehicle is entered into the blockchain node by a manufacturer.
- Manufacturer can handover the ownership to the dealer using smart contact as shown in Fig. 3.

4.2 Vehicle Sale/ Registration Workflow

Upon starting the smart contract, it may automatically send out requests for registration and insurance. The necessary information about the associated entity can be verified from the blockchain by the ROP and insurance agency and a general overview of the transaction is explained below:

- A vehicle selling contract can be started by the dealer.
- The registration and insurance requests can be sent out automatically by a smart contract.
- The required information regarding the vehicle and the customer can be verified from the blockchain by the ROP and insurance agency, who can then offer registration and insurance, as is shown in Fig. 4.

4.3 Vehicle Transfer Workflow

The vehicle transfer workflow typically involves the following steps:

- The first party can ask for a no objection certificate using smart contract. – A smart contract can be used to send out requests to government agencies for approval.

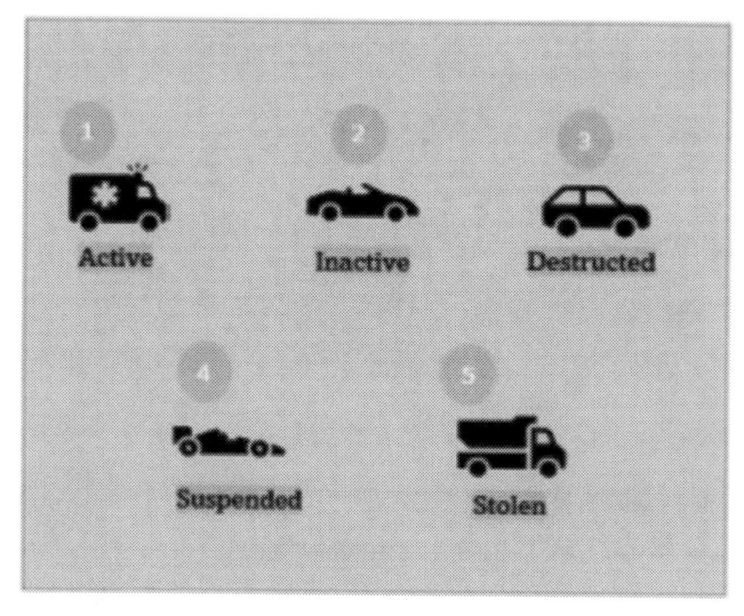

Fig. 2. Vehicle state

- After the details from Blockchain have been successfully validated, the relevant agencies may grant clearance and a no objection certificate, which may be utilized to obtain a new registration, as shown in Fig. 5.

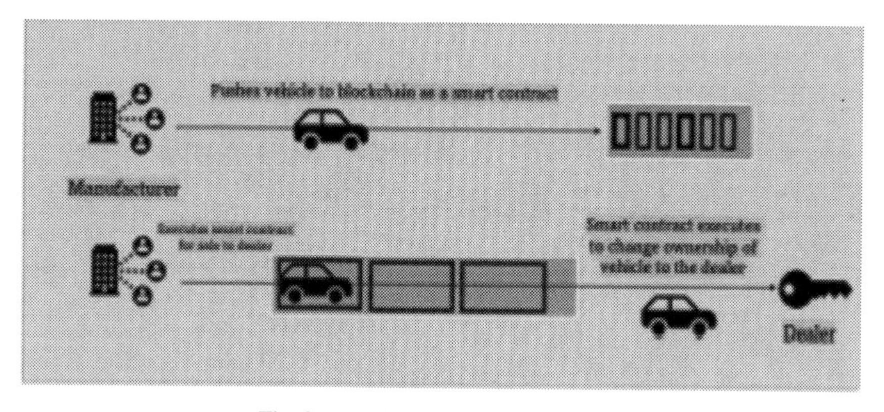

Fig. 3. Manufacturer-Dealer Workflow

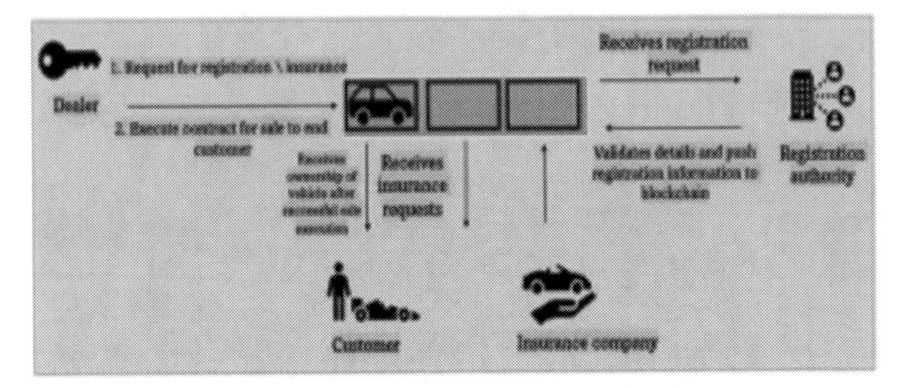

Fig. 4. Vehicle Sale/ Registration Workflow

4.4 Vehicle Resale Workflow

The vehicle resale workflow typically involves the following steps:

- Vehicle Owner executes the smart contract for the resale of a vehicle. – A smart contract can be used to send out requests to government agencies for approval.
- Upon validation, ownership will be shifted to the new purchaser, as shown in Fig. 6.

5 Testing and Evaluation

In order to draw a comparison between blockchain based system and existing paper-based system, we can deploy a proof-of-concept blockchain that encom passes a part of the process. The deployment in itself would be indicative of the merits that exist in opting for a blockchain system.

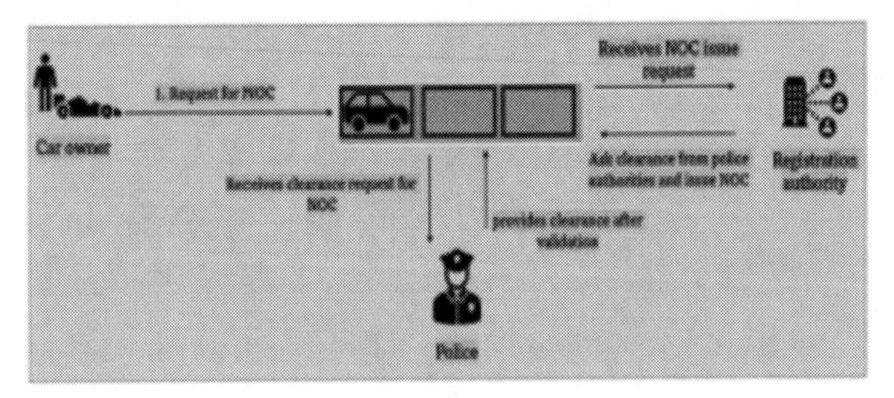

Fig. 5. Vehicle Transfer Workflow

For this purpose, the most tedious part of the workflow was set as the objective of implementation. The process in consideration is that of vehicle registration, which

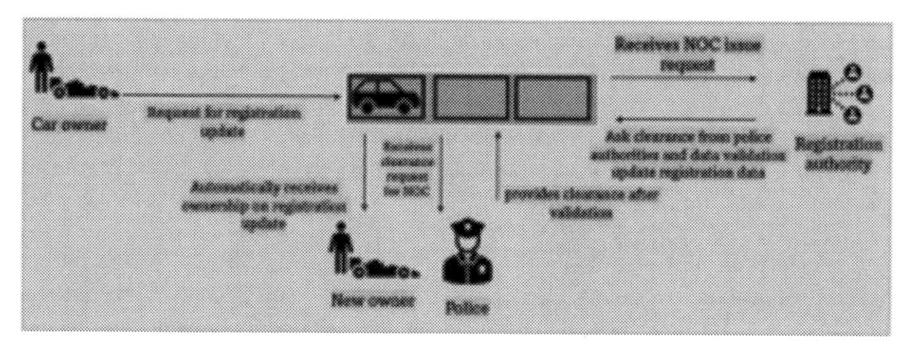

Fig. 6. Vehicle Resale Workflow

involves verification of records relating to different governmental bodies each of which maintains their records separately.

The process has four main steps in which different aspects related to citizens are validated using blockchain technology. The vehicles belonging to the citizens are linked through IDs. The process starts off when a citizen initiates a transaction in any upcoming block to perform vehicle registration. For this purpose, any node in blockchain can initialize a smart contract as well as the starting transaction. The smart contract automates the process involving all concerned governmental bodies, while the first transaction includes required details and data such as initiating citizen ID, signature indicative of request instance and can also include data about vehicles that are being associated with the registration. A vehicle registration testing scenario is shown in Fig. 7.

	ID	National Identity Authority	Residence Authority	BOLO Authority	Registration Authority
0	3241-2361	valid	inland	Nil	pending_renewal
1	6532-4634	Valid	Overseas	Nil	unregistered
2	2346-8358	valid	inland	Restricted	registered
3	7427-0783	expired	inland	BOLO	registered

Fig. 7. Vehicle Registration Testing Scenario

Each citizen might be linked to more than one vehicle, however in order to avoid unnecessarily complicated implementation an assumption is made that each ID is only associated with one vehicle. In practice, it can easily be differentiated via the data field contained in the transactions of this process. The four checks performed during registration are:

- Restrictions/BOLOs check
- Verification of valid identity
- Residency check
- Vehicle Registration Check

The first step in vehicle registration involves regulatory authorities verifying that initiating citizen does not have any BOLOs or restrictions against his name. For this purpose, the regulatory authority matches the ID of the citizen initiating the registration request with their database and returns the status of either restricted or unrestricted.

Concerned authority identifies that they have been called upon by simply keeping a track of whether their private key is compatible with the encrypted signature message found in transaction data. If a match gets found, it spurs the automated system of relevant organization into fulfilling its role, in the form of an automatically executing smart contract.

The second step involves a basic identity check of the citizen. This check is performed by querying the national identity record database for the citizens ID, and checking status of the respective ID card. If the ID returned from the query is either expired or invalid, then the transaction is considered as void and sent back to citizen in the upcoming mined block. However, if the ID check returns a positive result indicating that the citizen is a valid ID card holder then the process moves on towards a residency check.

A residency check is the third step of the registration process, and it deals with verifying that initiating citizen currently resides within the country. This check prevents various methods of fraud, however, it also calls into action the governmental departments that deal with passports and residences. This means that the more time gets spent on monitoring applications, the more wastage of skilled human resources might be occurring.

Finally, after a transaction clears all three of the previous stages with the help of smart contracts acting as liaisons between respective governing bodies, the vehicle registration authority finally accepts the transaction and updates the registration status. This step concludes the smart contract indicating its end of life.

The only other way in which the smart contract could have been concluded after deployment was if at any step the validation was declined. Other than successful completion or failure, the smart contract would remain immutable and transparent across all nodes. The primary recipients would be able to view the data contained in the contract and associated transactions, while all other nodes in the blockchain can monitor the process unfolding transparently as each governing body fulfils its validation duties.

6 Results and Discussion

The described blockchain was deployed successfully, with the results being computed under different scenarios of registrations. The following figures show the iterative nature of transactions unfolding through automation provided by smart contracts.

The first iteration of any blockchain framework always consists of creating a genesis block or block 1 of the chain. Transactions start being included in blocks mined after the first mining step. In the displayed results, each of the four test case citizens that were seen in the previous section initiate a request to register after the blockchain has been deployed. Alongside each registration request, a smart contract also gets deployed; smart contracts can be named or in case of no name being assigned a random string of alphanumeric characters gets assigned.

The Fig. 8 shows the smart contracts of all 4 requests being deployed, with their initiation transactions. The smart contracts having been deployed will wait until the next

Fig. 8. Smart Contracts being initiated

block to be mined until being automatically executed by the framework and advancing the validation request based on existing transactions.

Some requests that will fail validation will be returned to initiating nodes, which will be indicated from recipient field being replaced with citizen ID whereas successful validations will move to the next step in validation. This can be seen from following figure of next iteration in mining.

Fig. 9. Smart Contracts advancing transactions

The Fig. 9 indicates two of the requests being returned to senders, concluding their respective smart contracts as well while leaving only two intact. Signatures of requests remained consistent across the transactions, as coded in the contract, and the smart contracts retain their names as well.

One key detail also present in both figures is the data field, which appears encrypted to any node that does not have authority to view it but can be decrypted using private keys held by nodes that are addressed as recipients. This is accomplished by using public private key encryption that is commonly used in blockchain technology. Fig. 10

Fig. 10. Final transactions of test case results

and Fig. 11 show the final transactions of test case results and the resulting change in databases respectively.

The results conclude with one request resulting in update of registration status, while each of the three failed requesters were informed on respective stages where their validation could not succeed.

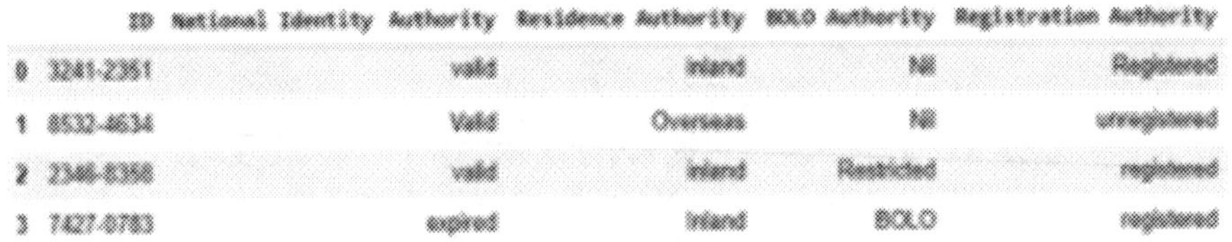

	ID	National Identity Authority	Residence Authority	BOLO Authority	Registration Authority
0	3241-2351	valid	inland	Nil	Registered
1	8532-4634	Valid	Overseas	Nil	unregistered
2	2346-8358	valid	inland	Restricted	registered
3	7427-0783	expired	inland	BOLO	registered

Fig. 11. Resulting change in Database

7 Conclusions

Immutability, Automation, Paperless Workflow, and Reduction of Manual Intervention are its primary benefits, according to the study conducted as well as the sample installation.

The blockchain that was established in particular demonstrated the important advantage of smart contracts the ability to eliminate paper in resource intensive, highly manual governmental procedures. The most important machinery of any state, the governmental bodies and regulatory authorities, may become more effective as a result of processes that once took weeks to complete being completed in a matter of days. Prior to the procedure, a framework that would stretch out and connect each of the bodies would need to be built up; however, once connected, the latency caused by inter-departmental cooperation would be reduced to seconds by networks alone. Also, the entire process would be completely accessible to all linked nodes while still protecting the privacy of the data provided by the persons making the requests. The two systems wouldn't continue to be similar in terms of overall system efficiency since they would undergo a revolutionary process that would result in an order-of-magnitude improvement.

Acknowledgements. The research leading to these results has received funding from the Research Council (TRC) of the Sultanate of Oman under the Block Funding Program.

References

1. Cocco, L., Pinna, A., Marchesi, M.: Banking on blockchain: costs savings thanks to the blockchain technology. Futur. Internet **9**(3), 27 (2017). https://doi.org/10.3390/fi9030025
2. Zohar, A.: Bitcoin: under the hood. Commun. ACM **58**(9), 104–113 (2015)
3. Mettler, M.: Blockchain technology in healthcare: the revolution starts here. In: Proceedings of the 2016 IEEE 18th International Conference on e-Health Networking, Applications and Services, pp. 1–3 (2016)
4. Crosby, M., Pattanayak, P., Verma, S., Kalyanaraman, V.: Blockchain technology: beyond bitcoin. Appl. Innov. **2**(6–10), 71–81 (2016)
5. Christian, J., Karim, A.: Blockchain applications in the energy sector: a system atic review. Renew. Sustain. Energy Rev. **101**, 143–174 (2019)
6. Kshetri, N.: Blockchain's roles in meeting key supply chain management objectives. Int. J. Inf. Manag. **39**, 80–89 (2018)
7. Siemens, G.: The role of blockchain in lifelong learning. In: Proceedings of the 8th International Conference on Learning Analytics and Knowledge, pp. 219–220. ACM (2018)
8. Abu-Samra, M., Hammouri, A.: Blockchain technology in the healthcare industry: a case study of Jordan. Int. J. Healthc. Inf. Syst. Inform. **16**(1), 38–53 (2021)
9. Al Hilali, R.A., Shaker, H.: Blockchain technology's status of Implementation in Oman: empirical study. Int. J. Comput. Digit. Syst. (2021)
10. Al Kemyani, M.K., Al Raisi, J., Al Kindi, A.R.T., Al Mughairi, I.Y., Tiwari, C.K.: Blockchain applications in accounting and finance: qualitative evidence from the banking sector. J. Res. Bus. Manag. **10**(4), 28–39 (2022)

Social Distancing and Face Mask Detection Using YOLO Object Detection Algorithm

Riddhiman Raguraman[1], T. S. Gautham Rajan[1], P. Subbulakshmi[1](✉), L. K. Pavithra[1], and Srimadhaven Thirumurthy[2]

[1] School of Computer Science and Engineering, Vellore Institute of Technology, Chennai, Tamil Nadu, India
{riddhiman.r2020,gauthamrajan.ts2020}@vitstudent.ac.in, {subbulakshmi.p,pavithra.lk}@vit.ac.in

[2] Thomas J. Watson College of Engineering and Applied Science, Binghamton University, State University of New York, Binghamton, NY, USA

Abstract. Due to the COVID-19 pandemic, there has been a huge impact worldwide. The transmission of COVID-19 can be prevented using preventive measures like social distancing and face masks. These measures could slow the spreading and prevent newer ones from occurring. Social distancing can be followed even by those with weaker immune systems or certain medical conditions. With the new normal into play, maintaining distance in social and wearing masks are likely to be followed for the next two years. This paper studies about maintaining distance in social and detection of masks using deep learning techniques. Several object detection models are used for detecting social distance. The inputs used are in the form of images and videos. With this system, the violations can be detected which will reduce the number of cases. In conclusion, the proposed system will be very efficient and can also be used to introduce newer preventive measures.

Keywords: COVID-19 · social distancing detection · wearing mask detection · CNN · YOLO · Faster R-CNN · SSD · AlexNet · Inception v3 · MobileNet · VGG

1 Introduction

Deep learning is a combination of artificial intelligence (AI) and machine learning (ML) that imitates the method of how humans gain knowledge. In recent years, more attention has been gained by deep learning in object detection and its applications. A research trend has been shown by deep learning in several classes of object detection and recognition in AI. Even on challenging datasets, outstanding performance has been achieved.

The processing of artificial intelligence (AI) is done by merging large data with intelligent algorithms in which the features and patters will be learnt automatically. Most of the AI examples are profoundly dependent on deep learning and natural language processing. AI is a wide-ranging field of study that comprises of several theories and methodologies. AI is divided into three subfields: Machine Learning, Neural networks

S. Aurelia et al. (Eds.): ICCSST 2023, CCIS 1973, pp. 66–79, 2024.
https://doi.org/10.1007/978-3-031-50993-3_6

and Deep Learning. By means of these technologies, computers could be trained for completing complex tasks with large datasets [30].”

Coronavirus disease (COVID-19) is a transmittable virus caused by the SARS CoV-2 virus [28]. The risk of being affected by COVID-19 is highest in crowded and poorly ventilated places. Outbreaks have been reported in crowded indoor spaces where people speak or sing loudly such as restaurants, choir rehearsals, fitness centres, nightclubs, offices and worship places [29].

Due to the COVID-19 pandemic, we have started taking precautionary measures such as social distancing and face masks. The aim of maintaining distance in social is to decrease the fast spread of the disease through breaking the chain of communicating and prevent additional ones from occurring (by WHO - World Health Organization). These measures would reduce the number of COVID-19 cases and also reduce the possibilities of being affected. One of the best ways of prevention or slowing down the transmission is to be well informed about the virus and the disease. By following social distancing by staying at least 1 metre apart from others, properly using a tight mask and regularly cleaning your hands [28]. Social distancing can be followed even by those with weak immune systems or certain medical conditions. Similarly, face masks can be worn in hazardous or polluted environment.

Manual checking of people following social distancing and people wearing face masks correctly is a long and tedious process. In order to monitored the violations with ease, the system can detect social distancing and facial mask using Deep Learning techniques. This paper also provides information on implementation of distancing from others and detecting the mask using YOLO v5 object detection model [3]. The sections are divided as follows: Sect. 2 provides information on the works which are related to the paper and similar researches carried out by other fellow researchers, Sect. 3 describes the various object detection that may be implemented for social distancing and detection of mask in detail. Section 4 provides information on how the implementation was carried out using YOLO v5 object detection model. Section 5 explains the outcomes obtained from the implementation. Finally, the conclusion and future scope of the paper is explained in Sect. 6.

2 Related Works

Since the announcement of the new COVID-19 virus, everyone was advised to wear face masks and follow social distancing. Hence, there has been several researchers, who were keen on finding a proper monitoring or detecting system for these new norms. Different types of object detection models and algorithms have been used.

There have been several applications for which the detection models have been proposed. Also, the type of inputs varies with each application. The inputs are in the form of images, videos and real-time videos.

Fan Zuo et al. in their paper, analysed several object detection models and found that YOLO v3 had better performance in comparison with others [1]. Similarly, X. Kong et al. in their paper, has compared YOLO v3 object detection model with Faster RCNN and Single-Shot detector (SSD) on the basis of Frame rate, Inference time and Mean Average Accuracy (mAP). The observations from the comparison were that the YOLO v3 was better and efficient than Faster RCNN and SSD [8].

Few researchers proposed new models and frameworks. This also includes combining two or more models and algorithms and forming a hybrid model. Shilpa Sethi et al. in their paper, has proposed a real time mask identification framework based on edge computing. This uses deep learning and includes video restoration, face detection and mask identification [6]. Similarly, Qin B et al. used a face mask wearing condition identification method by using image SRCNet [5]. SRCNet stands for super-resolution and classification networks.

Most of the distance calculations are calculated using the Euclidean distance for mula. X. Kong et al. and A. Gad et al. both have used Euclidean distance formula in their papers [2, 9]. Similarly, bird-eye is used for calibration. X. Kong et al. and A. Gad et al. both have used bird-eye for calibration in their papers [2, 9].

In the recent years, CNN models for example VGG16, AlexNet, ResNet-50 and InceptionV3 have been trained to achieve exceptional results in object detection [6]. With the help of the neural network structure, deep learning in object detection has been capable of self-constructing the object description and learning advanced features that cannot be obtained from the dataset. This has been the success of deep learning [4].

Raghav Magoo et al. proposes a model which can be used in embedded systems for surveillance applications. This model is a combination of one-stage and two-stage detectors. This model achieves less inference time and higher accuracy. An experiment was conducted using ResNet-50, MobileNet and AlexNet. From the experiment, it was observed that the proposed model obtains an accuracy of 98.2% when implemented with ResNet-50 [6].

B. Wang et al. in their paper proposes a two-stage model to identify the state of wearing a mask using a combination of machine learning methods. The first stage detects the mask wearing region of the person. This is on the basis of a transfer model of Faster RCNN and InceptionV2 structure. The second stage verifies the real face masks with the help of extensive methodology. The implementation is done by training a double stage model [7].

L. Jiao et al. in their paper have reviewed the various deep learning-based object detection methods. All the different methods have been explained and analysed in detail. This paper also lists the various applications and recent trends of object detection. This survey has been very helpful in the study of various deep learning techniques [12].

M. R. Bhuiyan et al. in their paper proposes a deep learning-based face mask detection method using YOLO v3 model. The authors have stated that "Experimental results that show average loss is 0.0730 after training 4000 epochs. After training 4000 epochs mAP score is 0.96. This unique approach of face mask visualization system attained noticeable output which has 96% classification and detection accuracy" [13].

M. Cristani in their paper has introduced the VSD (Visual Social Distancing) problem. The VSD analyses human behaviour with the help of video cameras and imaging sensors. The use of VSD beyond social distancing applications are also discussed in this paper [14].

M. Qian et al., in their paper has discussed about the social distancing measures for COVID-19. The author provides several information on social distancing other strategies. The author concludes that "social distance measure is the most effective practice to prevent and control the disease" [15].

S. Saponara et al., proposes an AI based system for classifying distance in social. This AI system can track people, detect social distancing and monitor body temperature with the images obtained from the thermal cameras [16].

Shashi Yadav proposes a computer vision based real time automated monitoring for social distancing and face masks. This approach uses computer vision and MobileNet V2 architecture. The process includes gathering and pre handling the data, development, training of the model, testing the model and model implementation [17].

Dzisi and Dei in their paper has discussed about the adherence to distancing in social and wearing mask in public transportation during COVID-19 in different countries. The paper mainly focuses on the situation in Ghana. A study on the people of Ghana about their compliance to face masks and social distancing in public transportation has also been discussed [18].

X. Peng et al., in their paper proposes a tracking algorithm in realtime for detection of face. Two steps of this approach: detection of face using RetinaFace and Kalman filter as tracking algorithm. Even though the proposed algorithm was not trained using datasets, it showed amazing results [19].

T. Q. Vinh and N. T. N. Anh in their paper propose a real-time face mask detector. This system uses Haar cascade classifier for face detection and YOLO v3 algorithm for identifying the face. The final output shows that the accuracy can achieve till 90.1% [20].

A. Nowrin et al., in their paper have studied about the various face mask detection techniques with different datasets. This paper also provides information on both ma chine learning and deep learning based algorithms which are used for object detection [21].

M. Z. Khan et al, have proposed a hotspot zone detection using computer vision and deep learning techniques. Experiments on different object detection algorithms have been carried out. Using confusion matrix, 86.7% accuracy was achieved for the whole human object interaction system [22].

J. Zhang et al., in their paper proposes a novel face mask detection framework namely Context-Attention R-CNN. This paper also proposes a new practical dataset which covers several conditions. Based on experiments, this proposed framework achieves 84.1% mAP on the proposed dataset [23].

S. Srinivasan et al., in their paper proposes an effective solution for person, distancing in social and detecting the mask using object detection and binary classifier based on CNN. This paper also studies the various face detection models. The final system achieves an accuracy of 91.2% [24].

A. Rahman et al., in their paper have analysed the various COVID-19 diagnosis methods which uses deep learning algorithms. The paper also provides information about the security threats to medical related deep learning systems [25].

M. Sharma et al., in their paper proposes an intelligent system for social distance detection using OpenCV. The process includes three steps: object detection, distance calculation, violation visualization [26].

K. Bhambani et al., in their paper has focused on providing a better solution for social distancing and face mask detection using YOLO object detection model on real time images and videos. The proposed model has obtained an accuracy of 94.75% mAP [27].

3 Relevant Methodologies

Several methodologies are available to be used for detection of social distance and wearing of mask. Each methodology has its own characteristics and differ from each other. A few of them are explained below.

3.1 Convolutional Neural Network (CNN)

CNN has evolved into a popular class of deep learning methods. It has also been drawing interest from a wide range of fields. CNN can automatically learn the three-dimensional orders of features and through using backpropagation algorithm. The CNN architecture can be described as numerous layers, particularly convolution layer, pooling layer and fully connected layers. It takes an image as input, assigns priority with learnable weights and biases to numerous object features and distinguishes between them (Fig. 1).

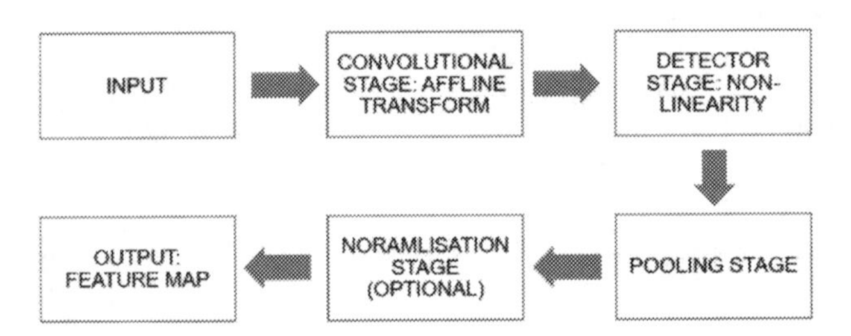

Fig. 1. Different stages in a typical CNN Layer.

The level of pre-processing required by CNN is significantly less in comparison with other classification algorithms. In comparison with its forerunners, CNN has the ability to identify key distinct features without any intervention of humans.

3.2 Object Detection

Object detection is associated with computer vision and image processing. It deals with feature detection of certain classes like humans, buildings or vehicles [12]. A traditional object detection algorithm can be divided into region selector, feature ex tractor and classifier [11]. Object detection comprises of both object classification and location regression [10] (Fig. 2).

The present deep learning-based object detectors are segregated into double stages namely, First-stage and second-stage detectors. Even though double-stage detectors obtain outstanding outcomes on various public datasets, they lag behind in low inference speed. On the other hand, one-stage detectors are fast and most preferred for several real-time object detection applications. On the whole, one-stage detectors have comparatively poor performance than two-stage detectors [10].

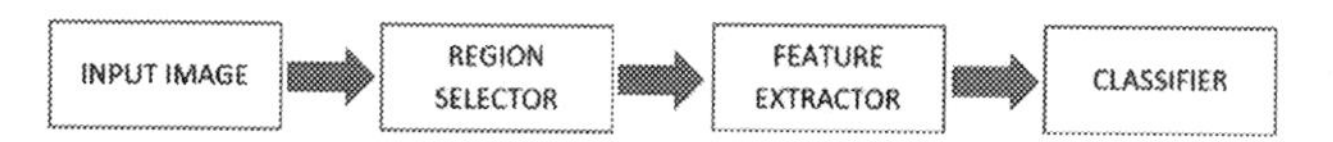

Fig. 2. Block diagram of a traditional object detection algorithm.

3.3 YOLO Object Detection Model

You Only Look Once (YOLO) is a popular object detection model used by research scholars around the world. YOLO is a CNN for executing the detection of objects in realtime. One of the advantages of YOLO is that it can be faster than other networks but still maintain the accuracy rate [29, 31] (Fig. 3).

Fig. 3. Block diagram of YOLO object detection model.

The YOLO model takes 45 fps (frames per second) to process images in real-time and the Fast YOLO model takes an extraordinary 155 fps to process images. The Fast YOLO model achieves twice the mAP of real-time conventional detection models. When generalising from natural photos to other domains like networks, the YOLO model outclasses existing detection models like R-CNN and DPM.

3.4 Faster R-CNN

Faster R-CNN is a combination of Fast R-CNN and RPN [6]. R-CNN represents Region-Based Convolutional Neural Network and RPN represents Region Proposal Network. To substitute the selection search algorithm, the Faster R-CNN uses RPN. This proposal generator is learned by using supervised learning methods. RPN is a fully convolutional network which uses a random sized images and generates various object detection proposal on each position of the feature map [10]. This allows cost free region proposals by integrating individual blocks of object detection in a single step. The individual blocks include feature extraction, proposal detection and bounding box regression [6].

3.5 Single-Shot Detector (SSD)

Initially, SSD identified for detecting objects using deep neural networks to solve computer vision problems. SSD will not reinitiate the features for the bounding box hypothesis [9]. RPN-based approaches like R-CNN require double stages. One stage for generating region proposal and another is for identifying object of each proposal. Therefore, two-shot approaches consume more time. In SSD, a single shot is enough to detect the various objects in an image. Hence, SSD is faster and time-efficient in comparison with the other two-shot RPN-based approaches [33].

3.6 AlexNet

AlexNet is a CNN architecture model which comprises a total of 8 layers along with weights. Among the 8 layers, the first 5 layers are convolutional layers and the remaining 3 layers are fully connected layers. The Rectified Linear Unit (ReLU) is used after every layer (both convolutional and fully connected) [34]. ReLU helps in preventing the computation growth required for operating the neural network. The next layer indiscriminately assigns inputs to zero with the occurrence rate of every stage. This helps to prevent overfitting [32]. The dropout method is used before or within the two fully connected layers. But on using this method, the time required for the network to converge increases [35].

3.7 Inception V3

Inception v3 is mainly used as assistance in image analysis and object detection. Inception v3 is the third version of the Inception CNN from Google. It comprises of 42 layers which slightly is higher than the v1 and v2 models. Inception v3 is an optimized edition of Inception v1. The main purpose of this version is to allow deeper networks, along with numerous parameters. In comparison with the parameters of AlexNet (60 million), Inception v3 has lesser parameters (25 million). In comparison with Inception v1, Inception v3 has a deeper network and higher efficiency. It is also less expensive.

3.8 MobileNet

MobileNet is a portable efficient CNN used in several applications. They are small in size, low latency and low powered models. Since it is light weighted, it has lesser parameters and higher classification accuracy. MobileNet comprises of depth-wise separable convolution layers which consists of a depth-wise convolution and a point wise convolution. Along with each depth-wise and point-wise convolution, a MobileNet includes a total of 28 layers. MobileNet also introduces two new universal hyperparameters: width and resolution multiplier. This permits developers to choose between latency and accuracy on the basis of the requirements.

3.9 Visual Geometry Group (VGG)

VGG or Visual Geometry Group is an advanced object detection and recognition model which supports up to 16 or 19 layers. As input, the VGG accepts 224×224 pixel RGB image. 3×3 convolutional layers are used. Also, a 1×1 convolutional filter is present. This acts as linear transformation for the input. The convolutional filter is then followed by a ReLU unit. The VGG is made up of three complete communicated layers. The initial layer comprises of 4095 channels each and the third layer comprises of 1000 channels. The hidden layers of VGG uses ReLU, as it is time efficient.

4 Implementation

Several papers with several methodologies and techniques of deep learning were studied. As per the study, it has been found that the social distancing and face mask can be detected efficiently by using CNN and YOLO object detection model. The latest version of the YOLO model, the YOLO v5 object detection model has been used.

Object detectors detect a corresponding label and a bounding box. In both the scenarios, bounding boxes are used for indication. All the codes were run in Google Colab. The inputs were mounted on Google Drive and imported into the Colab.

4.1 Face Mask Detection

Input is a custom dataset. The dataset consists of 50 images along with its corresponding labels to indicate the mask correctly. First step is to clone the repository from GitHub [36]. Then, all the required dependencies are installed. The inputs are mounted on Google Drive and imported into the Colab file (Fig. 4).

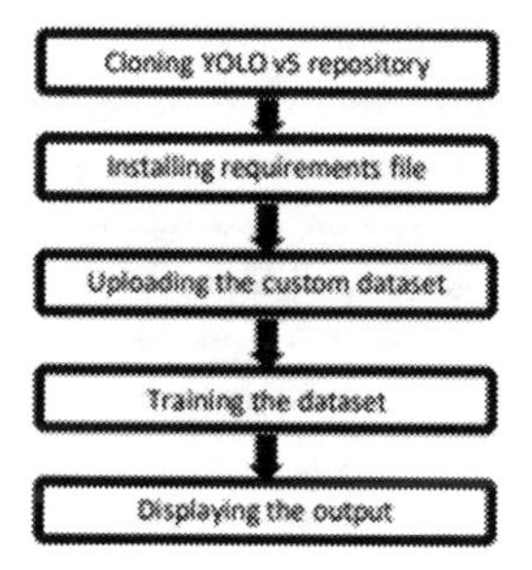

Fig. 4. Outline of Face Mask Detection

A configuration file is created to provide information about the dataset and its location to the model. The dataset path is provided for training and validation. Then the number of classes and name of the class are provided. The design of the model can be seen with all the layer informations. The model starts training and the results are logged. The models are trained for 50 epochs, tested and validated. PR and F1 curves are plotted. In addition to the custom data set, a video file is given as input, trained and validated (Fig. 5).

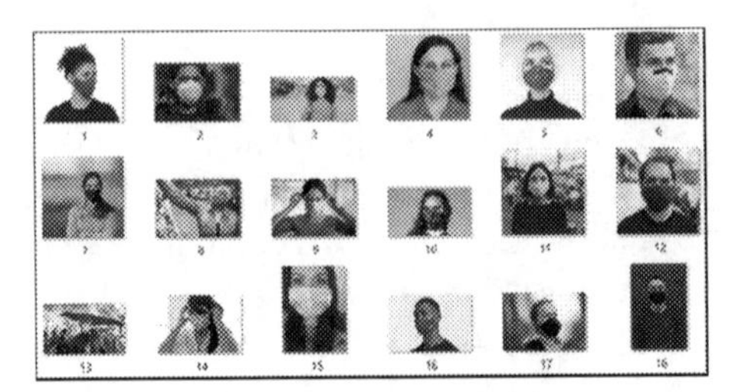

Fig. 5. Sample custom face mask dataset.

4.2 Social Distance Detection

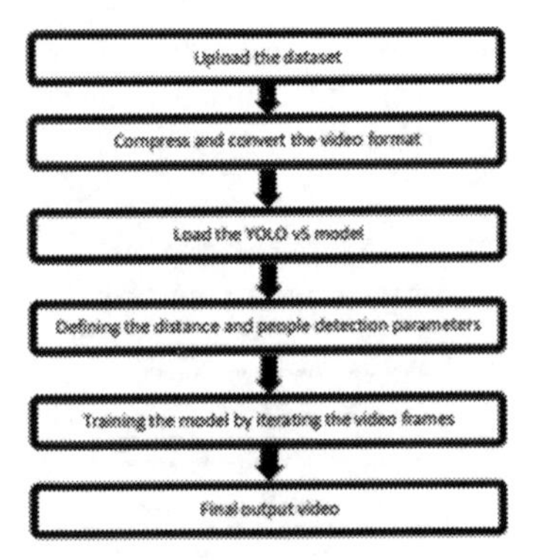

Fig. 6. Outline of Social Distancing Detection

The input is the processed video output of the face mask detection. This will use Euclidean distances. The ED between the centroids of all the regions projected will be computed (Fig. 6).

The Euclidean distance is expressed as

$$d = \sqrt{(x_2 - x_1)^2 + (y_2 - y_1)^2}$$

where,

(x_1, y_1) are the coordinates of one point.

(x_2, y_2) are the coordinates of the other point.

d is the distance between (x_1, y_1) and (x_2, y_2).

All the required libraries are imported. The drive is mounted and input is copied to the Colab file. Pre-processing of the video like compression and conversion of the video format is carried out. This YOLO model uses PyTorch library. With the help of the specified distance formula, the YOLO model detects people and draws rectangle boxes surrounding them.

Bounding boxes will be used to indicate the person and lines will be used to indicate close distance contacts. The colour of the bounding boxes will switch from green to red, if the social distancing norms are violated.

To detect people in a video, frame by frame iteration is carried out and at the end of the process, the output file is saved.

5 Results

The results of the implementation are analysed and found to be satisfactory. From the PR and F1 curve graphs from the face mask detection, the efficiency of the trained model can be understood. A PR curve is plotted by taking the values of Precision and Recall on the y-axis and x-axis respectively. The F1 curve is plotted by taking the values of the harmonic mean of both precision and recall. The average precision rate is found to be 60%, the average recall rate is found to be 74% and average accuracy mAP was found to be 62% (Figs. 7 and 8).

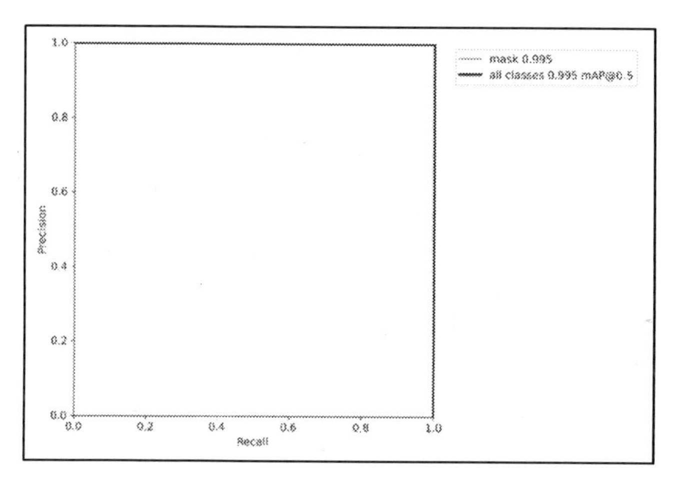

Fig. 7. Precision-Recall curve.

Figure 9 and 10 showcases the result outcome of the implementation. The red boxes detect the mask and the green box detects the social distance. Few limitations in the implementation are: uncontrollsed background and illumination. The camera angle and lighting conditions affect the detection of the person. Even in a diverse environment, problems occur in the people detection.

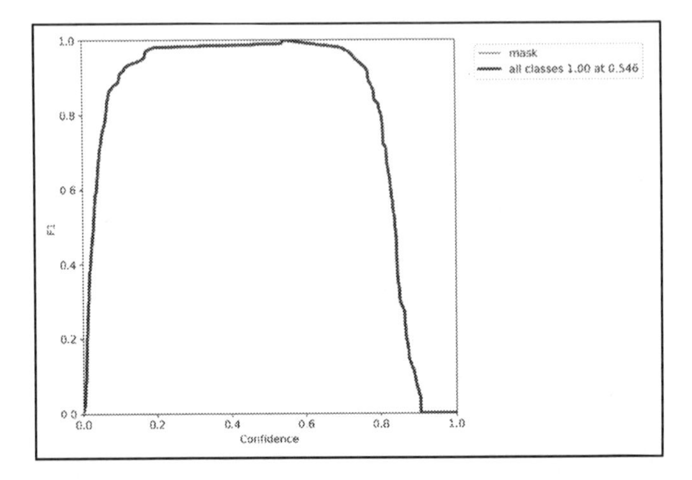

Fig. 8. F1 curve.

Fig. 9. Sample trained dataset with bounding boxes and mask prediction scores.

Fig. 10. A video frame of social distancing and face mask detection.

6 Conclusion and Future Scope

Even though the COVID-19 scenario has come to an end. New variants are still emerging around the world. To prevent further spread of the disease, everyone should follow the guidelines and norms advised by the WHO. The detection of social distancing and face mask is to provide an awareness to the people that the disease is still on the loose.

As future scope, the model can be trained to give comparatively better results. Real-time video can be taken as input and trained. To enhance the model further, temperature and cough detection can be added in addition to face mask and social distancing.

References

1. Zuo, F., et al.: Reference-free video-to-real distance approximation-based urban social distancing analytics amid COVID-19 pandemic. J. Transp. Health **21**. Science Direct (2021)
2. Gad, A., ElBary, G., Alkhedher, M., Ghazal, M.: Vision-based approach for automated social distance violators detection. In: 2020 International Conference on Innovation and Intelligence for Informatics, Computing and Technologies (3ICT). IEEE (2020)
3. Madane, S., Chitre, D.: Social distancing detection and analysis through computer vision. In: 2021 6th International Conference for Convergence in Technology (I2CT). IEEE (2021)
4. Hou, Y.C., et al.: Social distancing detection with deep learning model. In: 2020 8th International Conference on Information Technology and Multimedia (ICIMU). IEEE (2020)
5. Qin, B., Li, D.: Identifying facemask-wearing condition using image super resolution with classification network to prevent COVID-19. Sensors (Basel). MDPI (2020)
6. Sethi, S., Kathuria, M., Kaushik, T.: Face mask detection using deep learning: an approach to reduce risk of Coronavirus spread. J. Biomed. Inform. **120**. Science Direct (2021)
7. Wang, B., Zhao, Y., Chen, C.L.P.: Hybrid transfer learning and broad learning system for wearing mask detection in the COVID-19 Era. IEEE Trans. Instrum. Meas. **70**, 1–12. IEEE (2021)
8. Kong, X., et al.: Real-time mask identification for COVID-19: an edge-computing based deep learning framework. IEEE Internet Things J. **8**(21), 15929–15938. IEEE (2021)
9. Magoo, R., et al.: Deep learning-based bird eye view social distancing monitoring using surveillance video for curbing the COVID-19 spread. Neural Comput. Appl. **33**, 1–8. Springer (2021)
10. Wu, X., et al.: Recent advances in deep learning for object detection. Neurocomputing **396**, 39–64. Science Direct (2020)
11. Xiao, Y., et al.: A review of object detection based on deep learning. Multimed. Tools Appl. **79**, 23729–23791. Springer (2020)
12. Jiao, L., et al.: A survey of deep learning-based object detection. IEEE Access **7**, 128837–128868. IEEE (2019)
13. Bhuiyan, M.R., et al.: A deep learning based assistive system to classify COVID-19 face mask for human safety with YOLOv3. In: 2020 11th International Conference on Computing, Communication and Networking Technologies (ICCCNT). IEEE (2020)
14. Cristani, M., Bue, A.D., Murino, V., Setti, F., Vinciarelli, A.: The visual social distancing problem. IEEE Access **8**, 126876–126886. IEEE (2020)
15. Qian, M., Jiang, J.: COVID-19 and social distancing. J. Public Health **30**, 259–261. Springer (2022)
16. Saponara, S., Elhanashi, A., Gagliardi, A.: Implementing a real-time, AI-based, people detection and social distancing measuring system for Covid-19. J. Real-Time Image Process. **18**, 1937–1947. Springer (2021)

17. Yadav, S.: Deep learning based safe social distancing and face mask detection in public areas for COVID-19 safety guidelines adherence. Int. J. Res. Appl. Sci. Eng. Technol. **8**(VII), 1368–1375. Research Gate (2020)
18. Dzisi, E.K.J., Dei, O.A.: Adherence to social distancing and wearing of masks within public transportation during the COVID 19 pandemic. Transp. Res. Interdiscip. Perspect. **7**. Science Direct (2020)
19. Peng, X., Zhuang, H., Huang, G.-B., Li, H., Lin, Z.: Robust real-time face tracking for people wearing face masks. In: 2020 16th International Conference on Control, Automation, Robotics and Vision (ICARCV), pp. 779–783. IEEE (2020)
20. Vinh, T.Q., Anh, N.T.N.: Real-time face mask detector using YOLOv3 algorithm and haar cascade classifier. In: 2020 International Conference on Advanced Computing and Applications (ACOMP), pp. 146–149. IEEE (2020)
21. Nowrin, A., Afroz, S., Rahman, M.S., Mahmud, I., Cho, Y. -Z.: Comprehensive review on facemask detection techniques in the context of COVID-19. IEEE Access **9**, 106839–106864. IEEE (2021)
22. Khan, M.Z., Khan, M.U.G., Saba, T., Razzak, I., Rehman, A., Bahaj, S.A.: Hot spot zone detection to tackle COVID19 spread by fusing the traditional machine learning and deep learning approaches of computer vision. IEEE Access **9**, 100040–100049. IEEE (2021)
23. Zhang, J., Han, F., Chun, Y., Chen, W.: A novel detection framework about conditions of wearing face mask for helping control the spread of COVID-19. IEEE Access **9**, 42975–42984. IEEE (2021)
24. Srinivasan, S., Rujula Singh, R., Biradar, R.R., Revathi, S.: COVID-19 monitoring system using social distancing and face mask detection on surveillance video datasets. In: 2021 International Conference on Emerging Smart Computing and Informatics (ESCI), pp. 449–455. IEEE (2021)
25. Rahman, A., Hossain, M.S., Alrajeh, N.A., Alsolami, F.: Adversarial examples— security threats to COVID-19 deep learning systems in medical IoT devices. IEEE Internet Things J. **8**(12), 9603–9610. IEEE (2021)
26. Sharma, M.: Open-CV social distancing intelligent system. In: 2020 2nd International Conference on Advances in Computing, Communication Control and Networking (ICACCCN), pp. 972–975. IEEE (2020)
27. Bhambani, K., Jain, T., Sultanpure, K.A.: Real-time face mask and social distancing violation detection system using YOLO. In: 2020 IEEE Bangalore Humanitarian Technology Conference (B-HTC), pp. 1–6. IEEE (2020)
28. World Health Organization (WHO) – COVID-19. https://www.who.int/health-topics/coronavirus#tab=tab_1
29. World Health Organization (WHO) – Advice for the public: COVID-19. https://www.who.int/emergencies/diseases/novel-coronavirus-2019/advice-for-public
30. Artificial Intelligence. What it is and why it matters. https://www.sas.com/en_us/insights/analytics/what-is-artificial-intelligence.html
31. YOLOv3: Real-Time Object Detection Algorithm. https://viso.ai/deeplearning/yolov3-overview/
32. Keras API reference. https://keras.io/api/
33. Tsang, S.-H.: Review: SSD — Single Shot Detector (Object Detection). https://towardsdatascience.com/review-ssd-single-shot-detector-object-detection851a94607d11

34. AlexNet – ImageNet Classification with Deep Convolutional Neural Networks. https://neurohive.io/en/popular-networks/alexnet-imagenet-classification-withdeep-convolutional-neural-networks/
35. Alake, R.: What AlexNet Brought To The World Of Deep Learning. https://towardsdatascience.com/what-alexnet-brought-to-the-world-of-deep-learning46c7974b46fc
36. GitHub for YOLO v5. https://github.com/ultralytics/yolov5

Review on Colon Cancer Prevention Techniques and Polyp Classification

T. J. Jobin[1(✉)], P. C. Sherimon[2], and Vinu Sherimon[3]

[1] Lincoln University College Marian Research Centre, Marian College Kuttikkanam (Autonomous), Kuttikkanam, India
jtjoseph@lincoln.edu.my

[2] Arab Open University, Muscat, Sultanate of Oman

[3] University of Technology and Applied Sciences, Muscat, Sultanate of Oman

Abstract. Colorectal Cancer is the second largest type of life-threatening disease in humanity for a long time. The major causes of CRC are enlisted. Colonoscopy followed by the polypectomy during the procedure is the known method adopted for survival. Identifying the precancerous polyp is the challenging task as well as for disease-free survival. For this alone a few deep learning as well as profound learning strategies are recommended for the analysis, classification, and identification of the object that is generally in the form of a polyp. The detection of the presence of polyp for the given set of images materializes the success of the proposed process model that caters to the neural networks algorithm.

Keywords: CNN · CRC · RPN · Fast R-CNN · Faster R-CNN

1 Introduction

Most common name for colon cancer is colorectal cancer (CRC). The colorectal cancers over time are formed from adenomatous or precancerous polyps. Polyps are formed after certain changes happen to their cellular DNA. Colon cancer becomes more dangerous if any of the family members is diagnosed with colon or rectal cancer and also more dangerous if affected one have got liquor admissions, smoking and incendiary bowel malady. The four areas of the colon where the disease can be detected are the rising colon (moving up), transverse colon (moving over to the cleared out), descending colon (moving down), and sigmoid colon (moving back over to the appropriate). Major signs of colon cancer include altered bowel propensities, blood on or within the stool, unexplained iron deficiency, stomach or pelvic torment, weight loss, heaving, etc. Fecal immunochemical test (FIT), Guaiac-based faecal perplexing blood test (gFOBT), faecal DNA test, versatile sigmoidoscopy, colonoscopy, twice-separate barium bowel cleanses, and CT colonography are the new common screening tests (virtual colonoscopy). In the midst of the assurance plan, the number of tests experienced were Blood tests, Imaging tests, Biopsy, Expressive colonoscopy, proctoscopy. The higher frequency of Colon cancer has been connected to hereditary inclinations such as innate non-polyposis colorectal cancer (HNPC) or commonplace adenomatous polyposis (FAP). Other variables that causes

S. Aurelia et al. (Eds.): ICCSST 2023, CCIS 1973, pp. 80–88, 2024.
https://doi.org/10.1007/978-3-031-50993-3_7

lifestyle incorporate smoking, liquor utilization, ruddy meat slim down, moo vegetable and fibre admissions. For this reason, Convolutional Neural Systems (CNNs) are the most preferred for calculations including deep learning. The considerations related to colon cancer are partitioned into categories such as discovery, classification, segmentation, survival forecast, etc. Therapeutic imaging is a compelling application utilized within the early conclusion [1]. A special category of cancer that passes from parent to child is Genetic Non-Polyposis Colorectal Cancer (HNPCC). The colon cancer habitually happens in HNPCC families. In order to avoid colorectal cancer, customary colonoscopies are prescribed. Adenomatous polyps may turn to cancer but can be removed at the time of colonoscopy. In this situation, the surgical expulsion of the influenced, is as it were precise way to avoid CRC and to anticipate from repeating. Upon fruitful surgery, the slim down and way of life got to be controlled [2] A colon polyp that is formed as a result of clinging cells on the lining of the colon. Most colon polyps are secure. Colon polyps rarely exhibit symptoms. The primary method of treatment is through routine screening procedures like a colonoscopy since colon polyps that are discovered in the early stages can be safely and fully removed. The finest evasion for colon cancer is standard screening for and removal of polyps. There are 2 primary sorts of Colon Polyps: Hyperplastic polyps are small, create near to the conclusion of the colon, and don't turn into cancer. Adenomatous polyps impact more people. Most remain noncancerous (Fig. 1).

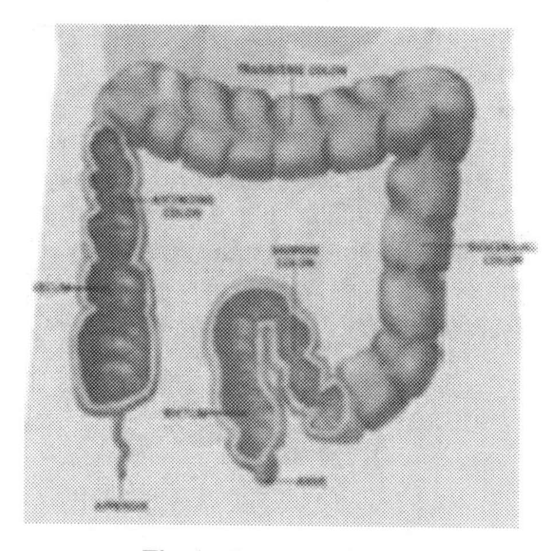

Fig. 1. Structure of colon

Polyps are common alterations to the intestinal lining. They can occur in a variety of locations throughout the digestive system, most frequently in the colon. They range in size from a fraction of an inch to many inches. They grow on a stalk and have a mushroom-like appearance. A family history of colon polyps raises the risk of polyps by colon cancer. A person is more likely than someone who has never had a polyp to develop an underused polyp in their inner long pull if they have a personal history of polyps or colon cancer. Hyperplastic polyp and adenoma are two common sorts. The hyperplastic polyp isn't at danger for cancer. For all colon cancers the adenoma is the forerunner. Polypectomy within the midst of colonoscopy may be an arranged outpatient technique. Conceivable complications, which are momentous, solidify concerning the tidy from the polypectomy region and opening of the colon. Gnawing the dust from the polypectomy area can be rapid or conceded for some of days; gaps rarely happen and can in a few cases be closed with clips in the midst of the colonoscopy, but other times require surgery to repair [3].

1.1 Objectives

The suggested system's goal is to find some techniques for polyp detection. The procedure discusses the recent developments in this area. Image segmentation procedures mainly targeted in deep learning could isolate the polyp from which dataset that contains images. The image segmentation from the images taken at the time of colonoscopy is considered for polyp detection. Here, the images with various qualities such as blurred, various sizes and different colours could be successfully used for the polyp identification. The values corresponding to the result will be displayed and are taken as the output.

2 Design

CVC-Clinic DB is an open-access dataset comprising 612 images from 31 colonoscopy categories with a determination of 384 $\times$ 288. This research will be carried out using the above dataset, which is obtained from kaggle.com. These images are mounted over Google Drive. The input images could be fed into the designed system, the feature learning is done thereafter and finally classification is done accordingly in which the images were in affected with polyp could be identified. The polyp determination from a set of images employs the deep learning model. Here the convolutional neural networks are used for object detection. Deep learning libraries such as Keras, TensorFlow and other common libraries including ImageDataGenerator are used. The training, validation, and testing of the data will be done with the help of Neural Networks Algorithm.

3 Setting and Participants

It is widely acknowledged that early detection and removal of colonic polyps can keep CRC at arm's length. Colonoscopy recordings of colonic polyps are dangerous because of the complicated environment of the colon and the variety of polyp's shapes. It is looked into whether the single shot detector framework has any potential for detecting polyps in colonoscopy records. SSD might be a one-stage method that uses a bolster-forward CNN to generate a set of fixed-size bounding boxes for each item from various feature maps [4] (Fig. 2).

The "resect-and-discard" and "diagnose-and-leave" approaches for reducing colorectal polyps have been proposed by the American Society for Gastrointestinal Endoscopy to cut down on the expenditures of unnecessary polyp excision and pathology evaluation. These guidelines' diagnostic boundaries are not always met in community hone. Endoscopy-related data has been created in order to correct this faulty execution. Endoscopy-related data has been created in order to correct this faulty execution.

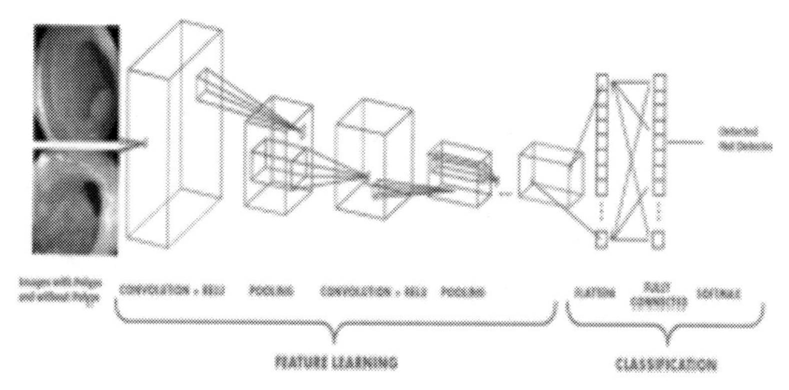

Fig. 2. Images having polyp or not, given as input and output generated

By combining significant learning calculations with AI models, it was possible to create virtually perfect adjustment frameworks that coordinate professional endoscopists' optical biopsies and outperform ASGE accepted edges. Considerations made recently suggest that the incorporation of AI in clinical training will soon take place [5].

The Focus U-Net establishes a few architectural modifications that aid in the formation of close-proximity skip associations and substantial oversight. The Crossbreed Central accident, a modern compound event work based on the Central incident and the Central Tversky accident, was designed to address the lesson of uneven picture division. The five open datasets (CVC-ClinicDB, Kvasir-SEG, CVC-ColonDB, ETIS-Laraib PolypDB, and EndoScene test set) that comprise images of polyps taken during an optical colonoscopy. Doing a clearing course of action after reviewing the Center U-Net on the CVC- ClinicDB and Kvasir-SEG datasets separately as well as on the combined dataset of all five open datasets. Use the Dice closeness coefficient (DSC) and Crossing point over Union (IoU) calculations to analyse show performance [6].

With some learning experiences, many computer-aided determination methodologies have been proposed for applications related to colonoscopy. Recently, basic learning calculations have been put up to break down show day thinking resistance into convolutional components, charting a significant potential for advancing the implementation of convolutional neural networks. When everything is taken into account, the majority of modern basic learning procedures progress from the high appearance complexity and organise computational weight. By using cross-channel common, this organisation can be made to memorise persuasive channel ideas. An additional collection of up to 2112 manually identified images from 1197 patients in a connected clinic using colonoscopy screening was made. From the CVC-ClinicDB, ETIS-Larib Polyp DB, and Kvasir-SEG data-set, more data tests were gathered [7].

The area out of polyps in endoscopic images is proposed using an unsupervised picture division technique. We propose a flexible Markov Random Field (MRF)-based framework based on the skeleton of an over-segmented image. As compared to normal tissues, the polyps or atypical regions exhibit fascinatingly unique surface and colour properties. The endoscopic images are initially over segmented into super pixels according to our methods. Inside the flexible MRF, neighbouring parallel design (LBP) and colour highlights are used for final refining. A test set of 612 images from 29 different video courses of action practically proves that the suggested method can precisely pinpoint polyp locations and get a brutal dice respect of 60.77%. A large number of planning photos are necessary for significant learning models [8].

Convolutional neural network (CNN) has been developed for independent disclosure of colorectal polyps in images taken during inaccessible colon capsule endoscopy, with a chance of risky progression to colorectal cancer. This version of CNN will be an improved version of ZF-Net, which combines pre-processing, trade learning, and data extension. The transmitted CNN is now being improved for a faster R-CNN to localise the area of images containing colorectal polyps. A database of 11,300 capsule endoscopy images from screening procedures are available here, along with images of standard mucosa (N = 6500) and colorectal polyps (any evaluate or morphology, N = 4800). This CNN received 98.0% accuracy, 98.1% affectability, and 96.3% specificity ratings [9].

Early diagnosis and treatment of colorectal polyps are essential for the success of early interventional drugs. The use of AI and ML techniques to analyse colonoscopy images has grown over time for early and precise locations of polyps and other colorectal types that deviate from the norm. Polyp classification and discovery approaches that are currently used are computationally real, restrict memory control, need extensive preparation, and have an impact on the optimization of hyper parameters. Because of this, they are inadequate for real-time applications and applications that require specific computer resources. The Dual Path Convolutional Neural Organize (DP-CNN) is a method that this research suggests for classifying polyp and non-polyp patches from colonoscopy images. The suggested method includes a sigmoid classifier and DP-CNN strategy for image update in order to reveal polyps in a useful way. The suggested orchestrate is prepared using the freely accessible CVC ClinicDB database, and it is tested on the ETIS-Laraib and CVC ColonDB datasets. The organization's testing pre cision on CVC ColonDB and ETIS-Laraib is 99.60% and 90.81%, respectively. Following are the execution metrics: exactness (100%), audit (99.20%), F1 score (99.60%), and F2 score (99.83%) on the CVC ColonDB database; and precision (89.81%), audit (92.85%), F1 score (91.00%) and F2 score (89.91%) on ETIS Laraib database [10].

A UNET structure with a spatial thought layer is suggested to improve the precision of polyp area division in colonoscopy videos. The CNN models for polyp division that are suggested in the writing are primarily set up using com mon incident capabilities like dice and parallel cross entropy incident. The seam learns inadequately in segmenting little measured polyps underneath the design of these incident capacities. The central Tversky incident can be used to understand this. Tests are run on a dataset that is completely open. It would seem that a UNET with a spatial thinking layer configured with a central Tversky in cident outperforms a regular UNet illustration configured with common mishap capabilities [11].

The colonoscopy may be a kind of robot Andon and on its tip there's a camera to procure pictures. This paper presents a thought pointed to move forward the rate of fruitful conclusion with a cutting edge picture data examination approach based on the speedier regional convolutional neural network (faster R-CNN). This unused approach has two steps for data examination: (i) pre-processing of pictures to characterize polyps, and (ii) joining the result of the pre-processing into the quicker R-CNN. Particularly, the pre-processing of The colonoscopy might resemble a robot Andon, with a camera for taking images on its tip. In this article, a novel picture data analysis method based on a quicker regional convolutional neural network is presented that aims to accelerate the rate of successful conclusion (faster R-CNN). This unutilized method consists of two processes for data analysis: (i) pre-processing images to identify polyps, and (ii) combining the pre-processing output into the faster R-CNN. In particular, it was anticipated that the pre-processing of colonoscopy would lessen the impact of specular reflections, which appeared in a made-progress image and were associated to the quicker R-CNN calculation [12].

It is appropriate to discuss R-CNN in this context and assess how well it performs while using various show day convolutional neural systems (CNN) as its highlight extractor for polyp disclosure and division. By adding more polyp images to the organising dataset, it examines how each highlight extractor is performing in order to determine whether more basic and advanced CNNs or a special dataset are needed for orchestrating in different polyp zones and divisions [13].

4 Methods

Artificial neural networks comprise of input, output, hidden layers. Their structure and activity reflecting the way that neurons of the human brain. Planning information is essential for neural networks to learn from and improve in accuracy over time. Once these learning calculations have been adjusted for precision, they can be used with computer science tools and experiences to quickly classify and cluster information. Weights are distributed when a choice of input layer has been made. These weights provide assistance in determining the significance of each particular variable, with higher weights contributing more significantly to the result than other inputs. Afterwards, all inputs are replicated by their respective weights before being added together. Whereas the abdication is passed via an endorsing work that picks a while later, the cede is passed through a sanctioning work that chooses a while after that [14] (Figs. 3, 4 and 5).

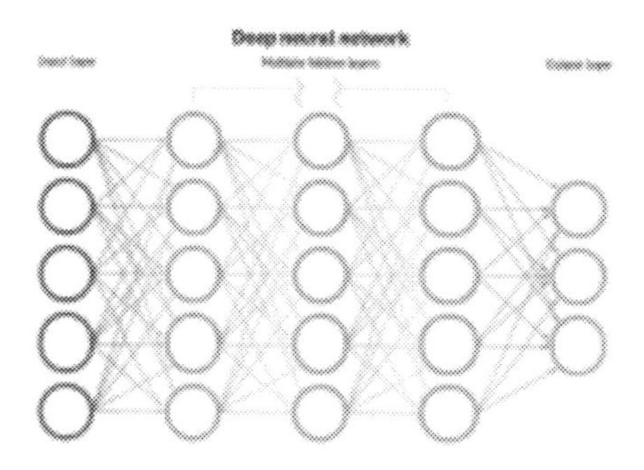

Fig. 3. Neural networks

With a multi-layered perceptron, the layers are connected to one another (MLP). By input layer, the input plans are gathered. The input plans may layout through the abdicate

$$\sum_{i=1}^{m} w_i x_i + bias = w_1 x_1 + w_2 x_2 + w_3 x_3 + bias$$

Fig. 4. Input Layer

$$output = f(x) = \begin{cases} 1 \text{ if } \sum w_1 x_1 + b \geq 0 \\ 0 \text{ if } \sum w_1 x_1 + b < 0 \end{cases}$$

Fig. 5. Output Layer

layers. It is argued that secured up layers extrapolate highlights with interior the input information that have prescient control with respect to the yields [15].

Here, various Neural Frameworks are used for different purposes, like picture classification where we utilize Convolution Neural Networks (CNN). Three Neural Networks (NN) layers are there.

Input Layer: This layer is used to give input to the model. The number of neurons corresponds to the data.

Hidden Layer: Through this layer input is accepted from the Input layer. The number of hidden layers depends on our model and that corresponds to the data size.

Output Layer: At that stage, the yield from the hidden layer is powered into a mathematical operation like sigmoid or softmax, which converts the yield of each course into the likelihood score of each class.

CNN can take in an input picture, and some aspect of the picture will be recognized to extract features from images.

CNN consists the described below:

The output layer may be a binary or multi-class label. The hidden layer contains the convolution layers, ReLU layers, pooling layers, and a fully connected neural network.

ANN made up of various neuron and is the combination of convolution and pooling layers. So here we have related Neural Network where the convolution and pooling layers can't perform classification [16, 17].

5 Results

The images containing polyps are considered in the model for detecting polyp in the form of objects. There will be a result corresponding to the input images given into system. Out of the images given to the process model those images that contain a polyp or not are distinguished. The process model successfully differentiated the so called images of the dataset mounted on to the google drive. The results obtained after running the neural network algorithm is as given below:

6 Conclusion and Implications

This inquiry could be a subjective investigation of the CRC and the examination of Convolutional Neural Network strategies taken after for the classification of threatening or generous tumours. The characteristics of different stages of the CRC cancer as well as

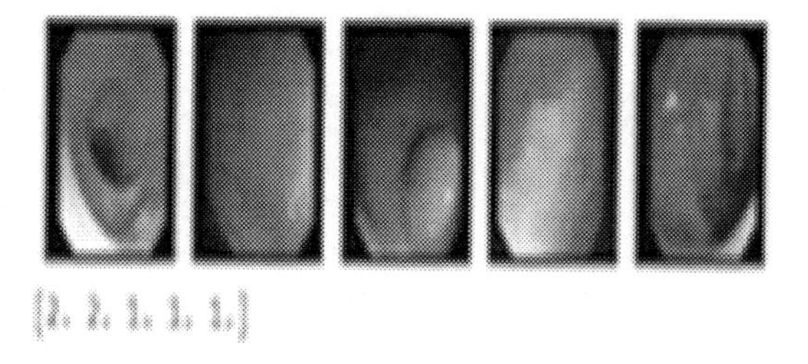

Fig. 6. Output after running the process model

the polyp characteristics utilizing various methodologies are highlighted in this overview. Long-standing time work will centre on the particular technique for the discovery of the CRC clutter. The identification of the polyp is done with the help of deep learning techniques. Real-time inference is not currently used for therapeutic purposes as the dataset is downloaded from the public Internet or other dataset sources.

References

1. https://my.clevelandclinic.org/health/diseases/14501-colorectal-colon-cancer
2. https://my.clevelandclinic.org/health/diseases/17097-hereditary-non-polyposiscolorectal-cancer-hnpcc
3. https://www.medicalnewstoday.com/articles/types-of-colon-polyps#contactinga-doctor
4. Liu, M., Jiang, J., Wang, Z.: Colonic polyp detection in endoscopic videos with single shot detection based deep convolutional neural network. IEEE Access **7**, 75058–75066 (2019). https://doi.org/10.1109/ACCESS.2019.2921027
5. Parsa, N., Rex, D.K., Byrne, M.F.: Colorectal polyp characterization with standard endoscopy: will Artificial Intelligence succeed where human eyes failed? Best Pract. Res. Clin. Gastroenterol. **52–53**, 101736 (2021). https://doi.org/10.1016/j.bpg.2021.101736
6. Yeung, M., Sala, E., Schönlieb, C.B., Rundo, L.: Focus U-Net: a novel dual attention-gated CNN for polyp segmentation during colonoscopy. Comput. Biol. Med. **137**(August) (2021). https://doi.org/10.1016/j.compbiomed.2021.104815
7. Yang, K., et al.: Automatic polyp detection and segmentation using shuffle efficient channel attention network. Alexandria Eng. J. **61**(1), 917–926 (2022). https://doi.org/10.1016/j.aej.2021.04.072
8. Sasmal, P., Bhuyan, M.K., Dutta, S., Iwahori, Y.: An unsupervised approach of colonic polyp segmentation using adaptive markov random fields. Pattern Recognit. Lett. **154**, 7–15 (2022). https://doi.org/10.1016/j.patrec.2021.12.014
9. Nadimi, E.S., et al.: Application of deep learning for autonomous detection and localization of colorectal polyps in wireless colon capsule endoscopy. Comput. Electr. Eng. **81** (2020). https://doi.org/10.1016/j.compeleceng.2019.106531
10. Nisha, J.S., Gopi, V.P., Palanisamy, P.: Biomedical signal processing and control automated colorectal polyp detection based on image enhancement and dual-path CNN architecture. Biomed. Signal Process. Control **73**(September 2021), 103465 (2022). https://doi.org/10.1016/j.bspc.2021.103465
11. Bengaluru, T., Bengaluru, T., Jayaramprash, T.B.: CNN based U-Net with modified skip connections for colon, no. Iccmc (2021)

12. Qian, Z., et al.: A New Approach to Polyp Detection by Pre-Processing of Images and Enhanced Faster R-CNN, vol. 21, no. 10, pp. 11374–11381 (2021)
13. Qadir, H.A., Shin, Y., Solhusvik, J., Bergsland, J., Aabakken, L., Balasingham, I.: Polyp detection and segmentation using mask R-CNN: does a deeper feature. In: 2019 13th International Symposium on Medical Information and Communication Technology (ISMICT), pp. 1–6 (2019). https://www.ibm.com/in-en/cloud/learn/neural-networks
14. https://www.sciencedirect.com/topics/computer-science/multilayer-perceptron
15. https://towardsdatascience.com/a-comprehensive-guide-to-convolutional-neuralnetworks-the-eli5-way-3bd2b1164a53
16. https://towardsdatascience.com/region-proposal-network-a-detailed-view1305c7875853
17. https://medium.com/smallfishbigsea/faster-r-cnn-explained-864d4fb7e3f8

Security Testing of Android Applications Using Drozer

Kamla AL-Aufi and Basant Kumar(✉)

Modern College of Business and Science, Muscat, Sultanate of Oman
{20202228,basant}@mcbs.edu.om

Abstract. Android applications are extensively utilized however, many of them include security flaws and malware that serve as entry points for hackers. Drozer is used in the current study to evaluate the security of four vulnerable Android applications namely AndroGoat, Diva, Insecure Bank v2 & BeetleBug. The information security knowledge of undergraduate students in Oman was evaluated, in the current study, using an online questionnaire. Where, A virtual emulation environment was employed to run a penetration test and examine the attack surfaces of four vulnerable Android applications. Participants in the study showed high level of security awareness when it comes to managing the applications and their permissions as well as posting personal information. The studied packages included a range of exporting components (Activities, content providers, services and broadcast receivers) that are not particularly covered by constraints, making them susceptible to hacking and data exploitation and potentially posing a security risk. Reducing attack surfaces in apps requires taking measures like defining permissions, executing authentication procedures during intents transition, securing databases, and cleaning data after usage. Using four unsecure Android applications, this study categorized Android vulnerabilities based on the OWASP mobile 2016 risks. This research is recognized as an adjunct model that security experts, researchers, and students may use to identify vulnerabilities and assure application security.

Keywords: Drozer · Drozer modules · Penetration testing · The Android package file (APK) · App components · Attack surfaces · AppUse VM

1 Introduction

In today's high-tech world, smartphones, tablets, and other portable electronic gadgets are increasingly indispensable for all daily activities such as communication, organization, and entertainment. Mobile phone vulnerabilities evolved rapidly as their use increased. Android operating system which is an open-source platform, is the most widely used operating system, not just among consumers but also among businesses, suppliers, and app developers of all kinds. It also has easy access to numerous free and notable apps from programs that support the Android OS [1]. The applications you use on your smart device are what give it any real value, no matter how much money you

S. Aurelia et al. (Eds.): ICCSST 2023, CCIS 1973, pp. 89–103, 2024.
https://doi.org/10.1007/978-3-031-50993-3_8

spent on it. There are several programs available, such as retail apps, dating apps, gaming apps, educational apps, and medical apps [2]. According to Statista (2022), Android users had a choice of 3.55 million applications, making Google Play the app store with the most available apps. Numerous firms work to protect apps, to identify vulnerabilities, combating cybercriminals, and prevent damaging data breaches [3]. However, it has been noted that the Android stack has security flaws. Synopsys conducted a study where they looked at the security weaknesses in 3,335 (paid and free) Android apps from 18 different categories that were available on the Google Play store. According to the survey, there were security flaws in 63% of the 3.3K applications. Top free gaming apps were the most vulnerable to attacks, with 96% of them having some kind of security flaws, while the second-most vulnerable category was "top grossing games" [4]. Security flaws in Android apps have been addressed and fixed using a variety of techniques. These detection methods are divided into categories, the main ones are static detection and dynamic detection tools. Static analysis examines static information available without executing the application, such as the executable file and source code. Dynamic analysis requires a program to be deployed to a sandbox or actual hardware by keeping tabs on the app while it's active [5]. Many organizations, institutions, and corporations utilize penetration testing to detect risks, security flaws, and viruses that might damage their database, applications, networks and systems. Penetration testing is an authorized attack on systems in order to identify vulnerabilities and improve security against hackers and attackers. Testing is important since it protects the organization's finances, foster customers trust, justifies the investment in IT security, determine how aware the organization's employees are, and aids in decision-making [6]. Drozer tool is one of several distinct penetration testing tools. Drozer is one of the best tools currently available for Android Security Evaluations which was developed by MWR Labs (MWR InfoSecurity). Dozer will be particularly beneficial because of its complexity and the way that analysis is done from the perspective of a malicious application that is installed on the same device. Yet, interaction is required in order to fully realize the tool's potential. Hence, comprehensive automation of this device won't be easy and may limit its potential. Numerous studies have examined the use of the Drozer tool to test security flaws in applications or other areas ([7–11]). Drozer, a ready-made framework for Android app analysis, was used by Hafiz et al. [12] to validate the tool's output. Using Drozer, they created unique intents and sent them to risky acts. It has been demonstrated that doing so allows attackers to alter user data and actually makes weak behaviors more evident. This study will demonstrate how to use Drozer to test Android apps for vulnerabilities to secure sensitive user data in virtual settings by emulating a hack test process on a virtual Android smartphone in an Ubuntu environment. The goal of this paper is to find out how much undergraduate students know about security so that they can use applications safely. Testing for security holes in four weak Android apps through emulation approach. Automate attack surface finding in target packets using the Drozer tool. OWASP's top 10 list of mobile risks were used to classify the vulnerabilities found in the test. With this paper, we specifically aim to give application developers, end users, students, researchers, and anyone else interested in application security a simple way to check how secure their applications are.

2 Methodology

In the current study two different approaches were applied: a qualitative approach using online questionnaire and an emulation of penetration testing approach.

2.1 Online Questionnaire

The online questionnaire was used to measure undergraduate student's security awareness and their knowledge of applications security. The questionnaire was divided into three axes including 1. Security of using mobile app stores (i.e., Apple and Google), 2. Evaluating the installed software tools and the security of personal information when using social media.

2.2 Emulation of Penetration Testing

The current study applied emulating Android application penetration testing to evaluate the surface vulnerabilities for application security. The AppUse VM which is a unified stand-alone environment for the Android Pentest Platform was chosen for this study because of its simplicity of use, low cost, and comprehensive range of features. This system is a free, one-of-a-kind platform for testing the security of mobile apps in the Android environment. It was created as a virtual machine for penetration testing teams looking for an easy-to-use, customized platform for testing the security of android apps, finding security flaws, and analyzing the apps traffic. AppUse is intended to be a weaponized penetration testing environment for Android and iOS apps [13, 14]. Drozer tool included in the AppUse VM was used to assess vulnerabilities in previously downloaded target application packages (AndroGoat, Diva, Insecure Bank v2 & Beetle-Bug). Table 1 describes the Drozer modules used in this study. Drozer modules used in this study for: 1. Retrieving Package Information, 2. Identifying the Attack Surface, 3. Identifying and Launching Activities, 4. Exploiting Content Providers, 5. Interacting with Services and 6. Listing broadcast receivers. Drozer examined the fundamental components of each app to assess the security of targeted packages namely 1) Activities (is the entry point for interacting with the user representing a single screen with a user interface), 2) Services (is a general-purpose entry point for keeping an app running in the background for all kinds of reasons), 3) Content providers (manages a shared set of app data that you can store in the file system, in a SQLite database, on the web, or on any other persistent storage location that your app can access) and 4) Broadcast receivers (is a component that enables the system to deliver events to the app outside of a regular user flow, allowing the app to respond to system-wide broadcast announcements) [15].

Table 1. Description of Drozer modules used in the current study

Commands	Description
list	This command displays a list of Drozer modules that can be executed in the current session
run app.package.list	list out all the packages installed in emulate or device
run app.package.info -a [package name] or run app.package.info −package [package name]	Retreving the information for particular package
run app. package.attacksurface [package name]	Exploring an app's potential vulnerabilities
run app.activity.info -a [package name]	to find out which activities in the package are exported
run app.activity.start −component [package name] [component name]	to exploit and invoke the activity
run app.provider.info -a [package name]	Obtain information from exposed Content providers
run scanner.provider.finduris -a [package name]	In order to get the content URIs for the chosen package
run app.provider.query [content URIs]	how to query a content provider and find out the sensitive information stored in it
run app.service.info -a [package name]	lists all exported services from the chosen package
run app.service.start −component [package name] [component name]	To start the service
run app.broadcast.info –a [package name] or run app.broadcast.info -a [package name] -i	Lists all exported broadcast receivers from the target package
run app.broadcast.send −component [package name] [component name]	To launch broadcast receivers in specific package

3 Implementation

In this study, testing will be performed to find vulnerabilities in the target applications pre-installed on the AVD, namely AndroGoat, Diva, Insecure Bank v2, and BeetleBug, utilizing the Drozer tool and Table 1-listed modules.

3.1 Retrieving Package Information

In order to find vulnerabilities in the components of Android applications and specifically the packages targeted in the study, i.e. to find information associated with a custom package (a specific application), we used the following Drozer module "run app.package.info

-a [package name]" or "run app. Package.info --package [package name]", which will fetch basic information about the package: permissions, configurations, shared libraries, and collection identifiers.

3.2 Identifying the Attack Surface

The second function is identifying vulnerabilities. Drozer helps limit the number of exporting activities, broadcast receivers, content providers, and services in the target applications. We Executed the following command "run app. Package. Attack surface [package name]" from the Drozer console in order to list the exported activities, exported broadcast receivers, exported content providers, and services We implemented the module for each target application package separately.

3.3 Identifying and Launching Activities

In order to define and launch the exported activities in a package we run two modules: First: To specify the exported activities in any package, the command "run app.activity.info -a [package name]" is run . Second: To start an exported activity, we execute the command "run app.activity.start --component [package name] [component name]".

3.4 Exploiting Content Providers

In order to identify the provider of the exported content in the target packages, we followed these steps: First, we executed the "Run app.provider.info -a [package name]" module.Second: Dozer's scanner module is used to collect target data. To get a list of URLs in this module, we first guessed the position of its component. We ran the module "run scanner.provider.finduris -a [package name]" to get the content URIs for the packages.Third: To query the provider for the critical information stored in the URL, we ran the "Run app.provider.query [content URIs]" module in the target packages.

3.5 Interacting with Services

Similar to activities, exported services can be called using Dozer modules. The module "run app.service.info -a [package name]" lists all services that have been exported from the chosen package. Then we run the module "run app.service.start – component [package name] [component name]" and observe what happens on the Android emulator.

3.6 Listing Broadcast Receivers

Android sends broadcasts when activities like booting or charging occur. Apps may send customized broadcasts to inform other apps of anything of interest (for example, some new data has been downloaded). Drozer provides modules to help obtain information about broadcast receivers .

In order to enumerate the exported broadcast receivers in the target applications, we ran the module "run app.broadcast.info –a [package name]". Then to start an exported broadcast receiver from any package, we execute the module "run app.broadcast.send – component [package name] [component name]" and note the changes on the Android emulator screen.

4 Results and Discussion

4.1 Questionnaire Results and Discussion

With the use of a questionnaire form, we were able to assess the level of security awareness among undergraduates as they downloaded apps. This method has performed the same goal in previous studies, for example, Abdulaziz [16], in his study measured the level of cybersecurity awareness of cybercrimes in the Kingdom of Saudi Arabia. Amita et al. [17] also analyzed smartphone security practices among undergraduate business students at a regional state university.

In the first axis, we found that females are more aware of how reliable apps in stores are than males are. On the other hand, there is a weakness in the security awareness of the importance of verifying the developer of the application before installing it for both sexes. While testing the second axis of undergraduate students' awareness of monitoring changes on their devices after installing applications, we found no significant gender differences. In the last axis, we found that males and females are equally aware of the privacy of the information they share on social media. This means that students aren't as aware of data privacy and data protection as they should be. This was the opposite of what we thought would happen, and it was also the opposite of what Philip et al. [18], found in their study, which showed that females are more willing to share their private data and information, while males are more careful to keep their privacy.

4.2 Vulnerability Testing Results

The following are the outcomes of the second recommended approach of the study, which is penetration testing and identification of vulnerabilities in four proposed vulnerable Android apps.

4.2.1 The Results of Retrieving Package Information

Figure 1 shows screenshots of the output generated by the module run "run app.package.info -a [package name]" in the target packets for this study. Figure 1 (A) illustrates the AndroGoat application package. It can access the Internet, write to, and read from external storage. Figure 1 (B) shows that the Diva application has package permissions (A). The Insecure Bank v2 application package in Fig. 1 (C) contains harmful access capabilities including send SMS, read call log, write to external storage, get accounts, read contacts, and more, as seen in the screenshot (C). The BeetleBug app bundle simply needs biometric authentication rights to view and validate Android biometric information, Fig. 1 (D). App permissions safeguard user privacy by limiting access to system data, contacts, audio recording, linked devices, and resources. According on the packages' results, these programs have security vulnerabilities due to authorization to access resources and data, which might lead to data loss, seizure, or detrimental application risks. Our result of getting package information and permission required via writing to external storage was M2 risk of the OWASP mobile top 10, consistent with Ashish et al. [19].

4.2.2 The Results of Identifying the Attack Surface

Figure 2 depicts the outcome of the target packages' command execution. Figure 2 (A) depicts the number of attack surfaces in an AndroGoat package, where the attack surfaces reflect the package's exported components. The package has two exported activities, one exported broadcast receiver and one exported service, and the package file (APK) is configured for debugging, as seen in the screenshot. The output of the Diva package command execution is shown in Fig. 2 (B). According to the test results, there are three debuggable exported activities, one exported content provider, and one exported service and package file (APK). As shown in snapshot Fig. 2 (C), an unsecured bank V2 package comprised 5 exported activities, 1 exported broadcast receiver, and 1 exported content providers as a consequence of running the create attack surfaces module. Moreover, the application package, like the previously stated packages, is debuggable. As illustrated in Fig. 2 (D), the BeetleBug package had only 15 exported activities and its APK file could not be modified. Moreover, the three packages (A, B, and C) are debuggable, which is a flaw. The APK file should not be debugged in any way. This means you can attach a Java debugger to a running program, parse it at runtime, set breakpoints, navigate through the code, gather variable values, and even change them.

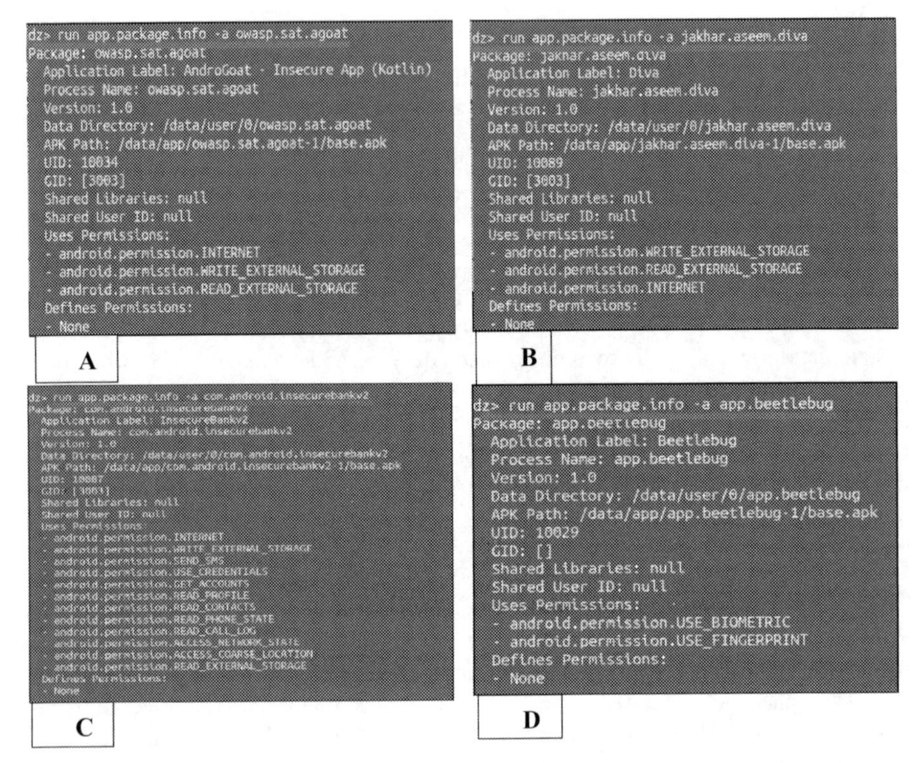

Fig. 1. Screenshots of retrieving packages information in the Drozer console for the target apps

4.2.3 The Results of Identifying and Launching Activities

The output in this task was the result of two modules being used in Drozer, the first module in order to accurately identify the activities resulting from the previous task for each package. The second command is to launch the activity. For AndroGoat package activity "owasp.sat.agoat.SplashActivity" is defined. The splash screen of the AndroGoat app appeared on the emulator screen, which is the launch screen of the application, which means that the splash activity is an exported activity to run the application. Accessing an application interface on the device without asking for an access token is an authentication vulnerability, as an attacker can request any service or gain unauthorized access to data. In terms of the OWASP Mobile Top 10 Vulnerabilities of 2016, this vulnerability is classified as Insecure Authentication: M4.

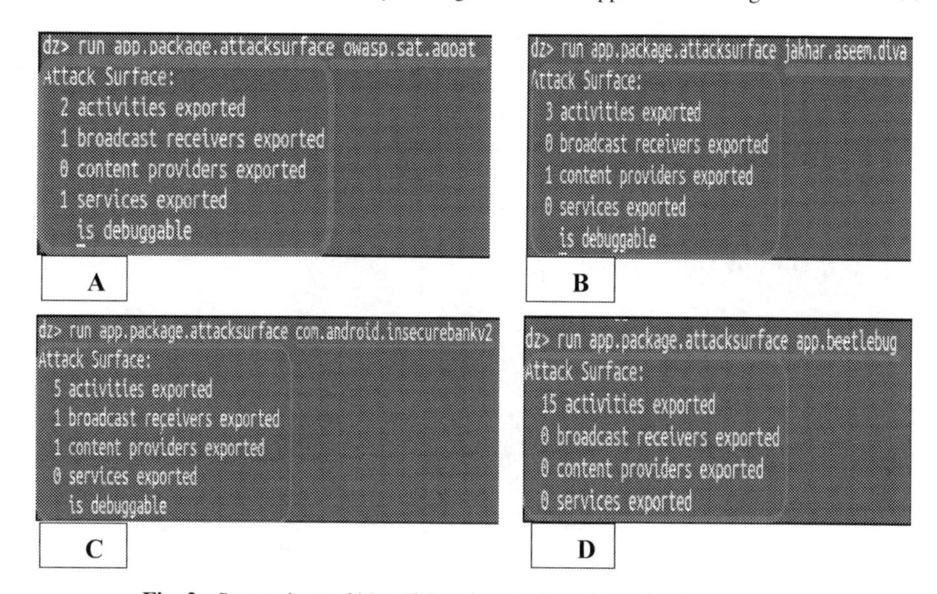

Fig. 2. Screenshots of identifying the attack surfaces for the target apps

Likewise, in the Diva package. When we launched the exported activity "jakhar.aseem.diva.APICredsActivity," a screen showed on the Android Emulator that displayed the API credentials as obtained from outside the app. As a result, there is an access control vulnerability that has been classified as Insecure Authentication: M4.

In the Insecure Bank v2 package in Fig. 3 the exported activity exploit is shown in the screenshot without any login verification. With additional capabilities, it is now possible to go directly to any required action without logging in. According to OWASP's 2016 list of the Top 10 Mobile Device Threats, this vulnerability—categorized as platform abuse—is categorized as an M1 risk. In addition, as can be seen from the screenshot, the app contained other exported activities such as. / PostLogin and./DoTransfer. These activities require authentication and have been exported, i.e., their execution can be called by authentication bypass and this risk is classified as M4: Insecure Authentication.

In the case of the BeetleBug package, it seems that the output of the "app.beetlebug.ctf.InsecureStorageSQLite" activity launched from the exported package activities was used to retrieve user credentials through insecure database storage. Although the vulnerability occurs when a user registers private credentials and logs into the application, the program immediately encrypts the credentials and stores them in a plain text database. Moreover, since the data is immediately encrypted upon recovery, the SQL injection vulnerability allows credentials to be recovered and exploited. This vulnerability has been classed as M2: Insecure Data Storage in OWASP's 2016 list of the top 10 mobile threats.

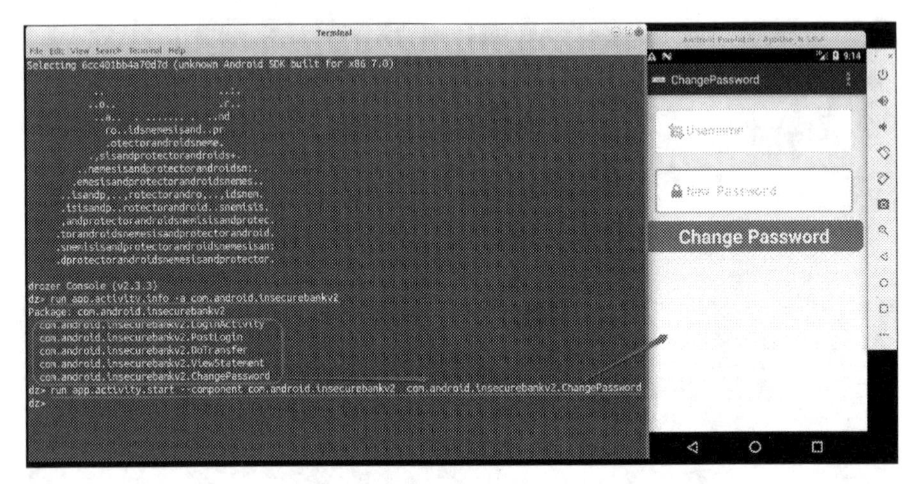

Fig. 3. Identifying and Launching Activities in Insecure Bank v2 package

4.2.4 The Results of Exported Content Providers

Figure 4 is the screenshot of the results of the "Run app.provider.info -a [package name]" module, which collects information about content providers in target packages. A content provider enables applications to communicate data in databases or files over a network. AndroGoat and BeetleBug packages do not have exported content providers, as seen in the screenshot. While the Diva package appeared, it was also present in the Insecure Bank v2 package. Note that the read and write permissions are set to null, indicating that the user may query the data storage through the content provider.

We get the result of querying the content provider of the Diva package using the Scanner module to specify the URI of the available content provider that you can access. The extracted data is a database, meaning you might be able to access sensitive information or exploit various flaws (SQL injection or path traversal). Most content providers are often used as a database interface. If you have access to it, you can extract, modify, insert, and remove data. This vulnerability was rated M2: Insecure Data Storage in OWASP's 2016 list of the top 10 mobile device threats. Similarly, we achieved a similar result while searching one of the URIs in the Insecure Bank v2 package, where the tested URI includes the data of the users who used the device.

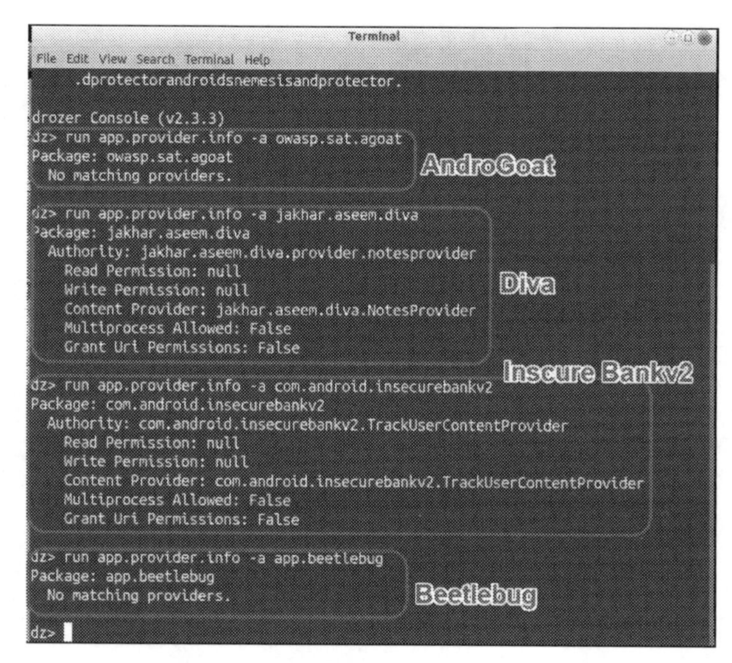

Fig. 4. Reading from content providers by Drozer console

4.2.5 Results of Interacting with Services

The services seem harmless and run in the background. However, they support other software components and may perform sensitive actions such as logging into an online profile, changing a password, or even supporting malicious processes by acting as a proxy for the host machine's system services. These must be taken into account while evaluating the application. If a potentially dangerous service does not require permissions, it may be misused to misuse users, elevation of privileges, or steal sensitive data [20]. The result of the enumeration of exported services in the target applications in this study is that all packages do not contain exported services except the AndroGoat package which contains one service exported without asking for permissions. The screenshot in Fig. 5 is the output for launching the exported service on Drozer. Note the notifications that appeared on the emulator screen after starting the service. Since the service runs in the background, we cannot interact with it on the interface. The outgoing service downloaded an invoice that may contain sensitive user data. This vulnerability was rated M4: Insecure Authentication in OWASP's 2016 list of the top 10 mobile device threats.

Fig. 5. Launching the exported service in AndroGoat package in Drozer Console

4.2.6 Results of Listing Broadcast Receivers

Figure 6 depicts the result of the module's implementation for recognizing the exported broadcast receivers in the target beams. Drozer found unsecured broadcast receivers in the vulnerable AndroGoat and Insecure Bank v2 apps, which are responsible for providing messages to users. In the AndroGoat package, the ShowDataReceiver broadcast receiver was used to display hidden sensitive user data after performing the "run app.broadcast.send —component [package name] [component name]" command. Notice the data shown as a notice on the emulator screen. According to the vulnerability research completed by Thomás and others in their study [21], this vulnerability is categorized as Reverse Engineering: M9 in terms of OWASP Mobile Top 10 vulnerabilities for 2016.

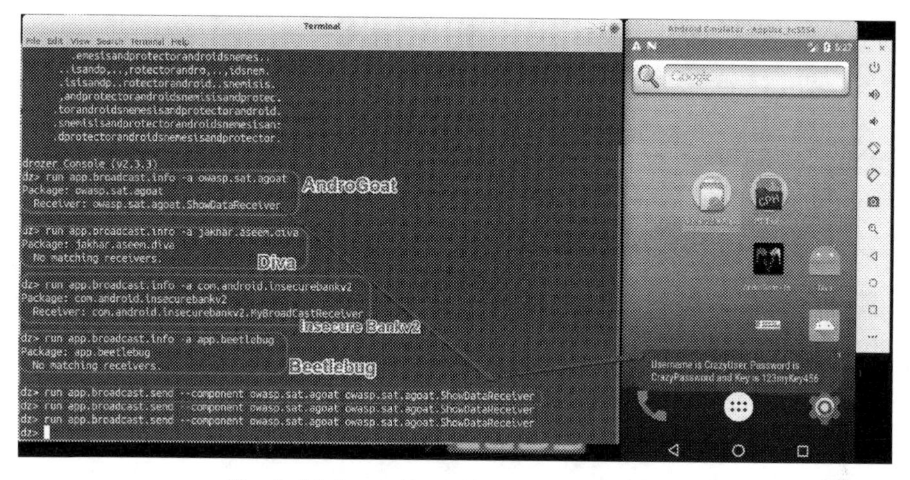

Fig. 6. Listing and launching broadcast receivers

5 Conclusions

Application security refers to the steps taken to prevent intruders from tampering with the source code. It is the process of improving any program by looking for, correcting, or avoiding vulnerabilities using innovative and diverse methodologies. To protect user data from tampering and illicit exploitation, to strengthen the role of security developers in verifying the security of programs before releasing them, and to raise public security awareness This study's goal was. Despite the security criteria imposed by Android on its apps and the processes followed by developers, criminal actors, and their illegal techniques to access user data and destroy mobile platforms are on the rise. As a result, it is critical to evaluate the application and its security before releasing or using it. Drozer, a dynamic detection tool, was used to check for application vulnerabilities. It has been used four legitimately vulnerable Android apps made specifically for training to ensure the security of users' personal information. Investigating and analyzing potentially vulnerable exported components and finding attack surfaces was the fundamental goal of this research. The results of the test for security flaws showed that the packages had several flaws. These included vulnerabilities related to debugging, application permissions, access issues, SQL injections, insecure authentication, unauthorized authorization, reverse engineering, and many more. After thorough analysis, these flaws have been placed in the OWASP Top 10 Mobile Risks category, making their status as an Approved Security Scale account more certain. This effort will provide a simple framework for students, researchers, and security professionals in the Sultanate of Oman to use in order to improve their security awareness while managing and deploying apps. This study may be improved in many ways, although it initially meets the necessary aims for beginners.

References

1. Adekotujo, A., et al.: A comparative study of operating systems: case of windows, Unix, Linux, Mac, Android and IOS. Int. J. Comput. Appl. **176**(39), 16–23 (2020). https://doi.org/10.5120/ijca2020920494
2. Vamsi, R., Jain, A.: Scientific and Practical Cyber Security Journal (SPCSJ), vol. 5, no. 3, pp. 1–10 (2021). ISSN 2587- 4667
3. Velu, V.K.: Mobile Application Penetration Testing: Explore Real-World Threat Scenarios, Attacks on Mobile Applications, and Ways to Counter Them. Packt Publishing, Birmingham (2016)
4. Ahmed, Arooj, and Arooj AhmedIam an experienced web content writer who loves to cover trendy and latest news. More than it I have written alot of blogs and Amazon listing. Writing is my passion so looking forward for better opportunities to fill the writing world with my o. "Data Reveals Android App Categories Most Vulnerable to Cyber Security Attacks." Digital Information World, 2 August 2021. www.digitalinformationworld.com/2021/08/data-reveal sandroid-app-categories.html?m=1
5. Wang, Z., et al.: Review of android malware detection based on deep learning. IEEE Access **8**, 181102–181126 (2020). https://doi.org/10.1109/access.2020.3028370
6. Al Shebli, H.M.Z., Beheshti, B.D.: A study on penetration testing process and tools. In: 2018 IEEE Long Island Systems, Applications and Technology Conference (LISAT) (2018). https://doi.org/10.1109/lisat.2018.8378035
7. Elsersy, W.F., et al.: Rootector: robust android rooting detection framework using machine learning algorithms. Arabian J. Sci. Eng. **48**(2), 1771–1791 (2022). https://doi.org/10.1007/s13369-022-06949-5
8. Hasan, U., et al.: User-centred design-based privacy and security framework for developing mobile health applications. In: Uddin, M.S., Bansal, J.C. (eds.) Proceedings of International Joint Conference on Advances in Computational Intelligence. Algorithms for Intelligent Systems, pp. 203–216. Springer, Singapore (2021). https://doi.org/10.1007/978-981-16-0586-4_17
9. Yankson, B., et al.: Security assessment for Zenbo robot using Drozer and MobSF frameworks. In: 2021 11th IFIP International Conference on New Technologies, Mobility and Security (NTMS) (2021). https://doi.org/10.1109/ntms49979.2021.9432666
10. Kohli, N., Mohaghegh, M.: Security testing of android based COVID tracer applications. In: 2020 IEEE Asia-Pacific Conference on Computer Science and Data Engineering (CSDE) (2020). https://doi.org/10.1109/csde50874.2020.9411579
11. Yang, Y., et al.: Research on non-authorized privilege escalation detection of android applications. In: 2016 17th IEEE/ACIS International Conference on Software Engineering, Artificial Intelligence, Networking and Parallel/Distributed Computing (SNPD) (2016). https://doi.org/10.1109/snpd.2016.7515959
12. Maqsood, H.M., et al.: Privacy leakage through exploitation of vulnerable inter-app communication on android. In: 2019 13th International Conference on Open Source Systems and Technologies (ICOSST) (2019). https://doi.org/10.1109/icosst48232.2019.9043935
13. Amiad. "Application Security." AppSec Labs, 31 May 2020. https://appsec-labs.com/
14. E-Spin. "Appuse Pro Mobile Pentesting: E-SPIN Group." E, E-SPIN, 17 April 2020. https://www.e-spincorp.com/appuse-pro-mobile-pentesting/
15. Ali, T., et al.: An automated permission selection framework for android platform. J. Grid Comput. **18**(3), 547–561 (2018). https://doi.org/10.1007/s10723-018-9455-1
16. Alzubaidi, A.: Measuring the level of cyber-security awareness for cybercrime in Saudi Arabia. Heliyon **7**(1) (2021). https://doi.org/10.1016/j.heliyon.2021.e06016

17. Chin, A.G., et al.: An analysis of smartphone security practices among undergraduate business students at a regional public university. Int. J. Educ. Dev. Inf. Commun. Technol. (IJEDICT) **16**(1), 44–61 (2020)
18. Nyblom, P., et al.: Risk perceptions on social media use in Norway. Futur. Internet **12**(12), 211 (2020). https://doi.org/10.3390/fi12120211
19. Sai, A.R., et al.: Privacy and security analysis of cryptocurrency mobile applications. In: 2019 Fifth Conference on Mobile and Secure Services (MobiSecServ) (2019). https://doi.org/10.1109/mobisecserv.2019.8686583
20. Makan. Android Security Cookbook. Packt Publishing (2013)
21. Borja, T., et al.: Risk analysis and android application penetration testing based on OWASP 2016. In: Rocha, Á., Ferrás, C., López-López, P.C., Guarda, T. (eds.) Information Technology and Systems. ICITS 2021. AISC, vol. 1330, pp. 461–478. Springer, Cham (2021). https://doi.org/10.1007/978-3-030-68285-9_44

Contemporary Global Trends in Small Project Management Practices and Their Impact on Oman

Safiya Al Salmi(✉) and P. Vijaya(✉)

Modern College of Business and Science, Bowshar, Muscat, Oman
{20202777,Vijaya.Padmanabha}@mcbs.edu.om

Abstract. Globalization and internationalization bring change in the business domain. Competition expands its roots in international dimensions and corporations are emerging. Organizations are now adopting new trends and accepting changes to win the market position. Due to global trends, the management of projects is now using different methods, tools, and practices to make their project successful. This study is exploring the impact of global trends on the management practices of small projects. A background study was done to analyze the past situation, moves in the traditional practices, and make a link with current global trends. The interview was conducted with the leaders of small projects to get an in-depth view of global trends and their impact on management practices. Highlighted global trends are digitalization, sustainability, diversity, people analytics, research and development, shared knowledge, and innovation. The second stage of research checks the awareness level of global trends and practices. Data was collected from the people working on small projects. These people are well aware of global trends and have a ready culture in them. Proper planning, supervision, and resources are needed to flourish small projects as these projects are a good contributor to Oman's economy and consider an important component of the business world. This study opens a new way for the research to explore industry-wise small project management practices and analyze their pattern of meeting global demand.

Keywords: Globalization · Internationalization · Global trends · Digitalization · Sustainability · Project management

1 Introduction

For decades, the area of project management or project organizing is gaining recognition among scholars and comes up with a range of different perspectives. (Dalcher, 2016; Hodgson and Cicmil, 2016; Svejvig and Grex, 2016; van der Hoorn, 2016; Walker and Lloyd-Walker, 2016). Governments are promoting small and medium projects and designing policiesin a way that gains the interest of the general public. This way leads to increased career development and a rise in job opportunities. Project work is generally undertaken by project-oriented organizations that use the projects to bring change. The main area is still in its

S. Aurelia et al. (Eds.): ICCSST 2023, CCIS 1973, pp. 104–112, 2024.
https://doi.org/10.1007/978-3-031-50993-3_9

infancy i.e. which global trends and future trends need to consider by projects to bring transformation and which set of knowledge, skills, attitude, and experience will be the requirement of the time rather than just make the career through projects. Project management certification (PMP) is a certification that still lacks many skills and areas needed by the project manager to handle the project effectively and efficiently in ongoing global trends (PMI, 2017). Project managers are excited to become part of the projects, but many challenges are coming on the way because of the topmost global trend "technology advancement" inclusive of artificial technology, virtual and augmented reality, big data, and the internet of things. Other than this trend climate change, demographic shifts and environmental wellbeing is also area that needs to handle in project management to witness successful outcomes. Trying to understand the impact of contemporary global trends in small project management practices in Oman, this study is important as it provides in depth knowledge of prevailing practices and expected happening.

2 Literature Survey

The main theme of this research and its literature review is to precisely look into the global change which raises and come up with the new and contemporary trends in business management that hit the globe and bring change in the practices of Small, medium, as well as huge organization. The reviewed literature also includes the impact of this hit on management practices specifically in the oil-dependent country Oman's small projects. This chapter discussed the important contribution of different authors discussing the modernization of Oman.

In different regions of the world, long-standing economies and economies that have just recently begun to grow are adapting to the changes brought about by the Fourth Industrial Revolution. To increase productivity over the next ten years, organizations must strike a balance between embracing emerging technology, making investments in human capital to train and develop them, and maintaining an environment that is open to new ideas. Companies must keep eye on ongoing change, access them, and handle it with a management tool to avoid any damages or losses. The process of internationalization and globalization raises the need to improve the management process to deal with socioeconomic changes. Old techniques and trends are now obsolete and replaced with new and effective techniques and tools to meet the demands of change and global trends. Even though artificial intelligence (AI) is now publicly accessible, the concept of AI is often associated with dystopian visions of the future in works of science fiction (Verma, S., Gustafsson, A. 2020). The Global Competitiveness Index 4.0 is made up of 12 components, some of which are institutions, infrastructure, the utilization of information and communications technology (ICT), macroeconomic stability, health, skills, product market, labor market, financial system, market size, business vitality, and innovation capabilities. Singapore is the most advanced technology of any country in the world, and the United States of America is in a league of its own when it comes to invention. The Index result and national scorecard serve as a yardstick to measure the current standing of a country to both its historical performance and the standing of other countries that fit into the same economic or geographical category as the country in question.

Project managers must be able to select the best way to implement changes smoothly, as failing to do so will make it impossible to achieve the desired level of increased productivity. Investing in people shouldn't be an afterthought anymore, as robust development and adaptability are necessary.

The most important details in this text are that a nation's capacity for innovation will not expand without other aspects being present, such as scientific publications, patent applications, research and development expenditures, and research institutes. Additionally, there has been a significant increase in market concentration as well as a rise in income inequality in both developed and developing nations, with the proportion of the income of the top decile increasing from 43% to 47% in the United States, 36% to 41% in China, and 35% to 35%) (Barkatullah and Djumadi, 2018). To recover productivity, one of the most essential aspects is a change toward an economic structure that is equal and kinder to the environment. To go ahead on a path that will secure development, shared prosperity, and long-term sustainability, politicians, business leaders, and global multilateral organizations work together to make choices that are both courageous and inventive.

3 Methodology

This study uses a qualitative approach to study the global trends in small projects in Oman. Data was collected via an interview and used to get in-depth knowledge of trending practices in the management of small projects. The qualitative approach is more towards understanding, interpreting, and observing the gathered data with a deep inside as well as an overall view. The quantitative research method is more structured and involves numerals and statistics. The mixed method is suited for understanding the behavior in both depth and breadth.

This study is a qualitative study. Purposive sampling is the best fit for this research, as it reflects the purpose of questions and allows for a broad range of different perspectives. The researcher chose to take only those individuals for interview who meet the mentioned criteria, such as project management, individual interviews, and data collection. Semi-structured interviews were conducted with randomly selected individuals who meet the study criteria, allowing participants to explain and elaborate on more information. Interviews are a time-consuming and expensive activity, but the most suitable approach is to draw a similar theme or pattern using the thematic approach.

The interview was conducted in October 2022 and was conducted with an open ended question along with a few close-ended questions to get the complete picture of global trends. The survey was conducted to assess the level of awareness regarding the highlighted trends of the small projects, project management tools used by small project managers, and their usefulness in terms of the project's success. To maintain ethical behavior, informed consent was taken on a prior basis and participation in the study was voluntary and free from any pressure. Limitations include researcher bias in drawing a theme, and the data is not numerical which will later on be tested statistically.

4 Results and Discussion

Oman, an oil-dependent country, progress significantly in economic and social transformation. Oil revenue is utilized to improve infrastructure, produce more jobs, and work for the support of social welfare. Although the recent oil crisis and other, global concerns demonstrate a straight multiplier effect on economic activities. The country is looking to diversify its economy to stabilize the ongoing circumstances. In Oman, people prefer government employment but due to the recent crisis, solely government was unable to meet the challenges. As a result, the government designed policies that support SMEs to share the burden of economic crisis, make the economy stronger, and strengthen business activities. Oman is one of the most improved nations in terms of success. This chapter of analysis supports the government in the alignment of small project goals with top trending practices of project management around the globe.

Countries in the future need access to their human capital and information infrastructure they hold, make some laws on the usage of one country's human capital used by the company of other countries run under the permit (Leal-Ayala, D., López-Gómez, C., and Palladino, M., 2018) (Table 1).

Table 1. Frequency of the interviewer 4.2 Analysis

Respondent's Level of Employment	Frequency	Percentage
CEO	6	30%
Managers	14	70%

4.1 Participants

4.1.1 Top Identified Trends Impacting on Project Management Practices of Oman

Little explanation about the academic purpose of the study to the interviewer, was given. After thematic analysis of the recorded interview, the following top main concerns impacting small project managers in consideration of global trends of project management practices.

- Digitalization
- Sustainability
- Research and development
- Human resource analytics or people analytics
- Diversity
- Shared Values
- Innovativeness and creativity
- Quality and risk management
- Integration
- Intellectual capital
- Others (easy entry, Metaverse) (Fig. 1)

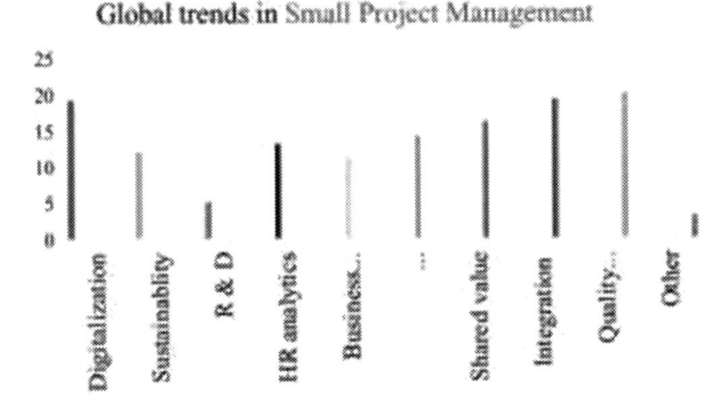

Fig. 1. Global trends in Small Project Management

Digitalization is an important aspect of all sizes of the firm in all sectors, as it reduces transaction cost and provides direct communication between two parties. It also helps small projects to smooth integration into the global market with broader trade services and operations, and provides timely access to inventory, resources, and training and recruitment channels. In Oman, the government has established the Government Innovation Initiatives to support and involve government organizations in innovative and creative work, and the latest technologies are supported by cloud computing, virtual application, block chain, the internet of things, and artificial intelligence. Technological advancement is quickly accepted in small projects due to its easy access and pocket friendly nature. Most managers focus on easy access to resources and workplace digital-ready culture, with little focus on innovation and integration.

Oman Technology Fund helps local businesses develop online auctions, teaching platforms, and marketplaces. Two commonly used internet services, Omantel and Ooredoo, are showing interest in 5G commercial partnerships. The national broadband strategy aims to provide broadband services in up to 90% of urban Oman with 75% penetration.

Sustainability is a global trend in small and large businesses, and Oman is a country that recognizes and adopts sustainable development goals. Implementation of these goals faces challenges, but Oman is progressing in the achievement of its theme "A Decent and Sustainable Life for All". Oman's major electricity production depends on crude oil and natural gas, leading to the thinking of transitioning Oman and constructing in green perspective. Oman suitability week was held in 2022 to highlight the Sultanate's commitment to uplifting its economy into the new green model. UN goals and target indicators are used to frame the agenda and investment related policies, but sustainability is still challenging and difficult to embed in the day-to-day practices of projects.

Small projects are recognizing the importance of sustainability and trying to reduce their ecological footprints, but face money and time constraints. Government awards green credentials based on energy efficient consumption practices and green disposable methods,

Quality management is very necessary for projects. It is possible via internal quality audits and management reviews whose main focus is continuous improvement in projects. This will bring a paradigm shift and cultural transition in the small projects (Silvius, 2017). Now researchers are exploring the sustainability in product base systems to product service-based systems which helps them to align and focus on people analytics. (Ali et al., 2016, Manzini, 1999, Johansson, 2002).

Diversity is a key trend in any business set up to manage the project efficiently. Cognitive diversity in the workforce relies on intellectual human capital and results in innovative teamwork. Now globally there is a trend of empowering teams across industries, from small regional firms to major multinationals, and in both emerging and developed markets (DK, 2022).

In addition, ethnic diversity is an acceptable business portfolio in all regions of the world but requires a high level of collaboration across the nation. Deep-level diversity is also highlighted by few as it impacts SMEs in Oman.

Oman's 2020 business development plan emphasizes the diversified business portfolio in its 2020 plan. Oman is focusing on diversifying and privatizing its small and big projects, including logistics, tourism, mining, manufacturing, and fisheries. The change of name from Muscat and Oman to the Sultanate of Oman reflects the unity of the country, and Oman is no longer among the ranks of lower-income nations in need of World Bank credit facilities. Analysis shows that cognitive and business diversity is the most important element in a small project to become a successful partner in the race. Positive aspects of diversity in small project management include mental flexibility, easy expansion, more attractive brand image, easy diversification of business and product line, innovative teams, powerful outcomes, and easy to respond to queries.

Research and development are an important part of businesses' survival, but it is often overlooked due to the lack of resources and cost-benefit. In Oman, government, semi-government, big enterprises, universities, and consultant agencies are dedicating investment to this activity, which raises economic growth and creates job opportunities, intellectual capital, national security, public health and well-being, and agricultural and environmental protection. The most important details in this text are that new products or projects always have the threat of failure, and it is the responsibility of the project manager to perform proper research before launching. In big organizations, proper budgets and resources can be allocated for this activity, but in small projects, these projects are to support the big organizations and launch after research of those big organizations. Small projects are reluctant to take on R&D activities due to their return's variation, and there is no legal obligation on small projects to conduct R&D. Additionally, three interviewers highlighted that the small projects are working for large projects so to do the research and explore more is only the task of larger projects.

People analytics is a global practice that is becoming increasingly popular. Globally, 70% of project heads are considering people analytics as a topmost priority, and organizations are taking time to develop people analytics capabilities. Big organizations categorize data analytics into three sub-categories: data and data management, analytics capabilities of the workforce and organization, and the model on which the organization is operating. In Oman, business managers are aware of it, but it is slow and continuous. Lack of data access is a challenge, leading to a new direction of data acquisition.

Additionally, organizations have fragmented human resources-specific technology, which makes the challenges compounded. The people analytics team in Oman relies on trust, ownership, and empowerment to keep them aligned with the organization. They rely on big data usage for decisions and planning ways to hold the employee for long.

People analytics is essential for small project management, as it allows project managers to cross- check their targets and retain customers for the long term. It is important to gather data, analyze it, and add value to projects. However, it is lacking to keep trend recording of customers, their feedback, pre-services, and post services. The overall impact is positive but takes time to get its proper place in small projects as it is in its early stage. This finding shows similar results to the studies of Ishaq & Matriano, 2020, ALSiyabi & Goel, 2020.

Creating Shared Value and Innovativeness. Global trends are focusing on shared values and innovation to create value in the economy. R&D is essential to bring feasible innovation to the organization, making it competitive and compatible in a global manner. Oman is one of the fastest developing countries, placed under the category of a high-income country, and top 50 countries in peace. It has a rich history and culture of exploration abroad and is categorized under the high-income economies and is placed in the top 50 leading nations in peace. Its economy is diversified and oil forms only a small fraction of the economic activities, which also include fishing, light manufacturing, and agriculture. The only successful rule from centuries is to do business on a model which is better than competitors and capture the economic value. The current environment is turbulent and requires project managers to align strategies and processes to meet global trends.

Risk management is the process of accessing and identifying business risk and planning to control it beforehand. In Oman, risk management companies are providing services to reduce the impact of disruption or extend their support to overcome the risk. Risk management strategies ensure the target's achievement and take actions and decisions proactively if something goes wrong. It is a continuous and essential project management part.

Risk management is an important part of small project management, as it helps to create awareness of positive and negative outcomes. It also enhances the intellectual capital of employees by providing on-the-job training and hands-on experience. Emerging global trends such as digitalization, environmental changes, complexity, and ambiguity in the business environment lead project managers to handle the project in a way that makes it competitive and compatible for a longer time. In the context of Oman, the government is providing continuous support to project managers.

Other than this, collaborative managers have expertise in all domains including risk and quality management. These findings contradict the findings of other studies which showed that risk managers are important or that project managers are key people to ensure the success of the project Gachie, Wanjiru. (2017), Akintoye and MacLeod (1997).

Global trends are important for small project managers in Oman, as they serve as a baseline to plan the project and align it with global practices. The main focus of the interviewee was on Technological and environmental change, which is the top leading global trend impacting Oman. Future targets should analyze the foundations of economic factors and explore opportunities from them via motivations. Experts should shape plans and authorize to construct attitudes and values. The most important details in this text are that mass scale business must adopt global practices and align goals accordingly, and that quality training is essential for small projects.

5 Conclusion

Today organizations are working and shaping according to the world where it is operating. Similarly, for long-term survival, organizations are transforming and updating themselves as per the requirements of future targets and trends. Then business managers must look into future trends of not only businesses but also the trends in politics, economics, and society and upgrade their management practices to make the organization sustainable. This study aims to understand the contemporary global trends and their impact on small project management practices in the sultanate of Oman. This study identified the top trends which impact Oman's economy and its management practices, hurdle, and efforts of the management of the small project to improve the weak areas. Furthermore, this study highlighted the challenges in the subdomain of global trends by interviewing the leaders and checking the level of awareness of identified trends impacting small projects.

Small projects are a big contributor to the economy if work successfully. The small project must keep its vision and mission aligned with global practices as per contemporary trends. Similarly, small project managers need to be very careful about planning, implementing, and accessing strategy not only in comparison with the competitors or designed criteria but with the global trends as well to remain successful in the long run. Top leading global trends in small project management must incorporate into vision, mission, and long-term goals. Incremental and radical innovation also need to be in the focus view of managers that lead the small firms to be in the race for global trends. Not only the small project managers are responsible, but the government of Oman also need to contribute their support towards small project and ensure growth which ultimately impacts Oman's economy.

This research is not the last but the least contribution towards the economy of Oman. This study analyses the significant impact of global trends on the practices of small project managers. As managers, are the main source of information regarding the impact of global trends with the little involvement of employees just to access their level of awareness on global trends? The future domain of the study must include the employees as the main source of the response. Another aspect is this study only targets the small project managers and their practices while future studies need to cover the aspect of the medium, and big organization practices. Sector-wise analysis and comparative analysis of small medium, and big firms and their transformational level due to globalization need to be checked to design future strategies. As Oman is welcoming foreign investors and become an attractive country, so research must be conducted to analyze and compare the practices and impact of a local and foreign firm, also need to study their alignment of the strategies with global trends and underlying mechanism which make them successful.

References

Azcona, D., Hsiao, I.H., Smeaton, P.M.I.: Project Management Job Growth and Talent Gap 2017–2027, PMI, Newtown Square, PA, US (2017)

Gachie, W.: Project risk management: a review of an institutional project life cycle. Risk Gov. Control Financ. Mark. Inst. **7**, 163–173 (2017). https://doi.org/10.22495/rgc7i4c1art8

Akintoye, A.S., Macleod, M.J.: Risk analysis, and management in construction. Int. J. Proj. Manag. **15**(1), 1–38 (1997)

Silvius, G.: Sustainability as a new school of thought in project management. J. Clean. Prod. **166**, 1479–1493 (2017)

Manzini, E.: Strategic design for sustainability: towards a new mix of products and services. In: Proceedings First International Symposium on Environmentally Conscious Design and Inverse Manufacturing (1999)

Johansson, G.: Success factors for integration of ecodesign in product development. Environ. Manag. Health **13**(1), 98–107 (2002)

Verma, S., Gustafsson, A.: Investigating the emerging COVID-19 research trends in the field of business and management: a bibliometric analysis approach. J. Bus. Res. **118**, 253–261 (2020)

Ishaq, Z.G., Matriano, M.T.: Future of people analytics: the case of petroleum development Oman. J. Stud. Res. (2020)

AL Siyabi, N.H., Goel, N.: The effectiveness of knowledge management and business intelligence for increasing financial performance. (A Case Study: Petroleum Development Oman). J. Stud. Res. (2020)

DK. (2022). Project Management. Penguin

Leal-Ayala, D., López-Gómez, C., Palladino, M.: A study on the digitalization of the manufacturing sector and implications for Oman. Austria: United Nations Industrial Development Organization (2018)

Bjorvatn, T., Wald, A.: Project complexity and team-level absorptive capacity as drivers of project management performance. Int. J. Proj. Manag. **36**(6), 876–888 (2018)

Schneider, B.R.: A comparative political economy of diversified business groups, or how states organize a big business. Rev. Int. Polit. Econ. **16**(2), 178–201 (2009)

Mata, N.N., Khan, F.H., Khan, K., Martins, J.M., Rita, J., Dantas, R.: Team diversity and project performance: role of trust and absorptive capacity in it industry. Acad. Strateg. Manag. J. **20**, 1–20 (2021)

Barkatullah, A.D.: Does self-regulation provide legal protection and security to ecommerce consumers? Electron. Commer. Res. Appl. **30**, 94–101 (2018). https://doi.org/10.1016/j.elerap.2018.05.008

Early Prediction of Sepsis Using Machine Learning Algorithms: A Review

N. Shanthi[1(✉)], A. Aadhishri[1], R. C. Suganthe[1], and Xiao-Zhi Gao[2]

[1] Kongu Engineering College, Perundurai, India
shanthi.cse@kongu.ac.in
[2] University of Eastern Finland, Kuopio, Finland

Abstract. With a high rate of morbidity as well as mortality, sepsis is a major worldwide health concern. The condition is complex, making diagnosis difficult, and mortality is still high, especially in intensive care units (ICUs), despite treatments. In fact, the most common reason for death in ICUs is sepsis. As sepsis progresses, fatality rates rise, making prompt diagnosis essential. Rapid sepsis detection is essential for bettering patient outcomes. To help with early sepsis prediction, machine learning models have been created using electronic health records (EHRs) to address this issue. However, there is currently little use of these models in ICU clinical practice or research. The objective is to use machine learning to provide early sepsis prediction and identification utilizing cutting-edge algorithms. For adult patients in particular, the creation of very precise algorithms for early sepsis prediction is crucial. Advanced analytical and machine learning approaches have the potential to predict patient outcomes and improve the automation of clinical decision-making systems when used in Electronic Health Records. The machine learning methods created for early sepsis prediction are briefly described in this paper.

Keywords: Sepsis diagnosis · Machine learning algorithms · Early Sepsis prediction · Electronic Health Records

1 Introduction

More than 250,000 people lose their lives to sepsis each year, a serious and possibly fatal condition brought on by infections. There are more than 1.5 million cases documented each year. Three stages of this sickness are typically referred to as sepsis, a condition called septic shock, and serious sepsis. One obvious symptom is a greater breathing rate of exceeding twenty breaths per minute. In order to address this medical conundrum, earlier studies looked into the application of different deep learning and machine learning strategies in building prediction models for sepsis detection. Individuals of at least 18 years of age are included in the sample data that was obtained, which was taken from patient health records. Researchers are continuously developing novel early sepsis prediction methods and concepts. It's vital to remember that these predictive models' effectiveness and accuracy depend on the choice of algorithms used.

S. Aurelia et al. (Eds.): ICCSST 2023, CCIS 1973, pp. 113–125, 2024.
https://doi.org/10.1007/978-3-031-50993-3_10

Systemic Inflammatory Response Syndrome (SIRS), a significant all-over inflammatory response, is usually connected to life-threatening medical conditions. It can be brought on by infectious or non-infectious sources. For many years, deter-mining the precise bacteria causing sepsis and quantifying immune-related variables have been essential steps in both diagnosing and treating the illness. Malaise, fever, chills, and leukocytosis are typical signs of septic patients, which leads doctors to order a blood culture investigation to see if there are any bacteria in the bloodstream [1]. The life threatening illness sepsis, which results from an inappropriate host response to infection, is characterized by organ failure [2].

Sepsis treatment is a huge burden on healthcare systems, with anticipated inpatient expenses in the United States exceeding $14.6 billion in 2017. One out of every four fatalities is caused by the more serious types of sepsis, such as severe sepsis or septic shock, which can be fatal. Sepsis frequently begins with the syndrome of systemic inflammatory response, which can progress to full-blown sepsis when a viral infection damages tissues across the entire body. The diagnostic criteria for SIRS are a temperature of at least 38 °C or less than 36 °C, a heart rate of at least 90 bpm, a respiratory rate of at least 20 bpm (or a blood vessel partial pressure of carbon dioxide beyond the typical range), and a white blood cell count of at least 12,000 bpm or less than 4,000 bpm (or more than 10% bands) [3].

Through the application of preprocessing, analysis of data, and artificial intelligence techniques to the enormous EHRs maintained in hospitals and healthcare systems, an information-driven strategy to predictive modeling in critical care is provided. Large volumes of clinical data can be accessed and analyzed thanks to electronic health records, which is especially useful when patients attend the emergency room [4]. Due to the increased use of EHRs, automated clinical decision-making and pre diction systems are now more practical in hospital settings. These systems have the potential to improve complicated syndrome monitoring and care. By converting a patient's medical information into clinically relevant data, such systems can provide warnings and recommendations to improve the treatment of patients. In the case of sepsis, an artificial intelligence learning-based sepsis detection system has been developed. Using individual medical records as input, this algorithm has demonstrated the ability to forecast sepsis, serious infection, and septic shock in patients for as long as four hours in advance. It is interesting that even when using only vital sign inputs in a mixed-ward data collection, the machine learning algorithm displayed accuracy in predicting as much as two days earlier the start of sepsis [5].

Programs for clinical decision support (CDS) are essential in helping doctors make diagnoses, write prescriptions, and manage patients. For doctors managing complex or difficult illnesses like sepsis, computerized CDS systems are extremely helpful [6]. Early confirmation of infection is made possible by early discovery of major sepsis cases, which is a considerable benefit [7]. Particularly significant evidence in support of bundled treatment techniques was discovered by a statewide group analysis of patients under the age of 18 who were treated in a range of healthcare settings, including emergency rooms, residential units, and intensive care facilities [8]. Sepsis can spread quickly when there are problems with antibiotic abuse and overuse along with compromised immune systems

[9]. As healthcare becomes more digitalized and the need of early sepsis identification and response is better understood, research activities in this area have expanded [10].

1.1 Description of Sepsis

For doctors treating patients with sepsis, the ability to forecast patient survival quickly is crucial because this condition can cause a patient's death in as little as one hour. Laboratory test results and hospital evaluation can provide beneficial data about the state of a patient, however they frequently arrive too late for clinicians to identify and address a serious risk of death. Machine learning provides a remedy by making patient survival predictions possible in a matter of minutes, especially when used with a limited number of easily accessible medical features [11]. Sepsis is more prone to develop among individuals in ICUs, underlining the significance of early diagnosis [12]. Machine learning-based classification systems that incorporate a range of readily available patient data may help with the early identification and forecasting of serious illnesses [13].

Improved algorithms for the early detection and treatment of severe diseases can result from ongoing physiological monitoring and analysis in ICU patients [14]. The potential to lower in-hospital mortality from major sepsis and septic shock exists with early sepsis identification [15]. Using machine learning, researchers have investigated the predictive validity of the physiological testing criteria suggested by the Surviving Sepsis Campaign for early sepsis identification [16]. Additionally, using machine learning techniques and high-resolution data from several organ systems, an early warning system has been created [17]. A patient's health status after being admitted to ICU is detailed in their EHR [18]. Due to the requirement to effectively analyze and integrate patient data from various unconnected sources, which can be a difficult and time-consuming task, sepsis can be difficult to diagnose [19]. Additionally, the likelihood of a patient surviving is considerably increased by prompt treatment for severe sepsis and septic shock [20].

Ideal Diagnostic Test for Sepsis. Given the difficulties in healthcare today and the need to influence clinical decision-making, the ideal system would have the following features to enable customized treatment notification:

1. Rapid detection.
2. Comprehensive Pathogen Coverage: It's critical to have the ability to identify bacteria, viruses, and fungi broadly.
3. Specimen Minimum Requirement: It is vital to use a modest specimen amount in a non-invasive way.
4. Early Antibiotic Intervention: The technology should be highly sensitive and specific, allowing the start of targeted antibiotic medication right away in the event of systemic inflammatory signs and symptoms.
5. Strong Detection in Contaminated Environments: It's critical to be able to identify pathogens among contaminants using polymicrobial detection across a wide range of pathogen loads.
6. Identification of Drug Tolerance: It ought to make drug tolerance easier to spot. 7. Recognition of Novel infections: A crucial prerequisite is the ability to recognize novel or previously undiscovered infections [1].

1.2 Challenges

Generalizability. It is challenging to evaluate current machine learning methods for sepsis prediction since there aren't enough standardized datasets. To solve this problem, use patient communities, medical factors, forecast sets (such as the clinical standards for sepsis, monitoring interval), and assessment measures to determine the applicability of these approaches [21].

Interpretability. However, many of them are complex, and clinicians are hesitant to use such "black-box" algorithms in practice since they outperform traditional evaluation methods and mathematical techniques in terms of prediction [21].

2 Methodology

2.1 Pathogenesis

The transition from mild sepsis to advanced sepsis and septicemia is triggered by a succession of pathogenic processes. An initial pro- and anti-inflammatory response that is neurohumoral in nature characterizes an infection. Beginning with the activation of neutrophils, monocytes, and macrophages, this response engages endothelial cells via a variety of pathogen recognition receptors. Essential treatments aim to re verse these infections in order to increase survival in situations of serious infection and septic shock [22]. For the past 20 years, sepsis has been acknowledged as a possible or confirmed infection [23].

2.2 Host Response

Evaluation of the body's reaction to illness is normally a speedier, albeit less precise, technique to identify sepsis than direct detection of pathogens. Assessing the host reaction over time, which may also include prognostic evaluations, is necessary for tracking the patient's response to treatment [24]. To aid in the early detection of sepsis, digital detection techniques constantly track patient information [25].

2.3 Analysis and Selection of Patients

Over thirty million individuals globally suffer from severe sepsis in a single year. Even though it has been shown that early sepsis diagnosis improves medical care, it is still a difficult problem in the field of medicine [26]. Information on people's health In prior studies, participants had to be over eighteen years. The trials either prohibited patients with underlying inflammatory conditions, such as Chronic Lupus Erythematosus, Crohn's illness, or Insulin-Dependent Diabetes, or those who were already undergoing chemotherapy, from responding [27]. Descriptions of the organ damage patterns seen in sepsis patients in critical care units have been published [28] to increase the ability to detect individuals in danger of sepsis based on their electronic health record data. For critically septic patients, intensive care unit management is frequently necessary, and infection-related problems account for 20% of all ICU ad missions [29].

Before individuals become dangerously ill, ambulance doctors frequently encounter sepsis patients early in their clinical course. It is yet unknown, nevertheless, whether blood pressure readings taken during prehospital and in-hospital clinical examinations for sepsis patients correlate [30]. The progression of patients from sepsis to advanced sepsis or septic shock has been studied recently using electronic medical records, laboratory results, and biological signs. This method permits prediction and the ability to anticipate changes in patient condition, permitting intense therapy to avoid tragic consequences [31]. Figure 1 depicts the procedure for detecting people with sepsis.

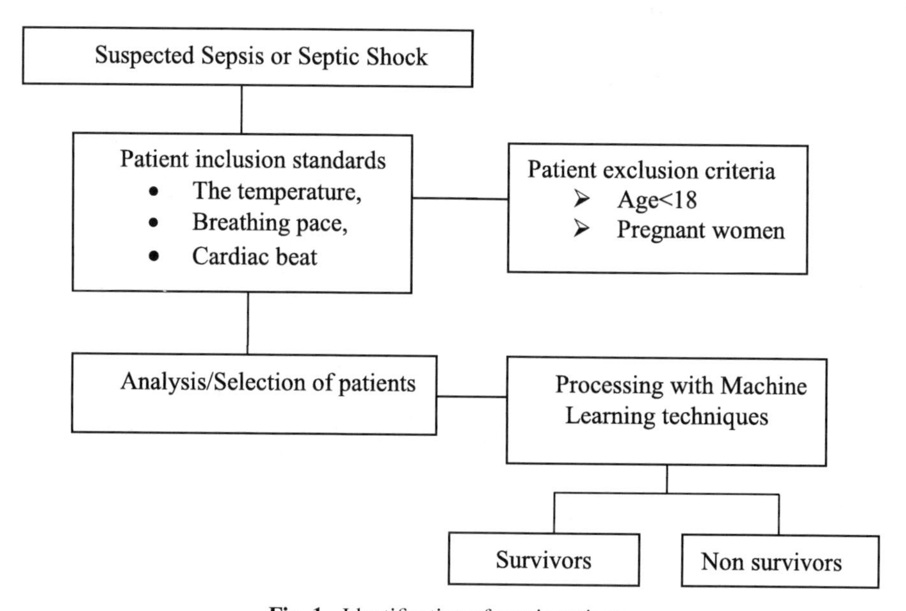

Fig. 1. Identification of sepsis patients

2.4 Collection of Data

Demographics, vital indicators (including cardiovascular health, heart rate, rate of respiration, and tympanic temperature), hematology (complete blood count), pharmaceutical chemistry (including urea, minerals, and blood sugar), and microbial condition will all be reviewed while evaluating patient data [27]. Patients were frequently evaluated if their temperature was between 36 and 38 °C, their heart rate exceeding 90 beats every minute, and their respiration rate surpassing 20 breaths each minute [32]. This medical data was gathered as part of the typical hospital workflow, and administrative information was also gathered as part of routine hospital operations. This data was extracted using SQL, a structured query language [33].

2.5 Data Imputation

Typically, all the data and values are gathered from hospital databases, such as adult patients' electronic health records. When values were missing, lost-one carries for ward imputation was used to fill in the gaps [34], and during a patient's hospital stay, a forward-fill technique was utilized to fill in the gaps. Sepsis will cause changes in the patient's metabolic system [35].

3 Model Design and Technique

Sepsis usually results from infection, frequently from an unidentified source [31]. There is currently no reliable, scientifically verified strategy for real-time sepsis pre diction in the elderly intensive care unit community [36]. Patients over the age of 18 hospitalized to the intensive care unit were the focus of retrospective research [32]. A foundation based on actual clinical practice is required for effective sepsis prediction [37]. Various monitoring technologies supply intensive care professionals with a significant amount of medical data and measures [38].

Sepsis results are better when it is diagnosed and treated early. Therefore, it was essential to evaluate how timely the Review of Systems score was in relation to established norms. The model's performance was evaluated 1, 3, 6, and 24 h after an index period, with the index time being the time at which each patient's first vital sign or laboratory result was entered into their electronic health record. A comprehensive clinical decision support system cannot be established without the integration of medical device connections across all types of ICUs [39].

A three-stage feature choice procedure was used to create and evaluate the model after the data were randomly separated into training and testing groups. Gradient boosting, a type of supervised machine learning, was used during this procedure, along with the examination of pre-existing sepsis screening models and consultations with subject-matter experts [33]. Through prompt action, timely early sepsis prediction can improve patient outcomes and lower the cost of late-stage sepsis infections [40]. Post-operative sepsis is becoming more commonplace [41]. Supervised models have the ability to categorize prospective clinical variables and provide more trust worthy predictions in comparison to current benchmark rule-based techniques [42].

To forecast hospital mortality, a number of severity grading systems and machine learning mortality prediction models have been created over the past few decades [43]. Table 1 provides a brief overview of the major machine learning algorithms for sepsis prediction.

Table 1. Taxonomy of Machine Learning Methods

ALGORITHM	DESCRIPTION	WORKING	ADVANTAGES
GRADIENT BOOSTING	Gradient boosting is a popular boosting technique in which, unlike Adaboost, which modifies training instance weights, each predictor corrects the errors of its predecessor by training on the residual errors from the prior predictor	The foundation of gradient boosting is the idea that combining the best next model with the prior models will reduce the overall prediction error. The important idea is to define target outcomes for the following model in order to minimize error	Unbeatable prediction accuracy, significant flexibility, and a lack of intensive data pre-processing are all ad vantages of gradient boosting
RANDOM FOREST	A classifier called Random Forest improves predicted accuracy by merging various decision trees, each built on a different subset of a dataset, and averaging their output	Two major steps make up the Random Forest algorithm. The random forest is first built by generating N decision trees. It then provides predictions for every tree produced in the first stage	A versatile technique, Random Forest can be used to address problems with regression and classification. It is exceptional at managing big datasets with high dimensionality, improving model precision, and reducing the overfitting problem
SUPPORT VECTOR MACHINE	Support Vector Machine can address classification and regression issues. However, classification jobs are where it is most frequently used	A Support Vector Machine creates the ideal separation hyper plane (typically a line in two dimensions) for the supplied points of data in order to optimize the distance between different categories	SVM functions well in cases with many dimensions. By using a portion of the training data known as support vectors in the decision function, memory is conserved. By enabling the creation of various kernel functions for decision functions, SVM also provides versatility

(continued)

Table 1. (*continued*)

ALGORITHM	DESCRIPTION	WORKING	ADVANTAGES
XG BOOST ALGORITHM	In Kaggle contests, the machine learning algorithm XGBoost has performed exceptionally well, especially for jobs involving structured or tabular data	Regression trees are regarded as weak learners in regression tasks. Each regression tree associates a single input data point with a leaf that has a continuous value or ranking	Regularization and efficient parallel preprocessing are two advantages of XGBoost

3.1 Gradient Boosting

A flexible machine learning technique called gradient boosting is used for both regression and classification tasks. It creates an ensemble by combining thousands of relatively straightforward decision trees. To improve a selected evaluation metric, various decision trees are iteratively blended in this procedure. A fresh decision tree is added to the current ensemble with each iteration while taking prior mistakes into account. Only six vital-sign occurrences from unprocessed Electronic Health Record sequences were used to create the GB-Vital model. Even though the generated features heavily aggregate the sequence information, little ordering information is still kept. The decision tree gradient-boosted classifier used to create this model allows each tree to split up to six times. It's a method, like decision trees [44], intended to strengthen learners who were initially weak.

3.2 Random Forest Model

Random Forest, the most modern machine learning algorithm employed, is comparable to classifiers created using decision trees. An ensemble method called Random Forest consists of many decision trees arranged in the shape of a "forest." It seeks to deal with the drawbacks of distinct decision trees. An independent subset of the initial training data is used to build each tree in the forest. Each tree divides nodes by evaluating a random subset of all accessible variables. Permutation importance is used in the Random Forest model to evaluate variable importance. It's important to keep in mind that permutation importance might not be able to tell apart associated variables. As a result, correlated variables with identical importance rankings in the variable space, such as the maximum respiration rate and initial respiratory rate, may exist [45].

3.3 Support Vector Machine

The supervised learning model Support Vector Machines (SVM) can solve classification and regression issues. They excel at resolving two-group classification issues where input vectors are frequently non-linearly transferred to a high-dimensional feature space. Although SVMs can be applied to classification and regression tasks, their main utility is in the former. SVM effectively generalizes because of its special ability to create a linear decision surface in a non-linear feature space. The kernel function of a Gaussian distribution is used in SVM models. SVM has the benefit of being less prone to overfitting [46].

3.4 XG Boost Algorithm

Gradient boosting and decision trees are the foundation of the ensemble machine learning technique known as XGBoost. It has a wide range of uses, including the solution of prediction issues involving unstructured data like text and images. Its adaptability allows it to handle issues in regression, grouping, rating, and custom prediction. It makes a point of addressing missing values in a few patient parameters, estimating these missing values during preprocessing using the Gaussian Mixture Model. The preprocessed data is subsequently processed using the Extreme Gradient Boosting (XGBoost) method [47]. The gradient boosting tool XGBoost functions as a decision tree. "Gain" is measured as the average proportional enhancement of production after the execution of a function to a model. "Coverage" describes how frequently the decision tree vertices that are separated based on a specific function are reached. How frequently a characteristic appears in a decision tree is indicated by its "frequency" [48].

Table 2 lists the many classifiers that were applied in various research between 2014 and 2020. These classifiers include Support Vector Machines (SVM), Gradient Boosting, Random Forest, and XG Boost. AUROC (Area Under the Receiver Operating Characteristic Curve) values ranging from 0.045 to 0.93 are reported in some studies, but not in others. Similar to this, the stated accuracy scores for several classifiers range from 0.73% to 81.6%, while some data are missing. The performance of classifiers varies, especially in the case of SVM, which provides two AUROC scores in a single study. Although the inclusion of recent studies in 2020 shows new patterns in classifier usage, there may occasionally be data gaps caused by certain experimental setups. Overall, the table emphasizes the need for careful classifier selection based on reported metrics by showcasing the varied areas of classifier performance in different research situations.

Table 2. Classifier Performance Comparison with AUROC and Accuracy Scores

AUTHOR	YEAR	CLASSIFIER	AUROC	ACCURACY
Abromavičius, V., et al. [46]	2020	XG BOOST	0.306	0.934
Lauritsen, S.M., et al. [26]	2020	GRADIENT BOOSTING	0.786	-
Wang, X., et al. [35]	2018	RANDOM FOREST	-	81.6
Rodriguez, A., et al., [45]	2020	SVM	0.61,0.58	84.3
Delahanty, R.J., et al. [33]	2019	GRADIENT BOOSTING	0.93	67.7
Gultepe, E., et al.. [3]	2014	SVM	0.045	0.73
Taylor, R.A., et al. [4]	2016	RANDOM FOREST	0.86	-
Giannini, H.M., et al. [49]	2019	RANDOM FOREST	0.88	0.73

The classifier complexity across several experiments is summarized in Table 3. Decision Trees (0.3 learning rate) and cross-validation are used by Delahanty et al. to lessen overfitting. Random Forest is used by Vellido et al. who emphasize multi-modality, complicated processes, and data privacy. Decision Trees with error-centered cross-validation are used by Taylor et al. According to Perng et al. SVM (0.5 learning rate) uses complexity to increase accuracy. Chic co and Jurman use Gradient Boosting (0.8 learning rate) and emphasize the difficulties brought on by a constrained understanding of sepsis. These observations highlight many classifier complexity factors, including as processes, learning rates, and particular complexities unique to each study.

Table 3. Complexity of Machine Learning Classifiers Used

AUTHOR AND YEAR	CLASSIFIER	PROCESS	LEARNING RATE	COMPLEXITY
Delahanty, R.J., et al. [33]	DECISION TREE	The probability of overfitting the model is reduced by using cross validation	0.3	complexity to reach a maximum depth of 3
Vellido, A., et al. [39]	RANDOM FOREST	Complexity includes Data privacy and anonymization issues, as well as data integration from potentially disparate sources	-	complexity includes prediction, and data multi-modality,
Taylor, R.A., et al. [4]	DECISION TREE	Centered on error re sults that have been cross-validated	-	complexity parame ter correlated with the smallest possi ble error
Perng, J.W., et al. [50]	SVM	The KNN, SVM, and RF patterns were more reliable than the more complex feature extrac tion algorithms	0.5	Complexity resulted in higher accuracy rates
Chicco, D. and G. Jurman [11]	GRADIENT BOOSTING	A more limited concept of sepsis adds to the ambiguity	0.8	high disease's complexity

4 Conclusion

In conclusion, various authors have addressed sepsis prediction from 2014 to 2020 using a range of machine learning classifier algorithms. High accuracy was attained by the SVM classifier used by Rodriguez, A., et al. outperforming other models, and this led to notable findings. Future studies should look on improving clinical uptake, optimizing algorithm design, and integrating interpretive methods into sepsis forecasts. Recent developments, such as the use of natural language processing to incor porate free text from provider documentation, show potential for enhancing machine learning predictions in sepsis and critical care scenarios. These advancements represent ongoing initiatives to improve sepsis prediction and support better patient outcomes.

References

1. Sinha, M., et al.: Emerging technologies for molecular diagnosis of sepsis. Clin. Microbiol. Rev. **31**(2), 10–1128 (2018)
2. Kok, C., et al.: Automated prediction of sepsis using temporal convolutional network. Comput. Biol. Med. **127**, 103957 (2020)
3. Gultepe, E., et al.: From vital signs to clinical outcomes for patients with sepsis: a machine learning basis for a clinical decision support system. J. Am. Med. Inform. Assoc. **21**(2), 315–325 (2014)
4. Taylor, R.A., et al.: Prediction of In-hospital Mortality in emergency department patients with sepsis: a local big data-driven. Mach. Learn. Approach. Acad. Emerg. Med. **23**(3), 269–278 (2016)
5. Barton, C., et al.: Evaluation of a machine learning algorithm for up to 48-hour advance prediction of sepsis using six vital signs. Comput. Biol. Med. **109**, 79–84 (2019)
6. Calvert, J., et al.: Cost and mortality impact of an algorithm-driven sepsis prediction sys tem. J. Med. Econ. **20**(6), 646–651 (2017)
7. Shimabukuro, D.W., et al.: Effect of a machine learning-based severe sepsis prediction algorithm on patient survival and hospital length of stay: a randomised clinical trial. BMJ Open Respir. Res. **4**(1), e000234 (2017)
8. Schonfeld, D., Paul, R.: Sepsis: an update on current improvement efforts. Current Treat. Options Pediatr. **6**(4), 366–376 (2020)
9. Asuroglu, T., Ogul, H.: A deep learning approach for sepsis monitoring via severity score estimation. Comput. Methods Programs Biomed. **198**, 105816 (2021)
10. Wulff, A., et al.: Clinical decision-support systems for detection of systemic inflammatory response syndrome, sepsis, and septic shock in critically ill patients: a systematic review. Methods Inf. Med. **58**(S02), e43–e57 (2019)
11. Chicco, D., Jurman, G.: Survival prediction of patients with sepsis from age, sex, and septic episode number alone. Sci. Rep. **10**(1), 17156 (2020)
12. Morrill, J.H., et al.: Utilization of the signature method to identify the early onset of sepsis from multivariate physiological time series in critical care monitoring. Crit. Care Med. **48**(10), e976–e981 (2020)
13. Desautels, T., et al.: Prediction of sepsis in the intensive care unit with minimal electronic health record data: a machine learning approach. JMIR Med. Inform. **4**(3), e5909 (2016)
14. Fairchild, K.D.: Predictive monitoring for early detection of sepsis in neonatal ICU patients. Curr. Opin. Pediatr. **25**(2), 172–179 (2013)
15. Westphal, G.A., et al.: Reduced mortality after the implementation of a protocol for the early detection of severe sepsis. J. Crit. Care **26**(1), 76–81 (2011)

16. Giuliano, K.K.: Physiological monitoring for critically ill patients: testing a predictive model for the early detection of sepsis. Am. J. Crit. Care **16**(2), 122–130 (2007)
17. Hyland, S.L., et al.: Early prediction of circulatory failure in the intensive care unit using machine learning. Nat. Med. **26**(3), 364–373 (2020)
18. Nakhashi, M., et al.: Early Prediction of sepsis: using state-of-the-art machine learning techniques on vital sign inputs. In: 2019 Computing in Cardiology (CinC). IEEE (2019)
19. Warttig, S., et al.: Automated monitoring compared to standard care for the early detection of sepsis in critically ill patients. Cochrane Database Syst. Rev. **6** (2018)
20. McCoy, A., Das, R.: Reducing patient mortality, length of stay and readmissions through machine learning-based sepsis prediction in the emergency department, intensive care unit and hospital floor units. BMJ Open Qual. **6**(2), e000158 (2017)
21. Li, X., et al.: A time-phased machine learning model for real-time prediction of sepsis in critical care. Crit. Care Med. **48**(10), e884–e888 (2020)
22. Nguyen, H.B., et al.: Severe sepsis and septic shock: review of the literature and emergency department management guidelines. Ann. Emerg. Med. **48**(1), 28–54 (2006)
23. Torsvik, M., et al.: Early identification of sepsis in hospital inpatients by ward nurses in creases 30-day survival. Crit. Care **20**(1), 244 (2016)
24. Gunsolus, I.L., et al.: Diagnosing and managing sepsis by probing the host response to infection: advances, opportunities, and challenges. J. Clini. Microbiol. **57**(7), 10–1128 (2019)
25. Evans, D.J.W., et al.: Automated monitoring for the early detection of sepsis in critically ill patients. Cochrane Database Syst. Rev. (2016)
26. Lauritsen, S.M., et al.: Early detection of sepsis utilizing deep learning on electronic health record event sequences. Artif. Intell. Med. **104**, 101820 (2020)
27. Sutherland, A., et al.: Development and validation of a novel molecular biomarker diag nostic test for the early detection of sepsis. Crit. Care **15**(3), 1–11 (2011)
28. Ibrahim, Z.M., et al.: On classifying sepsis heterogeneity in the ICU: insight using machine learning. J. Am. Med. Inform. Assoc. **27**(3), 437–443 (2020)
29. Hooper, M.H., et al.: Randomized trial of automated, electronic monitoring to facilitate early detection of sepsis in the intensive care unit*. Crit. Care Med. **40**(7), 2096–2101 (2012)
30. Smyth, M.A., Brace-McDonnell, S.J., Perkins, G.D.: Identification of adults with sepsis in the prehospital environment: a systematic review. BMJ Open **6**(8), e011218 (2016)
31. Kam, H.J., Kim, H.Y.: Learning representations for the early detection of sepsis with deep neural networks. Comput. Biol. Med. **89**, 248–255 (2017)
32. Calvert, J.S., et al.: A computational approach to early sepsis detection. Comput. Biol. Med. **74**, 69–73 (2016)
33. Delahanty, R.J., et al.: Development and evaluation of a machine learning model for the early identification of patients at risk for sepsis. Ann. Emerg. Med. **73**(4), 334–344 (2019)
34. Burdick, H., et al.: Effect of a sepsis prediction algorithm on patient mortality, length of stay and readmission: a prospective multicentre clinical outcomes evaluation of real-world patient data from US hospitals. BMJ Health Care Inform. **27**(1), e100109 (2020)
35. Wang, X., et al.: A new effective machine learning framework for sepsis diagnosis. IEEE Access **6**, 48300–48310 (2018)
36. Shashikumar, S.P., et al.: Early sepsis detection in critical care patients using multiscale blood pressure and heart rate dynamics. J. Electrocardiol. **50**(6), 739–743 (2017)
37. Schamoni, S., et al.: Leveraging implicit expert knowledge for non-circular machine learn ing in sepsis prediction. Artif. Intell. Med. **100**, 101725 (2019)
38. Hyland, S.L., et al.: Machine learning for early prediction of circulatory failure in the in tensive care unit. arXiv preprint arXiv:1904.07990 (2019)
39. Vellido, A., et al.: Machine learning in critical care: state-of-the-art and a sepsis case study. Biomed. Eng. Online **17**(Suppl 1), 135 (2018)

40. Baniasadi, A., et al.: Two-step imputation and adaboost-based classification for early prediction of sepsis on imbalanced clinical data. Crit. Care Med. **49**(1), e91–e97 (2020)
41. Yao, R.-Q., et al.: A machine learning-based prediction of hospital mortality in patients with postoperative sepsis. Front. Med. **7**, 445 (2020)
42. Shrestha, U., et al.: Supervised machine learning for early predicting the sepsis patient: modified mean imputation and modified chi-square feature selection. Multimedia Tools Appl. **80**, 20477–20500 (2021)
43. Awad, A., et al.: Early hospital mortality prediction of intensive care unit patients using an ensemble learning approach. Int. J. Med. Inform. **108**, 185–195 (2017)
44. Tran, L., Shahabi, C., Nguyen, M.: Representation learning for early sepsis prediction **45** (2019)
45. Rodriguez, A., et al. Supervised classification techniques for prediction of mortality in adult patients with sepsis. Am J Emerg Med, 2020
46. Abromavičius, V., et al.: Two-stage monitoring of patients in intensive care unit for sepsis prediction using non-overfitted machine learning models. Electronics **9**(7), 1133 (2020)
47. Parashar, A., Mohan, Y., Rathee, N.: Analysis of various health parameters for early and efficient prediction of sepsis. IOP Conf. Ser. Mater. Sci. Eng. **1022**, 012002 (2021)
48. Liu, R., et al. Early prediction of impending septic shock in children using age-adjusted Sepsis-3 criteria. medRxiv (2020)
49. Giannini, H.M., et al.: A machine learning algorithm to predict severe sepsis and septic shock: development, implementation, and impact on clinical practice. Crit. Care Med. **47**(11), 1485–1492 (2019)
50. Perng, J.W., et al.: Mortality prediction of septic patients in the emergency department based on machine learning. J. Clin. Med. **8**(11), 1906 (2019)

Solve My Problem-Grievance Redressal System

Sridhar Iyer(✉), Deshna Gandhi, Devraj Mishra, Rudra Trivedi, and Tahera Ansari

Department of Computer Engineering, Dwarkadas J Sanghvi College of Engineering, Mumbai 400056, India
sridhar.iyer@djsce.ac.in

Abstract. This study aimed to provide a platform that would be an interactive platform for the local common people to lodge their grievances and complaints. The grievances (such as potholes on the roads, garbage on the streets, sewage leakage, etc.) are raised exponentially with the rapid growth of the population. These grievances and complaints have severely. Government officials and developers worldwide strive for measures to control problems and complaints encountered by people in day-to-day life. The traditional grievance procedure or system is time-consuming, and people in remote areas frequently lack assistance. In existing grievances, redressal application complaints are lodged after registration through many forms; an alternative to that would be manually filing the complaint by visiting the official departmental office. To provide a recognized platform that enables proper communication between citizens and locally elected members, we proposed the application "Solve My Problem," which accepts complaints in a variety of different formats, such as text, images, or video based. The app is used to submit grievances and complaints, which are then displayed to the user's locally elected representative. In comparison to the existing grievance redressal system, the app doesn't need long procedures of form filing. It requires just a click of a photo of the issue. The proposed application then uses an ML model that has been trained on image datasets of common problems like pothole, sewers, garbage etc. for classification and extracts location content of the image to trace the location of the issue. The grievances are filtered based on the number of grievances in each area, which are then passed on to higher authorities. Using data visualization techniques that can be viewed by the authorities as well as by the citizen who filed the complaint, there will be a transparent process, and complaints can be tracked. Using the proposed application, the aim is to fast-track the process of filing grievances, connecting the citizen and the government and saving lives by fixing small problems.

Keywords: public complaints · grievances · ML model · GPS · tracking · API · authority

1 Introduction

People encounter numerous issues in emerging nations like India on a daily basis. A significant number of issues such as road potholes, littered streets, and sewage leaks are prevalent, leading to numerous accidents resulting in the loss of thousands of lives every

S. Aurelia et al. (Eds.): ICCSST 2023, CCIS 1973, pp. 126–142, 2024.
https://doi.org/10.1007/978-3-031-50993-3_11

year [1]. Unfortunately, these problems often persist for months or even years, as the current system for addressing them is inadequate. As many issues go unresolved, the lack of a suitable platform or procedure for addressing complaints simply makes matters worse. Even if a person does approach local authorities to report an issue, there is no guarantee that their complaint will be heard or documented.

A lack of a unified platform for public grievances makes it difficult to bring many problems to the attention of the relevant authorities, resulting in delays in their resolution. Furthermore, the current system does not allow for com plaints to be submitted in any format [2]. The complaint redressal system is more complex and tedious due to the following hierarchy:

- Lengthy registration and grievance submission procedures.
- Current system is not multilingual.
- Not able to provide a tracking feature for submitted grievances/complaints.

The proposed system "Solve My Problem" can overcome above listed gap which is present in the existing system can enable citizens to complaint against grievances they face every day. It is a one-click method for filing complaints that will be fed with images of various issues, including garbage spills, potholes, inadequate parking, open sewage, loose wires etc., along with their locations, to an ML model that will categorize the issues and allocate the work to responsible authorities. The gravity of these insignificant issues and their potential hazards to the country is illustrated by the following facts:

- According to the Supreme Court of India, accidents involving potholes result in over 3,600 fatalities annually. Authorities were not aware of the majority of these potholes since they were not reported. Using our "Solve My Problem" application will aid in lowering these figures by making the process of filing complaints transparent and simple.[3]
- People getting sick from rubbish is expanding rapidly, according to the report "Mismanaged waste" kills up to a million people a year globally. Most of the time, even when individuals wish to report a waste problem, they either lack the knowledge of who or where to contact, find it simple to complete a huge number of forms, are unconcerned with traceability, etc. All of these issues are resolved by the phrase "Solve My Problem". [4]
- In a similar vein, a surge in the number of common people dying as a result of undetected sewage leaks around the nation. A platform called "Solve My Problem" bridges the gap between locals and government to address these issues.

The app allows users to file complaints with the appropriate authorities by simply taking a photo of the issue. The app uses a trained machine learning model to automatically detect the problem, whether it be potholes, garbage, or sewage. This saves users the time and effort of manually identifying the problem and finding the correct authorities to report it to. The app provides a quick and easy way for individuals to help improve their community and hold officials accountable for addressing public issues.

The app also utilizes the HERE API [5] to automatically detect the location of the photo taken by the user. This allows for even more efficient tracking and management of the issues reported, as authorities can easily see where the problems are concentrated and prioritize their response accordingly. The use of location detection technology, combined with the app's other features, makes it an effective tool for addressing community issues and improving the quality of life for all members of the community.

In addition to its user-friendly features, the app also includes an admin web page that allows officials to keep track of the complaints that have been filed. This allows them to efficiently manage and address the issues reported by users. The combination of user-friendly reporting and effective tracking makes the app a valuable tool for improving public infrastructure and holding official's account able.

Implementing data analysis and data visualization "Solve My Problem" offers a dashboard to acquire visual reports of their region on a weekly, monthly, and annual basis using various graphs and data visualization techniques that can be viewed by both the authorities and regular users to ensure a transparent process and the ability to track complaints. The elected local authority will have access to a list of problems within their area, categorized by type and quantity. More severe issues can then be escalated to higher authorities.

2 Literature Review

There are many workings, characteristics, structures, etc. that offer a wide range of grievance redressal systems. A compliant redressal system (using GPS), com plaints to the appropriate government departments (using the KNN algorithm), a smart e-grievance system (that set up relational databases linking lists of grievances to line lists of issues faced in a particular region), and compliant tracking (used for self-reporting of grievances by people through mobile phone apps or SMS text messages) are all based on technology, software, detecting strategies, use of cases, and data-storing methodology.

As seen in (Table 1), we have researched multiple IEEE/Research articles on complaint handling procedures, we have come to the conclusion that technology is crucial to every component. Location tracking applications still need to be improved in order to reach desired goals in a time-conscious manner, despite the fact that there have been major technology improvements to help with grievance resolution.

The process for handling complaints is the most crucial element of any administration. Yet, a lot of websites and apps have laborious and time-consuming processes. The Government of India's current grievance redressal mechanism adheres to the same conventional grievance filing procedure. The primary characteristic of such applications is grumbling about having to register through several forms and from various government agencies.

The grievance redressal system is that it can help improve the accountability of government agencies and officials. By providing a channel for complaints and grievances, the system enables citizens to hold public officials and agencies accountable for their actions or in actions.

The field survey and papers also gave us an insight into the requirements of the target demographic. It showed us that the people needed a quick and easy to-use grievance redressal system that directly connects the common people to the local authorities. It should not require lengthy procedures and extensive red tape.

The findings of the field survey revealed that people are willing to take action, but they require the appropriate tools to do so. Research papers have also indicated that no

Table 1. Research Gaps in IEEE papers on existing grievance redressal systems

Sr.no	Paper	Publisher	Research Gaps Identified
1	Smart E-Grievance System For Effective Communication In smart Cities [1]	IEEE	1) Not able to detect location from where the issue is delivered 2) Limited to address the road and light related problems
2	Public complaint service engineering based on good governance principles [6]	IEEE	1) Limited area coverage for addressing the problems 2) The proposed application is Android based, cannot run on IOS devices Limited number of problem are addressed
3	GPS based complaint redressal system [7]	IEEE	1) Not Updating the location as safe after the grievance is solved 2) Unable to prioritization by area 3) No list of grievances is displayed (just displayed on the maps with marker)
4	Determining citizen complaints to the appropriate government departments using KNN algorithm [8]	IEEE	1) No prioritization of complaints 2) Co-ordinate are not accurate as GPS enabled location detection is not been used 3) No filtration or area-related target could be found

existing model or application incorporates a prioritizing algorithm to determine which complaints should be addressed first.

The authorities need to have such an algorithm to allocate their resources effectively. Furthermore, the current models lack the use of geo-tagging, which is a simple way to extract the location where a photo is captured.

So finally, we can determine that the outcome of the literature and field survey was that there is a need for a solution to the pre-existing problem.

3 Problem Statement

The problem statement associated with this subject revolves around the ethical challenges that must be factored in and the issues that demand careful deliberation during the planning, development, and implementation of such tools. Technical limitations, limited public acceptance, time-consuming processes, and a conventional approach to handling complaints all represent potential barriers to their effectiveness.

The problem that is quite evident living in any metropolitan area seems to be the deterioration in the city's infrastructure. There is a huge number of potholes, misplaced dividers, bad roads, open gutters, and loose wiring that are ever-increasing by the day. But this is the problem that the government should solve.

The problem that the people of India face every day is the breaking down of their city's infrastructure. What should be a clean, safe and green ecosystem is in reality a dangerous place filled with potholes, loose wires and garbage everywhere. Failure to promptly address these problems can result in injuries to the general populace, and in more severe cases, even fatalities. E-grievances contribute to heightened transparency, efficiency, and improved monitoring of city management's effectiveness.

Nonetheless, the manner in which online complaints, opinions, and suggestions from citizens are managed may differ from traditional approaches. Our field survey revealed that individuals are inclined to take action when equipped with appropriate tools. Research papers have also highlighted the absence of a current model featuring a prioritization algorithm for determining which complaints to address first. The authorities severely require such an algorithm to know where to focus their resources at. The current models also do not use geo-tagging.

The problem that is quite evident living in any metropolitan area seems to be the deterioration in the city's infrastructure. There is a huge number of potholes, misplaced dividers, bad roads, open gutters, and loose wiring that are ever-increasing by the day. But this is the problem that the government should solve. We aim to provide information about these problems to the government via the common people. The problem we are solving is the delayed repairing and solving of issues in the city's infrastructure.

The proposed system can solve the issues of the common public while helping the government in doing its job. It will be used in good use to notify the government of the issues in their city. The app directly informs the relevant department for solving the problem. For example, instead of troubling the government's office for small issues like potholes, it will inform the road repair department of the BMC.

4 Existing Systems

The existing system for addressing grievances is limited to web-based and android applications, which only allow users to file complaints of certain types pre-determined by the government. The process for filing a complaint is still traditional, requiring the user to fill out a form and submit their complaint in written format. Other complaints are not accepted and must adhere to the specified formats.

The current systems include message content filtering to prevent the delivery of undesired words. Users have the choice of either typing their complaint in the provided text box or uploading a PDF following the app's format. The registered complaint is then transferred from one department to another as they seek a resolution.

As seen in (Table 2), the existing systems share many research gaps within themselves. There has not been any update in the technology, which Solve My Problem aims to do. The existing systems have not yet implemented many basic features like complaint tracking and transparency.

The current system presents the following limitations: It only allows for the submission of specific types of issues. The government portal exclusively supports the submission of grievances or complaints falling into the following categories:

Table 2. Comparative Analysis of Existing Systems and Solve My Problem

Features	Existing Systems	Solve My Problem(SMP)
Registration Pro cess	Citizens have to register their complaints through phone, email, letters, or in-person visits, which can be time-consuming and complicated	SMP offers a quick and automated registration process through a mobile application. Citizens can register complaints with a few clicks and submitting a picture of the problem
Complaint Filing Process	Citizens have to file complaints manually, which can be tedious and time-consuming	SMP offers an easy and hassle free complaint filing process by clicking a picture of the issue and submitting it through the app. For instance, a citizen can file a complaint about a pothole by clicking a picture and uploading it
Complaint Detection	Complaint detection and classification is done manually, which can lead to errors and delays	SMP uses machine learning algorithm to detect and classify com- plaints accurately and quickly. The app can recognize various is- sues like garbage accumulation, potholes, open sewage, etc. and assign them to the local authorities. For instance, if a citizen submits a picture of a pothole, the app will automatically detect the issue and send it to the BMC
Complaint Tracking	Complaint tracking is limited, with citizens having limited visibility on the status of their com- plaints	SMP offers real-time tracking of complaints through the app and website, providing citizens with updates on the status of their complaints. Citizens can view the progress of their complaints and know when they have been resolved
Transparency and Accountability	Transparency and accountability are limited, with citizens having limited visibility on the actions taken by the authorities on their complaints	SMP offers improved transparency and accountability through real-time tracking and a detailed feedback mechanism. Citizens can see the progress of their complaints and hold the authorities accountable for any delays or issues

- Matters under legal consideration.

- Personal and familial disputes.
- Grievances related to the Right to Information (RTI).
- Concerns that may have implications for the nation's territorial integrity or diplomatic relations with other countries.

There are many prevalent systems in place to tackle the problem that we have taken up. There is even an official government app [9] that helps in solving the problems faced by the citizens. But there are many drawbacks that we have found in them. We observed that the system lacks the capability to determine the precise location of the problem, making it impossible to accurately mark it on a satellite map.

We have also observed that the existing system is limited to recognizing only road problems like potholes and misplaced dividers. Another drawback found is that the location once marked with an issue is not unmarked and put as "Safe" once the issue has been solved. This way the people are unaware of the status of the issue. And lastly, we have seen that there is no prioritizing of the issues that are displayed. The local authorities are forced to pick a grievance at random and work towards the rest.

5 Proposed System

The commuter who faces grievances (such as Potholes on the roads, Garbage on the streets, sewage leakage, etc.) can submit a grievance through "Solve My Problem" which is an android-based application. There are two signs up one for the public other one for higher authority. Then users first and only once can shortly sign up using public sign-up for submitting grievances through our proposed application. Location can be automatically fetched using GPS. Once the user is logged in then he/she fills the grievance to solve the issue they faced in daily life. He/She need can fill the grievance by giving the texts, videos and images of the issue which will be automatically classified whether it is porthole, savage or garbage. Once the grievance is filled successfully the grievance is shown in your reports section. Afterwards, they can view the status of the grievance and keep track of their reported grievances.

Higher authority can also make a sign-up to the application using higher authority sign-up. The relevant local authority responsible for addressing grievances can access the system by entering their ID and password. After logging in, they will have the ability to view complaints sorted based on the number of grievances reported in a particular region and prioritize those that require immediate resolution. Our proposed system "Solve My Problem" is a one click grievance reporting system that allows users to re- port various issues, such as garbage spills, potholes, no parking areas, and stray dogs, by uploading pictures and location details. ML model will then classify the problems into their respective categories and assign tasks to the relevant authorities. Through the use of data analysis and visualization techniques, the system offers a dashboard that provides visual reports of the area on a weekly, monthly, and annual basis. Both the authorities and users can access these reports, ensuring transparency and allowing complaints to be tracked efficiently.

To get a better understanding of how the product works, an architectural design (Fig. 1) of our system is made. This system design depicts the flow control of the application. It illustrates the different functionalities of the app. It gives us a brief understanding of how each module will interact with the other.

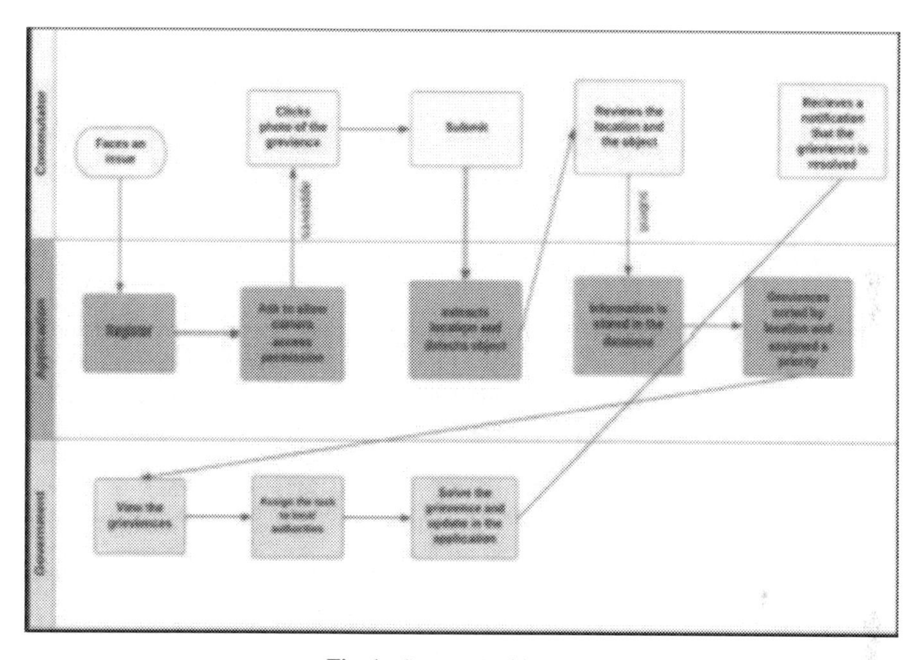

Fig. 1. System Architecture.

The proposed system overcomes the shortcomings of the outdated mobile technology employed in the existing public complaint system. It achieves this by reimagining the system with a commitment to upholding the principles of good governance, including accountability, transparency, and active citizen participation. The proposed provides data visualization to get visual reports of their area weekly, and monthly using data visualization techniques that can be viewed by both authorities as well as normal users so there will be a transparent process and complaints can be tracked.

The Proposed System can serve a very small part of the population. But it can be divisible into many different segments once scaled. A separate database for each small-scale city when brought together can lead to the speedy resolution of an entire district. If implemented in a major part of a city, it is capable of fixing all the problems as fast as the local authorities repair the issues. When seen in a bigger picture, we are left with safer roads, cleaner cities and better infrastructure.

That brings opportunity for new development. We can also implement it on a small scale like an office complex. The grievance could be reported to the management of the building. Issues like broken furniture, ill-managed washrooms etc. could all be solved with ease. A data-flow diagram (DFD) is a graphical representation that depicts the movement of data within a system or process, often found in information systems. The

DFD level 0 diagram (Fig. 2.) shows the inputs, outputs, and processes involved for each entity. DFD is more concernedwith the overall flow of data in a system.

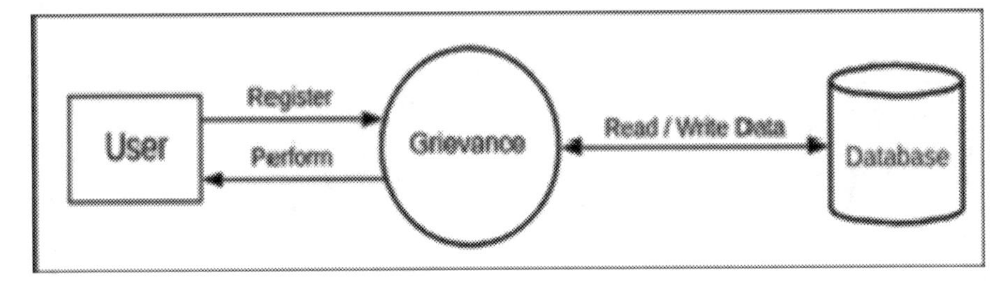

Fig. 2. DFD level 0

The DFD Level 1 diagram (Fig. 3.) identifies various tasks the user performs briefly. After logging in, users can submit a grievance to address any issues they encounter in their daily lives. The grievance can be submitted by uploading an image of the issue, which will then be automatically classified such as potholes, sewage, or garbage. Once the grievance has been successfully submitted, it will appear in the user's reports section. Users can then monitor the status of their grievances and keep track of their reported issues.

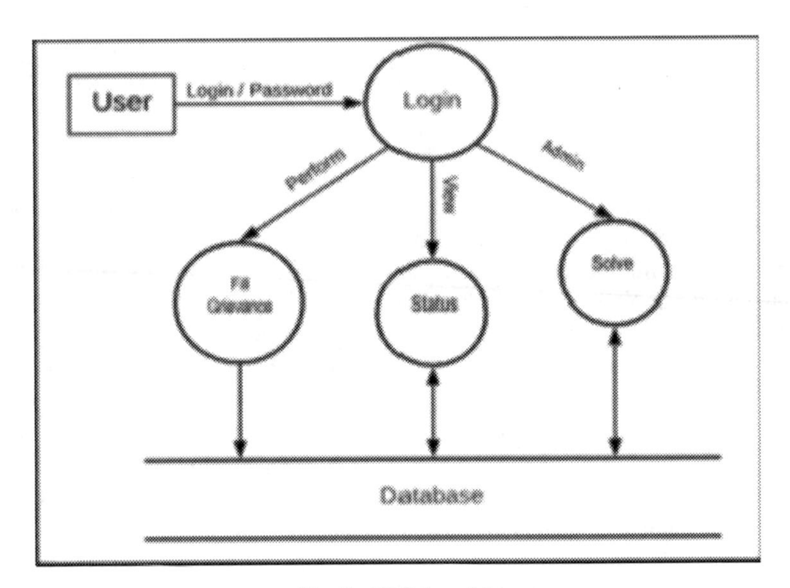

Fig. 3. DFD Level 1.

The proposed systems cater to the needs of ordinary citizens in the locality, local governing bodies, and higher-level authorities within the grievance redressal system.

6 Implementation

Registration is the process of creating a new account in the grievanceredressal system. As we see in (Fig. 4), the user is required to provide their personal information such as name, email, password, and phone number during the registration process. This information is used to create a unique account for the user in the system, which is required to access the features of the system.

Once the user has registered, they can log in to the system using their email and password. The user is required to enter their email and password on the login page to access the system. After logging in users can edit the details which they submitted while registering. After logging in, the user can view the dashboard where they can see the different features of the grievance redressal system, including the option to file a new grievance. The user can also view their previously filed grievances, their status, and any updates on their grievance.

The notification for GPS access popped up. Once the user enables GPS access the location of the user is automatically traced. In the proposed system "Solve my Problem", grievances submitted by users, such as potholes, garbage, and sewage, are shown to the authorities for solving them.

The system uses image classification through a machine learning model to automatically classify the type of grievance based on the submitted image. Once classified, the grievance is visible to the respective authorities who can take the necessary actions to solve the issue. This approach not only simplifies the grievance submission process for users but also enables the authorities to prioritize and address grievances efficiently.

The admin panel page (Fig. 5), shows the local authorities an in-detail description of the grievance that has been submitted by the users. It is displayed in the form of cards that have the date of submission, the location of the issue and the category under which the issue is detected. It also has a solved/unsolved button that can help the authorities keep track of the issues that have been solved. Once the issues have been solved, the button is clicked and a notification is sent to the citizen.

The visual representation page in (Fig. 5) in the grievance redressal system shows the number of grievances submitted by users that fall into different categories such as garbage, sewage, and potholes. This page provides a graphical representation of the total number of grievances in the system, along with the number of grievances in each category. The graph can be viewed by both authorities and normal users to get a better understanding of the current status of the grievances in their area. This visualization can help authorities to allocate resources more efficiently and focus on addressing the most pressing issues first (Fig. 6).

The ML model needs to be trained to classify images uploaded by users into categories such as garbage, pothole or sewage. Therefore to train the ML model we use a well-constructed pre-defined convolutional neural network called ResNet50.

ResNet50 is 50 layers deep and can be rigorously trained on huge datasets. ResNet50 has already been trained on a huge dataset of ImageNet that helps it classify between different images of dog, cat, pens etc. In (Fig. 7) we see the architecture of the convolutional neural network and how it parses the input to give output.

The training of the model involves several iterations known as epochs. In each epoch, the model receives the training data, processes it, and then adjusts its parameters based

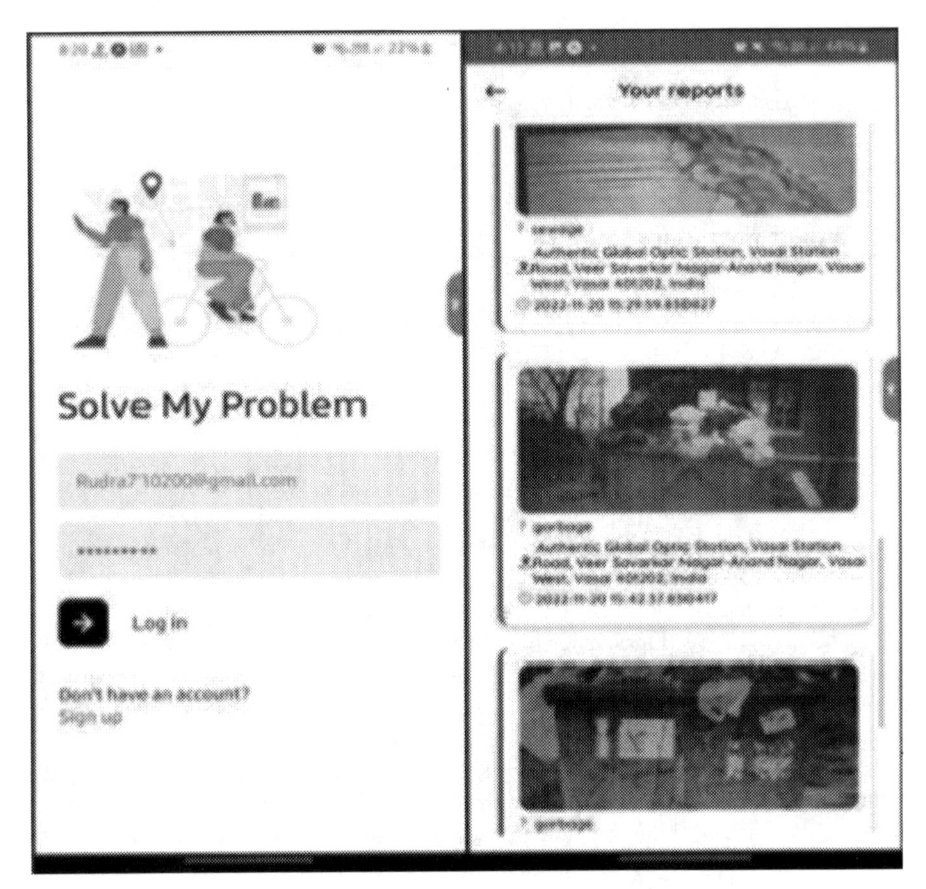

Fig. 4. GUI of Users logging in and submitting grievances.

on the feedback received. This is done to reduce the error rate and increase accuracy. We have trained the model for a 100 epochs as seen in (Fig. 8). The number of epochs required for training a model depends on the size and complexity of the dataset, as well as the performance of the model during the training process.

The dataset used for training the ML model consists of a large number of images of garbage, potholes, and sewage. These images are labelled with the appropriate category so that the ML model can learn to classify new images based on their features. This type of training is called supervised training. We have scraped huge datasets from google images as seen in (Fig. 9). The dataset should be large enough to cover a wide variety of situations and conditions and should be representative of the types of images that are likely to be uploaded by users.

When a user uploads an image of a problem (such as garbage, pothole, or sewage) for redressal, the image may not always be clear or in the appropriate format for the ML model to classify it correctly. In such cases, image transformation techniques are applied to pre-process the image and improve its quality.

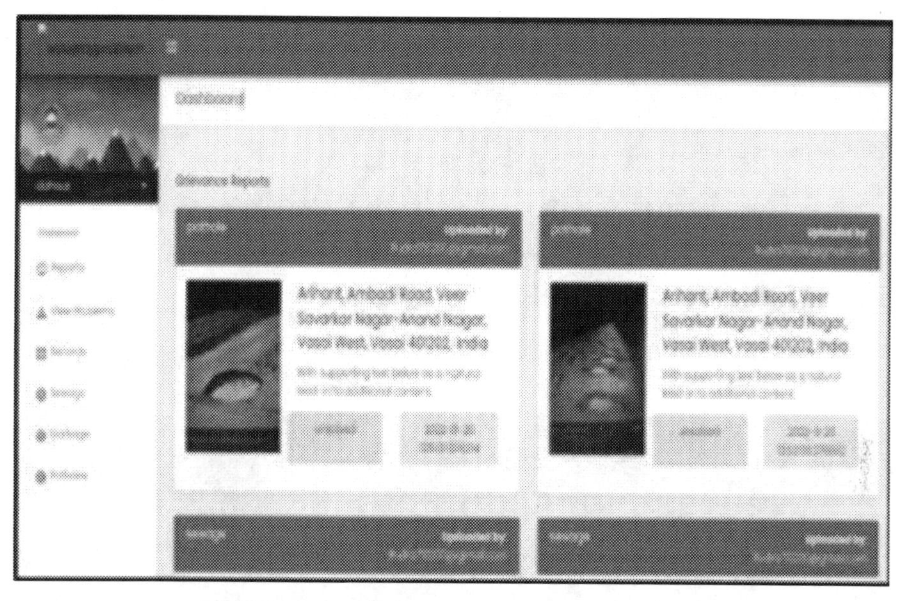

Fig. 5. Admin Panel (Grievances submitted by users).

Fig. 6. Visual Representation of total grievances.

Some of the image transformation techniques that can be used include resizing, cropping, rotation, and noise reduction as done in (Fig. 10). For example, resizing can be used to adjust the size of the image, while cropping can be used to focus on the relevant

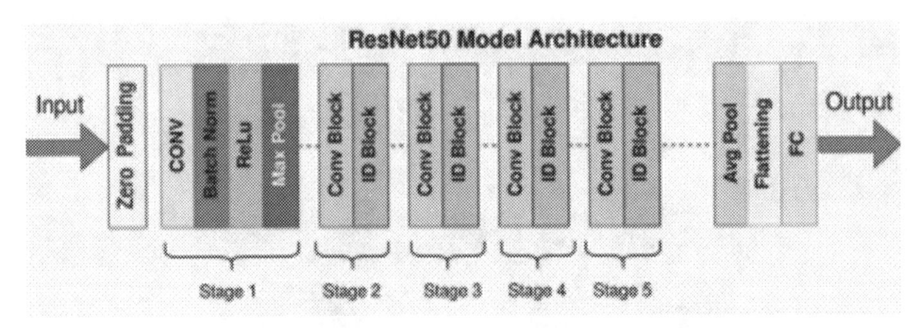

Fig. 7. ResNet50 Architecture

Fig. 8. A sample of epochs that have been executed.

part of the image. Rotation can be used to orient the image in the correct position, while noise reduction can be used to remove any unwanted noise from the image.

After applying the appropriate image transformations, the image is passed through the ML model, which then classifies it as one of the categories (garbage, pothole, sewage). By applying image transformation techniques, the system can improve the accuracy of the classification and ensure that the grievances are addressed more effectively.

Once this procedure is completed, we test the ML model by uploading multiple images and testing its accuracy such that they can be addressed and handled by the authority. To assess the precision of a grievance redressal system employing a Machine Learning (ML) model for image classification, an evaluation can be conducted using a confusion matrix. The confusion matrix, as shown in (Fig. 11), serves as a tool to gauge the ML model's effectiveness in accurately categorizing various types of issues depicted in images, such as potholes, garbage spills, and sewage problems.

Once the ML model has been trained on the dataset, it can be used to classify new images uploaded by users. The model should be tested on a separate test dataset to ensure

Fig.9. Dataset

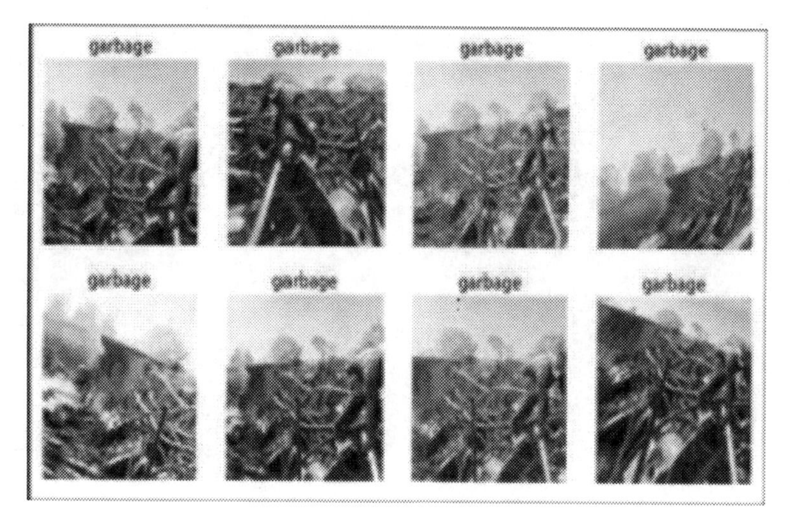

Fig. 10. Transformation

that it is accurate and reliable in classifying images into the appropriate categories. The accuracy of the model can be further improved by fine-tuning the model using additional training data or by adjusting its parameters based on user feedback.

As seen in (Fig. 12), the prediction phase of the ML model for classifying images in the grievance redressal system involves feeding the trained model with new, unseen images to classify. The predictions made by the model can then be used to prioritize and

Confusion matrix
garbage
pothole
sewage
Actual
23
0
0
0
25
0
1
0
20
garbage
pothole
sewage
Predicted

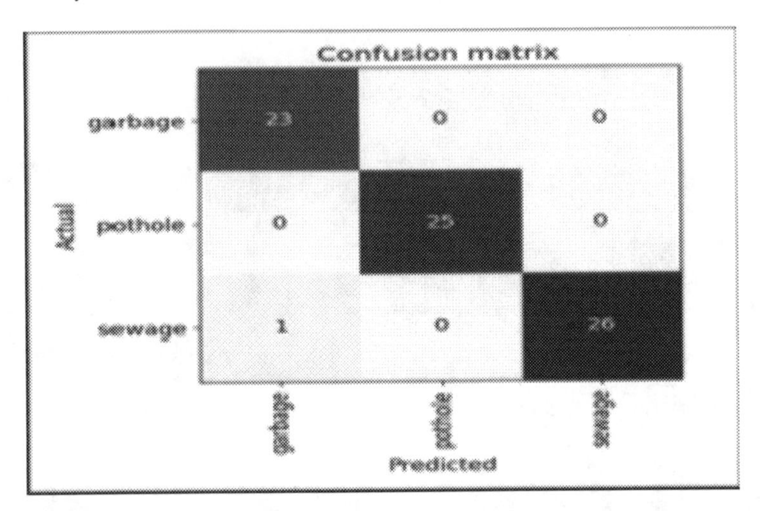

Fig. 11. Confusion Matrix

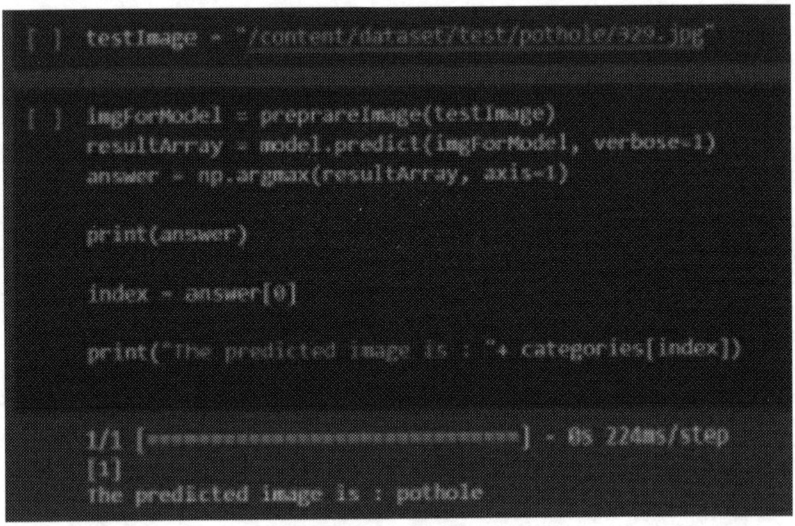

[ ] testImage = "/content/dataset/test/pothole/929.jpg"
[ ] imgForModel = preprareImage(testImage)
resultArray = model.predict(imgForModel, verbose=1)
answer = np.argmax(resultArray, axis=1)
print(answer)
index = answer[0]
print("The predicted image is : "+ categories[index])
1/1 [==============================] - 0s 224ms/step
[1]
The predicted image is : pothole

Fig. 12. Prediction

assign tasks to the respective authorities based on the nature of the problem reported by the users (Fig. 13).

After Modelling, Testing and predicting, we successfully implemented the ML model for the classification of images as sewage, garbage, and pothole. This model helps the authority to classify the grievance faced and submitted by the commuter of the system. Take appropriate action and update the status of the received grievance when they are taken and the grievance fixed by higher authority.

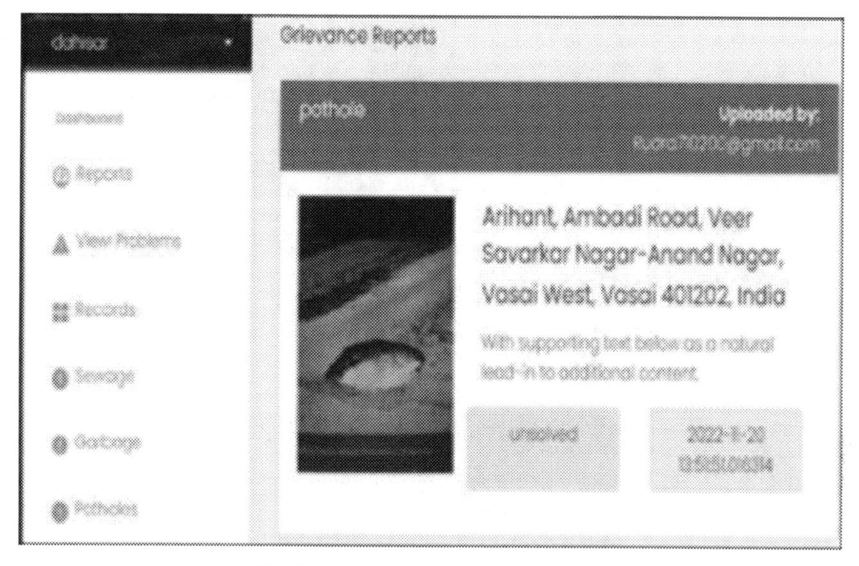

Fig. 13. Implementation of ML model

7 Conclusion

The "Solve My Problem" grievance resolution system aims to overcome the constraints of the outdated mobile technology employed in the current public complaint system. It has been revamped in accordance with the principles of good governance to offer a user-friendly and interactive platform for individuals to report daily-life issues encountered during their commute. The use of an ML model for image classification helps in automatically classifying grievances such as potholes, sewage, and garbage, which are visible to the authorities and are sorted based on the number of grievances faced in a specific region. The proposed system integrates all types of complaints in a single platform and can be further enhanced to improve its functionality. Hence, this system holds the promise of rectifying multiple shortcomings in the current grievance redressal system and establishing a transparent and efficient mechanism for addressing public grievances.

References

1. Over 5,000 people died in road accidents due to potholes during 2018–2020 (no date) Moneycontrol. https://www.moneycontrol.com/news/india/over-5000-people-died-in-road-accidents-due-to-potholes-during-2018-2020-9069791.html. Accessed 15 Dec 2022
2. "Smart E-Grievance System for Effective Communication in Smart Cities." Smart E Grievance System for Effective Communication in Smart Cities | IEEE Conference Publication | IEEE Xplore. ieeexplore.ieee.org/document/8537244. Accessed 15 Dec 2022
3. 1481 deaths, 3103 injured: Pothole-related accidents on the rise, says report (2022) English. Mathrubhumi. https://english.mathrubhumi.com/news/kerala/1–481-dead-3–103-injured-union-govt-report-shows-pothole-relatedaccidents-on-increase-1.8187782#:~:text=According%20to%20the%20report%20titled,1471%20and%203064%20in%202020. . Accessed 15 Dec 2022

4. Mismanaged waste 'kills up to a million people a year globally' (2019). https://www.theguardian.com/environment/2019/may/14/mismanaged-waste kills-up-to-a-million-people-a-year-globally#:~:text=Mismanaged%20waste%20is%20causing%20hundreds,problem%2C%20a%20report%20has%20found. Accessed 15 Dec 2022
5. Map Data | Static Map API | Platform | HERE (2022). https://www.here.com/platform/map-data. Accessed 16 Dec 2022
6. "Public Complaint Service Engineering Based on Good Governance Prin ciples." Public Complaint Service Engineering Based on Good Gover nance Principles | IEEE Conference Publication IEEE Xplore. ieeexplore.ieee.org/abstract/document/8002470. Accessed 15 Dec 2022
7. "GPS Based Complaint Redressal System." GPS-Based Complaint Redressal System IEEE Conference Publication IEEE Xplore. ieeexplore.ieee.org/document/6967558. Accessed 15 Dec 2022
8. Tjandra, S., Warsito, A.A.P., Sugiono, J.P.: Determining citizen complaints to the appropriate government departments using KNN algorithm. In: 2015 13th International Conference on ICT and Knowledge Engineering (ICT & Knowledge Engineering 2015), Bangkok, Thailand, pp. 1–4 (2015). https://doi.org/10.1109/ICTKE.2015.7368461. Accessed 15 Dec 2022
9. CPGRAMS-Home (2022). https://pgportal.gov.in/. Accessed 16 Dec 2022

Finite Automata Application in Monitoring the Digital Scoreboard of a Cricket Game

K. Kaushik, L. K. Pavithra(✉), and P. Subbulakshmi

School of Computer Science and Engineering, Vellore Institute of Technology, Chennai, India
kaushik.k2020@vitstudent.ac.in, {pavithra.lk, subbulakshmi.p}@vit.ac.in

Abstract. Cricket is one of the most viewed sports in the world and tracking of this game is very important throughout the gameplay. The application of automata in sports is least explored and this paper proposes the use of automata theory in cricket. The paper sheds light upon the potential application of finite automata concept in cricket which proposes proper tracking flow of the game. The major focus of the paper is shown upon ball-tracking and on-strike batsman tracking with each tracking system being explained in detail using the state diagrams. With the help of Deterministic Finite Automata (DFA), it is possible to achieve the flow of track and thus, helps in efficient tracking of the sport. It also minimizes the rate of committing the tracking errors during the gameplay.

Keywords: Automata Theory · Ball tracking · Cricket · Deterministic Finite Automata · On-strike batsman · State Diagrams

1 Introduction

Cricket is claimed to be the second most viewed sports in the world with the viewership range of more than 2.5 billion. The game has evolved various changes from its origin and the laws of the game have been improved by times. Unlike other games, Cricket involves lot of tracking. A small error in the tracking system can affect the entire gameplay and might lead to wrong decisions/results. Hence, it is necessary to track the game accurately.

Finite Automata (FA) is used to effectively recognize the patterns and can determine the sequence of flow based on the input. It minimizes the uncertainty of the flow of any state machine. The use of FA in cricket will help in minimizing the tracking errors which in-turn ensures proper tracking flow of the game. This paper concentrates on ball-tracking and on-strike batsman tracking. The dashboard of the game which is projected during the live matches provides the information about the on-strike batsman and the runs conceded in the over. Therefore, it is necessary and important to display the correct information on the screen for the spectators.

S. Aurelia et al. (Eds.): ICCSST 2023, CCIS 1973, pp. 143–158, 2024.
https://doi.org/10.1007/978-3-031-50993-3_12

2 Literature Review

Roger Caillois proposed the classification of games into four categories namely agon, alea, mimicry and ilinx [1]. The games that involve competition between teams or individuals fall under the agon category. Alea category of games depend upon probability and luck. The games that involve action of imitations comes under the category of mimicry and ilinx categorical games are the ones that aim to alter reality through quick motion. Cricket falls under the Agon category as per Caillois classification as it is a sport of competition between two different teams. David Best concentrates on elucidating the problems associated with aesthetic education [2].

Gurram S has proposed a work on the design of finite state diagrams for sports in which the winner of the game is predicted based on the rules that are set for a particular sport [3]. The finite automata concept is involved and thus, the work focuses on the games that follow only sequential events. The state diagrams that are proposed for each sport in the work clearly predicts the path of the winner based on the input in each stage. Badminton, Shot Put, Hurdles, Long Jump, Pacman, Track, Bowling and Weightlifting are the games that were considered in the paper and all of these games fall under the Agon category as per the Caillois classification.

Lee et al. have proposed a work on the recognition of events in the basketball game with the help of DFA [4]. The proposed technique in the work accordingly identifies the events generated by tracking the players and the referees. The paper is more concerned about the actions and motions of the players to determine the next flow for designing the state automata diagrams.

Potechin et al. proposed a non-deterministic finite automata model for finding the length of the words that are accepted [6]. In the paper, the complexity of acyclic NFA acceptance problem was examined and the authors have highlighted an algorithm using matrix multiplication which determines all input lengths that are accepted in case of the unary NFA. The paper also stated a method of reducing the triangle detection problem to unary acyclic NFA which in-turn improved the algorithm for acceptance. Further, the authors provide a link between strong exponential-time hypothesis and acceptance of NFA in the work.

Ransikarbum et al. proposed a new agent-based affordance model framework called “Affordance Based Finite State Automata” (AFSA) for highway lane driver system [7]. The paper makes a comparative analysis between the real driving data and an agent-based simulation data with the help of state diagrams. A correlation and statistical study have been performed to understand the behaviour of driver agents in their proposed work.

The work done by Gribkoff highlights the various applications of DFA in various domains [8]. Initially, the basic concepts of DFA were covered in the paper and then the applications of the automata were discussed. In this work, a detailed explanation of each application of DFA was provided with the help of state diagrams. The vending machine was represented as a finite automata model, which has followed certain specific parameters to control the flow of the process. The behaviour of Pac-Man Ghost in the game was also represented as a finite automata model in the paper. The Transmission Control Protocol (TCP) was represented as DFA followed by which the implementation of finite automata concept for auto-complete in Apache Lucene Open-Source Search engine was proposed.

Krishnan et al. presents a paper on fake identifier algorithm in online social networks with the help of DFA [9]. A comparative study was made with other existing manual validation methods to identify the performance level. The proposed algorithm performs well in most of the cases but it generates long regular expression in certain conditions. Hence, the author suggests to combine the proposed algorithm with neural networks or other classification techniques. Eilouti et al. used Finite State Automata (FSA) to describe the ways to map the basic shapes together into meaningful systems [10]. The concept of FSA was used for describing the sequence of element grouping and the paper has shown that FSA can help to derive and represent families of composition that contains geometric elements.

The paper proposed the implementation of home automation system using Finite State Machine (FSM) model [11]. Each module has been represented in the form of finite diagram and based on certain criteria, the flow of the automata is determined. Theft detection at door and window, security check at the entrance, fire alarm, heater, air-conditioner, water tank and luminosity are the modules that are covered in the work. All the modules that are mentioned have been implemented in the FSM model except water tank which is not FSM-based. Each module has been implemented in Verilog code on Xilinx platform for which Spartan3 FPGA (Field Programmable Gate Array) board of device XC3S1000L was used.

Slamet et al. proposed a system to convert Latin to Sundanese which helps to learn the script [12]. The system accepts Latin script as the input, splits it into syllable using the Finite State Automata algorithm with the Non-Deterministic Finite Automata (NDFA) model and converts it to Sundaneses script. The state diagrams have been used to represent the transition of separation of syllable and the basic set of rules for writing and pattern of Sundanese syllables were followed in the work. Also, a web-based application for preserving, facilitating, proving, and testing the FSA in recognizing the Sundanese script has been developed for the research work.

The paper deals with the online pulse deinterleaving with the help of PRI pattern information [13]. The pulse streams are represented in the form of transition diagrams and an automaton-based deinterleaving method for a single emitter was proposed in the work. The method has been extended for multiple emitters as well. The paper illustrates the regular grammar for streams with repetitive structures and in the end, the performance of the proposed method is analysed.

The paper proposed an approach of recognizing daily human activities based on Probabilistic Finite State Automata for which a new indicator called "normalised likelihood" was introduced [14]. The paper also addresses the robustness and the computational complexity of the proposed indicator. An experiment has been done in a living lab (smart flat) and the obtained results were discussed in the work. The activities done for the preparation of hot beverage is considered for explaining their work with the help of the automaton.

The paper proposed the relationship between the automata theory and linear congruence [15]. The authors have tried solving problems of Chinese Remainder Theorem (CRT) using the cartesian product of finite automata theory and the solution of the CRT is the acceptable strings of DFA. The paper also states that finite automaton can be used to identify the residue classes.

The paper proposed a model for an automatic fruit vending machine with the help of finite state automata [16]. The model uses non-deterministic finite state automata and the process from selection of fruit till the choice of payment was represented in the transition diagram. The paper also covered the use case diagram, activity diagram and fruit vending machine interface design.

Egyir et al. proposed a design for the cricket game using FPGA [17]. The work focused on creating a functioning simulation of cricket scorecard which included runs scored, present count of balls, etc. Verilog Hardware Description Language (HDL) has been used for the programming and Digital Schematic Tool (DSCH) was used for validation. The push buttons of the board were used for getting the user input and the various components used in the circuit assisted the process.

Ali et al. encourages activity-based learning of Push Down Automata (PDA) and hence, proposed the methodology for the same [18]. The authors have used different tools and software like thoth, Formal Language and Automata Environment (FLUTE), Java Formal Languages and Automata (JFLAP), Deus Ex Machina (DEM), Automata Tutor v3, Automata online tool, etc. and have developed strategies to simulate and practice the concept. The paper proposed an algorithm to design DFA for a regular language with prefix strings [19]. The proposed algorithm has been represented in a flow-chart and each step of the algorithm has been described in detail. Further, the authors have tried to simplify the lexical analysis process of compiler design in the work.

Vayadande et al. proposed GUI-based visualization tool for simulation and testing of DFA [20]. The simulator proposed helps to construct DFA by addition of states, marking of a specific state as a final state and checking of string if it can be accepted for the constructed DFA model. The model was limited to DFA and it cannot perform tasks on NFA.

3 Methodology

This paper concentrates upon tracking of ball in an over and tracking of on-strike batsman after a ball is bowled in an over. The rules and laws of cricket are the basis for this work and the finite automata diagrams has been designed and discussed in this paper based on it. The following section describes the tracking methodology of the batsman on-strike after a ball is bowled in an over by the bowler.

3.1 Tracking of On-Strike Batsman

In an innings of a match, there can be only 2 players from the batting side to play the game on the ground. A new batsman enters the gameplay when one of the present batsmen in the ground gets out/injured. The batsman who is about to face the ball from the bowler is known as on-strike batsman and the other batsman who is standing on the bowler's end is known as non-striker. There is a chance of change in the strike of the batsman after a ball is bowled based on several conditions.

The above defined concept can be enforced into finite automata diagram (i.e., Fig. 1, 2, 3, 4 and 5). Let the state of on-strike batsman be represented as S_0 state and non-strike end batsman be represented as S_1 state. If the on-strike batsman does not score any run in the ball, then it is considered as a dot ball. So, there will not be any change in the position and the state transition does not occur as shown in Fig. 1(a), exception if the ball bowled is a last ball since the positions of the batsman will be inter-changed after an over is completely bowled by the bowler according to the rules of this sport. For better understanding of each scenario, we ignore the over-up condition in all the upcoming possibilities that is about to be addressed and will consider the ball is not bowled in the final ball of the over.

If the on-strike batsman secures a single run, then there happens a swap in the position where the non-strike end batsman comes to the striker crease and the on-strike batsman moves to the bowler's end. The same is followed for 3 runs and 5 runs in case of running between the wickets. If the batsman secures 2 runs or 4 runs in a ball from running between the wickets, then there is no change in the position. In general, odd runs scored from running between the wickets makes a change of strike while securing even runs does not change the position of the batsman. When the batsman hits a boundary (four or six), the strike is not switched as shown in Fig. 1(e) and Fig. 1(f) respectively.

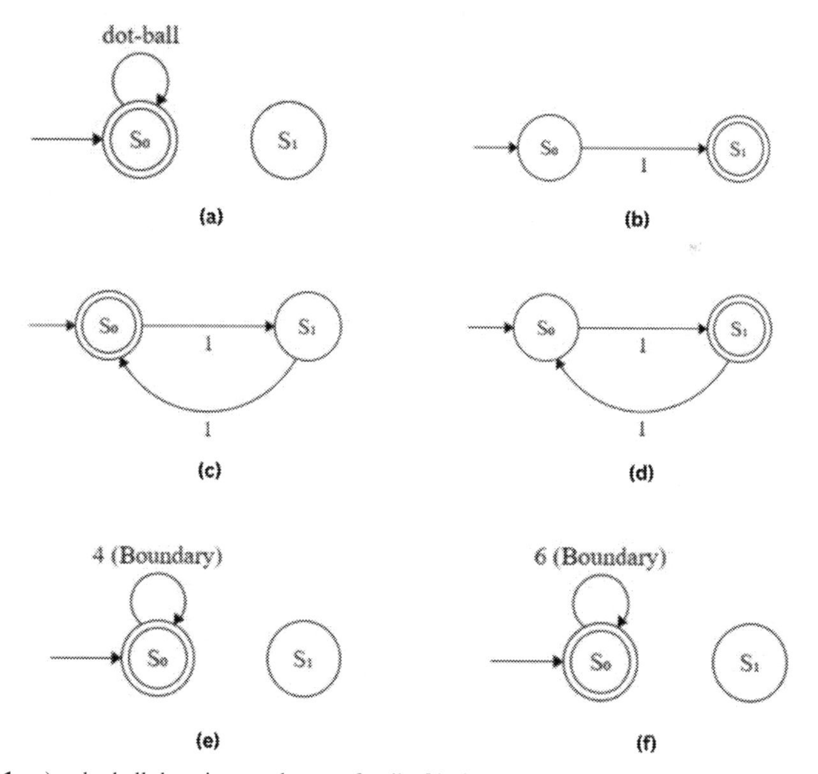

Fig. 1. **a)** a dot ball denoting no change of strike **b)** single run is scored and the state is moved from S_0 to S_1 state **c)** S_0 has scored two runs and the batsman keeps up the strike **d)** three runs scored which makes a shift in strike position as indicated using two concentric circles **e)** four runs scored in the form of a boundary which enforces no change of strike **f)** six runs scored through a boundary and no change of strike

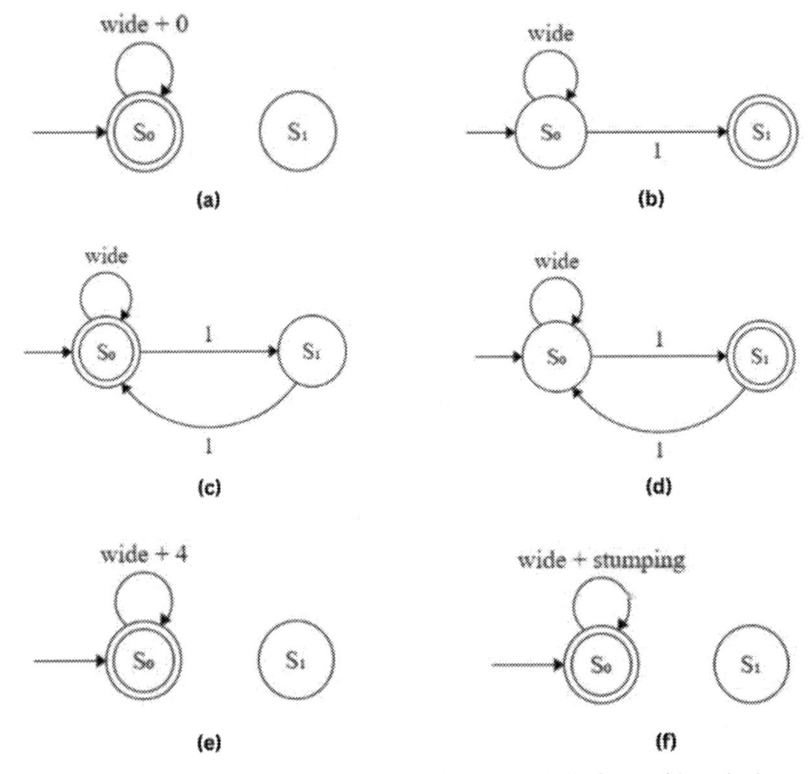

Fig. 2. **a)** a wide ball with no extra runs scored **b)** a wide ball along with a single run **c)** a combination of wide ball with two extra runs **d)** three runs are scored by the batsman in a wide delivery **e)** combination of a wide and four extra runs **f)** a stumping in a wide ball

When the bowler bowls a wide delivery, the strike is held with the same batsman if the batsman does not involve in scoring extra runs. If, in case, the batsman scores run by running between the wickets, then the change of strike happens based on the previously discussed scenarios. The same principle appears as mentioned before but the only change here is that a wide delivery is considered as not a fair delivery by the bowler and hence, the ball must be re-bowled and also extra run(s) is added to the batting team accordingly. The possible scenarios to happen when the bowler bowls a wide delivery is depicted in Fig. 2. Also, there is a possibility for the on-strike batsman to get out from stumping in a wide delivery and the respective state diagram is shown in Fig. 2(f). This case usually happens when the bowler is a spinner.

As discussed in previous case, the same flow happens when the bowler bowls a no-ball. For the batting team, the difference between a wide and a no-ball is the maximum runs that can be secured by them from that particular ball and the possibility of wicket loss. In case of wide delivery, there is no chance for batting team to score six runs over the boundary and also, there is a possibility of wicket loss for the batting team. But, in case of no-ball, it is possible to score six runs over the boundary (shown in Fig. 3(f))

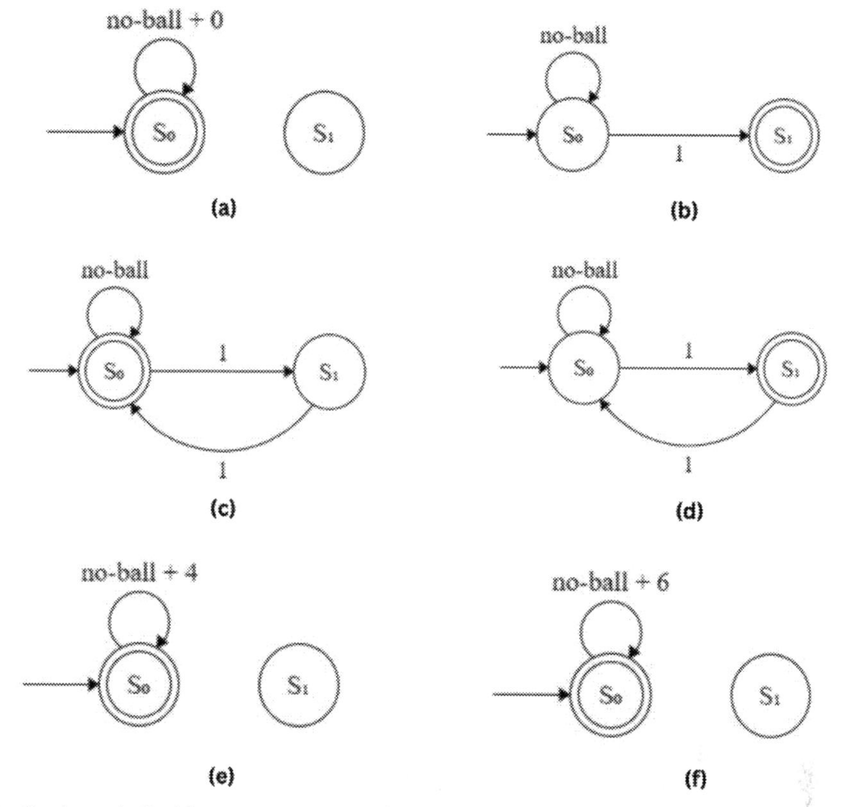

Fig. 3. **a**) a no-ball with no extra runs scored **b**) a no-ball along with a single run **c**) a combination of no-ball with two extra runs **d**) three runs scored by the batsman in a no-ball delivery **e**) combination of a no-ball and four extra runs **f**) six runs scored over the boundary in a no-ball delivery

and the batting team does not lose any batsmen, even if the bowler takes a wicket in that particular ball, exception for the runouts.

If the batsman is out, then the position of strike is moved to the new batsman who arrives at the striker crease. In case of run-out, the batsman's position is determined based upon the end where the run-out takes place. If the run-out is made at non-striker end, then the new batsman goes to non-strike end position. If the run-out is made at striker-end, then the new batsman holds the strike of upcoming ball.

When dealing with over-up condition, the position of strike differs. For example, if the bowler bowls a dot ball in the final delivery of an over, then the strike is swapped between the batsmen as depicted in Fig. 5(a). Suppose, if the batsman scores a single run in the final ball, then the strike for the next ball of new over is held up with the same batsman who faced the final ball of the current over. In general, for any happenings in a ball, the state diagrams will be same unless it is a last ball of the over and if it is the final ball of the over, then the swap of strike occurs.

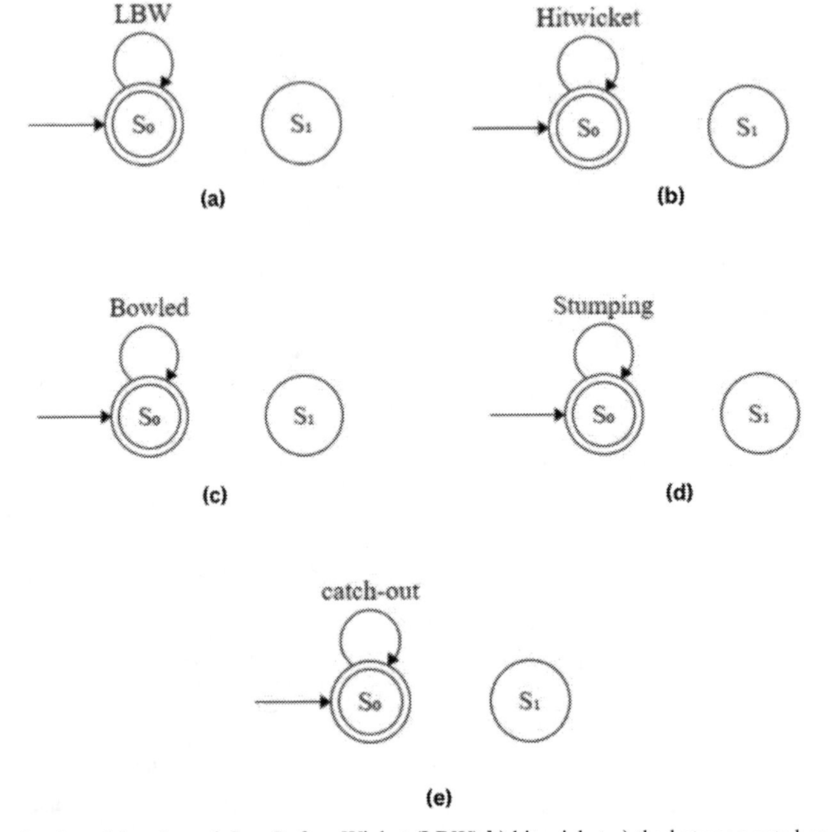

Fig. 4. **a)** a wicket through Leg Before Wicket (LBW) **b)** hit-wicket **c)** the batsman gets bowled by the bowler **d)** stumping **e)** catch-out

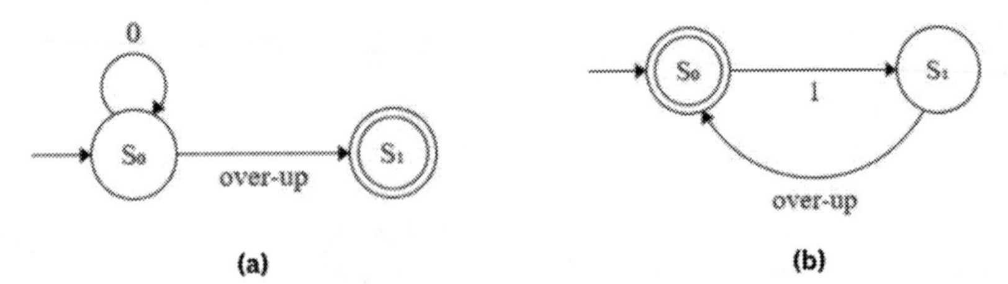

Fig. 5. **a)** a dot ball in the final ball of the over which enforces an over-up that makes a shift from the striking position of batsman for the next ball of new over **b)** a single run is scored in the final ball of the over which enforces an over-up that makes shift in striking position back to the same batsman for the first ball of new over

3.2 Ball Tracking in an Over

An over consists of six legal deliveries. In order to complete an over, the bowler must bowl all the six balls adhering to the rules of cricket. If the bowler fails to bowl a legal delivery, the ball must be re-bowled and if the runs are scored during those balls, then those runs are awarded to the batting team accordingly as per the game laws. Thus, there is possibility for a bowler to bowl more than 6 balls in an over unless he bowls legally.

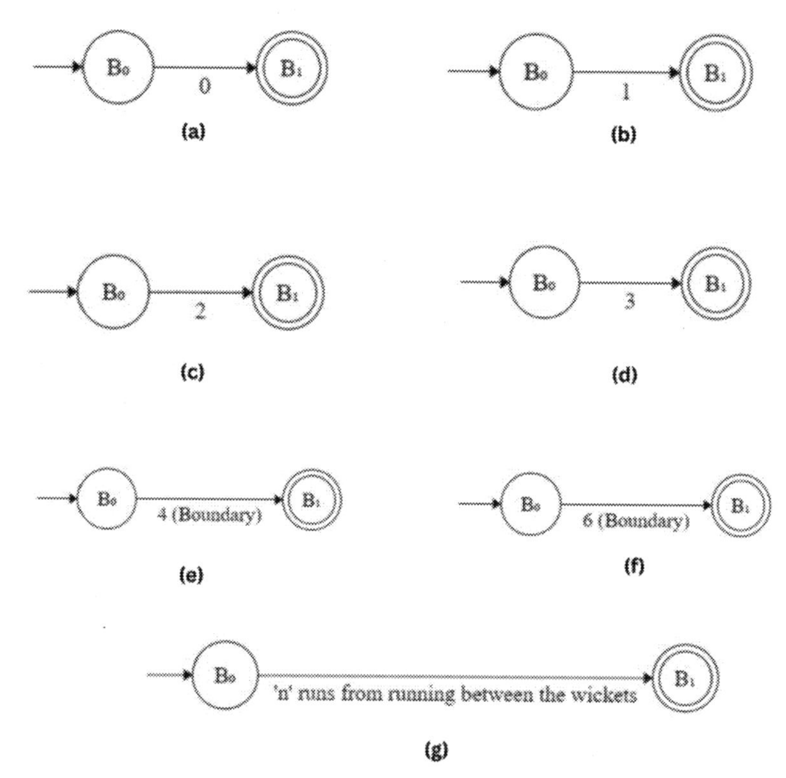

Fig. 6. **a)** a dot ball **b)** single run **c)** two runs **d)** three runs **e)** four runs (boundary) **f)** six runs (boundary) **g)** 'n' runs scored from running between the wickets

The finite automata concept can be extended to ball tracking technique in cricket. Let the starting state of the ball be B0 and subsequent states of the ball be B1, B2, B3, B4, B5 and B6. When the bowler bowls a legal ball, the ball is considered and the state is moved to the subsequent state. In Fig. 6, there has been a transition in the state position in all the cases since all the possibilities are considered "legal" as per the laws of cricket. The probability of runs that can be scored from running between the wickets is vast and is clearly mentioned in the tracking of on-strike batsman concept. In general, for any runs scored from running between the wickets in a legal delivery, the ball is counted and the state transition takes place as shown in Fig. 6(g).

When the bowler bowls a wide delivery, there is no state transition happening from B0 to B1 as the ball is considered as "illegal" and the extra runs scored in that delivery is added with a penalty run accordingly to the batting team's score. Figure 7 illustrates the finite state diagrams of ball tracking for the possibilities of a delivery along with a wide. As previously mentioned, the batsman is possible to get out in a wide delivery

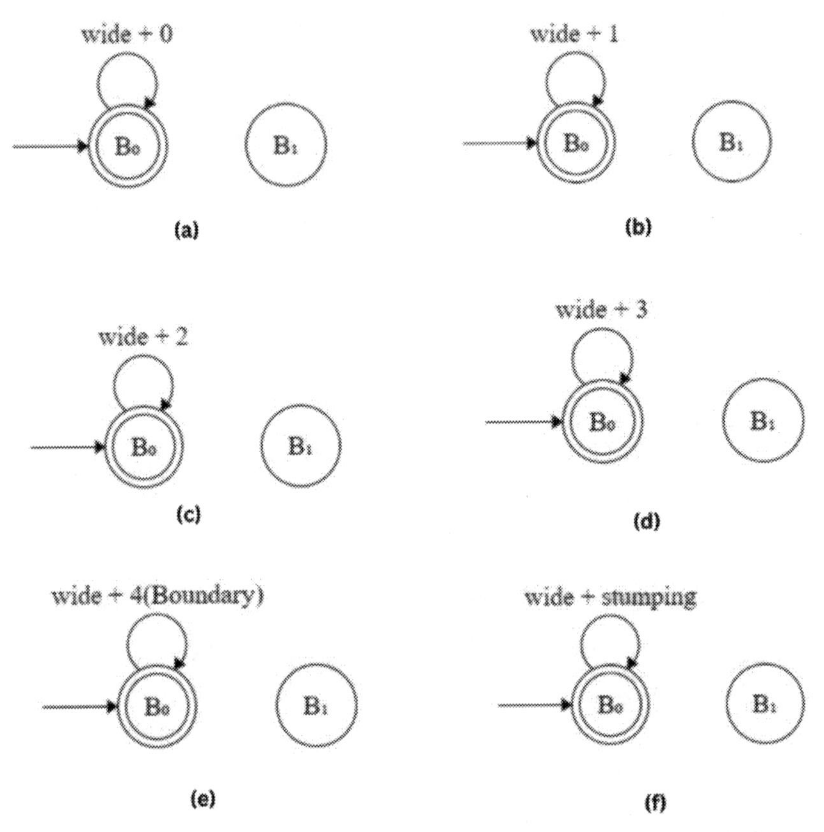

Fig. 7. **a)** a wide ball with no extra runs scored **b)** a wide ball along with a single run **c)** a combination of wide ball with two extra runs **d)** three runs scored by the batsman in a wide delivery **e)** combination of a wide and four extra runs (boundary) **f)** a stumping in a wide ball

from stumping which is depicted in Fig. 7(f). In such case, the ball is not counted as one of the over whereas the extra run and the stumping is considered.

Similarly, the same concept applies when dealing with a no-ball delivery. The state transition does not happen and ball is not taken into consideration as it fails to satisfy the legal criteria of a delivery. Hence, the bowler has to re-bowl that particular ball of the over. However, the extra runs scored in a no-ball delivery is added to the total score of batting team like the case of wide delivery. The possible scenarios along with a no-ball is shown in Fig. 8. In case of a runout in a no-ball delivery, the wicket is counted and the batsman is considered as out. But the ball is still not taken into consideration and it has to be re-bowled by the bowler in this case as well. This is shown in Fig. 8(g) where the state transition does not take place. Also, other than a runout in a no-ball delivery, all other chances of wicket taking possibilities including stumping, are not considered as a wicket and hence, it is not counted as shown in Fig. 8(h).

In Fig. 9, the finite state diagrams are shown for the case of ball-tracking in a wicket taking delivery (through bowled, LBW, catch, stumping and hit-wicket). It is to be noted

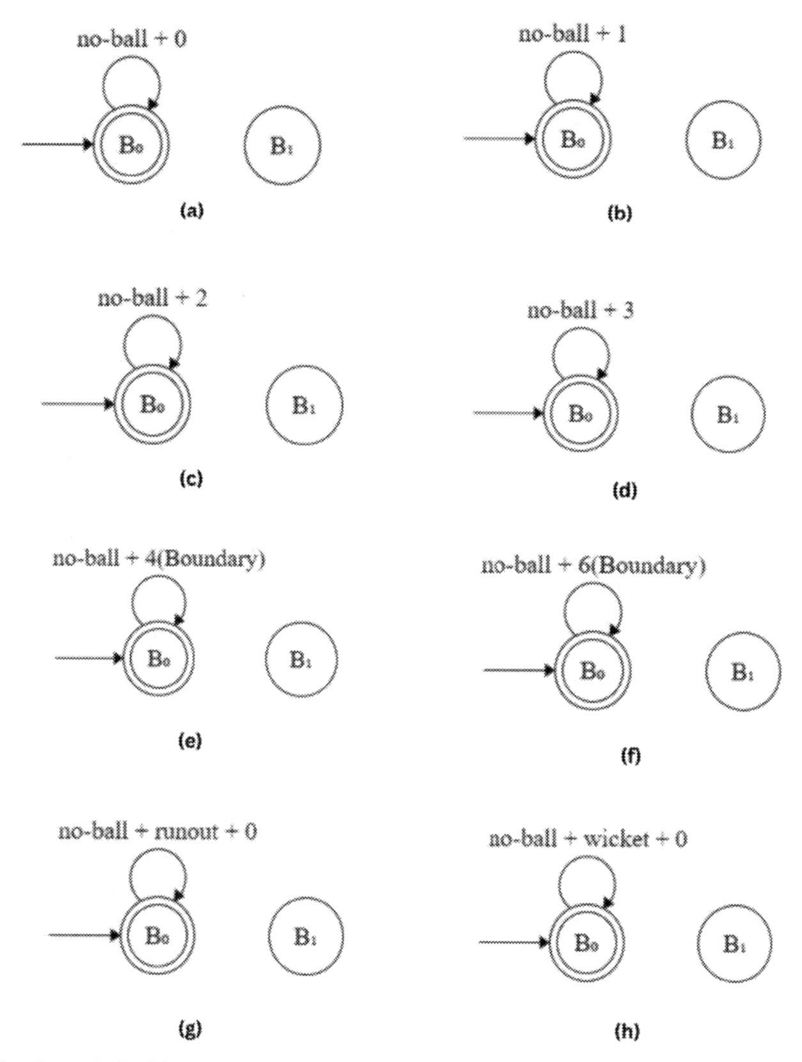

Fig. 8. **a)** a no-ball with no extra runs scored **b)** a no-ball along with a single run **c)** a combination of no-ball with two extra runs **d)** three runs scored by the batsman in a no-ball delivery **e)** combination of a no-ball and four extra runs **f)** six runs scored over the boundary in a no-ball delivery **g)** a runout in a no-ball delivery with no extra runs scored **h)** a wicket in a no-ball delivery with no extra runs scored

that ball transition happens in all wicket taking deliveries excluding no-ball and wide deliveries. In case of a runout, the runs scored if any and a wicket is considered as per the laws. This is depicted in Fig. 10.

It is necessary to provide proper input for proper flow of finite diagrams. The automata diagrams for ball tracking in an over and on-strike batsman tracking of the game is explained in detail but providing a wrong input to the automata can yield wrong results.

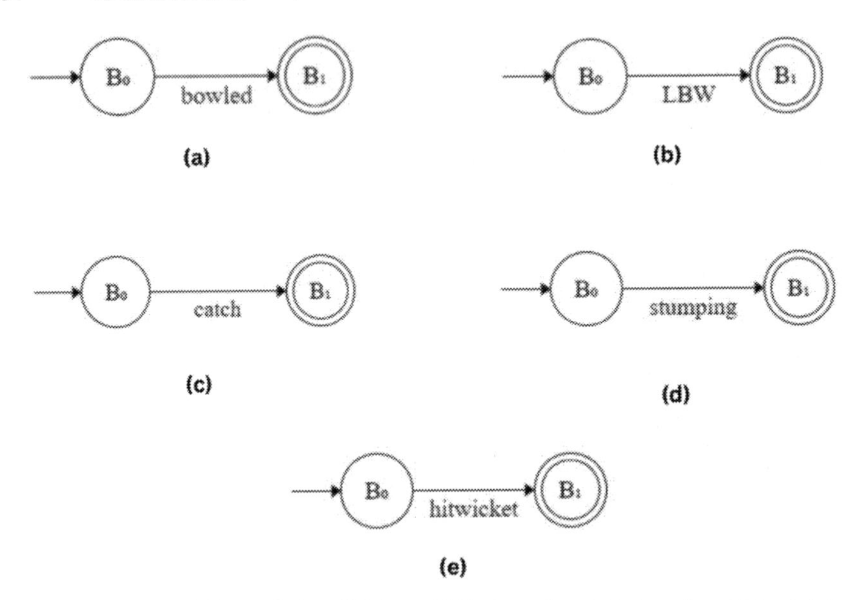

Fig. 9. **a)** bowled **b)** Leg Before Wicket (LBW) **c)** catch-out **d)** stumping **e)** hit-wicket

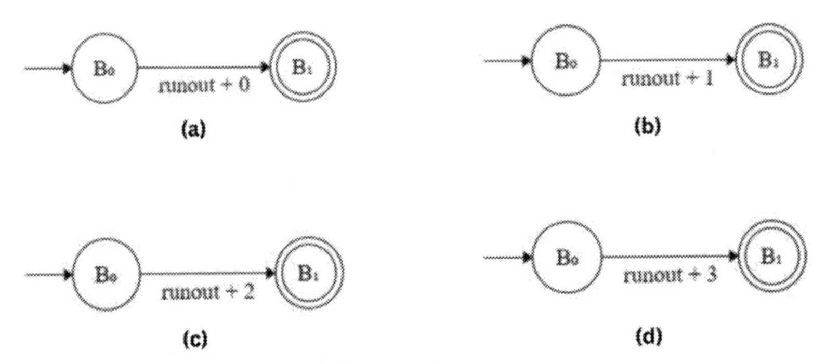

Fig. 10. **a)** runout with no extra runs scored **b)** runout with a single run scored **c)** two runs scored with a runout in the same ball **d)** three runs scored in a runout ball

4 Results/Discussion

With the help of advanced technology, it is important and necessary to efficiently handle the tracking system of the game. But there are few cases where the current existing methodology has failed to handle it. In a match of ICC Men's T20 World Cup 2022 between Australia and Afghanistan, only 5 deliveries were bowled in a particular over instead of 6 legal deliveries which became a controversial issue after the match. In the fourth ball of that controversial over, the batsman goes for two runs initially and then runs for a single after an overthrow by the fielder, which makes three runs from that delivery. However, the system considered two runs for fourth delivery and three runs for fifth delivery, though the fifth ball has not been bowled by the bowler. This has been failed to be noticed by the umpires, on-field, and off-field players as well as the team managements. This is an umpiring error and the supply of wrong input to the tracking software system has resulted in the wrong gameplay decision.

During the live broadcast, the dashboard graphics that is used to project the game statistics, which are manually handled for input supply till date, has also failed to monitor the innings properly because of wrong input supply to the model by the operators. This has led to the consideration of 'un-bowled' delivery as 'bowled' delivery in the over. Most of the other dashboard monitoring software used by various websites for score projection has just ignored the 6th ball and resumed its operation, without indicating the error. This implies that the digital scoreboard monitoring and projection process is still not efficiently handled for the game. Supply of wrong input to the tracking system will result in wrong decisions and this match incident is an example. Also, tracking models developed by the websites do not follow the ethics of the gameplay flow as it is reflected by the ignorance of the last ball in that particular over.

In this paper, the focus was shown in the ball-tracking and on-strike batsman tracking. The combination of both the tracking systems together will help in better accuracy of gameplay track. The accuracy of the automata can be achieved only if the input supplied to the state diagram is proper. Hence, it is essential to provide correct input to the state diagrams for proper functioning.

An example is shown in Fig. 11 which shows the combination results of ball-tracking and on-strike batsman tracking in automata diagrams for an over and the respective process flow is shown in Table 1. In this assumed example, the over consists of 8 deliveries where two of the deliveries fail to meet the requirements for a legal ball. The bowler gives single run during the first delivery of the over. In the second ball, the bowler concedes two runs. The third ball of the over is a dot ball which makes the batsman to score no run from the delivered ball. During the fourth delivery of the over, the bowler bowls a wide delivery which makes him/her to re-bowl the ball again though penalty runs are counted. In the fifth delivery, the batsman bowls a no-ball and the batsman runs for a single. This adds extra penalty runs to the batting team. This over involves a wide delivery followed by a combination of no-ball and a single run between 3rd and 4th ball. Hence, there is a self-transition in B3 state in case of ball tracking automaton. In the sixth and seventh delivery of the over, the bowler gives a six and a four respectively. During the eighth ball, the bowler takes a wicket through LBW. This makes the end of the over. Other than B3 state, all other states (excluding B6 since it is the final state of the over) have a transition to subsequent states. The transition in ball-tracking happens when the bowled ball is a legal delivery as per the cricket laws which were discussed under the methodology section in this paper.

The following content deals with on-strike batsman tracking of the over for the given example. In the first ball of the over, the batsman S_0 scores a single run which leads to a change in striking position to batsman S_1 as indicated using two circles around the state S_1. In the second ball, the batsman S_1 scores two runs and holds up the strike with him/her for the next ball. The third ball is a dot ball and thus, no swapping occurs. The fourth delivery is a wide ball with no extra runs scored by the batsman. Hence, the position of strike does not change and no transition takes place. Also, the ball is not counted and it must be re-bowled. In the fifth delivery, the bowler bowls a no-ball which makes the bowler to re-bowl the ball again and this time, the batsman S_1 secures one

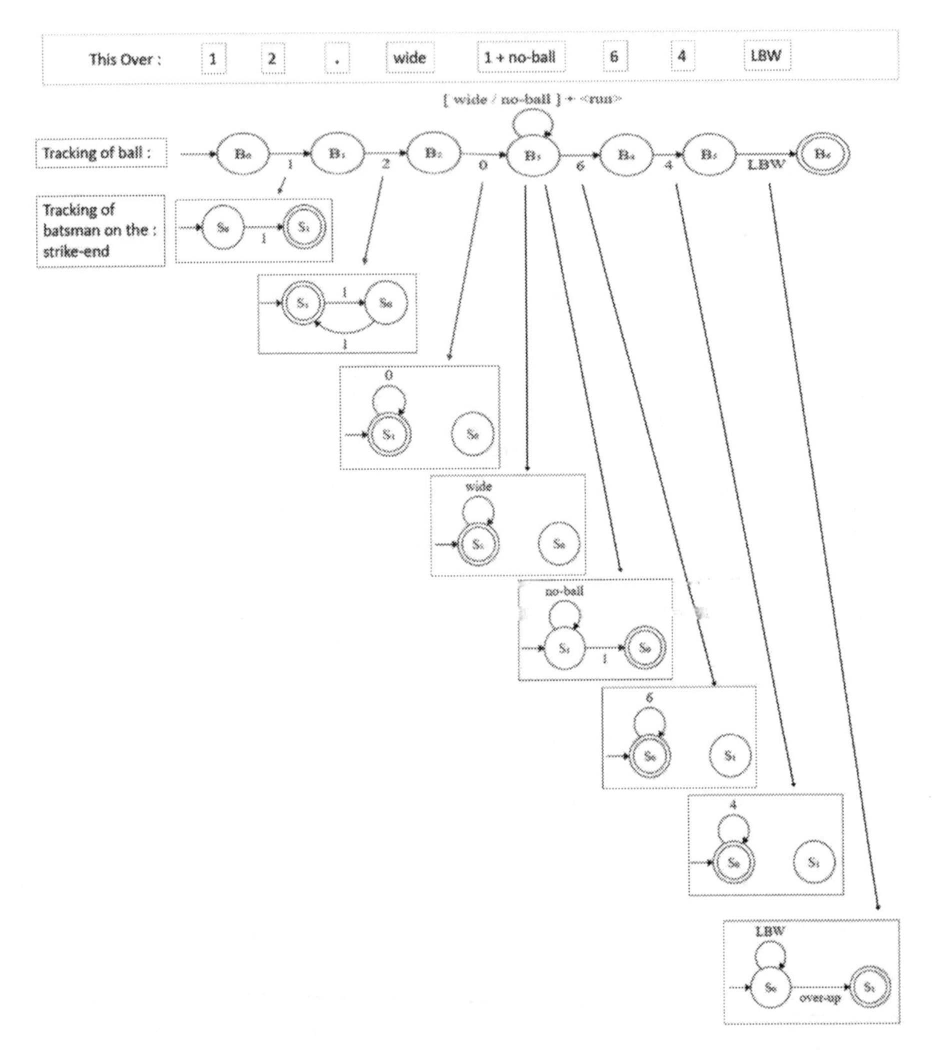

Fig. 11. An example to illustrate the application of finite automata in cricket along with the state diagrams with the combination of ball tracking in an over and on-strike batsman tracking in an over

run from running between the wickets. A no-ball with no extra runs does not change the striking positions but since, an extra run is scored by the batsman, there is a change in the position of strike to batsman S_0.

In the sixth delivery, the batsman hits the ball over the boundary and scores 6 runs. Thus, no transition change occurs. In the seventh delivery, the batsman hits the ball for four runs and keeps up the strike with him which is indicated using the self-transition in automata diagram for the respective ball. In the eighth delivery, the bowler bowls the last ball to complete his over. This time, the batsman S_0 becomes out because of LBW. Hence, the new batsman who arrives to the play will be taking up the striking position

Table 1. Process flow for the illustrated example

After n^{th} ball (n values)	On-Strike Batsman	Runs Scored
0 (initial)	S0	-
1	S1	1
2	S_1	2
3	S_1	0
3	S_1	Wide = {1}
3	S_0	No-ball + 1 = {2}
4	S_0	6
5	S_0	4
6	S_0	Wicket
Over-up	Now, S_1 comes to strike	At the end of the over: 16–1

as S_0 state. However, since the wicket was taken in the last ball of the over, the batsmen will be swapping their positions. Hence, the strike goes to batsman S_1 for the upcoming first ball of the new over.

5 Conclusion

In this paper, the concept of finite automata has been implemented in tracking of a cricket game and it has been explained in accord with the rules of cricket. The tracking of the balls bowled in an over and tracking of on-strike batsman position in the ground has been explained using the automata concepts. The application of finite automata concept can further be extended to other tracking scenarios of cricket or other sports in future works. The major highlight of this paper is to provide a detailed application of automata theory in cricket and it has been done with the help of state diagrams.

References

1. Caillois, R.: Man, Play, and Games. University of Illinois press, Champaign (2001)
2. Best, D.: Art and sport. J. Aesthetic Educ. **14**(2), 69–80 (1980)
3. Gurram, S.: Classification of sequential sports using automata theory. New Approaches Exerc. Physiol. **2**(4), 100–134 (2020)
4. Lee, J., Lee, J., Moon, S., Nam, D., Yoo, W.: Basketball event recognition technique using deterministic finite automata (DFA). In: 2018 20th International Conference on Advanced Communication Technology (ICACT), pp. 675–678. IEEE, Korea (South) (2018)
5. The Laws of Cricket, https://www.lords.org/mcc/the-laws-of-cricket. Accessed 10 Nov 2022
6. Potechin, A., Shallit, J.: Lengths of words accepted by nondeterministic finite automata. Inf. Process. Lett. **162**, 105993 (2020)
7. Ransikarbum, K., Kim, N., Ha, S., Wysk, R.A., Rothrock, L.: A highway-driving system design viewpoint using an agent-based modeling of an affordance-based finite state automata. IEEE Access **6**, 2193–2205 (2017)

8. Gribkoff, E.: Applications of deterministic finite automata. UC Davis, pp. 1–9 (2013)
9. Krishnan, P., Aravindhar, D.J., Reddy, P.B.P.: Finite automata for fake profile identification in online social networks. In: 2020 4th International Conference on Intelligent Computing and Control Systems (ICICCS), pp. 1301–1305. IEEE, India (2020)
10. Eilouti, B., Vakalo, E.G.: Finite state automata as form-generation and visualization tools. In: 1999 IEEE International Conference on Information Visualization (Cat. No. PR00210), pp. 220–224. IEEE, UK (1999)
11. Vandana, S.S., Abinandhan, K., Ramesh, S.R.: Implementation of smart home using finite state machine model. In: 2022 7th International Conference on Communication and Electronics Systems (ICCES), pp. 45–49. IEEE, India (2022)
12. Slamet, C., Gerhana, Y.A., Maylawati, D.S., Ramdhani, M.A., Silmi, N.Z.: Latin to Sundanese script conversion using finite state automata algorithm. In: IOP Conference Series: Materials Science and Engineering, vol. 434, no. 1, p. 012063. IOP Publishing, Indonesia (2018)
13. Liu, Z.M.: Online pulse deinterleaving with finite automata. IEEE Trans. Aerosp. Electron. Syst. **56**(2), 1139–1147 (2019)
14. Viard, K., Fanti, M.P., Faraut, G., Lesage, J.J.: Recognition of human activity based on probabilistic finite-state automata. In: 2017 22nd IEEE International Conference on Emerging Technologies and Factory Automation (ETFA), pp. 1–7. IEEE, Cyprus (2017)
15. Dutta, M., Saikia, H.K.: Automata in Chinese remainder theorem. Commun. Math. Appl. **13**(1), 183–197 (2022)
16. Kurniawan, O., Ismaya, F., Gata, W., Putra, J.L., Nugraha, F.S.: Application of the finite state automata concept in applications fruit vending machine simulation. Jurnal Mantik **6**(2), 1467–1474 (2022)
17. Egyir, R.J., Devendorf, R.P.: Design of a Cricket Game Using FPGAs (2020)
18. Ali, L., Sahawneh, N., Kanwal, S., Ahmed, I., Elmitwally, N.: Activity based easy learning of pushdown automata. In: 2022 International Conference on Cyber Resilience (ICCR), pp. 1–8. IEEE, United Arab Emirates (2023)
19. Singh, R., Goyal, G.: Algorithm design for deterministic finite automata for a given regular language with prefix strings. J. Sci. Res. **66**(2), 16–21 (2022)
20. Vayadande, K.B., Sheth, P., Shelke, A., Patil, V., Shevate, S., Sawakare, C.: Simulation and testing of deterministic finite automata machine. Int. J. Comput. Sci. Eng.Comput. Sci. Eng. **10**(1), 13–17 (2022)

Diabetes Prediction Using Machine Learning: A Detailed Insight

Gour Sundar Mitra Thakur[1](✉), Subhayu Dutta[2], and Bratajit Das[2]

[1] Department of Artificial Intelligence and Machine Learning, Dr. B. C. Roy Engineering College, Durgapur, West Bengal, India
gour.mitrathakur@bcrec.ac.in

[2] Department of Computer Science and Engineering, Dr. B. C. Roy Engineering College, Durgapur, West Bengal, India

Abstract. Diabetes often referred to as Diabetes mellitus is a general, continuing and deadly syndrome occurring all over the world. It is characterized by hyperglycemia which occurs due to abnormal insulin secretion which results in an irregular rise of glucose level in the blood. It is affecting numerous people all over the world. Diabetes remained untreated over a long period of time, may include complications like premature heart disease and stroke, blindness, limb amputations and kidney failure, making early detection of diabetes mellitus important. Now a days in healthcare, machine learning is used to draw insights from large medical data sets to improve the quality of patient care, improve patient outcomes, enhance operational efficiency and accelerate medical research. In this paper, we have applied different ML algorithms like Logistic Regression, Gaussian Naive Bayes, K-nearest neighbors, Support Vector Machine, Decision Tree, Random Forest, Gradient Boost, AdaBoost and Multi Layered Perceptron using Artificial Neural Network on reduced PIMA Indian Diabetes dataset and provided a detailed performance comparison of the algorithms. From this article readers are expected to gain a detailed insight of different symptoms of diabetes along with their applicability in different ML algorithms for diabetes onset prediction.

Keywords: Diabetes mellitus · Classification · Machine Learning · Onset Prediction · feature engineering

1 Introduction

Diabetes often referred to as Diabetes mellitus is a metabolic disease that causes high blood sugar. The hormone insulin moves sugar from the blood into the cells to be stored or used for energy. With diabetes, our body either isn't able to produce enough insulin or can't effectively use the insulin it does make. If diabetes remains untreated for a longer period of time it can damage the nerves, eyes, kidneys, and other organs [3]. Chronic diabetes conditions include Type 1 diabetes, Type 2 diabetes and Gestational diabetes.

S. Aurelia et al. (Eds.): ICCSST 2023, CCIS 1973, pp. 159–173, 2024.
https://doi.org/10.1007/978-3-031-50993-3_13

- *Type 1* diabetes is an autoimmune disease where the immune system attacks and destroys the cells in the pancreas, where insulin is made.
- In *Type 2* diabetes the body either doesn't make enough insulin or the body's cells don't respond normally to the insulin. Up to 95% of people with diabetes have Type 2 and it usually occurs in middle-aged and older people.
- In *Gestational diabetes* there is high blood sugar during pregnancy. However, gestational diabetic patients have higher risk of developing Type 2 diabetes later on in life.

In 2014, 8.5% of adults who were 18 years and above had diabetes. Whereas in 2019, diabetes was the major cause of 1.5 million deaths and 48% of total deaths due to diabetes occurred before the age of 70 years. In 2015, 1.6 million diabetic patients died and 2.2 million further deaths were attributed to high blood glucose levels in 2012. Another 4,60,000 kidney disease deaths were caused by diabetes, and raised blood glucose causes around 20% of cardiovascular deaths. By 2030 diabetes will become the 7th major illness that will cause deaths in the global population [14]. Diabetes may include complications like premature heart disease and stroke, blindness, limb amputations and kidney failure. This is exactly why early detection of onset diabetes is important so that the patients can get aware of their condition and they can act before it gets worse and avoid the much serious side effects that can decrease the quality of life.

A technique called, onset prediction, incorporates a variety of machine learning algorithms, feature engineering and statistical methods that uses current and past data to find knowledge and predict future events. By applying machine learning algorithm on healthcare data, significant decisions can be taken and predictions can be made. Onset prediction aims at diagnosing the disease with best possible accuracy, enhancing patient care, optimizing resources along with improving clinical outcomes. Machine learning is considered to be one of the most important artificial intelligence features that supports development of computer systems having the ability to acquire knowledge from past experiences with no need of programming for every case. Machine learning algorithms are widely used in predicting diabetes, and they get preferable results. Decision tree, Logistic regression, support vector machine, ensemble techniques are some of the popular machine learning methods in medical field, which has grateful classification power. Neural network using multi layered perceptron is a recently used popular machine learning method, which has a better performance in many aspects [13].

Different Researchers in the previous times have proposed various Machine Learning (ML) based techniques in order to predict the onset of Diabetes. Dritsas et al. [5] have utilized the information gathered from a clinical study conducted at Waikato Environment for Knowledge Analysis (WEKA) in University of Waikato, New Zealand, wherein various Machine Learning models, such as Bayesian Network, Naive Bayes, Support Vector Machine(SVM), Logistic Regression, Artificial Neural Network (ANN), K-Nearest Neighbours (KNN), J48, Random Forest, Random Tree, Logistic Model Tree, Reduced Error Pruning Tree, Rotation Forest, AdaBoostM1 and Stochastic Gradient Descent were selected in order to evaluate their prediction performance. Applying SMOTE with 10-fold cross-validation, the Random Forest and K-Nearest Neighbour (KNN) outperformed the other models by resulting with an accuracy of 98.59%. Authors like Ahmad et al. [1] had applied four different ML algorithms (decision tree, artificial neural network, K-nearest neighbors and deep neural network) to predict patients with or without type 2 diabetes mellitus and had obtained an accuracy of 90% and above, and in some special cases an accuracy 100% also. Nadeem et al. [14] tried to improve the research on prediction of diabetes and have used Artificial Neural Networks (ANN) and General Regression Neural Networks (GRNN) to create algorithms for the diagnosis of diabetes. The GRNN approach have achieved an 80% prediction accuracy which is an improvement compared to the Radial Basis Function (RBF) and Multi-Layer Perceptron (MLP) based techniques. Few authors like Maniruzzaman et al. [12] conducted a rigorous study and took help of LR model which demonstrated that 7 factors out of 14 as age, Body Mass Index(BMI), education, systolic BP, diastolic BP, direct cholesterol, and total cholesterol are the risk factors of diabetes. The overall ACC of ML-based system is 90.62%. The combination of LR-based feature selection and RF-based classifier gives 94.25% ACC and 0.95 AUC for K10 protocol. Naz et al. [15] tried to introduce a new way to predict diabetes and thus used Deep Learning approach and achieved an accuracy of 98.07%, which can be used for further development of the automatic prognosis tool. The accuracy of the Deep Learning approach can be enhanced by including the omics data for prediction of the onset of disease. Islam et al. [9] have analyzed the dataset with Logistic Regression Algorithm, Naive Bayes Algorithm and Random Forest Algorithm. After applying tenfold Cross- Validation and Percentage Split evaluation techniques, Random forest has been found having best accuracy on this dataset. Authors like Gadekallu et al. [7] tried a completely different approach and used Firefly algorithm which was used for the purpose of dimensionality reduction. This reduced dataset was then fed into the deep neural network (DNN) and it generated classifications results with enhanced accuracy. Aminian et al. [2] performed a direct comparison of IDC Risk Scores with RECODe models on a separate cohort of nonsurgical patients and it yielded in better performance of new models. Islam et al. [8] used two Ensemble Machine Learning algorithms namely Bagging and Decorate. Bagging classified the types of diabetes 95.59% accurately, whereas Decorate classified 98.53% accurately. Some Authors like Sneha et al. [19] tried to improve the predictiotion of diabestes and thus have only used decision tree algorithm and Random forest which had given the highest specificity of 98.20% and 98.00% and respectively holds best for the analysis of diabetic data. Support vector machine(SVM) and Naive Bayes techniques give the accuracy of 77.73% and 73.48% respectively from the existing method and the proposed method improves the accuracy of the classification techniques too. Authors like Larabi et al. [11] have used Reduced Error Pruning Tree and had obtained an accuracy of 74.48%. Upon using Multivariate Regression prediction model, the accuracy further increased to 97.17%. Faruque et al. [6] have used four popular machine learning algorithms, namely Naive Bayes (NB), Support Vector Machine (SVM), K-Nearest Neighbor (KNN) and C4.5 Decision

Tree (DT), on the adult population data to predict diabetic mellitus and found that C4.5 decision tree achieved higher accuracy as compared to other machine learning techniques cited above. Authors like Birjais et al. [4] have also done rigorous study on the Diabetes dataset and have used different Machine Learning techniques such as Gradient Boosting, Naive Bayes and Logistic Regression and have obtained an accuracy of 86% for the Gradient Boosting,77% for Naive Bayes and 79% for Logistic Regression. Authors like Sisodia et al. [17] have done a rigorous research taking in account the PIMA Indian Diabetes dataset wherein, he used three machine learning classification algorithms namely, Decision Tree, Support Vector Machine and Naive Bayes. Experimental results determined an overwhelming adequacy of the designed system with an achieved accuracy of 76.30% using the Naive Bayes classification algorithm. Patil et al. [16] have tried to make a change to the existing research on the PIMA Indian Diabetes Dataset and have used eight different Machine Learning algorithms namely K Nearest Neighbors (KNN), Logistic Regression, Support Vector Machine(SVM), Gradient Boost, Decision tree, MLP, Random Forest and Gaussian Naïve to predict the population who are most likely to develop diabetes and the studies had concluded that Logistic Regression and Gradient Boost classifiers achieved a higher test accuracy of 79 % when compared to other classifiers. Swapna et al. [20] used a Deep Learning approach where she analysed input HRV signals employing deep learning architectures of CNN, LSTM and its combinations and achieved a high accuracy value of 95.7% employing CNN 5-LSTM architecture with SVM using 5-fold cross-validation. Kandhasamy et al. [10] used four different machine learning classifiers which is no less than J48 Decision Tree, K-Nearest Neighbors(KNN), Random Forest and Support Vector Machines and found that the decision tree J48 classifier achieved a higher accuracy of 73.82% when compared to other three classifiers.

From the above literature survey it can be found that in most of the researches PIMA Indian dataset with eight features has been used. In this article initially we have analysed the features used in PIMA Indian dataset [18] and calculated the correlation matrix with all the features. We then removed two features with lowest correlation with the diabetes outcome and then applied different ML algorithms in the reduced dataset. Finally we compared the performances of ML algorithms in the actual dataset with that when applied in the reduced dataset. In Sect. 2 we have briefly introduced different symptoms used for diabetes prediction. Section 3 gives detailed feature analysis and identification of the features which are less relevant in diabetes prediction. In Sect. 4 we have given detailed comparison of the performances of different ML algorithms when applied in the original dataset as well in the reduced dataset with 6 features. Finally Sect. 5 gives the conclusion.

2 Identification of Symptoms for Diabetes Prediction

The "Pima Indian diabetes" dataset is used in the training and testing of the various machine learning and deep learning models. This dataset is originally taken from the National Institute of Diabetes and Digestive and Kidney Diseases which has 769 records and 8 features. The main objective of the dataset is diabetes onset prediction, based on certain diagnostic measurements included in the dataset.

The dataset contains 8 features which are as follows:

- *Pregnancies* - Diabetes can cause serious problems during pregnancy for women and their developing embryo. If not controlled/taken care of at an very early stage, it may lead to birth defects(such as high birth weight or hypoglycemia) and several serious complications for the women. So, pregnancy become most important factor for diabetes prediction.
- *Glucose* - When concentration of blood glucose increases rapidly in human body, than the required normal necessary level due to defective insulin secretion by the body or its impaired biological effects, it leads to a condition medically known as Hyperglycemia which is one of the main/primitive factor for causing diabetes.
- *Blood Pressure* (mm HG) - High Blood Pressure, often referred to as Hypertension is a common condition that affects the arteries of the bodies. Since the force of the blood pushing against the artery walls is consistently high, it has to work harder to pump blood for the proper functioning of the rest of the body organs. Blood pressure does not have too much impact on diabetes.
- *Skin Thickness* (mm) - Skin thickness is primarily determined by collagen content and increase in insulin-dependent diabetes mellitus (IDDM). Skin thickness for patients aged between 24–38 years is found to be 66IDM and it is investigated whether it is correlated with long-term glycemic control and the presence of certain diabetic complications and it is found that widespread connective tissue brings changes in diabetes mellitus.
- *Insulin* (mu U/mL) - Insulin is a hormone produced in the Pancreas of human body which is responsible for the control of glucose in the bloodstream at any given point of time. It helps the blood sugar to enter the body cells so that it can be used for energy. People with type1 diabetes are unable to produce insulin and are unable to control their blood glucose level.
- *Body Mass Index (BMI)* (kg/m^2) - When BMI of a person goes above his/her personal threshold and blood sugar levels of the body increase, it triggers the onset of type 2 diabetes which may lead to damaged blood vessels and increased risk of developing heart problems and circulatory diseases.
- *Diabetes Pedigree Function* - It is a summarized score that indicates the genetic predisposition of the patient for diabetes as extrapolated from the patient's family record for diabetes.
- *Age* (years)- Age is a vital and most important risk factor for diabetes as it is directly proportional to the risk of complications especially for type-2 diabetes. This is the reason why older adults prone to diabetes have a higher risk of heart attack and stroke.

3 Feature Analysis

In machine learning the main goal of feature selection techniques is to find the best set of features that allows one to build useful models for accurate prediction. Various methods of feature selection techniques are applied in the dataset like correlation matrix, density plot, etc. A correlation matrix is a table that displays the correlation coefficients for different variables in a dataset. The matrix in Fig. 1 depicts the correlation between outcome (dependent variable) and independent variables like Glucose, BMI, pregnancies and age as compared to other independent variables.

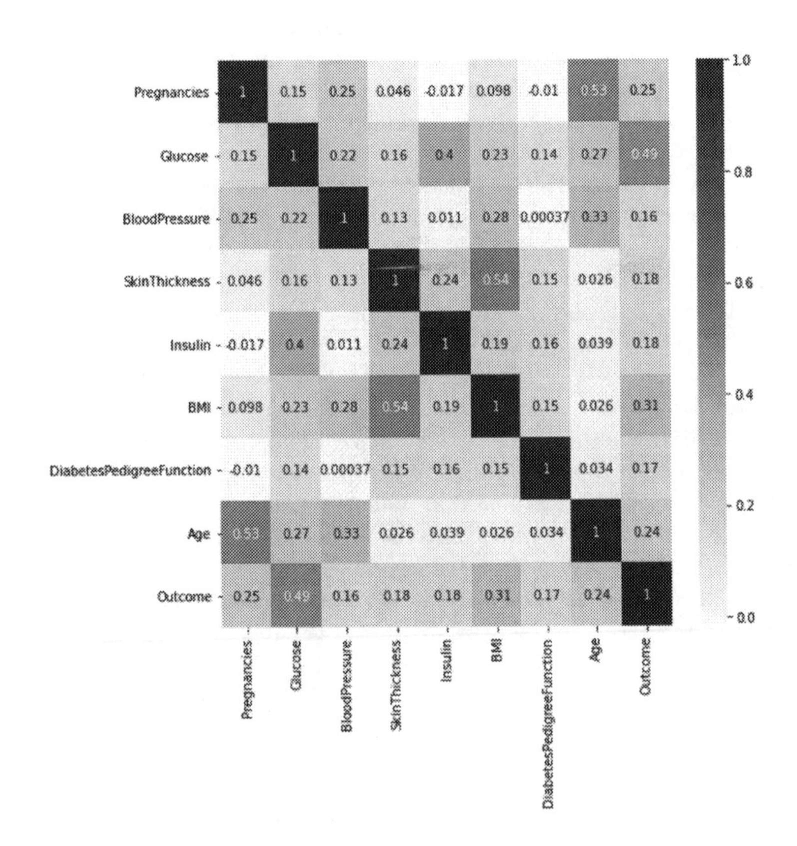

Fig. 1. Correlation Matrix of outcome dependent variable and other independent variables

By visualising the distribution of the nine variables in the dataset by plotting a histogram in Fig. 2 which provides some interesting insights into the data. It depicts that most of the data was collected from young people, between the age group of 20–30 years and the distribution for BMI, Blood Pressure, and Glucose concentration is normally distributed. The distribution of outcome shows that approximately 65% of the population don't have diabetes, while the remaining 35% have diabetes.

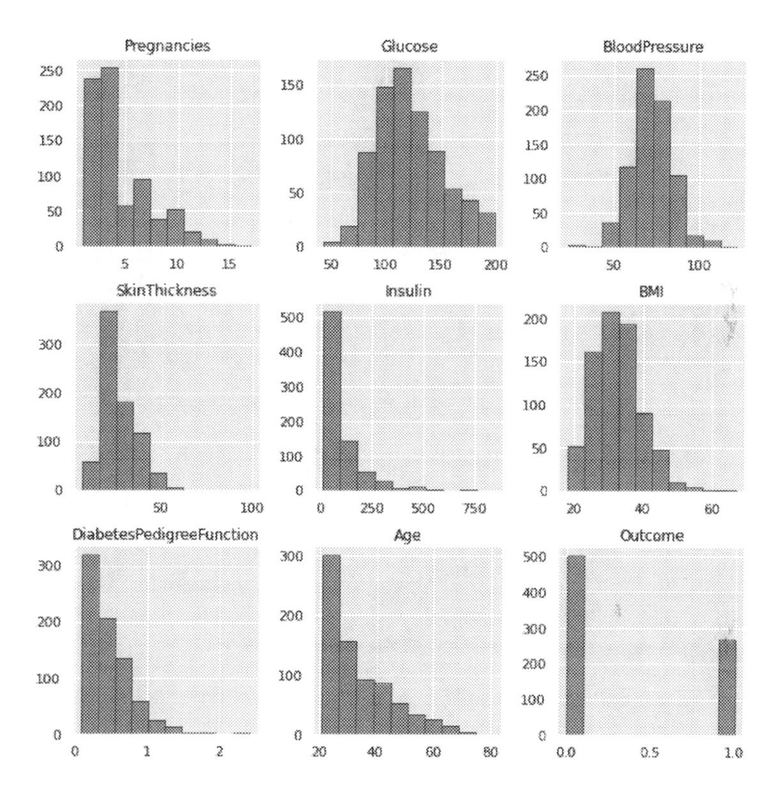

Fig. 2. Histogram plot of independent variables

By visualising the density plots in Fig. 3, it depicts that the people with diabetes tend to have higher blood glucose level, higher insulin, higher BMI, and are older. Whereas the other variables such as Blood Pressure, Diabetes Pedigree Function and skin thickness have no significant difference in the distribution between diabetic and non-diabetic patients.

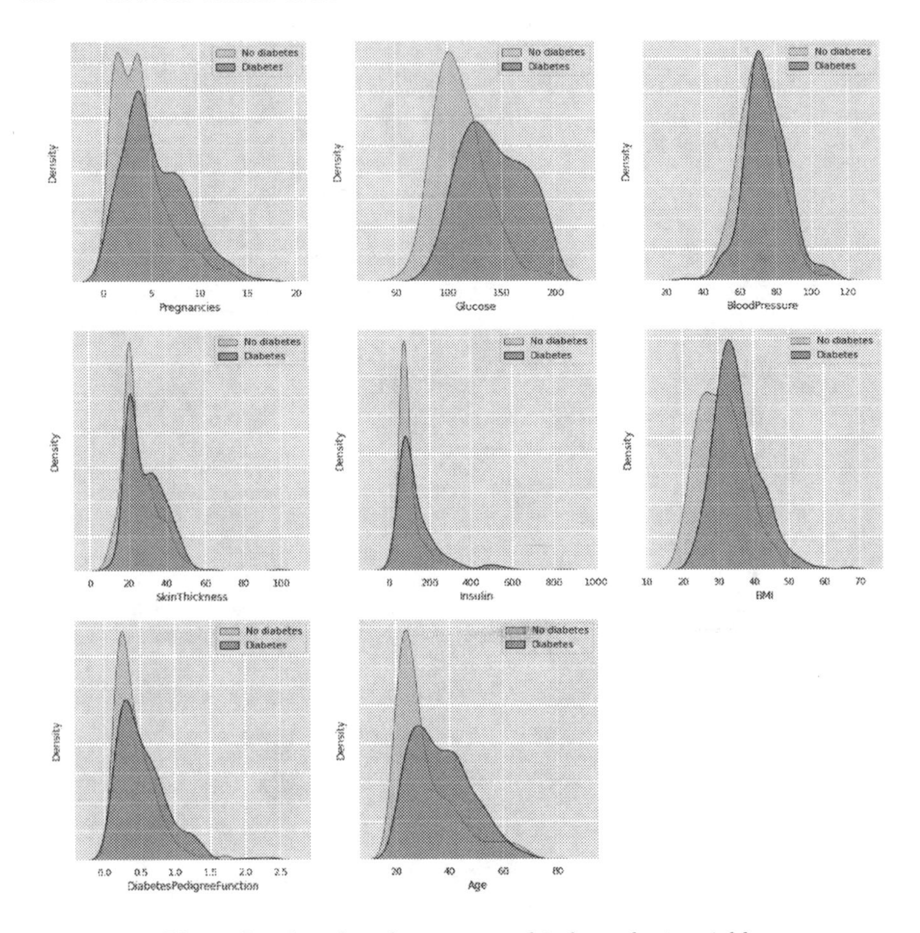

Fig. 3. Density plot of outcome and independent variables

Thus, from correlation matrix and density plots it is observed that the **Glucose**, **Insulin**, **BMI**, **pregnancies** and **Age** variables are strong predictors for diabetes whereas **Blood Pressure**, **Diabetes Pedigree Function** and **Skin Thickness** are poorer predictors for diabetes. Above shows some graphical analysis of the features along with correlation and identification of two features i.e. skin thickness and diabetes pedigree function which can be removed to further reduce the complexity of algorithms. So we will apply the machine learning algorithms on the dataset with all features and the dataset with reduced features (Diabetes Pedigree Function and Skin Thickness features are removed).

4 Comparative Analysis of Different ML Algorithms in Diabetes Onset Prediction

In the experimental studies the dataset is divided into two parts, one with dataset with all features and other with dataset with reduced features (Diabetes Pedigree Function and Skin Thickness features are removed). Then, the dataset is partitioned between 70–30% for training and testing purpose respectively. Data cleaning is done in the dataset to fill up the null values and data values are centered around the mean with a unit standard deviation using Standard Scaler which is an important part of feature engineering. Now both the datasets is compared in terms of accuracy, train and test score, mean absolute error, root mean squared error training time, precision, recall, F1 score and ROC by applying different machine learning algorithm such as Logistic Regression, Gaussian Naïve Bayes, K-nearest neighbors, Support Vector Machine, Decision Tree, Random Forest, Gradient Boost, AdaBoost and Multi Layered Perceptron.

Table 1 shows the comparison of various algorithms in terms of Accuracy, train and test score, mean absolute error, root mean square error and training time in the dataset containing all the features and Table 2 shows the comparison of various algorithms in the dataset with reduced features. It is seen that the accuracy of dataset with all the features is 1–2% more than that of dataset with reduced feature which means accuracy is comparably same in both the datasets. Training time is also less in reduced feature dataset as compared to the dataset with all the features which in turn decreases the time complexity of all algorithms in case of reduced feature dataset.

Table 1. Comparison in terms of Accuracy, Train Score, Test Score, Training time in the dataset containing all features

Algorithm	Accuracy	Train Score	Test Score	MAE	RMSE	Train Time
Logistic Regression	0.7965	0.7854	0.7923	0.2142	0.4629	0.0487
Gaussian Naive Bayes	0.7543	0.7822	0.7512	0.2662	0.5159	0.0081
K-nearest neighbors	0.7134	0.7785	0.7839	0.2662	0.5159	0.0964
Support Vector Machine	0.7791	0.7798	0.7812	0.2207	0.7687	0.0492
Decision Tree	0.7698	0.7734	0.7654	0.2207	0.4698	0.0875
Random Forest	0.7383	0.7743	0.7754	0.2727	0.5222	0.1875
Gradient Boost	0.7397	0.7532	0.7691	0.2727	0.5222	0.0937
AdaBoost	0.7345	0.7842	0.7962	0.2727	0.5222	0.2175
Multi Layered Perceptron	0.7346	0.6523	0.6513	0.2951	0.5376	0.9816

Table 2. Comparison in terms of Accuracy, Train Score, Test Score, Training time in reduced feature dataset

Algorithm	Accuracy	Train Score	Test Score	MAE	RMSE	Train Time
Logistic Regression	0.7857	0.7736	0.7857	0.2101	0.4793	0.0359
Gaussian Naive Bayes	0.7337	0.7687	0.7792	0.2576	0.5082	0.0062
K-nearest neighbors	0.7337	0.7687	0.7792	0.2576	0.5082	0.0716
Support Vector Machine	0.7792	0.7792	0.7823	0.2387	0.7512	0.0310
Decision Tree	0.7792	0.7687	0.7687	0.2345	0.4623	0.0505
Random Forest	0.7277	0.7687	0.7792	0.2798	0.5246	0.1538
Gradient Boost	0.7272	0.7687	0.7792	0.2798	0.5246	0.0852
AdaBoost	0.7277	0.7687	0.7790	0.2798	0.5246	0.1969
Multi Layered Perceptron	0.7272	0.6498	0.6455	0.2732	0.5376	0.9298

The graph in Fig. 4 depicts the comparison of both the datasets in terms of accuracy. It is seen that Logistic Regression, Support Vector Machine and Decision Tree algorithm achieved the highest accuracy in around 77–79% in both the datasets. In K-nearest neighbors and Decision Tree algorithm dataset with reduced features achieves a rise in accuracy as compared to dataset with all features. In Support Vector Machine algorithm both the dataset has same accuracy. And in all other algorithms accuracy of dataset with all features is more as compared to dataset with reduced feature. The graph in Fig. 5 depicts the comparison of both the datasets in terms of training time. Training time is less in reduced feature dataset as compared to the dataset with all the features. But in case of Multi Layered perceptron the time taken by both the dataset is more than all algorithms because training an neural network involves using an optimization algorithm to find the weights to best map inputs to outputs using various activation functions, loss functions and optimizers.

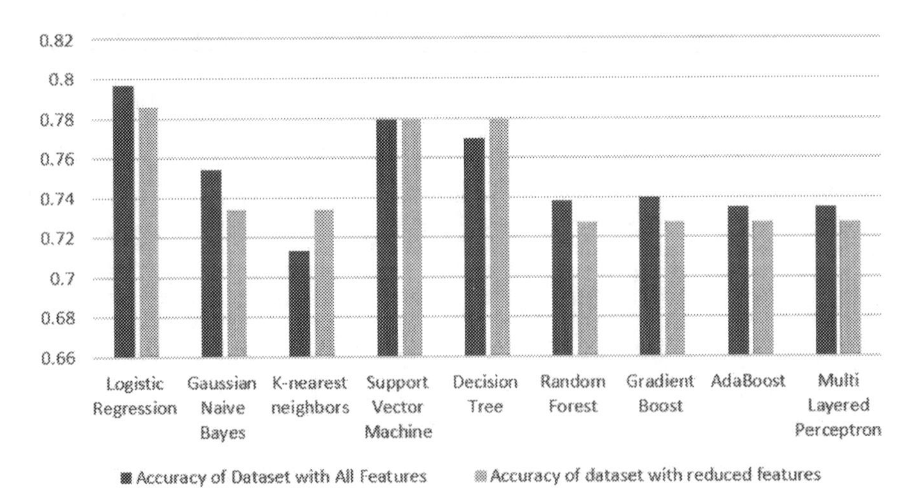

Fig. 4. Comparison of both the datasets in terms of accuracy

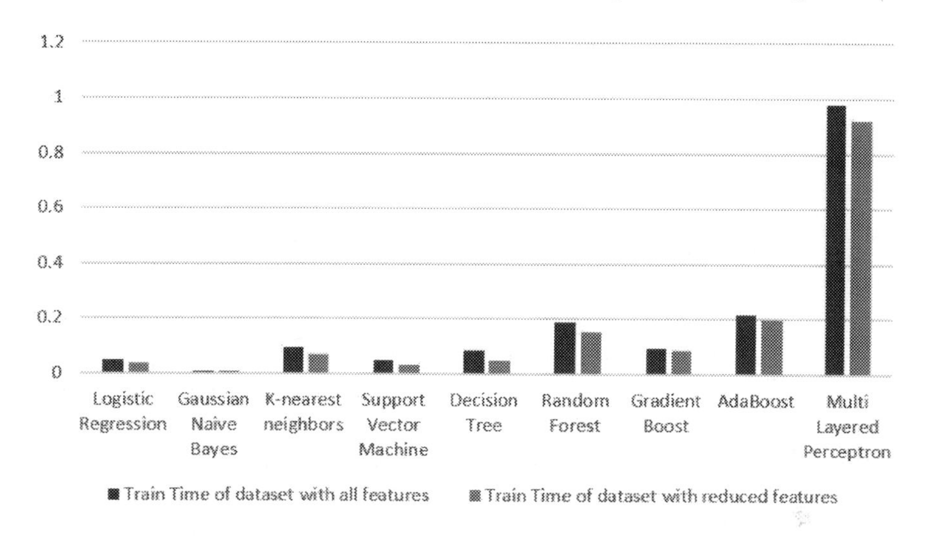

Fig. 5. Comparison of both the datasets in terms of time

Table 3 and 4 shows the comparison of algorithms for precision, recall, F1 score and ROC which is calculated using the confusion matrix in both the datasets. Precision shows that how many of the positive predictions are correct, recall shows that how many of the positive cases the classifier correctly predicts over all positive cases in the data, F1 score is calculated by taking harmonic mean of precision and recall. It is seen that in both the datasets the precision, recall, f1 score and roc is comparably same.

Table 3. Comparison in terms of Precision, Recall and F1 Score in dataset containing all features

Algorithm	Precision	Recall	F1 Score	ROC
Logistic Regression	0.78	0.78	0.77	0.75
Gaussian Naive Bayes	0.71	0.74	0.75	0.72
K-nearest neighbors	0.72	0.71	0.73	0.71
Support Vector Machine	0.75	0.79	0.76	0.76
Decision Tree	0.78	0.75	0.78	0.73
Random Forest	0.72	0.75	0.73	0.71
Gradient Boost	0.74	0.73	0.74	0.70
AdaBoost	0.73	0.74	0.73	0.72
Multi Layered Perceptron	0.74	0.73	0.74	0.63

Table 4. Comaparison in terms of Precision, Recall and F1 Score in reduced feature dataset

Algorithm	Precision	Recall	F1 Score	ROC
Logistic Regression	0.77	0.78	0.76	0.75
Gaussian Naive Bayes	0.72	0.73	0.73	0.71
K-nearest neighbors	0.72	0.73	0.73	0.70
Support Vector Machine	0.77	0.78	0.76	0.75
Decision Tree	0.77	0.78	0.76	0.72
Random Forest	0.71	0.73	0.72	0.70
Gradient Boost	0.71	0.74	0.72	0.73
AdaBoost	0.71	0.73	0.71	0.70
Multi Layered Perceptron	0.73	0.73	0.73	0.62

Figure 6 and 7 shows the overall comparative analysis of algorithms in terms of precision, recall, F1 score and ROC in both the datasets which depicts that precision, recall, F1 score and ROC is comparably same in both the datasets which in turn provides greater accuracy in predicting the population who are most likely to develop diabetes on Pima Indians Diabetes data. From the graph it is noticed that logistic regression, support vector machine and decision tree algorithm has the higher precision, recall, f1 score and roc than the other algorithms in both the datasets.

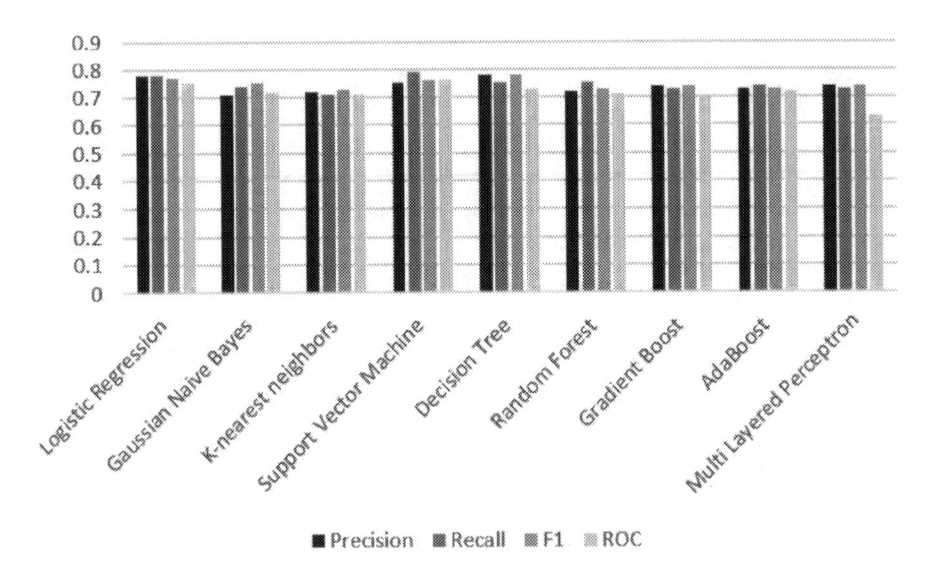

Fig. 6. Comparison in terms of precision, recall, F1 score and ROC in the dataset containing all features

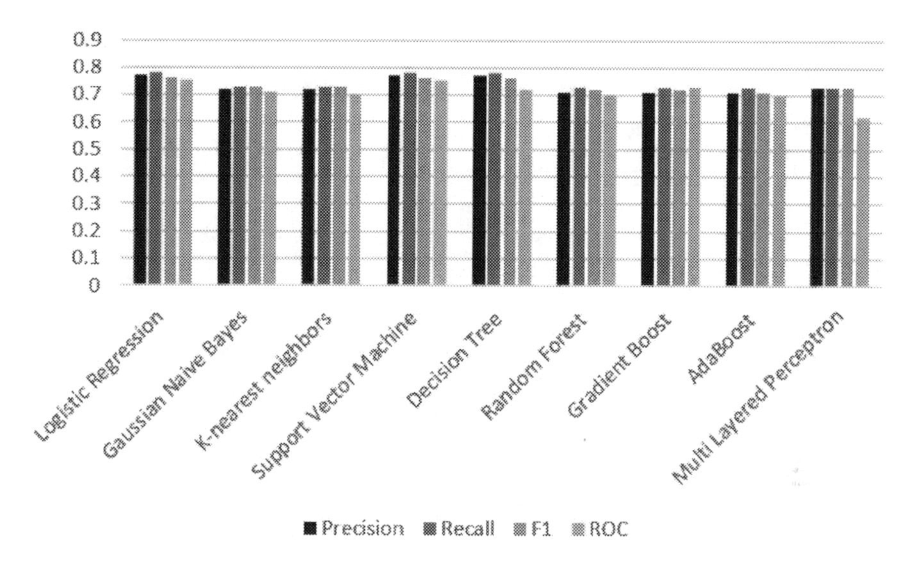

Fig. 7. Comparison in terms of precision, recall, F1 score and ROC in the dataset containing reduced features

5 Conclusion

In this paper, we have done comparitive analysis by executing nine different machine learning algorithms namely Logistic Regression, Gaussian Naive Bayes, K-nearest neighbors, Support Vector Machine, Decision Tree, Random Forest, Gradient Boost, AdaBoost and Multi Layered Perceptron to predict the population who are most likely to develop diabetes on Pima Indian diabetes data. The performance measurement is compared in terms of Test Accuracy, train score, test score, mean absolute error, root mean square error, training time, precision, recall, f1 score and roc obtained from the dataset containing all features and dataset containing reduced feature (Diabetes Pedigree Function and Skin Thickness features are removed). Here the studies conclude that Logistic Regression achieve higher test accuracy of 79% in dataset containing all features and 78% in dataset with reduced features than other classifiers. Other algorithm like Support Vector Machine and Decision Tree has achieved an accuracy of more than 76% in both the datasets. And the training time is also less in reduced feature dataset as compared to the dataset with all the features which in turn decreases the time complexity of all algorithms in case of reduced feature dataset. Another important point is that we cannot achieve an accuracy of more than 90% using this traditional machine learning algorithms. Further, we plan to recreate our study by introducing the intelligent machine learning algorithms applied to a large collection of real life dataset. This study can be used as a best classifier for predicting diabetes.

References

1. Ahmad, H.F., Mukhtar, H., Alaqail, H., Seliaman, M., Alhumam, A.: Investigating health-related features and their impact on the prediction of diabetes using machine learning. Appl. Sci. **11**(3), 1173 (2021)
2. Aminian, A., et al.: Predicting 10-year risk of end-organ complications of type 2 diabetes with and without metabolic surgery: a machine learning approach. Diabetes Care **43**(4), 852–859 (2020)
3. Bastaki, S.: Diabetes mellitus and its treatment. Dubai Diabetes Endocrinol. J. **13**, 111–134 (2005)
4. Birjais, R., Mourya, A.K., Chauhan, R., Kaur, H.: Prediction and diagnosis of future diabetes risk: a machine learning approach. SN Appl. Sci. **1**(9), 1–8 (2019)
5. Dritsas, E., Trigka, M.: Data-driven machine-learning methods for diabetes risk prediction. Sensors **22**(14), 5304 (2022)
6. Faruque, M.F., Sarker, I.H., et al.: Performance analysis of machine learning techniques to predict diabetes mellitus. In: 2019 International Conference on Electrical, Computer and Communication Engineering (ECCE), pp. 1–4. IEEE (2019)
7. Gadekallu, T.R., et al.: Early detection of diabetic retinopathy using PCA-firefly based deep learning model. Electronics **9**(2), 274 (2020)
8. Islam, M.T., Raihan, M., Akash, S.R.I., Farzana, F., Aktar, N.: Diabetes mellitus prediction using ensemble machine learning techniques. In: Saha, A., Kar, N., Deb, S. (eds.) ICCISIoT 2019. CCIS, vol. 1192, pp. 453–467. Springer, Singapore (2020). https://doi.org/10.1007/978-981-15-3666-3_37
9. Islam, M.M.F., Ferdousi, R., Rahman, S., Bushra, H.Y.: Likelihood prediction of diabetes at early stage using data mining techniques. In: Gupta, M., Konar, D., Bhattacharyya, S., Biswas, S. (eds.) Computer Vision and Machine Intelligence in Medical Image Analysis. AISC, vol. 992, pp. 113–125. Springer, Singapore (2020). https://doi.org/10.1007/978-981-13-8798-2_12
10. Kandhasamy, J.P., Balamurali, S.: Performance analysis of classifier models to predict diabetes mellitus. Procedia Comput. Sci. **47**, 45–51 (2015)
11. Larabi-Marie-Sainte, S., Aburahmah, L., Almohaini, R., Saba, T.: Current techniques for diabetes prediction: review and case study. Appl. Sci. **9**(21), 4604 (2019)
12. Maniruzzaman, M., Rahman, M., Ahammed, B., Abedin, M., et al.: Classification and prediction of diabetes disease using machine learning paradigm. Health Inf. Sci. Syst. **8**(1), 1–14 (2020)
13. Mujumdar, A., Vaidehi, V.: Diabetes prediction using machine learning algorithms. Procedia Comput. Sci. **165**, 292–299 (2019)
14. Nadeem, M.W., Goh, H.G., Ponnusamy, V., Andonovic, I., Khan, M.A., Hussain, M.: A fusion-based machine learning approach for the prediction of the onset of diabetes. In: Healthcare, vol. 9, p. 1393. MDPI (2021)
15. Naz, H., Ahuja, S.: Deep learning approach for diabetes prediction using PIMA Indian dataset. J. Diabetes Metab. Disord. **19**(1), 391–403 (2020)
16. Patil, R., Tamane, S.: A comparative analysis on the evaluation of classification algorithms in the prediction of diabetes. Int. J. Electr. Comput. Eng. **8**(5), 3966 (2018)
17. Sisodia, D., Sisodia, D.S.: Prediction of diabetes using classification algorithms. Procedia Comput. Sci. **132**, 1578–1585 (2018)

18. Smith, J.W., Everhart, J.E., Dickson, W., Knowler, W.C., Johannes, R.S.: Using the ADAP learning algorithm to forecast the onset of diabetes mellitus. In: Proceedings of the Annual Symposium on Computer Application in Medical Care, p. 261. American Medical Informatics Association (1988)
19. Sneha, N., Gangil, T.: Analysis of diabetes mellitus for early prediction using optimal features selection. J. Big Data **6**(1), 1–19 (2019)
20. Swapna, G., Vinayakumar, R., Soman, K.: Diabetes detection using deep learning algorithms. ICT Express **4**(4), 243–246 (2018)

Empirical Analysis of Resource Scheduling Algorithms in Cloud Simulated Environment

Prathamesh Vijay Lahande and Parag Ravikant Kaveri(✉)

Symbiosis Institute of Computer Studies and Research, Symbiosis International (Deemed University), Pune, India
{prathamesh.lahande,parag.kaveri}@sicsr.ac.in

Abstract. The cloud environment is a collection of resources providing multiple services to the end-users. Users submit tasks to this cloud computing environ ment for computation purposes. Using statically fixed resource scheduling algo rithms, the cloud accepts these tasks for computations on its Virtual Machines (VM). Resource scheduling is considered a complex job, and computing these challenging tasks without external intelligence becomes a challenge for the cloud. The key objective of this study is to compute different task sizes in the Work flowSim cloud simulation environment using scheduling algorithms Max – Min (MxMn), Minimum Completion Time (M.C.T.), and Min – Min (MnMn) and later compare their behavior concerning various performance metrics. The exper imental results show that all these algorithms have dissimilar behavior, and no method supplies the best results under all performance metrics. Therefore, an in telligence mechanism is required to be provided to these resource scheduling al gorithms so they can perform better for all the performance metrics. Lastly, it is suggested that Reinforcement Learning (RL) acts as an intelligence mechanism, enhances the resource scheduling procedure, and makes the entire process dynamic, enhancing cloud performance.

Keywords: Cloud Environment · Empirical Analysis · Performance · Reinforce ment Learning · Time

1 Introduction

The cloud computing platform is an on-demand availability of computing services pro vided by the cloud to the end-user [18]. The cloud uses its Virtual Machines (VMs) and resource scheduling algorithms to compute the tasks submitted for computations [2, 6]. To provide superior services, the main challenge for the cloud is to process these tasks as efficiently as possible without any delays [12, 19]. For the cloud to function smooth computation of tasks, its scheduling algorithms are its backbone [13]. With ac curate resource scheduling, the cloud can compute more tasks; otherwise, the latency gradually grows in the cloud environment [20]. Therefore, it is important to observe the behavior and perform its empirical analysis study (EA). The primary objective of this

research study is to compute Cybershake Tasks (CT) using the VMs by using the sched uling algorithms Max – Min (MxMn), Minimum Completion Time (M.C.T.), and

S. Aurelia et al. (Eds.): ICCSST 2023, CCIS 1973, pp. 174–182, 2024.
https://doi.org/10.1007/978-3-031-50993-3_14

Min – Min (MnMn) This paper is roughly categorized into the following phases: the initial phase consists of an experiment conducted in the WorkflowSim [17] platform where tasks are computed in four scenarios where each scenario computes CT - 30, CT - 50, CT - 100, and CT -1000 tasks on VMs in the series 5, 10, 15, …, 50, respectively; the second phase consists of performing an EA on the results obtained from the first phase concerning performance metrics such as Average Start Time (A.S.T.), Average Com pletion Time (A.C.T.), Average Turn Around Time (A.T.A.T.), and Average Cost (A.C.); the last phase includes introducing a mechanism of intelligence to the cloud to improve its resource scheduling mechanism.

Figure 1 depicts the architecture of the flow of the experiment conducted.

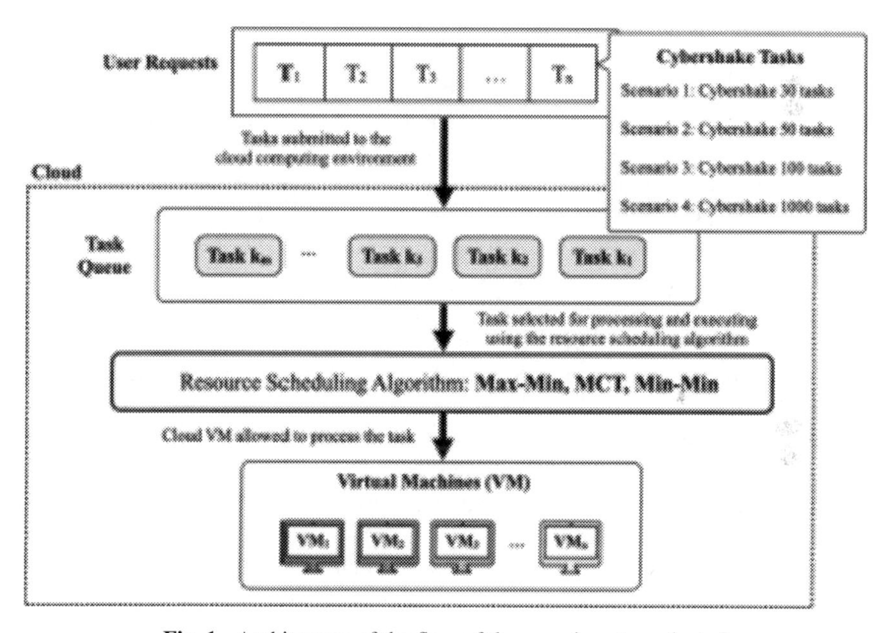

Fig. 1. Architecture of the flow of the experiment conducted

2 Literature Review

Resource scheduling is a significant criterion for making the cloud output the best results. Several researchers have tried to focus on this area and provided multiple solu tions to this NP-hard resource scheduling problem. The authors of this paper have pro posed a method to lower the usage of resources while making sure the satisfactory rate of the user is attained in the internet of things application [1]. The RL algorithms are implemented to enhance the resource scheduling issue in the cloud environment [3]. The researchers have implemented the RL technique to propose a fog scheduling method in the Internet of Things platform to solve various problems in it [2]. To im prove cost required by the cloud service providers, a scheduling technique is presented according to the deep

RL [13]. To focus on resource scheduling and failures of tasks, a scheduling method based on the deep RL is used in this research paper [4]. The task scheduling issue of the cloud applications is focused on in this research paper, where a method is presented to improve task scheduling and reduce the computation cost [6]. The mathematical concept of the Knapsack algorithm is used to provide a VM relocation method to enhance the energy effectivity and density of VMs [10]. To manage the VM resources, a deep elastic resource provisioning agent is used to obtain automated elasticity to schedule the VM resources, which gains better rewards compared with other approaches [11]. To enhance performance concerning finish time, breaching rates, and VM resource use, a task-scheduling solution based upon the fuzzy algorithm of self-defense is proposed [7]. The researchers have presented a dynamic resource scheduling policy using the RL method with fuzzy rules to reconfigure the system and boost the rewards and penalties that occurred [20].

A metaheuristic dynamic virtual machine allocation framework is proposed to im prove the scheduling process of tasks in the cloud system [5]. When this framework is implemented in a simulation environment, the results show that it optimizes the cloud. The researchers have studied the fault-tolerance VM placement issue, and by proving this problem as an NP-hard, they have presented an integer linear programming method to provide an optimal solution for this fault-tolerance VM placement issue [8]. The researchers have used an RL approach to optimally select the best placement strategy for each VM while considering its load [9]. The researchers have executed tasks using the existing algorithms to compare their behavior concerning cost, latency, and overall output [14]. To balance the Quality of Service (QoS) and power consumption in the cloud environment, an RL-based resource management algorithm is presented [15]. The results depict that this presented algorithm provides improved results than existing methods. To obtain more cloud reliability, an enhanced resource scheduling method is presented in this research paper, which uses adaptive RL combined with queuing theory [12]. The results from this proposed method show the efficiency in terms of task processing, consumption rate, and latency. The RL is used to enhance resource manage ment and improve the overall success rate of task execution while reducing the com plexities of computations [16].

3 EA of the Results and Their Implications

The empirical analysis (EA) of the results is presented in four sub-sections. The Subsects. 3.1, 3.2, 3.3 and 3.4 provide the EA concerning A.S.T., A.C.T., A.T.A.T., and A.C., respectively. Terminologies used are: **PRMTR**: Parameter; **LNR-RGRN-EQN**: Linear Regression Equation; **RGRN-LNE-SLPE**: Regression Line Slope; **RGRN-Y INTRCPT**: Regression Y-Intercept; **RLSHP**: Relationship; **N (-ve):** Negative, **P (+ve)**: Positive.

3.1 EA Concerning A.S.T

Table 1 depicts the EA concerning the A.S.T. for executing CT - 30, CT - 50, CT - 100, and CT - 1000 tasks respectively.

Table 1. EA of MxMn, M.C.T., and MnMn algorithms concerning Average Start Time (A.S.T.).

	CT - 30			CT – 50			CT – 100			CT - 1000		
PRMTR	MxMn	M.C.T	MnMn	MxMn	M.C.T	MnMn	MxMn	M.C.T	MnMn	MxMn	M.C.T	MnMn
LNR-RGRN -EQN	y = − 1.3x + 177.3	y = − 1.0x + 175.5	y = − 0.7x + 173.1	y = − 5.9x + 213.1	y = − 4.9x + 206.2	y = − 3.3x + 194.2	y = − 23.8x + 379.0	y = − 20.4x + 353.3	y = − 14.8x + 311.0	y = − 257.1x + 2559.2	y = − 243.7x + 2420.5	y = − 139.6x + 1462.2
RGRN-LNE-$_{SLPE}$	−1.3031	−1.0604	−0.7225	−5.9201	−4.9869	−3.355	−23.898	−20.488	−14.823	−257.14	−243.75	−139.67
RGRN-$_{Y\text{-}INTRCPT}$	177.32	175.58	173.11	213.16	206.28	194.26	379.04	353.39	311.06	2559.2	2420.5	1462.2
RLSHP	N (−VE)	N (−VE)	N (−VE)	N (−VE)	N (−VE)	N (−VE)	N (−VE)	N (−VE)	N (−VE)	N (−VE)	N (−VE)	N (−VE)
R^2	0.2727	0.2727	0.2727	0.3198	0.3144	0.3164	0.4084	0.3910	0.3745	0.6542	0.6544	0.6532
Results	MxMn ≈ M.C.T. ≈ MnMn			M.C.T. > MnMn > MxMn			MnMn > M.C.T. > MxMn			MnMn > MxMn > M.C.T		

From Table 1, the following observations regarding the Average Completion Time (A.S.T.) can be made:

- ↑ VM = ↓ A.S.T.: With the increase in VMs in each scenario, the A.S.T. gradually decreases for all the algorithms.
- MxMn ≈ M.C.T. ≈ MnMn: The performance of all algorithms is similar for executing 30 tasks.
- Performance (M.C.T.) > Performance (MnMn) > Performance (MxMn) for executing 50 tasks.
- Performance (MnMn) > Performance (M.C.T.) > Performance (MxMn) for executing 100 tasks.
- Performance (MnMn) > Performance (MxMn) > Performance (M.C.T.) for executing 1000 tasks.

3.2 EA Concerning A.C.T

Table 2 depicts the EA concerning the A.C.T. for executing CT - 30, CT - 50, CT - 100, and CT - 1000 tasks respectively.

Table 2. EA of MxMn, M.C.T., and MnMn algorithms concerning Average Completion Time (A.C.T.).

	CT - 30			CT – 50			CT – 100			CT - 1000		
PRMTR	MxMn	M.C.T	MnMn	MxMn	M.C.T	MnMn	MxMn	M.C.T	MnMn	MxMn	M.C.T	MnMn
LNR-RGRN -EQN	y = − 1.2x + 204.6	y = − 1.0x + 202.8	y = − 0.7x + 200.4	y = − 5.9x + 246.3	y = − 4.9x + 239.4	y = − 3.3x + 227.4	y = − 23.8x + 414.1	y = − 20.4x + 388.5	y = − 14.8x + 346.1	y = − 256.9x + 2580.6	y = − 243.4x + 2440.8	y = − 139.6x + 1462.2
RGRN-LNE-$_{SLPE}$	−1.2992	−1.0566	−0.7186	−5.9187	−4.984	−3.3518	−23.895	−20.486	−14.818	−256.99	−243.47	−139.67
RGRN-$_{Y\text{-}INTRCPT}$	204.63	202.89	200.41	246.35	239.46	227.44	414.17	388.53	346.18	2580.6	2440.8	1462.2
RLSHP	N (−VE)	N (−VE)	N (−VE)	N (−VE)	N (−VE)	N (−VE)	N (−VE)	N (−VE)	N (−VE)	N (−VE)	N (−VE)	N (−VE)
R^2	0.2727	0.2727	0.2727	0.3199	0.3144	0.3164	0.4085	0.3911	0.3745	0.6534	0.6528	0.6532
Results	MxMn ≈ M.C.T. ≈ MnMn			M.C.T. > MnMn > MxMn			MnMn > M.C.T. > MxMn			MnMn > M.C.T. > MxMn		

From Table 2, the following observations regarding the Average Completion Time (A.C.T.) can be made:

- $\uparrow$ VM = $\downarrow$ A.C.T.: With the increase in VMs in each scenario, the A.C.T. gradually decreases for all the algorithms.
- MxMn $\approx$ M.C.T. $\approx$ MnMn: The performance of all algorithms is similar for executing 30 tasks.
- Performance (M.C.T.) > Performance (MnMn) > Performance (MxMn) for executing 50 tasks.
- Performance (MnMn) > Performance (M.C.T.) > Performance (MxMn) for executing 100 tasks.
- Performance (MnMn) > Performance (M.C.T.) > Performance (MxMn) for executing 1000 tasks.

3.3 EA Concerning A.T.A.T

Table 3 depicts the EA concerning the A.T.A.T. for executing CT - 30, CT - 50, CT - 100, and CT - 1000 tasks respectively.

Table 3. EA of MxMn, M.C.T., and MnMn algorithms concerning Average Turn Around Time (A.T.A.T.).

	CT - 30			CT – 50			CT – 100			CT - 1000		
PRMTR	MxMn	M.C.T	MnMn	MxMn	M.C.T	MnMn	MxMn	M.C.T	MnMn	MxMn	M.C.T	MnMn
LNR-RGRN-EQN	y = 0.0039x + 27.302	y = 0.0038x + 27.303	y = 0.0039x + 27.303	y = 0.0015x + 33.188	y = 0.003x + 33.177	y = 0.0033x + 33.175	y = 0.003x + 35.135	y = 0.0021x + 35.142	y = 0.0043x + 35.124	y = 0.0107x + 22.892	y = 0.0109x + 22.889	y = 0.0106x + 22.888
RGRN-LNE-SLPE	0.0039	0.0038	0.0039	0.0015	0.003	0.0033	0.003	0.0021	0.0043	0.0107	0.0109	0.0106
RGRN-Y-INTRCPT	27.302	27.303	27.303	33.188	33.177	33.175	35.135	35.142	35.124	22.892	22.889	22.888
RLSHP	P (+VE)	P (+VE)	P (+VE)	P (+VE)	P (+VE)	P (+VE)	P (+VE)	P (+VE)	P (+VE)	P (+VE)	P (+VE)	P (+VE)
R^2	0.2727	0.2727	0.2727	0.1863	0.3881	0.3814	0.1961	0.1324	0.4986	0.9976	0.9997	0.9996
Results	MxMn $\approx$ M.C.T. $\approx$ MnMn			MxMn > MnMn > M.C.T			M.C.T. > MxMn > MnMn			MxMn > MnMn > M.C.T		

From Table 3, the following observations regarding the Average Turn Around Time (A.T.A.T.) can be made:

- $\uparrow$ VM = $\uparrow$ A.T.A.T.: With the increase in VMs in each scenario, the A.T.A.T. gradually increases for all the algorithms.
- MxMn $\approx$ M.C.T. $\approx$ MnMn: The performance of all algorithms is similar for executing 30 tasks.
- Performance (MxMn) > Performance (MnMn) > Performance (M.C.T.) for executing 50 tasks.
- Performance (M.C.T.) > Performance (MxMn) > Performance (MnMn) for executing 100 tasks.
- Performance (MxMn) > Performance (MnMn) > Performance (M.C.T.) for executing 1000 tasks.

3.4 EA Concerning A.C

Table 4 depicts the EA concerning the AC for executing CT - 30, CT - 50, CT - 100, and CT - 1000 tasks respectively.

Table 4. EA of MxMn, M.C.T., and MnMn algorithms concerning Average Cost (AC).

	CT - 30			CT – 50			CT – 100			CT - 1000		
PRMTR	MxMn	M.C.T	MnMn	MxMn	M.C.T	MnMn	MxMn	M.C.T	MnMn	MxMn	M.C.T	MnMn
LNR-RGRN-EQN	y = 0.0115x + 627.7	y = 0.0115x + 627.7	y = 0.0116x + 627.7	y = 0.0044x + 751.4	y = 0.0089x + 751.37	y = 0.0098x + 751.36	y = 0.0089x + 759.48	y = 0.0061x + 759.5	y = 0.0129x + 759.45	y = 0.032x + 127.69	y = 0.0259x + 127.73	y = 0.0318x + 127.68
RGRN-LNE-$_{SLPE}$	0.0115	0.0115	0.0116	0.0044	0.0089	0.0098	0.0089	0.0061	0.0129	0.032	0.0259	0.0318
RGRN-$_{Y\text{-}INTRCPT}$	627.7	627.7	627.7	751.4	751.37	751.36	759.48	759.5	759.45	127.69	127.73	127.68
RLSHP	P (+VE)	P (+VE)	P (+VE)	P (+VE)	P (+VE)	P (+VE)	P (+VE)	P (+VE)	P (+VE)	P (+VE)	P (+VE)	P (+VE)
R^2	0.2727	0.2727	0.2727	0.1869	0.3880	0.3805	0.1947	0.1327	0.5002	0.9977	0.8501	0.9996
Results	MxMn ≈ M.C.T. ≈ MnMn			MxMn > MnMn > M.C.T			M.C.T. > MxMn > MnMn			M.C.T. > MxMn > MnMn		

From Table 4, the following observations regarding the Average Cost (A.C.) can be made:

- ↑ VM = ↑ A.C.: With the increase in VMs in each scenario, the A.C. gradually increases for all the algorithms.
- MxMn ≈ M.C.T. ≈ MnMn: The performance of all algorithms is similar for executing 30 tasks.
- Performance (MxMn) > Performance (MnMn) > Performance (M.C.T.) for executing 50 tasks.
- Performance (M.C.T.) > Performance (MxMn) > Performance (MnMn) for executing 100 tasks.
- Performance (M.C.T.) > Performance (MxMn) > Performance (MnMn) for executing 1000 tasks.

4 Improving Resource Scheduling Using Intelligence Mechanism

This section includes proposing an intelligence mechanism for improving the Resource Scheduling mechanism. Table 5 represents the performance comparison of the experiment conducted.

From Table 5, we can observe that when the CT - 30 tasks are processed and executed, the MxMn, M.C.T., and MnMn resource scheduling algorithms behave similarly. Therefore, any one of them can be implemented for smaller task sizes. When the CT - 50 tasks are submitted for processing, the M.C.T. algorithm provides the optimal output concerning the A.S.T. and A.C.T., whereas the MxMn algorithm provides the optimal results concerning the A.T.A.T. and A.C. With CT - 100 tasks, the MnMn algorithm can be opted concerning A.S.T. or A.C.T., whereas the M.C.T. algorithm can be opted concerning the A.T.A.T. and A.C. With a large task size of CT - 1000, the MnMn algorithm gives an optimal performance concerning the A.S.T. and A.C.T., whereas the

Table 5. Performance comparison of Resource Scheduling Algorithms

Parameter	CT - 30	CT – 50	CT – 100	CT – 100
A.S.T	MxMn. ≈ M.C.T. ≈ MnMn	1. M.C.T 2. MnMn 3. MxMn	1. MnMn 2. M.C.T 3. MxMn	1. MnMn 2. MxMn 3. M.C.T
A.C.T	MxMn. ≈ M.C.T. ≈ MnMn	1. M.C.T 2. MnMn 3. MxMn	1. MnMn 2. M.C.T 3. MxMn	1. MnMn 2. M.C.T 3. MxMn
A.T.A.T	MxMn. ≈ M.C.T. ≈ MnMn	1. MxMn 2. MnMn 3. M.C.T	1. M.C.T 2. MxMn 3. MnMn	1. MxMn 2. MnMn 3. M.C.T
A.C	MxMn. ≈ M.C.T. ≈ MnMn	1. MxMn 2. MnMn 3. M.C.T	1. M.C.T 2. MxMn 3. MnMn	1. M.C.T 2. MxMn 3. MnMn

MxMn and M.C.T. algorithms provide the best performance in terms of A.T.A.T. and A.C., respectively.

Therefore, we can observe that no single resource scheduling algorithm can be im plemented and used across all the scenarios and performance parameters. The resource scheduling algorithm that outputs the best results in a certain performance parameter fails to deliver similar results in other performance parameters. The primary explana tion for this is that these scheduling algorithms are initially fixed statically and imple mented for all the performance parameters. Hence, there is a need for an intelligence mechanism that will dynamically select a resource scheduling algorithm according to the current situation and a certain performance parameter. The Reinforcement Learning (RL) [1, 2] method is famous for making any system dynamic and improving its over all performance. The working mechanism of this RL method is feedback based, and the system enhances itself over time. This RL technique can be applied to the cloud system to enhance the resource scheduling mechanism and optimal output results. When the RL technique is used to optimize cloud performance, the cloud will learn initially. How ever, using the right feedback mechanism, the cloud will adapt to choose the best re source scheduling algorithm for each given situation time.

5 Conclusion

Cybershake tasks of various sizes were computed using the scheduling algorithms MxMn, M.C.T., and MnMn to compare their behavior under various performance pa rameters. This research study includes experimenting in the cloud simulation of Work flowSim, where these tasks were computed in four different scenarios. From this con ducted experiment, obtained results, and empirical analysis (EA) performed on the re sults, it has been noticed that no resource scheduling algorithm outperforms under all the scenarios and all the performance parameters. The primary cause for this is that the cloud system initially fixes these resource scheduling algorithms and uses the same algorithm to compute all its tasks. To make this entire process dynamic and provide consistent results under all performance parameters, the Reinforcement Learning (RL) method is proposed. Unlike other Machine Learning (ML) techniques, the working of RL is completely based on feedback. Also, the main advantage of embedding RL in the cloud is that the cloud will not require any past or labeled data. Using RL, the cloud will initially learn and slowly adapt to the external conditions in its environment. Over time, the cloud, with appropriate feedback, will learn and opt for a specific scheduling algorithm concerning the current scenario. Over time, with a series of feedbacks, the cloud will perform the resource scheduling process effectively, thereby becoming ca pable of computing more tasks, ultimately improving the overall cloud performance.

References

1. Anoushee, M., Fartash, M., & Akbari Torkestani, J.: An intelligent resource management method in SDN based fog computing using reinforcement learning. Computing (2023)
2. Shahidani, F.R., Ghasemi, A., Haghighat, A.T., Keshavarzi, A.: Task scheduling in edge-fog-cloud architecture: a multi-objective load balancing approach using reinforcement learning algorithm. Computing **105**, 1337–1359 (2023)
3. Shaw, R., Howley, E., Barrett, E.: Applying reinforcement learning towards automating energy efficient virtual machine consolidation in Cloud Data Centers. Inf. Syst. **107**, 101722 (2022)
4. Zheng, T., Wan, J., Zhang, J., Jiang, C.: Deep reinforcement learning-based workload scheduling for Edge Computing. J. Cloud Comput. **11**(1), 3 (2022)
5. Alsadie, D.: A metaheuristic framework for dynamic virtual machine allocation with opti mized task scheduling in Cloud Data Centers. IEEE Access **9**, 74218–74233 (2021)
6. Swarup, S., Shakshuki, E.M., Yasar, A.: Task scheduling in cloud using deep reinforcement learning. Procedia Comput. Sci. **184**, 42–51 (2021)
7. Guo, X.: Multi-objective task scheduling optimization in cloud computing based on Fuzzy Self-defense algorithm. Alexandria Eng. J. **60**(6), 5603–5609 (2021)
8. Gonzalez, C., Tang, B.: FT-VMP: fault-tolerant virtual machine placement in cloud data centers. In: 2020 29th International Conference on Computer Communications and Net works (ICCCN) (2020)
9. Caviglione, L., Gaggero, M., Paolucci, M., Ronco, R.: Deep reinforcement learning for multi-objective placement of Virtual Machines in cloud datacenters. Soft Computing 25(19), 12569–12588, 2020
10. Han, S.W., Min, S.D., Lee, H.M.: Energy efficient VM scheduling for Big Data Processing in cloud computing environments. J. Ambient Intell. Humanized Comput. (2019)
11. Bitsakos, C., Konstantinou, I., Koziris, N.: Derp: A deep reinforcement learning cloud system for elastic resource provisioning. In: 2018 IEEE International Conference on Cloud Computing Technology and Science (CloudCom) (2018)
12. H. A. M. N. Balla, C. G. Sheng, and J. Weipeng.: Reliability enhancement in cloud com puting via optimized job scheduling implementing reinforcement learning algorithm and queuing theory. 2018 1st International Conference on Data Intelligence and Security (ICDIS), 2018
13. Cheng, M., Li, J., Nazarian, S.: DRL-cloud: deep reinforcement learning-based re source provisioning and task scheduling for cloud service providers. In: 2018 23rd Asia and South Pacific Design Automation Conference (ASP-DAC) (2018)
14. Madni, S.H.H., Abd Latiff, M.S., Abdullahi, M., Abdulhamid, S.I.M., Usman, M.J.: Performance comparison of heu ristic algorithms for task scheduling in IaaS cloud computing environment (2017)

15. Zhou, X., Wang, K., Jia,W., Guo, M.: Reinforcement learning-based adaptive resource management of differentiated services in geo-distributed data centers. In: 2017 IEEE/ACM 25th International Symposium on Quality of Service (IWQoS) (2017)
16. Hussin, M., Hamid, N.A.W.A., Kasmiran, K.A.: Improving reliability in resource management through adaptive reinforcement learning for distributed systems. J. Parallel Distrib. Comput. **75**, 93–100 (2015)
17. Chen, W., Deelman, E.: WorkflowSim: a toolkit for simulating scientific workflows in distributed environments. In: 2012 IEEE 8th International Conference on E-Science (2012)
18. Armbrust, M., et al.: A view of cloud computing. Commun. ACM **53**(4), 50–58 (2010)
19. Dillon, T., Wu, C., Chang, E.: Cloud computing: issues and challenges. In: 2010 24th IEEE International Conference on Advanced Information Networking and Applications (2010)
20. Vengerov, D.: A reinforcement learning approach to Dynamic Resource Allocation. Eng. Appl. Artif. Intell. **20**(3), 383–390 (2007)

A Recommendation Model System Using Health Aware- Krill Herd Optimization that Develops Food Habits and Retains Physical Fitness

N. Valliammal(✉) and A. Rathna(✉)

Department of Computer Science, Avinashilingam Institute for Home Science and Higher Education for Women, Coimbatore, India
vallinarayancbe@gmail.com, rathu2001dhanam@gmail.com

Abstract. Major contributors to a healthy lifespan are regular physical activity and a nutritious diet. Although ageing is linked to decreased muscular function and endurance, these factors can be positively altered by regular physical activity, especially exercise training at all ages and in both sexes. Additionally, older people's altered body composition and metabolism, which can happen to even highly trained athletes, can be somewhat offset by active people's increased exercise metabolic efficiency. As a result, sportsmen are frequently cited as examples of good ageing, and their outstanding physical ability serves as a proof to what is still achievable as people age. This is due to their physical activities and the intake of food with high nutrient content. This paper provides a nutrient recommendation to the individual to improve their food habits and maintain the physical condition using the proposed Recommendation system using the Health Aware- Krill Herd Optimization (Recsys-HA-KHO). The Resys utilize HA-KHO to provide an efficient recommendation to consume nutritious food based on their Physical Activity (PA). The results obtained from the proposed HA-KHO is compared with other optimization algorithms based on the parameters likeprecision, recall, F1 score, Area Under Curve (AUC) and Discounted Cumulative Gain (DCG). The results show that the proposed HA-KHO obtained highest precision of 93.54% and the recall value of 97.32% .

Keywords: Health aware- krill herd optimization · nutrition · physical activity · recommendation system · sports

1 Introduction

Sport is characterized as an activity that depends on participants' physical fitness and is connected to athletes' athletic performance and health. It is possible to highlight nutrition, food, body composition, hydration, supplements, and doping among these indicators [1]. Post-workout nutritional advice is essential for the efficiency of recovery

S. Aurelia et al. (Eds.): ICCSST 2023, CCIS 1973, pp. 183–199, 2024.
https://doi.org/10.1007/978-3-031-50993-3_15

and adaptation processes because nutrition is one of the pillars of athletic performance. In order to maximize adaptive responses to different mechanisms of exhaustion, an efficient recovery plan between workouts or during competition is therefore necessary [2]. This improves muscle function and raises exercise tolerance. In order to boost physical working capacity during the pre-competition period and produce better results in sporting competitions, an athlete's diet should be linked with their training regimen [3]. The anaerobic and aerobic energy systems are both necessary for the outstanding speed, power, endurance, and explosiveness that sports players must possess. To power these systems, one must consume enough calories and carbohydrates [4, 5]. A fundamental procedure is regarded to be an effective intervention to restore an athlete's physical fitness through regimen and diet monitoring, prompt admission, and the intake of food components of a certain quality and amount [6, 7]. Inadvertent undereating, information gaps, food insecurity, time restraints, restricted dietary practices, and overt eating disorders like anorexia nervosa are just a few of the diseases that can cause these deficiencies [8]. The Athlete Diet Index (ADI) was created especially for elite athletes and is a valid and trustworthy diet quality assessment instrument. It quickly assesses athletes' typical dietary consumption and compares results to recommendations for sports nutrition and the general population [7, 9]. The most notable distinction from the user's perspective is that meal recommendations are highly relevant to users' health [10, 11]. The nutrition system based on gluten-free diets has a significant benefit in providing a balanced diet to sports person [12]. Therefore, an ideal food recommendation system should self-adaptively build a tradeoff between personalized food preference/interest and personalized nutrition/health requirement [13, 14]. Hence, this paper provides with an improvised nutrition recommendation system using the Krill herd algorithm.

The major contributions of the paper are listed below,

(a) The proposed HA-KHO Recsys attributes an efficient recommendation system to provide a balanced diet to the user based on their physical activities
(b) The HA-KHO is employed with the real-time dataset to evaluate the efficiency of the proposed frame work.
(c) The proposed Recsys is compared with the existing methods in the related works based on the value of precision, recall, F-1 score, AUC and DCG.

2 Related Works

Meal Recommender System for Athletes and Active Individuals by Mustafa, Norashikin (2020)

- The developed web iDietScore that provides a diet plan to every individual user through the mobile which helped them in computing the balanced diet. (16)

A Novel Time-aware Food recommender-system based on Deep Learning and Graph Clustering by Rostami, M., Oussalah, M. and Farrahi, V (2022)

- They introduced a novel time aware recommendation system which was based on two phases such as recommendation based on the content and the user by deep learning and Clustering. (!7)

A food recommender system considering nutritional information and user preferences by Toledo, R. Y., Alzahrani, A. A., & Martinez, L. (2019)

- They introduced a food recommendation system on the nutritional data that provides day-to-day plan for the nutrient meal on the basis of user need. (!8)

Kiriakos Stefanidis et.al [18] have introduced a recommendation system to certify the food plan to provide healthy diet for the user. A qualitative layer is used to assess the suitability of ingredients, while a quantitative layer is used to combine meal plans in the recommendation system. Using an ontology of rules amassed by nutrition experts, the first layer is implemented as an expert system for fuzzy inference, and the second layer is an optimization approach for creating daily meal plans based on goal nutrient values and ranges. Moreover, the system creates an appropriate recommendation for macronutrients and provides a balanced diet for the users. But, the system is suited for global varieties and not for traditional food varieties.

Jieyu Zhang et.al [19] introduced a Many-Objective optimization (MaOO) based method of making recommendations that has been established to offer a fair and organized way of handling food suggestion chores. In the suggested suggestion technique, four important goals user choice, diet pattern, food nutritional values, and food diversity have all been taken into account at once. The presented recommendation task was solved using three Pareto-based algorithms, and extensive experiments using real-world data sets were carried out to confirm the viability of the suggested MaOO-based recommendation system. Although, the system does not provide specified recommendation for the patients suffered from various types of diseases.

Resma, et.al [20]. Introduce a new method to resolve multilevel thresholding using a bio-inspired computing paradigm such as Krill Herding Optimization (KHO) algorithm. The Krill Herd Optimization algorithm represents one of the innovative clustering intellectual methods depend on the modelling of Krill independent grouping behaviour and attitude.

3 Proposed Method

RecSys (Recommender System), is a frame work utilized for the recommendation based on a multi objective optimization problem using Krill Herd (KH) optimization, is introduced in this section. RecSys's recommendations for the end user are demonstrated byitems or groups of items in this framework. According to this method, it is conceivable to develop recommendations that are I geared toward achieving a particular well-being goal that the user has chosen and (ii) take into account the preference of the user. The structure for generating recommendation provide a balance between the user's tastes and the kinds of food and exercise the user should take into account for achieving a goal. The process of recommendation system is represented in Fig. 1.

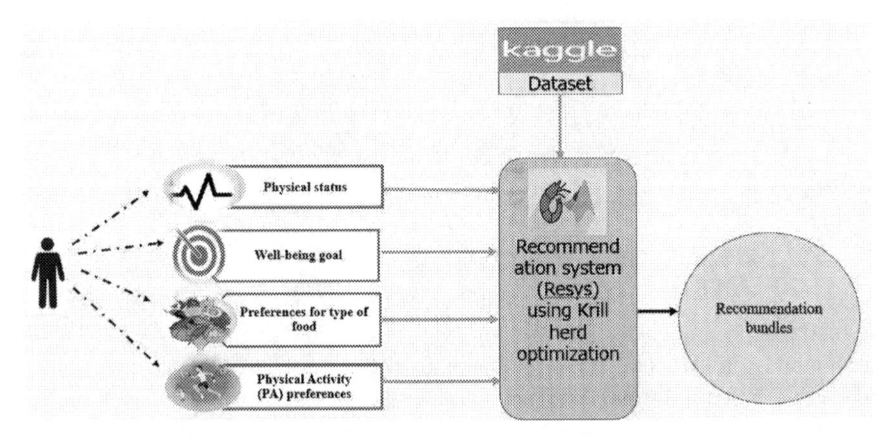

Fig. 1. The process in nutrition recommendation system

3.1 Architecture of RecSys

Through an evolutionary optimization process, RecSys uses the input of the user (attributes, well-being objectives, and nourishment and preferences for workout) to recommend meal and physical activity (PA) bundle which is customized to individual. RecSys' primary inputs are broken down into four categories of data related to user.

(i) **Physical condition and exercise routines**: Information about the user's age, gender, and physical measurements, as well as how often they exercise.
(ii) **Preferences for food categories**: How much the user like particular components, particular food varieties, etc. A numerical scale, such the 5-point Likert scale, can be utilized for this. This element can be utilized in a variety of ways due to its adaptability. For instance, it could concentrate on a particular nutritional need like veggies, fruitarians, or persons with definite sensitiveness.
(iii) **Preferences for various forms of exercise**: Data describing the user's level of preference for various PA types. It can be instigated focused on a programmed

corporal doing, just like the previous component. For instance, aquatic sports or ball-based athletics. A numerical scale based on minimum-maximum values, related to the prior component, is used to evaluate the user preferences.

(iv) **The aim of wellbeing**: a goal that the user selects from a list of predefined goals that is targeted at a particular area of well-being that has to be addressed. Depending on the specific aims of the model built upon this broad framework, the goals can be set for dealing with either general conditions (weight reducing) or specific ones (such as diseases).

The framework cooperates with the source of data which includes less data onnutriment and PA while the evolutionary suggestion process is underway. The compiled user interactions and response across interval, previous user preferences are also required. The data source can be any easily available database (Kaggle), as long as it has the information required to carry out the evolutionary recommendation, according to the framework's decoupled architecture. The framework generates a K-item suggestion list that is customized to the user's tastes and the selected well-being objective.

3.2 Krill Herd Algorithm for Optimization

The Krill is known as best species which lives in large group around marine habitat. The Krill herd algorithm is inspired from the swarm of krill which normally lives in a group, but when they are visible to the predators like seals, birds etc. they might kill the individual krill from the group and leads to decrease in density of the krill. Increasing krill density and obtaining food are the two fundamental objectives of the multi-objective process of krill herding [20]. This technique is taken into consideration in the current work to suggest a new metaheuristic algorithm for resolving global optimization issues. Krill are drawn to areas of higher food concentration due to their density-dependent attraction, which eventually causes them to herd around minimal population. Here, each krill tends to locate themselves to the best region to mingle with high density population and availability of food.

When the individual krill is removed by the predators, and relies as the reason in decreased density of Krill and separates the krill population in search of food.So, the position of every individual krill is considered based on dependent of time by the three steps mentioned below,

- Motion by individual krill
- Activity based on foraging
- Random diffusion

The overall equation of KH algorithm on dimensional space based on decision is represented using the Eq. (1). $^{dX}\underline{i}$

$$\frac{dx_i}{dt} = N_i + F_i + D_i \tag{1}$$

where, the motion of an individual krill is denoted as N_i, the foraging motion is denoted as F_i, and the random diffusion is denoted as D_i.

3.2.1 Motion by Individual Krill

The individuals present in the swarm of krill tries to present among the high density and locate in search of their food. The motion of krill in the particular direction is computed from the density of krill in the local swarm, density of the krill in the target swarm, density of the krill in repulsive swarm. This is mathematically represented using the Eq. (2) mentioned below,

$$N_i^{new} = N^{max}\alpha_i + \omega_n N_i^{old} \tag{2}$$

where,

$$\alpha_i = \alpha_i^{local} + \alpha_i^{target} \tag{3}$$

In the Eq. (2), the maximum speed of the individual krill is represented as N^{max}, the weight of the motion performed by individual krill in range [0,1] is denoted as ω_n and the final motion of the krill is denoted as N_i^{old}. In Eq. (3), the action of the local swarm of krill is denoted as α_i^{local} and the direction to the target shown by the best among the krill swarm is denoted as α_i^{target}.

In the local search, the neighbor's action as attractive and repulsive among the individuals is represented using the Eq. (4)

$$\alpha_i^{local} = \sum_{j=1}^{NN} K_{i.j} X_{i.j} \tag{4}$$

The value of $K_{i.j}$ and $X_{i.j}$ is represented in the below Eq. (5).

$$K_{i.j} = \frac{K_i - K_j}{K^{worst} - K^{best}} \text{ and } X_{i.j} = \frac{x_j - X_i}{\|x_j - X_i\| + \varepsilon} \tag{5}$$

The best and worst fitness values of the individuals are represented as K^{best} and K^{worst} respectively. The fitness value of the individual krill present in the swarm is represented as K_i and the fitness value of the neighboring krill is represented as K_j. The related positions of the krill are denoted as X and the total count of neighboring krill is denoted as NN. A small positive value, $\varepsilon$is summed to the denominator to neglect the singularities.

The direction of the target's effect on the best individual present in the krill swarm is represented using the Eq. (6).

$$\alpha_i^{target} = C^{best} K_{i,best} X_{i,best} \tag{6}$$

The effective coefficient of the individual present in the krill swarm with best value of fitness is represented using the Eq. (7).

$$C^{best} = 2\left(\text{rand} + \frac{I}{I_{max}}\right) \tag{7}$$

where, the random values among 0 and 1 is denoted as rand, the iteration and the maximum count of iterations are represented as I and I_{max} respectively.

For choosing the neighbor, Several methods can be applied. For example, a neighborhood ratio is used to defined the number of the closest krill individuals. Using the certain practices of the krill individuals, a sensing distance (d_s) could be determined by the surroundings for a krill individual (as shown in Fig. 2) and the neighbors should be identified.

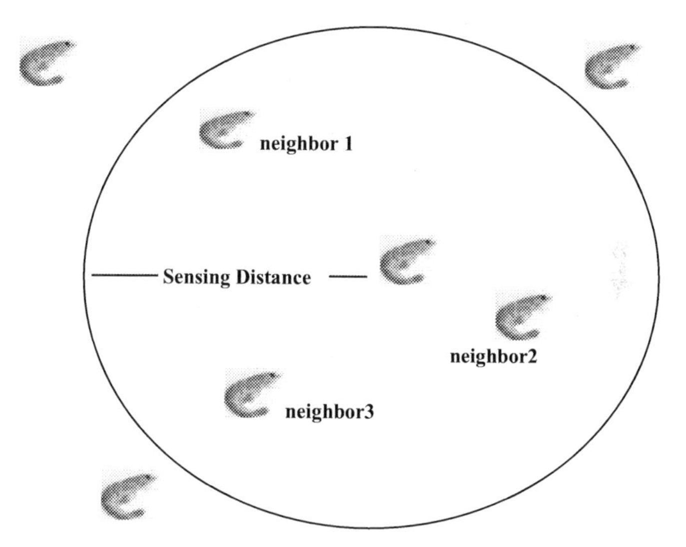

Fig. 2. A schematic presentation of the sensing ambit

The sensing distance of every individual krill is computed using heuristic methods and in the proposed work, the sensing distance is calculated using the formula in Eq. (8)

$$d_{s,i} = \frac{1}{5N} \sum\nolimits_{j=1}^{N} \|X_i - X_j\| \tag{8}$$

where, the distance for sensing the ith individual present in the krill swarm is represented as $d_{s,i}$ and the count of individuals present in the krill is denoted as N.

3.2.2 Activity Based on Foraging

The motion of foraging is based on two parameters such as location of the food and experience obtained from searching food. The following Eq. (9) is used to represent the motion of the ith individual present in the krill swarm.

$$F_i = V_f \beta_i + \omega_f F_i^{old} \tag{9}$$

where, $\beta_i = \beta_i^{food} + \beta_i^{best}$ and the speed of foraging is denoted as V_f. . The inertial weight of the motion during foraging in the range [0,1] is represented as ω_f. The food which appears attractive for the swarm is denoted as β_{ifood} and the best fitness of the individual krill in the swarm is denoted as β_i^{best} .

The middle of the food should be identified initially to formulate the attraction for food and the middle of the food can't be determined but it can be predictable. Here, the food center for the individual krill is represented using the formula mentioned below in Eq. (10).

$$X^{food} = \frac{\sum_{i=1}^{N} \frac{1}{K_i} X_i}{\sum_{i=1}^{N} \frac{1}{K_i}} \tag{10}$$

So, the attraction of food for the ith individual krill is represented in Eq. (11) as,

$$\beta_i^{food} = C^{food} K_{i,food} X_{i,food} \tag{11}$$

where the coefficient of food is denoted as C^{food}. Since the attraction for food in the swarm of krill gets reduced, the coefficient of the food is distinguished in Eq. (12) as,

$$C^{food} = 2\left(1 - \frac{I}{I_{max}}\right) \tag{12}$$

3.2.3 Physical Diffusion

The diffusion occurs in the individuals of the krill population is a process which takes place in a random manner. Here, the physical diffusion motion is determined in terms of maximum speed of diffusion and vector based in random direction. The linear reduction in the random speed based on time and works is determined using the Eq. (13) represented below,

$$D_i = D^{max}\left(1 - \frac{I}{I_{max}}\right)\delta \tag{13}$$

where, the maximum speed of diffusion is denoted as D^{max} and the random directional vector between the range $[-1,1]$ is represented as δ.

3.2.4 Process in Krill Herd Algorithm During Movement

The position vector of the individual krill between the range $[t, t+\Delta t]$ is attributed on the basis of parameters related to movement of the krill swarm. According to the mentioned range, the movement of the individual among the swarm of Krill is determined using the Eq. (14).

$$X_i(t+\Delta t) = X_i(t) + \Delta t \frac{dx_i}{dt} \tag{14}$$

where Δt is the constant which relies on space for search and the value of Δt is attainedusing the equation represented in the Eq. (15)

$$\Delta t = C_t \sum_{j=1}^{NV} (UB_j - LB_j) \tag{15}$$

The total count of variables is denoted as NV, upper and lower bounds present in the jth variable is denoted as UB_j and LB_j respectively. The $C_t$is known as a constant among the range [0,2]. Moreover, when the C_t values are low, the individuals present in the krill swarm attains an opportunity to perform space search.

3.3 Genetic Operators

The overall performance of the algorithm is improvised using the mechanisms based on genetic reproduction. The performance gets improvised using two stages such as mutation and crossover combined with krill herd algorithm.

3.3.1 Crossover

The operator called crossover is generally utilized to provide an efficient and effective optimization. The cross over operator is usually controlled using the probability of the crossover, C_r, actual crossover. The crossover operator performs crossover in two phases such as binomial and exponential. The crossover is performed for individual elements d by creating a distributed random number uniformly among the range [0,1] and it is represented using the Eq. (16)

$$X_{i,m} = \begin{cases} X_{r,m} \text{ rand }_{i,m} < C_r \\ X_{i,m} \text{ else} \end{cases} \tag{16}$$

where, $C_r = 0.2K_{i,best}$ and by utilizing the obtained probability from the crossover operator, the fitness value gets improvised.

3.3.2 Mutation

The process of mutation has significant role in providing the probability of mutation (Mu) and the efficient scheme for mutation is provided using the formula represented in Eq. (17).

$$X_{i,m} = \begin{cases} X_{gbes,m} + \mu\left(X_{p,m} - X_{q,m}\right) \text{ rand }_{i,m} < M_u \\ X_{i,m} \text{ else} \end{cases} \tag{17}$$

where $M_u = 0.05/K_{i,best}$ and μ is known as the constant which lies among the range [0,1]. By utilizing the obtained probability from the mutation operator, the fitness value gets improvised.

3.4 Recommendation System (Recsys) Using KHO Algorithm

The major steps in KHO algorithm is utilized for evolution in the architecture of proposed Recsys and the recommendation of the individual are initialized among various search spaces. The created individual should equal with the preference of the user, within the boundaries of the defined input, and correlate among the food and physical activity (PA).

3.4.1 Dataset in the Architecture of Recsys

In architecture Recsys, the inputs are based on the physical status of the person, his exercising habits and goal of well-being. Here, the data is collected from the Kaggle dataset where the data about food preference on various types of food items and physical workouts distributed to various tasks are collected [21]. The data from the Kaggle is used to find the preference of the user at the collaborative stage for filtering and output of the Recsys is computed for the individual based on bundle recommendations. The data from the Kaggle on food preference can be accessed using the link as follows: https://www.kaggle.com/datasets/vijayashreer/food-preferences.

3.4.2 Creating Individuals

In the proposed architecture, there are about four types of food items and to attribute diverse and non-repeating recommendations, there are two type of food in each meal. The types of meal include solitarychief food and three lateral foods. An assumption is made that the individual has three bundles, (k = 3).

A related parameter that correlates among the meal and physical activity is used to assist the user to inform the calorie intake on a day. The health and nutrition based predictions are performed based on the Basal Metabolic Rate (BMR) which is represented for male and female in Eq. (18) and (19) respectively.

$$\text{BMR(male)} = 655.1 + (9.563 \times \text{ weight }) + (1.850 \times \text{ height })(4.676 \times \text{ age }) \quad (18)$$

$$\text{BMR(female)} = 66.5 + (13.75 \times \text{ weight }) + (5.003 \times \text{ height })(6.755 \times \text{ age }) \quad (19)$$

The average BMR for male and female is given as 1600–1800 and 1400–1550 cal/day respectively.

Where, the weight is calculated by kilograms, age is years and height is measured in centimeters. The product of BMR and the Level of Physical Activity (LPA) gives the value of Total energy expenditure (TEE). The TEE is evaluated using the formula represented in Eq. (20).

$$\text{TEE} = \text{BMR} \times \text{PAL} \quad (20)$$

For example, if the person reduced the calorie intake to 551.26 in day-to-day TEE, he will lose his about 500g of the total weight in 7 days approximately. The TEE value is utilized in determining the preference of the user according to therequirements based on nutrition value.

3.4.3 Selection

The two people are compared based on their aptitude scores, and the person with the highest aptitude is chosen to be a part of the new population. Until a new child population is generated that is the same size as the parent population, the process is repeated N times. The population's selection pressure is directly influenced by the tournament size T. The best parent population member is expected on average, T = 2 offspring in the new population given and it is utilized for tournament of size T = 2, while median parent population is expected on average, T /2 = 1 offspring and the lowest member is assured to have none. The best individual from each group is produced as offspring from the population to prove that better solutions are obtained from the process of reproduction.

3.4.4 Crossover and Mutation in Recsys

At the element level of each bundle, there are stochastic processes for crossover and mutation to take place. Each element is chosen in the crossover and mutation processes using the following procedure: The following steps are taken on individual bundle: (ii) the bundle is selected with probability $p0 = 0.9$; (ii) one of the bundle's elements—main food, side food, or physical activity is chosen with probabilities of (0.2, 0.6, and 0.2; and (iii) if Side is chosen, each sub-element is chosen with probability 0.5, allowing for the probability of choosing multiple sides.A copy of parent A is made as the offspring, and using the crossover procedure mentioned above, each element chosen for parent B is then introduced into the offspring. The element or sub-element chosen is substituted by the comparable element in surrounding user when a mutation occurs after "similar" neighbors have been identified via collaborative filtering on user preferences.

3.4.5 Directing Recsys Using Collaborative Filtering Method

The filtering based on collaborative method is generally utilized in recommending preference for the user. For instance, if users u_a and u_b are similar and u_b has given an item $X_j$that u_a hasn't seen yet a positive rating or like, $X_j$is suited to recommendu_a. As a result, the suggested model integrates a method that finds users that are similar to the target consumers using collaborative filtering in KH algorithm.

To obtain efficient Recsys, the combined similarity $Sim(u_a, u_b)$among the user is computed based on (a) preference for the food, (b) preference of the physical activity, (c) physical status of the user and (d) goal of well being. Assume that $FT = \{ft_1, ft_2, \ldots\}$ is considered as non-empty set which contains finite number of food types. The vector for the preference of $u_a$for various types of food is denoted as $P_a^{ft} = \left[P_a^{ft_1}, P_a^{ft_2}, \ldots P_a^{ft|FT|}\right]$. The similarity among the users u_a, u_b is formulated using the Eq. (21).

$$sim^{ft}(u_a, u_b) = 1 - d\left(P_a^{ft}, P_b^{ft}\right) \tag{21}$$

where $d\left(P_a^{ft}, P_b^{ft}\right)$ is known as the distance among two vectors u_a and u_b.

The non-empty set of finite type of physical activities are represented as $AT = \{at_1, at_2, \ldots\}$ and the vector for the preference of $u_a$for the physical activities are represented as $P_a^{at} = \left[P_a^{at_1}, P_a^{at_2}, \ldots P_a^{at|AT|}\right]$. The similarity among the users u_a, u_b based on physical activities are formulated using the Eq. (22).

$$sim^{at}(u_a, u_b) = 1 - d\left(P_a^{at}, P_b^{at}\right) \tag{22}$$

The physical condition of the user is computed based on the parameters such as height of the individual, individual's age, weight of the individual and based on gender. Initially, the consideration is taken as $status_a = [\text{weight (kg), height (cm), age (years), gender}]$. The physical status of two users with similar levels of expenditure rate of calories and the correlation between the consumption of food and the requirement of physical activities should be similar in the bundle. Based on the value obtained from Eq. (20), the similarity among the physical stage of the user is represented in the below Eq. (23).

$$sim^{at}(u_a, u_b) = 1 - \frac{|TEE_a - TEE_b|}{TEE_{max} - TEE_{min}} \tag{23}$$

At final stage, a function η_w is utilized to associate three similarities based on the preference of user, preference to the physical activities and physical conditions of the user. The combination of mentioned parameters is represented in Eq. (24).

$$sim(u_a, u_b) = \eta_w\left(sim^{ft}(u_a, u_b), sim^{at}(u_a, u_b), sim^{st}(u_a, u_b)\right) \tag{24}$$

where the vector to specify the significance on preference of food, preference for the physical activities and physical condition of the user is represented as w. At final stage. The goal of well-being is computed for two users who are seems to be nearest neighbors. The equation for computing the goal of well-being is represented in Eq. (25)

$$sim'(u_a, u_b) = \begin{cases} \sqrt{sim(u_a, u_b)} when\ equal\ goal\ of\ well\ being\ among\ u_a\ and\ u_b \\ sim(u_a, u_b)\ otherwise \end{cases} \tag{25}$$

where the range of $sim(u_a, u_b)$ lies among the range [0,1] and $\sqrt{sim(u_a, u_b)} \geq sim(u_a, u_b)$. The data regarding the preference of the user is utilized to assist the operator for the process of mutation and helps to provide recommendations based on space of search.

3.5 Evaluation of Fitness Value

The fitness function of the proposed Recsys is employed based on three parameters such as restrictions to the healthy food, restriction for physical activity and restrictions based on diversity and consistency.

- The healthy food restrictions are utilized in computing the nutrient content such as carbs, proteins, fat, fiber etc. in the food.
- The restrictions to physical activities is based on the recommendation time in PA where the user tends to spend time in specific exercise. The metabolic equivalent is considered as the value for reference to compute the combination of food and physical activities of the user goal.
- Restriction on consistency and diversity is computed based on two ways, the evaluation of diversity with a single meal and evaluation of diversity within the individual. The values of restrictions are computed in terms of size of serving and the proportion of nutrients present in each meal.

The above-mentioned restrictions are evaluated to attribute the fitness function, the set of restrictions is evaluated as $\rho = \{\gamma_{hf}, \gamma_{pa}, \gamma_{cd}\}$ The fitness value FF is calculated according to the user u_i accompanied to achieve the goal ζ.

is determined on the basis of ρ and the preference of the individual ϑ. The fitness function is evaluated using the formula mentioned in Eq. (26).

$$FF = FF(\vartheta, \zeta) = \lambda(\vartheta \cdot F(\gamma_{hf}), F(\gamma_{pa}), F(\gamma_{cd}) \tag{26}$$

4 Results and Analysis

The results and discussion of the HA-KHA is explained in this section. The design and implementation of efficient Recsys is performed using python toolkit version 3.7. The features of the system during the analysis have i5 processor with RAM capacity of 6 GB. Here, the analysis of the proposed algorithm is based on two types of analysis such as quantitative and qualitative analysis. The quantitative analysis is performed for various optimization algorithms such as Firefly Optimization (FFO), Particle Swarm Optimization (PSO), Ant Colony Optimization (ACO), Krill Herd Optimization (KHO) and the proposed Health-Aware Krill Herd Optimization (HA-KHO). In qualitative analysis the comparison of the proposed Recsys using HA-KHO is performed with various recommendation systems provided in the literature.

4.1 Quantitative Analysis

In quantitative analysis, the performance of the proposed HA-KHA algorithm is compared with the existing optimization algorithms such as Firefly Optimization (FFO), Particle Swarm Optimization (PSO), Ant Colony Optimization (ACO), Krill Herd Optimization (KHO). The performance of the proposed HA-KHA is evaluated based on performance metrics such as precision, recall, F1 score, Area Under Curve (AUC) and Discounted Cumulative Gain (DCG).

Precision

The precision is defined as ratio of correctly recommended nutrition to the sum of correct and wrong recommended values of nutrition. The precision value is represented in Eq. (27).

$$precision = TP/(TP + FP) \tag{27}$$

where *TP* refers to true positives and *FP* refers to false positives.

The value of precision obtained from various algorithms including proposed HA-KHA is represented in Fig. 3.

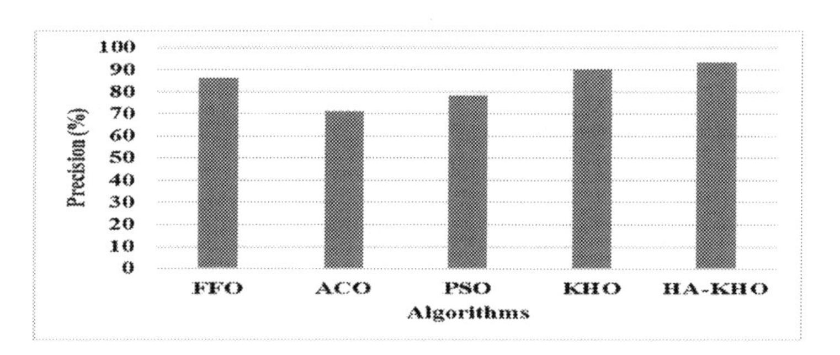

Fig. 3. Graphical representation of precision values

Recall

The ratio of recommended nutrient value attained by user to sum of correctly missed nutrients and the count of correctly recommended nutrients. The recall value is represented in Eq. (28).

$$Recall = TP/(TP + FN) \tag{28}$$

The recall value obtained from various existing algorithms including HA-KHA is represented in Fig. 4.

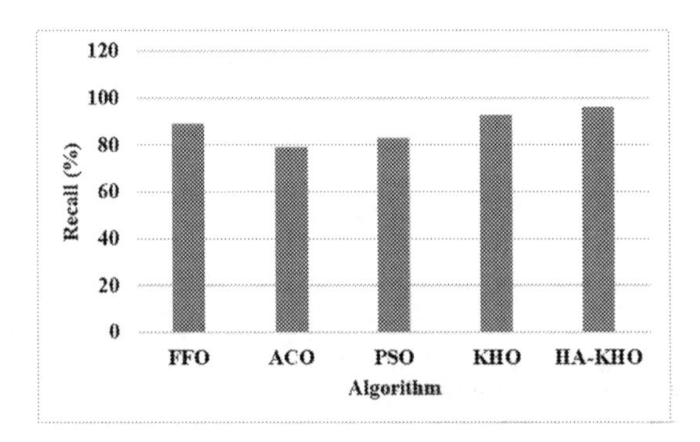

Fig. 4. Graphical representation of recall value

F1-score

The average obtained from recall and precision is defined as F1 score. It is the mean value obtained when the values of recall and precision gets closer to each other. The value of F1 score is considered as high when the values of precision and recall gets high. The F1 score is computed using the following Eq. (29).

$$\text{F1 score} = \frac{2(\text{ precision } * \text{ Recall })}{\text{Precision } + \text{ Recall}} \tag{29}$$

The F1 score obtained from various existing algorithms including HA-KHA is represented in Fig. 5.

Area Under Curve (AUC)

The AUC is defined as the area lies among the curve and the axis. The area may be fully above the axis or fully below the axis or combined form of both below and above axis. The area under the curve respect to x axis is computed using the formula in Eq. (30).

$$\mathrm{A} = \int_{a}^{b} \mathrm{f(x)} \cdot \mathrm{dx} \tag{30}$$

The AUC value obtained from various existing algorithms including HA-KHA is represented in Fig. 6

Discounted Cumulative Gain (DCG)

DCG is defined as the parameter to measure the quality of algorithms based on ranking.

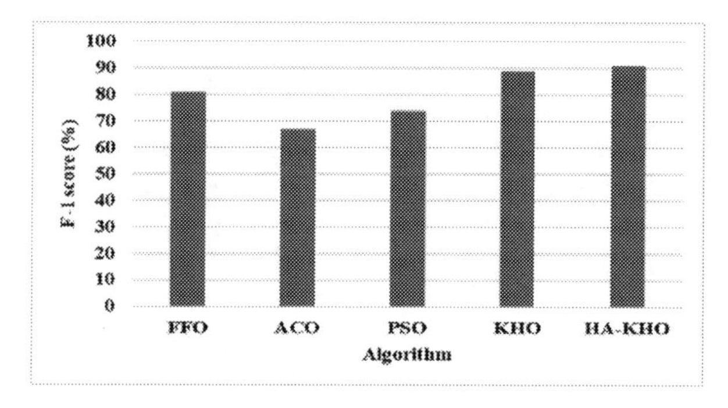

Fig. 5. Graphical representations of F1 score

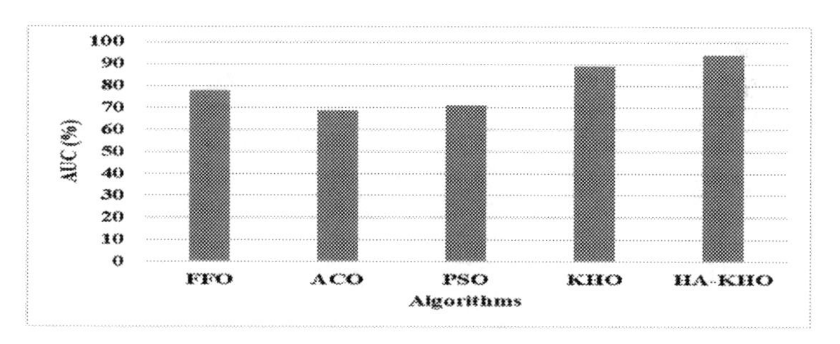

Fig. 6. Graphical representations of AUC

It is utilized in problems based on retrieval of information. The DCG is measured using the formula represented in Eq. (31).

$$\mathrm{NDCG} = \sum\nolimits_{i=1}^{n} (\mathrm{rel})_i \tag{21}$$

The DCG value obtained from various existing algorithms including HA-KHA is represented in Figure 7.

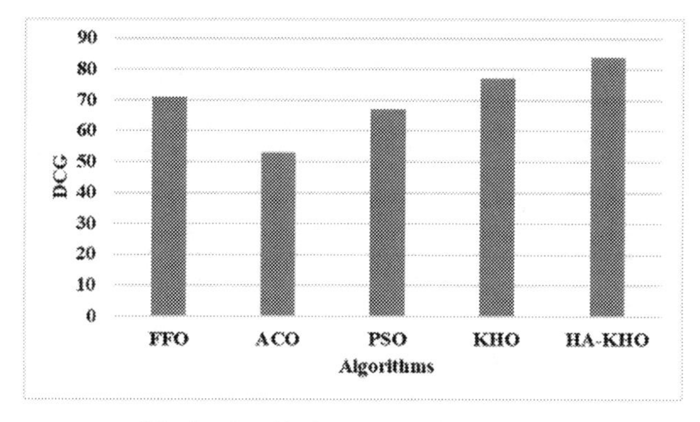

Fig. 7. Graphical representations of DCG

4.2 Qualitative Analysis

The qualitative analysis the performance of the proposed HA-KHO Rec sys is compared with the existing recommendation systems such as Time-Aware Food Recommendation system (TA-F Recsys) [16], Nutrition Recommendation system (Resys) [17], Knowledge based Food Recommendation system (KB- F Recsys) [17] and the proposed HA-KHO Recsys. The performance of the proposed HA-KHA Recsys is evaluated based on performance metrics such as precision, recall, F1 score, Area Under Curve (AUC) and Discounted Cumulative Gain (DCG). The comparative results for the mentioned methods for mentioned parameters are provided in Table.1

Table 1. Comparative Table

Methods	Precision (%)	Re-call (%)	F1 score (%)	AUC (%)	DCG (%)
TA-F Recsys [16]	89.32	91.41	85.65	90.34	80.46
Nutrition Resys (Resys) [17]	85.55	88.21	81.74	87.63	78.29
KB-F Recsys [18]	91.26	93.45	88	92.61	82.73
HA-KHO Resys	93.54	97.32	91.02	95.62	87.18

5 Conclusion

The basic assumption of this study is feasible to establish a stabilityamong three key proportions (i) the preference of the user, (ii) the needs of the user, and (iii) the goal of the user. This balance is achieved by the Recsys for recommendations in health-related domains. The HA-KHO algorithmic approach, which the framework defines, institutes the stability of these three constituentsby a multi-objective optimization. The framework's capability to create vastrecommendations in meal-exercise bundles as opposed to immutable items enables more individualized and dependable recommendations for the user. Moreover, the HA-KHA is utilized in recommendation system to provide an efficient recommendations compared with other methods such as TA-F Recsys [16], Nutrition Resys (Resys) [17], KB-F Recsys [18]. In future, the proposed recommendation can be improvised based on recommending balanced diet to the patients related to medical applications.

References

1. Roberts, B.M., Helms, E.R., Trexler, E.T., Fitschen, P.J.: Nutritional recommendations for physique athletes. J. Hum. Kinet. **71**(1), 79–108 (2020)

2. Hannon, M.P., Close, G.L., Morton, J.P.: Energy and macronutrient considerations for young athletes. Strength Cond. J. **42**(6), 109–119 (2020)
3. Strasser, B., Pesta, D., Rittweger, J., Burtscher, J., Burtscher, M.: Nutrition for older athletes: focus on sex-differences. Nutrients **13**(5), 1409 (2021)
4. Kim, J., Kim, E.-K.: Nutritional strategies to optimize performance and recovery in rowing athletes. Nutrients **12**(6), 1685 (2020)
5. Baranauskas, M., Jablonskienė, V., Abaravičius, J.A., Stukas, R.: Actual nutrition and dietary supplementation in Lithuanian elite athletes. Medicina **56**(5), 247 (2020)
6. Martínez-Sanz, J.M., Menal-Puey, S., Sospedra, I., Russolillo, G., Norte, A., Marques-Lopes, I.: Development of a sport food exchange list for dietetic practice in sport nutrition. Nutrients **12**(8), 2403 (2020)
7. Malsagova, K.A., et al.: Sports nutrition: diets, selection factors, recommendations. Nutrients **13**(11), 3771 (2021)
8. Holtzman, B., Ackerman, K.E.: Recommendations and nutritional considerations for female athletes: health and performance. Sports Med. **51**(1), 43–57 (2021)
9. Kerksick, C.M., et al.: ISSN exercise & sports nutrition review update: research & recommendations. J. Int. Soc. Sports Nutr. **15**(1), 38 (2018)
10. Vermeulen, T.F., Boyd, L.A., Spriet, L.L.: Dietary macronutrient and micronutrient intake over a 7-day period in female varsity ice hockey players. Nutrients **13**(7), 2262 (2021)
11. Capling, L., et al.: Diet quality of elite Australian athletes evaluated using the athlete diet index. Nutrients **13**(1), 126 (2020)
12. Książek, A., Zagrodna, A., Słowińska-Lisowska, M.: Assessment of the dietary intake of high-rank professional male football players during a preseason training week. Int. J. Environ. Res. Public Health **17**(22), 8567 (2020)
13. Min, W., Jiang, S., Jain, R.: Food recommendation: framework, existing solutions, and challenges. IEEE Trans. Multimedia **22**(10), 2659–2671 (2019)
14. Devi, K.R., Bhavithra, J., Saradha, A.: Diet recommendation for glycemic patients using improved K-means and Krill-Herd optimization. ICTACT J. Soft Comput. **10**(03), 2096–2101 (2020)
15. Mustafa, N., et al.: iDietScore^{TM}: meal recommender system for athletes and active individuals. Int. J. Adv. Comput. Sci. Appl. **11**(12), 269–276 (2020)
16. Rostami, M., Oussalah, M., Farrahi, V.: A novel time-aware food recommender-system based on deep learning and graph clustering. IEEE Access **10**, 52508–52524 (2022)
17. Toledo, R.Y., Alzahrani, A.A., Martinez, L.: A food recommender system considering nutritional information and user preferences. IEEE Access **7**, 96695–96711 (2019)
18. Stefanidis, K., et al.: PROTEIN AI advisor: a knowledge-based recommendation framework using expert-validated meals for healthy diets. Nutrients **14**(20), 4435 (2022)
19. Zhang, J., Li, M., Liu, W., Lauria, S., Liu, X.: Many-objective optimization meets recommendation systems: a food recommendation scenario. Neurocomputing **503**, 109–117 (2022)
20. Resma, K.B., Nair, M.S.: Multilevel thresholding for image segmentation using Krill Herd Optimization algorithm. J. King Saud Univ.-Comput. Inf. Sci. **33**(5), 528–541 (2021)
21. Kaggle dataset for food preference. https://www.kaggle.com/datasets/vijayashreer/food-preferences

Video Summarization on E-Sport

Vani Vasudevan(✉), M. R. Darshan, J. V. S. S. Pavan Kumar, Saiel K. Gaonkar, and Tallaka Ekeswar Reddy

Department of Computer Science Engineering, Nitte Meenakshi Institute of Technology, Bengaluru, India
{vani.v,1nt19cs063.darshan,1nt19cs087.pavankumar, 1nt19cs166.saiel,1nt19cs198.ekeswar}@nmit.ac.in

Abstract. Video summarizing is a useful technique for extracting the most important information from long videos. The "Video Summarization on E-sport" technique, which hasn't been extensively used in the esports industry yet attempts to automatically create highlights from e-sport game footage. This program, which users may input an e-sport game into, can produce summarized movies that capture the most thrilling parts of the game without the need for specialized equipment or experienced video editors. The suggested approach leverages AI, machine learning, and certain Python modules to produce highlights automatically, as opposed to the conventional way of making summarized movies, which is labor and time intensive. Users may save important time and effort by using the technology to automatically summarize key game moments by analyzing audio and text cues. The future of e-sports content creation may be greatly impacted by this technology, which has the potential to revolutionize the way e-sport game footage are summarized.

Keywords: Artificial Intelligence · Machine Learning · Video Summarization · e-sport · Python · Audio · text cues

1 Introduction

Modern storage and digital media technologies make it simple to record and store large volumes of video. Many videos are uploaded every minute to video-sharing websites like YouTube, Twitch, Loco. To be consistent in such platforms one needs to regularly upload the content which consumes too much time. Indian gamers are predicted to increase from 507 million in 2021 to 700 million in FY25, growing at a compound annual growth rate (CAGR) of 12%. According to a brand-new Dentsu report named Gaming Report India 2022 [1].

A video summary that condenses the entire video into a shorter format is being worked on as a solution to these problems. Large video collections can be quickly viewed using the technique of "video summarization," which also boosts the effectiveness of content indexing and access. Either skim the video or choose the key frames that best convey the information to create the summary. Select shot segments of the video are

S. Aurelia et al. (Eds.): ICCSST 2023, CCIS 1973, pp. 200–212, 2024.
https://doi.org/10.1007/978-3-031-50993-3_16

used for the summary, and the key frames are determined using the audio energy levels and text [2–4]. When selecting segments, make sure they reflect reality the entire video while also engaging the viewer. Advances in technology have expanded the options for multimedia content, particularly video, and the ability to quickly search through large video archives is essential. Video summarization is a promising technology with the goal of creating a shorter video clip or video poster using as much semantic content from the source video as possible. As video collections grow, it becomes increasingly difficult to effectively browse between and within videos. Video summarization is one solution to this problem. This study presents a dataset with frame-based relevance score labels to support research on query-controllable video summary and facilitate further investigation. Results show that text-based queries improve the management of video summaries and increase the efficiency of the algorithm.

In today's world, video data is now pervasive. Most raw videos are quite long and filled with unnecessary details. The amount of video data that users must now consume has significantly increased as a result. Artificial intelligence and Machine learning can do functions like learning and problem solving. Through AI, computers can simulate reasoning using logic and math and in machine learning, it uses mathematical models of data to help a computer learn without direct instruction. Which makes the computers to learn and improve based on experience, it will keep improving until the model's accuracy is high enough to finish the task completely. This enables us to predict trends and patterns, which will helped in many fields like Recommendation engines, Speech recognition, Video summarization, natural language understanding, Image and video processing, Sentiment analysis. Machine Learning and AI enables companies to discover valuable insights, improve data integrity and reduce human error, increases operational efficiency which reduces cost. As AI and machine learning started to be used in almost all fields, it got a very good response in video summarization, because people prefer to watch the highlights instead of entire video. This reduces work efforts while editing if only highlights are needed.

In this paper the approach is to summarize an e-sport gameplay video by live recording the gameplay along with audio and then the recorded file is sent for fragmentation where required number of frames are generated per second and these frames are later sent for text recognition, From the recorded file the audio format of the file is sent for audio classification based on the energy levels. Through text recognition we extract the frames required for highlighting the events and through audio classification we highlight the major occurrences from the required gameplay video. Also, a website is developed using ReactJS to download the windows application which can be used globally.

2 Literature Review

The use of a computer vision system has a basic objective of detecting moving objects and about its performance measure which is quite insufficient on few systems mainly because of the key reason which is environmental fluctuations which causes the detection of moving objects harder. The motion detection software used in this paper [5] allows us to detect movement around an item or a visual area. In this paper we have used OpenCV machine learning library for the detection of moving objects. Here we have taken a webcam that takes photos at frequent intervals. If two consecutive photos have no similarity, then they are deleted to free up memory but if there's some change then both the photos are saved.

Digital video fraud is referred to as the deliberate replacement of digital video for production. The most common way to manipulate digital video is temporary trickery, which includes frame sequence tricks like drop, insertion, rearrangement, and loops. By simply changing the temporary feature of the video the deception is not reflected in a single image research strategy; therefore, there is a need for digital forensics methods that perform temporary video analysis to detect such deception. We have implemented the python code for frame fragmentation in Jupiter notebook which requires a basic system which can run codes on Jupiter IDE. This research paper addresses about how the original video can be forged by adding duplicate video files or by deleting few frames from the video and still making it look the original video which is hard to find out with our naked eyes.

The new technology that is being used for frame fragmentation and uses three tier strategy for fragmenting the video, it talks briefly about how that method [6] works and we use this frame fragmentation technique in our project where we fragment the screen recorded video and give the frames per second and save all the frames in one folder. Talking about the technology, we use Three-tier architecture strategy for frame fragmentation and tools what we use is Jupiter notebook where we run our python code for frame fragmentation. To achieve the flexible and efficient transfer of files to the video system on demand, this paper proposed a three-stage split strategy that provided a seamless way to split HTTP live streaming video and larger file.

The four stages of video text recognition are typically detection, localization, extraction, and recognition [7]. In general, the detection step separates the text area from the non-text area. Using Sobel and canny masks, edge detection (i.e., horizontal, and vertical edge detection) is carried out on grayscale images. Each character has strong edges, regardless of font or color, making it possible to detect characters with different fonts and colors using our system, which uses edges as its primary feature. The precise text line boundaries are determined during the positioning step. We will extract the correct text using a stroke filter after the extraction step has removed the background from text lines so that only text pixels are left for recognition. Optical character recognition (OCR) software can be used to complete the recognition step [8]. OCR generates patterns by segmenting blocks of text into individual characters. Additionally, we use audio-based classification, where we have access to every audio cue from every frame of the video. Based on the audio intensity, we summarize a lengthy video. The module accepts a video of a sporting event as input and performs breaking the video into manageable chunks so that it can be processed. The audio is taken out of the video and divided into separate audio files by the module. The audio is then time-domain and high frequency analyzed.

OCR is the process of reading text from scanned materials (like pictures) and converting it into computer-friendly text that has been encoded. OCR makes digitization easier since the document may be scanned, processed, and the text extracted from it saved in an editable format, like a word document. Human involvement may be required to rectify certain items that were not accurately scanned because the procedure may not be 100% accurate. A dictionary or even Natural Language Processing (NLP) can be used to fix errors. According to recent research, the accuracy of the English language may reach up to 93% when written by hand and 98% when typed. On the other side, Arabic text extraction likewise achieves a recognition and translation rate of above 90%. Based on these findings [9], OCR was chosen as the top technique for converting paper-based records into digital ones.

This research proposes a novel role-based framework for goal event recognition that makes advantage of the semantic structure of football games. The ball is often thrown back to the centre after a goal scene, and the camera may zoom in on the player, show the excitement of the crowd, replay the goal scene, or show a mix of these things. As a result, by examining the sequences of the aforementioned roles, the objective event's occurrence will be discernible.The framework that is suggested in this paper consists of four main procedures: 1-detection of game critical events using audio channel, 2-detection of shot boundary and shots classification, 3-selection of candidate events based on shot type and presence of goalmouth in the shot, and 4-detection of restarting the game from the centre of the field.

By leveraging cues like a rise in sound intensity and the presence of a goal bar in the shot, this approach selects the potential shots. The goal events are then identified by looking at the subsequent shoots and locating visual indicators of goal scoring, such as the audience shot and the replay shot. In order to identify goal occurrences in a soccer game video, this study offers a framework. We extract video events for this purpose by using both the visual and audio channels. This framework consists of many primary phases, and within each of these levels, we tie certain traits to more advanced semantic ideas. In the first stage, we identify the key events in the video using fundamental audio attributes. In the following step, we utilise dissimilarity features to identify shot boundaries before categorizing these images using fundamental visual cues. Finally, in the following stage, we look for instances when the audience shot and close shot coexist, which is a sign of significant video sequences. The last phase is detecting the goalmouth and game start framework using the relevant units that have been produced. These procedures [10] can be used to locate goal events and summarize the film.

Merits	Demerits
1. With easily accessible software and equipment, screen recording or uploading a video is a simple and quick procedure. 2. Using current video processing libraries and methods, frame fragmentation may be accomplished rapidly and automatically 3. To aid with the creation of an effective summary, the most crucial portions of the video may be identified using text recognition and audio-based recognition 4. By removing the need to view the complete film in order to understand the key points, video summary helps conserve time and resources 5. A well-done video summary may be helpful in a variety of industries, including sports, journalism, and education	1. The effectiveness of the text recognition and audio-based recognition algorithms, which can be difficult to develop and may need a lot of computer power, strongly influences the quality of the summary. The most crucial elements of the video may not be included in the summary if the algorithms are not reliable 2. The most significant portions of the video may not always be adequately represented by text-only and audio-based recognition methods. Visual clues including changes in setting, body language, and facial emotions might be overlooked 3. The process of summarizing the video could miss some of the subtleties and specifics, which would mean that context and information would be lost 4. Some sorts of videos, including artistic or experimental ones that do not have a clear narrative or aim, could not be good candidates for video summary 5. Using a video without seeing the complete material may raise copyright and legal concerns, especially in professions like journalism and documentary filmmaking where authenticity and integrity are vital

3 Proposed System

3.1 Flow Diagram

Figure 1 displays the architecture for the "Video Summarization on E-sport" project. The system comprises various modules such as audio cues and player killed. The video is analyzed using these modules to extract relevant information and produce the highlights. The steps involved are further outlined.

So, the aim of this project is to give the highlighted video for an e-sport game from the entire gameplay based on the event mentioned. This is done by initially live recording the gameplay along with game audio and is saved in a folder in the local system. This video is later sent for frame fragmentation where we fragment the frames on number of frames mentioned, here we have taken 10 frames per second, and we do the text recognition on each frame to keep the important ones and discard which are not needed. We also use the recorded game audio where we check the intensity of the audio file and whenever the intensity is more than the threshold mentioned it will divide the audio file accordingly and we extract the frames which have higher intensity. Finally, we use this frame and the audio clips to stitch, and we get a summarized video.

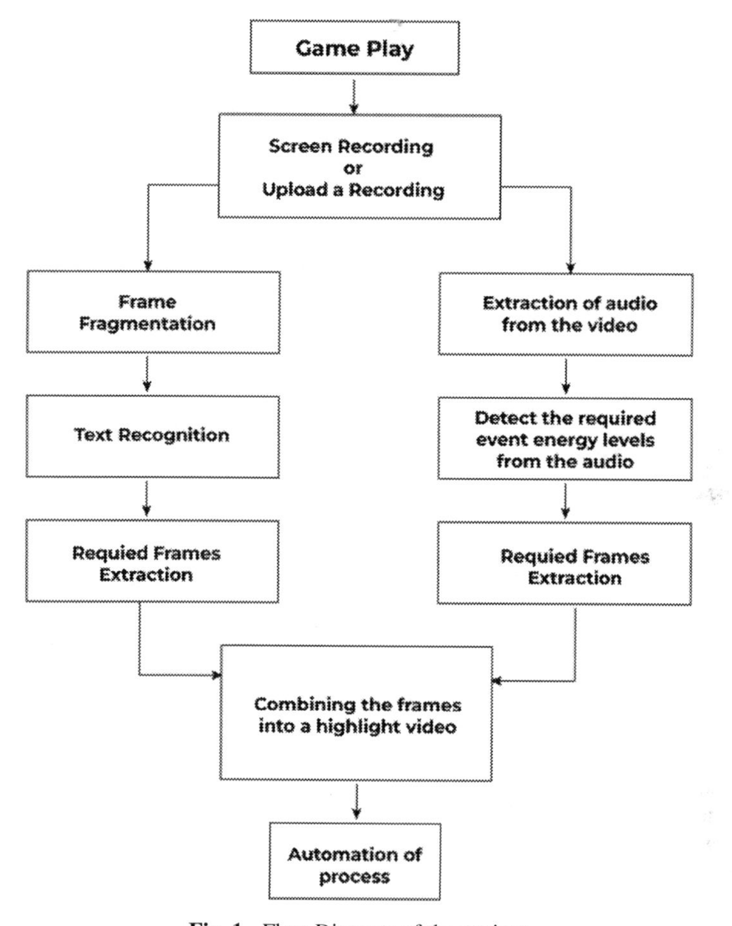

Fig. 1. Flow Diagram of the project

1. *Screen Recording:*

   In this module, we will be utilizing the tkinter and PIL libraries for the execution of GUI and Image processing functionalities. A User interface was also developed through the windows application where a button is provided to implement the screen recording process. The process involves a function record_screen() where we use the imageTK, Image, ImageGrab packages from the PIL library to capture images and use OpenCV for the creation of a video writer [11] which stores these images until a termination event occurs through a special key, given as q here.
2. *Frame fragmentation:*

   This model features the OpenCV and datetime functionality as major portion. We implement three functions in this module where the first function format_timedelta() returns the frame duration attached with each frame. The second function get_saving_frames_durations() arrange the list of durations in an order and the third main() function involves the intake of the file and then using openCV packages to achieve the frames from the recorded video.

3. *Text Recognition:*
   This is the major module for our project since it is this module where the events required for the highlights of the gameplay are decided. The require text is detected and is extracted from the frames using openCV and pytessearct packages which helps in the textual and image processing of the frames from which we obtain the text required for classifying [12] the necessary and unnecessary frames.
4. *Highlighting based on Audio:*
   In this module, we create a copy for the audio format of the video file and then pass it for processing. The audio is broken down into chunks, each having a certain energy level. We now specify a threshold value for this energy level based on the event which we want to be highlighted. Now based on the threshold value and each chunk value, we extract the required phases which are meant to be useful. The extracted audio bits are combined and then reverted to their video format [13] which provides the summarized part.
5. *Frame stitching:*
   Since the useful frames are now detected we use the same openCV packages for the combining of the images or frames by deciding the length of the event which the user may want to watch. We use the ImageTk, image etc. packages so that we read each required frame and now combine these frames by considering the timestamp for each frame and keeping a set of 10,20,30 s duration starting a bit early from those timestamps.

3.2 Algorithmic Steps

1. *Gameplay Video Input:* This module accepts the video of an e-sport game as input.
2. *Live Screen Recording with Audio:* This module records the live gameplay of an e-sport game, including the game audio.
3. *Audio Cues Processing*:
   (a) *Audio Extraction:* This module separates the audio from the video and saves it as individual audio files.
   (b) *Short-term Energy Computation:* The audio signals are analyzed using both time domain and frequency domain techniques to assess their time and frequency components, respectively.

4 Implementation

Figure 2 shows the data flow of screen recording. Initially when the user runs the program a tab will open in a new window which has the button to start the live recording of the screen and also it displays a message telling to press button "e" on keyboard to stop the live recording. So whenever the user wants to record the gameplay, he/she can click on the record button to start recording and after the gameplay is recorded along with the game audio, than user need to switch back to the tab and click on button "e" to stop the recording.

Figure 3 shows the tab in the windows screen which has option to start and stop recoding the live gameplay.

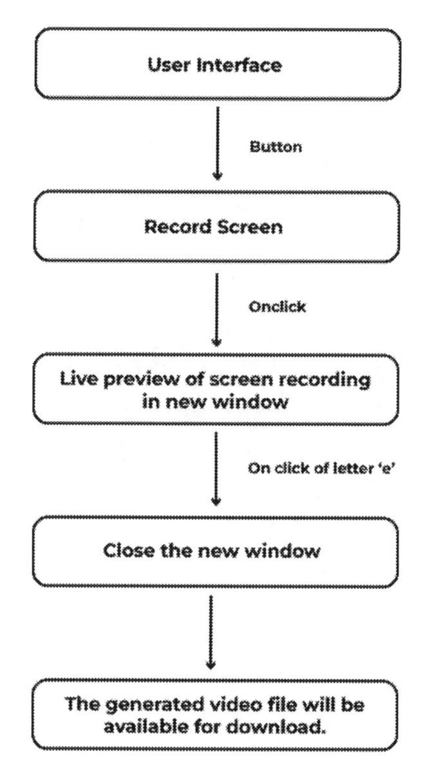

Fig. 2. Data flow of screen recording

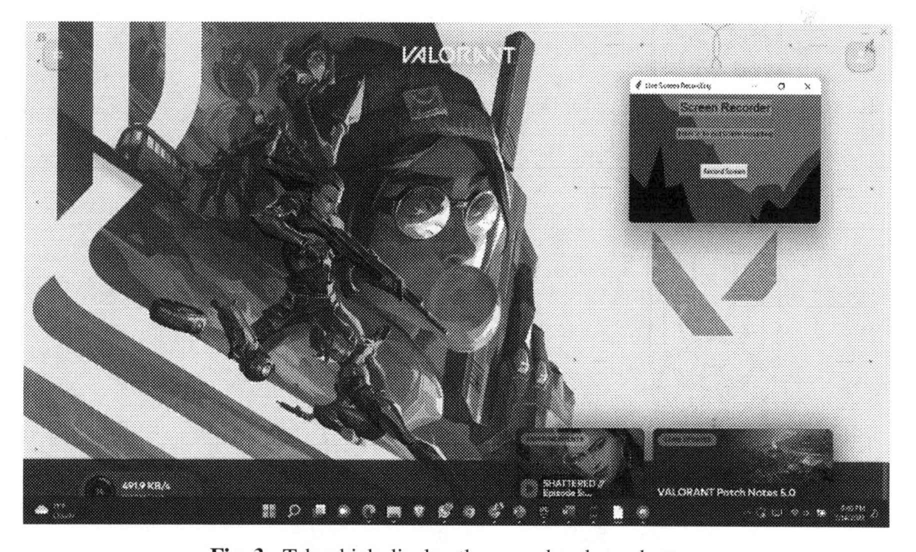

Fig. 3. Tab which display the record and stop button.

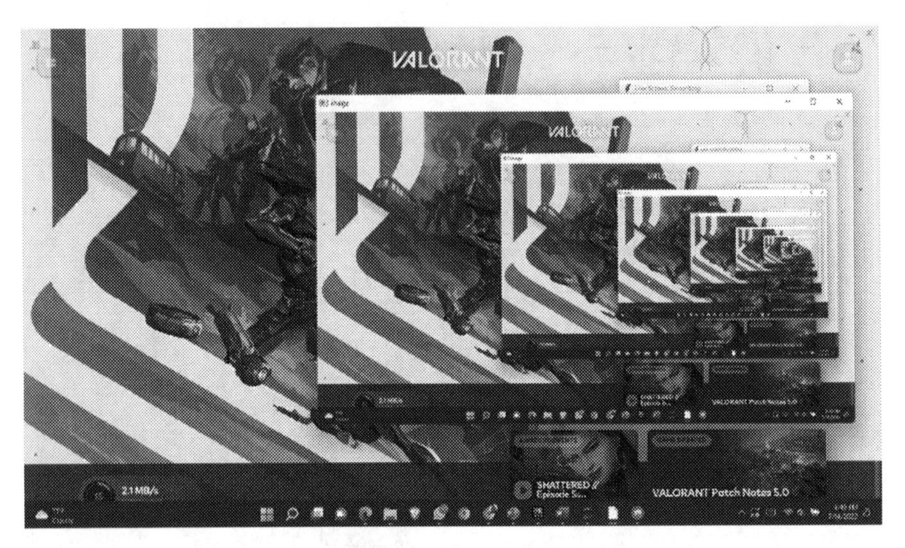

Fig. 4. Tab confirming the live recording of the screen.

Figure 4 confirms the live screen recoding of the tab that is being displayed by showing multiple tabs in loop confirming the recording of the screen.

Figure 5 displays the data flow of the frame fragmentation where we are fragmenting the video on 10 frames per second.

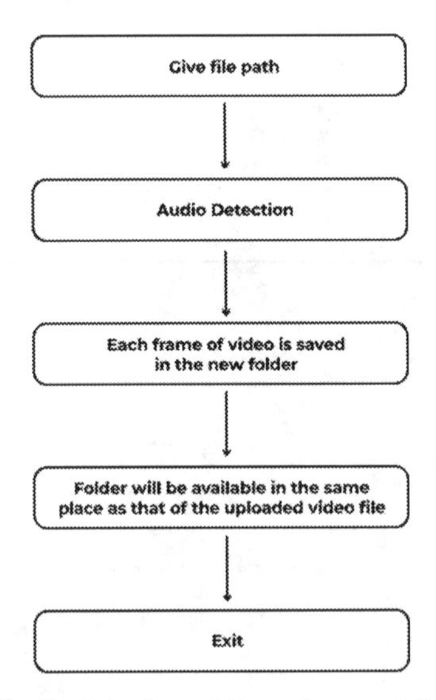

Fig. 5. Data flow of frame fragmentation

Initially in the program we will give the path of the video file that need to be fragmanted, then we run the program the entire video will be fragmented based on the number of frames per second we have given in the program. Those frames will be saved in a new folder also this folder will be located in the same place as that of the uploaded video file.

Figure 6 displays the folder where the fragmented frames are being downloaded in the local system with the frame fragmented at the rate of 10 frames per second.

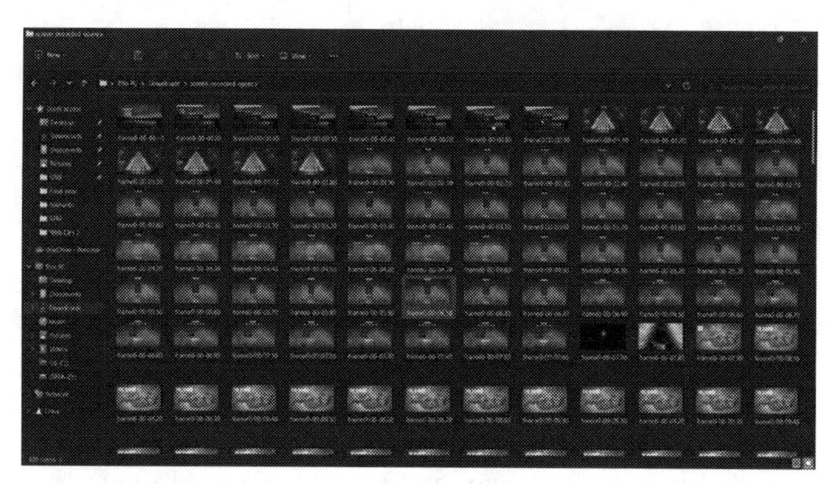

Fig. 6. Tab displaying the fragmented frames in the folder downloaded in local system

Here we can see all the fragmented frames in a single new folder which is also located in the same folder as that of video.

Figure 7 below showcases the time series domain of the energy chunks divided from the audio file.

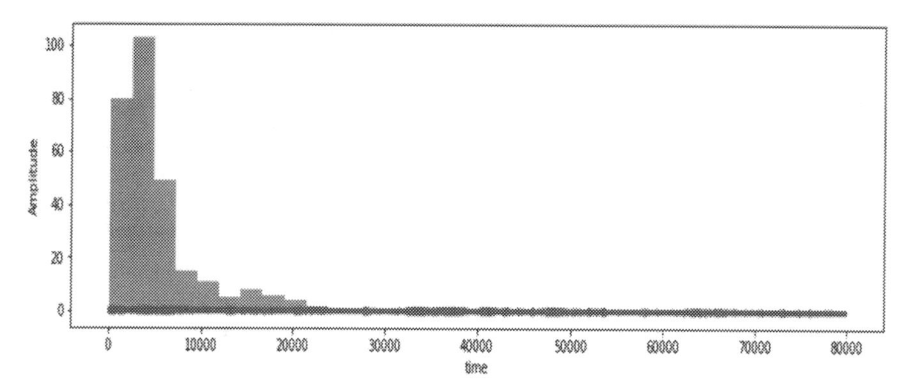

Fig. 7. Visualization of energy chunks

Figure 8 below showcases text recognition of selected area which is achieved through pytesseract library that is highlighted by a green rectangular box in the image.

Starting with the python package librosa which is used for video and audio analysis, we provide the audio format to the program which then uses librosa for getting the

Fig. 8. Image of selected text recognition.

duration and loading of the file. We break the audio file into chuncks and then proceed with highlight extraction. Now we use Ipython library for the analysis of each chunk by calculating the energy levels with one another. We visualise the analysis made using matplotlib numpy and pandas library for the statistical measures. Now to extract the important event highlight we put a threshold value and extract the frames for the required value which will be either greater or lower than the threshold [14]. Now we have the pieces of highlight for the required video, from those we now combine the audio and revert it back to the video format that now showcases the highlighted events from the entire video source using the moviepy python library

5 Results

The "Video Summarization on E-sport" technique that is being suggested uses a straightforward yet efficient method to extract highlights from e-sport game footage. To begin with, the system records the gameplay screen and saves it as a video file. The audio is then taken out of the video clip in order to pinpoint crucial moments in the gameplay.

This is accomplished by breaking up the audio into smaller segments based on a predetermined threshold intensity. The algorithm then recognizes the time periods when the audio surpasses this threshold and makes use of these crucial time periods to identify crucial gaming occurrences. The algorithm creates highlights to encapsulate the game's most thrilling moments whenever the occurrences are identified.

With the use of the Pytesseract library, we were also able to recognize the text in the chosen frame, and anytime a kill occurs and appears on the screen, that text will be identified, and those frames will be retrieved for the video summary.

Because it is automated, this method is quite effective and requires little assistance from humans. The time and effort needed for video editing and content generation can be greatly reduced by the technology's capacity to recognize noteworthy occurrences in real-time and automatically produce highlights. The process of producing entertaining and educational e-sport game video summaries is expedited as a consequence.

6 Conclusion

E-sports players and digital content developers spend numerous hours compiling gaming highlights and editing them into a compiled video. This procedure is frequently time-consuming and tiresome, thus the "Video Summarization on E-sport" technology that has been developed is a game-changer. This system can automatically create highlights of an e-sport match using artificial intelligence and machine learning techniques, saving both content producers and player's considerable time.

Another important benefit of the system is how simple it is to use. It generates a video output that content providers may publish directly, requiring no further editing. The highlights were produced using image processing and neural networks, while the system's original architecture was based on audio-based concepts. The end result is a highly automated and effective method that generates precise and interesting gaming highlights.

Additionally, this system may be created to produce performance highlights for specific players. With the help of this tool, gamers may show off their best moments to an audience without spending many hours on video production and editing. Overall, this technology can completely transform the way e-sports material is produced and make it easier and more effective for all parties involved.

References

1. https://www.techcircle.in/2022/12/02/professional-gamers-to-cross-700-mn-markin-india-by-2025-report
2. Vasudevan, V., Gounder, M.S.: A systematic review on machine learning-based sports video summarization techniques. In: Kumar, B.V., Sivakumar, P., Surendiran, B., Ding, J. (eds.) Smart Computer Vision. EAI/Springer Innovations in Communication and Computing. Springer, Cham (2023). https://doi.org/10.1007/978-3-031-20541-5_1
3. Vasudevan, V., Gounder, M.S.: Advances in sports video summarization – a review based on cricket videos. In: Advances and Trends in Artificial Intelligence. From Theory to Practice: 34th International Conference on Industrial, Engineering and Other Applications of Applied Intelligent Systems, IEA/AIE 2021, Kuala Lumpur, Malaysia, July 26–29, 2021, Proceedings, Part II, PP. 347–359. Springer, Heidelberg (2022). https://doi.org/10.1007/978-3-030-79463-7_29
4. Haq, H.B.U., Asif, M., Ahmad, M.B.: Video summarization techniques: a review. Int. J. Sci. Technol. Res. **9**(11), 146–153 (2020). ISSN 2277-8616
5. Mishra, S., Verma, V., Akhtar, N., Chaturvedi, S., Perwej, Y.: An intelligent motion detection using OpenCV. J. Sci. Res. Sci. Eng. Technol. **9**, 51–63 (2022). https://doi.org/10.32628/IJSRSET22925
6. Li, X., Wang, L., Cui, J., Zheng, B.: A new fragmentation strategy for video of HTTP live streaming. In: 2016 12th International Conference on Mobile Ad-Hoc and Sensor Networks (MSN), pp. 86–89 (2016). https://doi.org/10.1109/MSN.2016.022

7. Shetty, S., Devadiga, A., Chakravarthy, S., Varun, K.A.: Ote-OCR based text recognition and extraction from video frames, pp. 229–232 (2014). https://doi.org/10.1109/ISCO.2014.7103949
8. Dange, B.J., Kshirsagar, D.B., Khodke, H.E., Gunjal, S.N.: Automatic video summarization for cricket match highlights using convolutional neural network. In: 2022 International Conference on Smart Technologies and Systems for Next Generation Computing (ICSTSN), Villupuram, India, pp. 1–7 (2022). https://doi.org/10.1109/ICSTSN53084.2022.9761338
9. Jayoma, J.M., Moyon, E.S., Morales, E.M.O.: OCR based document archiving and indexing using pytesseract: a record management system for DSWD caraga, Philippines. In: 2020 IEEE 12th International Conference on Humanoid, Nanotechnology, Information Technology, Communication and Control, Environment, and Management (HNICEM), Manila, Philippines, pp. 1–6 (2020). https://doi.org/10.1109/HNICEM51456.2020.9400000
10. Bayat, F., Moin, M.S., Bayat, F., Mokhtari, M.: A new framework for goal detection based on semantic events detection in soccer video. In: 2013 8th Iranian Conference on Machine Vision and Image Processing (MVIP), Zanjan, Iran, pp. 184–189 (2013). https://doi.org/10.1109/IranianMVIP.2013.6779975
11. Currell, G.: The use of screen-capture video as a learning resource. In: New Directions in the Teaching of Physical Sciences, pp. 37–40 (2016). https://doi.org/10.29311/ndtps.v0i3.415
12. Li, L. -J., Li, J., Wang, L.: An integration text extraction approach in video frame. In: 2010 International Conference on Machine Learning and Cybernetics, Qingdao, China, pp. 2115–2120 (2010). https://doi.org/10.1109/ICMLC.2010.5580492
13. Ajmal, M., Ashraf, M.H., Shakir, M., Abbas, Y., Shah, F.A.: Video summarization: techniques and classification. In: Bolc, L., Tadeusiewicz, R., Chmielewski, L.J., Wojciechowski, K. (eds.) ICCVG. LNCS, vol. 7594, pp. 1–13. Springer, Heidelberg (2012). https://doi.org/10.1007/978-3-642-33564-8_1
14. Chu, W.T., Chou, Y.C.: On broadcasted game video analysis: event detection, highlight detection, and highlight forecast. Multimedia Tools Appl. **76**, 9735–9758 (2017). https://doi.org/10.1007/s11042-016-3577-x

SQL Injection Attack Detection and Prevention Based on Manipulating the SQL Query Input Attributes

R. Mahesh[1(✉)], Samuel Chellathurai[1], Meyyappan Thirunavukkarasu[2], and Pandiselvam Raman[3]

[1] Sree Saraswathi Thyagaraja College, Pollachi, India
aummahesh@gmail.com, samuel.c@stc.ac.in
[2] Alagappa University, Karaikudi, India
meyyappant@alagappauniversity.ac.in
[3] Alagappa Government Arts College, Karaikudi, India

Abstract. SQL injection refers to one of the types of database attacks for web applications. The database security is compromised when wild card characters, malicious code, or malicious SQL query string are injected into the database. These changes in syntax and semantic allow the attacker to gain access to sensitive information and manipulate the database. Various techniques have been developed to detect and prevent this type of attacks. In this article, we proposed an method for preventing and detecting SQL injection. This method manipulates the SQL query input parameters and determining the distance between query strings. This method satisfies static query and dynamic also.

Keywords: SQL Injection · Database Security · Anomaly Detection · Intrusion Prevention · Intrusion Detection · Runtime Monitoring

1 Introduction

In the digital age, we are dealing with information in different formats and transmitting it all over the world. Web applications are one of the main sources of access to data from any location. The work model of the web application retrieves the information from the database and displays it in the web pages. Attackers steal the information by using different types of attack models. SQL injection is a familiar attack model that compromise the database security without any additional use of system resources. In this paper, we have proposed a method for prevention and detection of SQL injection using special string replacement algorithm and the Levenshtein distance method. Chapter 2: Types of SQL injection attacks Chap. 3: Work model of SQL injection Chap. 4: Proposed algorithm with an example Chap. 6: Implementation process Chap. 7: Solution for static query structure Chap. 8: Moving to the next stage of this research

S. Aurelia et al. (Eds.): ICCSST 2023, CCIS 1973, pp. 213–221, 2024.
https://doi.org/10.1007/978-3-031-50993-3_17

2 SQL-Injection Attacks

The attacker inserts characters or keywords into the SQL input parameters that are built using the web page's input fields. It alters a legal query's logical operation. [1]. There is no limitation in the attack models that do not include restricted attack techniques like tautologies, exploiting untyped parameters, stored procedures, excessively elaborative error messages, union queries, and piggy-backed queries.

3 Work Model of SQL Injection

Here, we discuss SQL injection attacks with an example. In the business model of the web application, a few lines of code construct the SQL query with the value of input fields. The database server executes the constructed query and then the result sends to the web application. Let us consider the input attribute values are "admin" and "admin#123", then the legitimate query dynamically gets the syntax structure like

Q0="select data from login where userid='admin' and password='admin#123"

The constructed query Q0 retrieves the information from the database and sent back to the web application. The application layer of the web application renders the information on the web page. The attacker injects the malicious code and wild card characters into genuine query for example

Q1= select data from login where userid=" or 1 =1 - and password="

The syntax of above query is acceptable but it is not correct logically. The attacker divert the purpose of query through malicious code injection. Compromised query Q1 retrieve the information of all users.

4 Related Work

Researchers have contributed numerous strategies for identifying SQL injection. Code is written by developers to look for input parameters' keywords that contain harmful code and attack patterns. In order to prevent the injection, Stephen W. Boyd and Angelos D. Keromytis [2] created a mechanism that concatenates the random keywords with an appropriate query. For this strategy to work, developer and proxy server skills are required. The detection method based on depth-first traversal of the non-deterministic finite automata was introduced by William Halfond et al. in their publication [5]. to locate the problem area and to parse the strings in the NDFA's SQL query. With the exception of stored procedures and runtime SQL query generation, this security architecture meets both static and dynamic query analysis. A approach based on the parse tree of the SQL statement prior to construction with user input and that obtained after including input has been proposed by G. Buehrer et al. [5]. A technique that satisfies both analysis and dynamic analysis has been put forth by Y.-W. Huang [4]. Sanitizing chores are carried out, and the vulnerable area of code is where the sanitizing routine is inserted. The prerequisites for converting contaminated input into trusted output must be stored. A remedy for preventing SQL injection in stored procedures has been offered by KeWei [12]. Static and run-time validation are satisfied by this solution. Create a graph to represent the query dependencies, then store them in a table. While scanning the query execution, check the query dependencies. if the DFA flow graph is satisfied. Angelo Ciampa [17] applied a heuristic approach. Generates attacks by crawling web applications and storing attack patterns in a database. Injection detection is based on stored attack patterns. R.A. McClure [7] provided a DLL-based solution. The DLL monitors SQL syntax and eliminates possible SQL syntax problems, spelling errors, and data type mismatches. Information from each table and query details must be integrated and compiled. This method only provides a static analysis solution. William G.J. Halfond and Alessandro Orso [9] proposed a method for a syntax-aware evaluation process that consists of fragmenting queries and preserving keywords and operators. Evaluate SQL queries based on the order of syntactic sequences. This method is difficult to implement because it requires collecting operator keywords, operations, and sequences that correspond to the SQL syntax. Frank S. Rietta [11] established his method for detecting SQL injection at the application view layer. This detection method fragments the data and monitors access frequency compared to normal data observed in the past. This method is not reliable when implemented in large-scale he web applications. Gregory T. Buehrer [8] implemented tree concepts to parse SQL queries and compared tree nodes with predefined parsed tree lengths and dynamic query string lengths that are parsed at runtime. Did. This method requires longer processing time. Debasish Das [16] approached a dynamic query pattern matching based detection system. Legitimate requests are collected and saved in an XML formatted file. At runtime, SQL queries are converted to XML and checked for a predefined structure for legal queries in the master XML file. Not all web applications behave so dynamically. Can be implemented in a single web application. Inyong [18] proposed a new method. Remove input attributes from SQL queries and measure the length of SQL queries. However, this only satisfies string input parameters, not numeric input parameters. Y. Kosuga [13] analyzed the semantics and syntax of SQL queries to detect SQL injections. The purpose of creating a repository is to collect normal queries compared to those generated by attacks. Sruthi Bandhakavi [14] proposed a method to detect SQL injections by dynamically determining the query structure intended by the programmer for each input and comparing it with the structure of the actual output query. did. Romil Rawat [15] described protection measures for his web applications against SQL injection. Muyang liu.et.al [22] provided a method to detect vulnerabilities in his SQLi using his natural language processing-based DeepSQLi tool for test case generation. The DIAVA method [21] is a new traffic-based his SQLIA detection and vulnerability analysis framework. Send alerts instantly. Inspect network traffic for SQL operations and multi-stage regular expression models. An LSTM

[20]-based method that solves the overfitting problem by comparing generated positive samples with insufficient positive samples. Debabrata kar.et.al [19] used node measurement centrality to train a support vector mechanism and token approach graph for SQL injection detection. This method removes the value of the SQL query attribute from the web page when passing the parameter and compares it to the specified value. This method [19] uses a combination of static and dynamic analysis that removes the value of a SQL query from a web page when a parameter is passed and compares it to the specified value. This method uses a combination of static and dynamic analysis.

5 Proposed Work

The input fields of the web form or HTML input controls are the gateway where the attackers inject the malicious code. A web application needs either a mechanism or a method to monitor the input values for detecting the SQL injection attack. The proposed method detects the injected SQL vulnerabilities and prevents data breaches against SQL injection. The first step is to apply the string replacement algorithm to replace the special string constraints (STR, NUM) instead of the value of the input parameters of the SQL query. The special string STR is used to replace the input parameter that is a string data type and NUM is used for numeric data type input parameters. The detection method explains with an example given below.

*LQ1= "select * from user_account where user_id="'+usrid+"' and password="'+passwd+"''*

In LQ1, there are two input parameters such as 'usrid' and 'passwd'.

Here LQ1 is a legitimate query to access the user table by allowed users. We use the algorithm which replaces special string constraints instead of every input parameter. Before replacement, the entire SQL query rewrites into string format including input parameters. The following structure of SQL query string assign to replace the special string constraint.

*string SQ1="Select * from user_account where user_id='+usrid+'*
and password='+apsswd+"' ;

After replacement process, the following restructured query is derived

*RSQ1= Select * from user_account where userid=STR and password=STR*

5.1 Proposed Algorithm for Replacing Special String Constraints Instead of Input Parameter

function(query)
input sql query

```
    extract substring from "where" keyword to end of query statement
    while(upto substring length)
      if input parameter=string pattern start and end single quotes.
        replace( string input parameter,STR)
        form new query
      else if input parameter=Numeric pattern
        replace( numeric input parameter,NUM)
        form new query
      end if
    end while
    return formated string
end function
```

5.2 Levenshtein Method

Levenshtein distance (LD) is a method that measure the similarity between source string (s) and the target string (t). Here LD is represented the function of measuring the distance. The function LD returns the value of distance d. If distance d is equal to 0, the requested query is legitimate. If distance d is not equal to 0, the requested query is not legitimate SQL query

Detecting Legitimate Query. The following structure of SQL query string is a genuine query string.

*string SQ1="Select * from user_account where user_id='+usrid+' and password='+apsswd+' " ;*

After replacement process, the following restructured query is derived

*RSQ1= Select * from user_account where user_id=STR and password=STR*

In the next process, we pick the original SQL query(LQ1) and replace the special string characters which is bind with single quotes instead of input parameter values of the SQL query(LQ1).

*LQ1="select * from user_account where user_id='sami05' and password= '123#12"'*

Input attribute value of User_id is'sami05' and password value is "123#12"

We uses the algorithm which replace the special string value instead of input parameter values. The following statements is obtained after replace.

*LQ1= "select * from user_account where user_id='sami05' and password= '123#12"'*

*RSQ2="select * from user_account where user_id=STR and password=STR*

Finally, we get two restructured query in given below

RSQ1-By replacing the special string constraint instead of input parameters
*RSQ1= Select * from user_account where user_id=STR and password=STR:*
RSQ2-By replacing the special string keyword instead of input attribute values
*RSQ2="select * from user_account where user_id=STR and password=STR*

Finally, we have to applied the levenstein method for measuring the distance of two string,
d= LD(RSQ1,RSQ2);
if d=0
SQL query is legitimate
if d !=0
SQL query is not legitimate

Detecting Malicious SQL Query. After applying the string replacement algorithm, the following SQL query obtained

*RSQ1= Select * from user_account where user_id=STR and password=STR:*

then the user pass the input values of userid and password.

*LQ1="select * from user_account where user_id=" OR 1=1–' and password= 'abc"'*

The following SQL query obtained after string replacement algorithm.

*RSQ2="select * from user_account where user_id=STR OR 1=1–' and password=STR"*
d= LD(RSQ1,RSQ2);
#The value of d is not equal to 0
if d=0
SQL query is legitimate
if d !=0
SQL query is not legitimate

Numeric Input Attribute Value. After apply the algorithm for replacing the special constraint "NUM" instead of string input parameter The following SQL query obtained

*RS1= select * from user_account where customer_id=NUM*

then the user pass the input value of customer_id. We get

*LQ1= "select * from user_ account where customer_ id=123 or 1=1"*

After apply the algorithm for replacing the special constraint instead of numeric input attribute values The following SQL query obtained

*RS2="select * from user_ account where customer_ id=NUM or 1=1"*

Finally, we get two restructured query in given below

RSQ2 - By replacing the special string constraint instead of input attribute values
*RSQ2="select * from user_ account where customer_ id=NUM or 1=1"*

d= LD(RSQ1,RSQ2);
#The value of d is not equal to 0
if d=0
    SQL query is legitimate
if d !=0
    SQL query is not legitimate

6 Implementation

The proposed method can be experimented with web applications. This experiment involves his two different query execution methods: Dynamic versus static query analysis. The proposed algorithm has been applied to a real web application. A compiled collection of web applications that are cited online and used in many research papers. To implement the proposed method, we used the cited web application. The results of our work satisfy the detection and prevention of various types of attacks, including tautologies, piggyback queries, union queries, untyped parameters, stored procedures, and alternative inference encodings. Table 1 shows the performance of SQL-Injection detection and prevention compared to existing methods. The performance of the proposed method is studied on web applications. The recommended results are shown in Table 2.

Table 1. Comparison of prevention and detection techniques for various SQL injection attacks. tables.

Detection Method	Code Modification	Injection Detection	Attack Prevention
AMNESIA [5]	Not Needed	Automatic	Automatic
JDBC-checker [3]	Not Needed	Automatic	Source code adjustment
SQLCheck [10]	Not Needed	Partially Automatic	Automatic
SQLGuard [6]	Needed	N/A	Automatic
SQL DOM [7]	Needed	Automatic	Automatic
Proposed Method	Not Needed	Automatic	Automatic

Table 2. Performance of SQL Injection Detection in Various Applications. tables.

Web Apps	Proposed Method	Detection(%)	SQL DOM	Detection(%)	AMENSIA	Detection(%)
Employee directory	247/247	100	3937/3937	100	280/280	100
Events	87/87	100	3605/3605	100	260/260	100
Classifieds	319/319	100	3724/3724	100	200/200	100
Portal	288/288	100	3685/3685	100	140/140	100
Bookstore	366/366	100	3473/3473	100	182/182	100

7 Conclusion and Future Work

We introduced techniques to prevent SQL injection attacks in web applications. The proposed method satisfied both static and dynamic analysis. We implemented the considered application without evidence of false-positive results. This is an effective method that has stopped over 1500 attacks. It is a very reliable and lightweight application that can be implemented in many different types of applications, including desktop applications, web applications, and web services. The proposed method satisfied both string and numeric input type parameters. Future research will explore techniques to track attacks within data packets by scanning the port layer. Consider how to design web application environment-agnostic components that incorporate database software to protect applications from SQL injection attacks.

References

1. Anley, C.: Advanced SQL Injection In SQL Server Applications. Next Generation Security Software Ltd, White Paper (2002)
2. Boyd, S.W., Keromytis, A.D.: SQLrand: preventing SQL injection attacks. In: Jakobsson, M., Yung, M., Zhou, J. (eds.) ACNS 2004. LNCS, vol. 3089, pp. 292–302. Springer, Heidelberg (2004). https://doi.org/10.1007/978-3-540-24852-1_21
3. Gould, C., Su, Z., Devanbu, P.: JDBC checker: a static analysis tool for SQL/JDBC applications. In: Proceedings of the 26th International Conference on Software Engineering (ICSE), pp. 697–698 (2004)
4. Huang, Y.-W., Yu, F., Hang, C., Tsai, C.-H., Lee, D.T., Kuo, S.-Y.: Securing Web application code by static analysis and runtime protection. In: Proceedings of the 12th International World Wide Web Conference (WWW 2004), pp. 40–52 (2004)
5. Halfond, W.G., Orso, A.: AMNESIA: analysis and monitoring for NEutralizing SQL-injection attacks. In: ACM, ASE 2005, November 7–11 (2005)
6. Buehrer, G., Weide, B.W., Sivilotti, P.A.G.: Using parse tree validation to prevent SQL injection attacks. In: Proceeding of the 5th International Workshop on Software Engineering and Middleware, pp. 106–113 ACM (2005)
7. McClure, R., Krger, I.: SQL DOM: compile time checking of dynamic SQL statements. In: Proceedings of the 27th International Conference on Software Engineering (ICSE 2005), pp. 88–96 (2005)
8. Buehrer, G., Weide, B.W., Sivilotti, P.A.: Using parse tree validation to prevent SQL InjectionAttacks, SEM 2005. In: Proceedings of the 5th international workshop on Software engineering and middleware, pp. 106–113 (2005)

9. Halfond, W.G., Orso, A., Manolios, P.: Using positive tainting and syntax-aware evaluation to counter sql injection attacks. In: ACM-SIGSOFT, pp. 175–185 (2006)
10. Su, Z., Wassermann, G.: The essence of command injection attacks in web applications. In: Conference Record of the 33rd ACM SIGPLAN-SIGACT Symposium on Principles of Programming Languages, pp. 372–382 (2006)
11. Rietta, F.S.: Application layer intrusion detection for SQL injection, ACM-SE 44. In: Proceedings of the 44th annual Southeast regional conference, pp. 531–536 (2006)
12. Wei, K., Muthuprasanna, M., Kothari, S.: Preventing SQL injection attacks in stored procedurce. In: Software Engineering Conference, pp. 18–21 (2006)
13. Kosuga, Y., Kono, K., Hanaoka, M., Hishiyama, M., Takahama, Y.: Syntactic and semantic analysis for automated testing against SQL injection. In: Proceedings of the Computer Security Application Conference, pp. 107–117 (2007)
14. Bandhakavi, S., Bisht, P., Madhusudan, P., Venkatakrishnan, V.N.: CANDID: preventing SQL injection attacks using dynamic candidate evaluations. In: Proceedings of the Computer Security Application Conference, pp. 12–24 (2007)
15. Rawat, R., Dangi, C.S., Patil, J.: Safe guards anomalies against SQL injection attacks. Int. J. Comput. Appl. **22**(2), 11–14 (2011)
16. Das, D., Sharma, U., Bhattacharyya, D.K.: An approach to detection of SQL injection attack based on dynamic query matching. Int. J. Comput. Appl. **127**(14), 15–24 (2010)
17. Ciampa, A., Visaggio, C.A., Di Penta, M.: A heuristic-based approach for detecting SQL-injection vulnerabilities in Web applications. In: SESS 2010: Proceedings of the 2010 ICSE Workshop on Software Engineering for Secure Systems, pp. 43–49 (2010)
18. Lee, I., Jeong, S., Yeo, S., Moon, J.: Novel method for SQL injection attack detection based on removing SQL query attribute values. Math. Comput. Model. **55**(1–2), 58–68 (2012)
19. Kar, D., Panigrahi, S., Sundararajan, S.: SQLiGoT: detecting SQL injection attacks using graph of tokens and SVM. Comput. Secur. **60**, 206–225 (2016)
20. Li, Q., Wang, F., Wang, J., Li, W.: LSTM-based SQL injection detection method for intelligent transportation system. IEEE Trans. Veh. Technol. **68**(5), 4182–4191 (2019)
21. Gu, H., et al.: DIAVA: a traffic-based framework for detection of SQL injection attacks and vulnerability analysis of leaked data. IEEE Trans. Reliab. **69**(1), 188–202 (2020)
22. Liu, M., Li, K., Chen, T.: DeepSQLi: deep semantic learning for testing SQL injection, ISSTA 2020. In: Proceedings of the 29th ACM SIGSOFT, pp. 286–297 (2020)

Comparative Analysis of State-of-the-Art Face Recognition Models: FaceNet, ArcFace, and OpenFace Using Image Classification Metrics

Joseph K. Iype(✉) and Shoney Sebastian

Department of Computer Science, CHRIST (Deemed to be University), Bengaluru, Karnataka 560001, India
joseph.iype@mca.christuniversity.in,
shoney.sebastian@christuniversity.in

Abstract. In recent years, facial recognition has emerged as a key technological advancement with numerous useful applications in numerous industries. FaceNet, ArcFace, and OpenFace are three widely used techniques for facial identification. In this study, we examined the accuracy, speed, and capacity to manage variations in face expression, illumination, and occlusion of these three approaches over a period of five years, from 2018 to 2023. According to our findings, FaceNet is more accurate than ArcFace and OpenFace, even under difficult circumstances like shifting lighting and facial occlusion. Also, during the previous five years, FaceNet has shown a significant improvement in performance. Even while ArcFace and OpenFace have made significant strides, they still lag behind FaceNet in terms of accuracy. Therefore, based on our findings, we conclude that FaceNet is the most effective method for facial recognition and is well-suited for use in high-stakes applications where accuracy is crucial.

Keywords: Facial recognition · FaceNet · ArcFace · OpenFace · Comparison · Performance · Accuracy · Speed · Facial expression · Lighting · Occlusion · Improvement · High stakes applications

1 Introduction

Recent years have witnessed a substantial increase in the use of facial recognition technology, which has a wide range of practical uses in everything from social media to security systems. Several potent models have been created to take on the challenge of facial identification with the development of deep learning and computer vision techniques. The most well-liked facial recognition models are FaceNet, ArcFace, and OpenFace.

In this work, we do a real-time application comparison analysis of these three models. We employ a dataset of 100 photographs of well-known individuals from diverse fields, with each subject appearing in two photographs that were shot five years apart. This dataset is used to evaluate the models' capacity to identify people despite changes in their looks over time.

S. Aurelia et al. (Eds.): ICCSST 2023, CCIS 1973, pp. 222–234, 2024.
https://doi.org/10.1007/978-3-031-50993-3_18

Applying the FaceNet, ArcFace, and OpenFace models to the dataset allows us to assess each model's performance in terms of accuracy, speed, and adaptability to changes in facial expression, illumination, and occlusion. Our attention is on how well they work in a real-time environment, which is crucial for applications like security and surveillance systems.

The purpose of the work is to shed light on the advantages and disadvantages of these models for face recognition in real-time applications. In addition to determining which model performs best overall, we also want to determine which models are best suited for particular use cases. The findings of this study, particularly for those working on real-time applications, should be of great interest to researchers and practitioners in the field of facial recognition.

2 Problem Statement

The growing use of facial recognition technology in a variety of industries, including security systems, social media platforms, and virtual reality applications, has made it a significant topic of research in recent years. FaceNet, ArcFace, and OpenFace are just a few examples of facial recognition models that have been built and have produced good accuracy results on benchmark datasets. To ascertain whether a model performs better in real-world circumstances, comparative research is required.

The goal of this study is to assess the efficacy of FaceNet, ArcFace, and OpenFace on real-time facial recognition tasks using a dataset of well-known people from around the globe. 100 photos of each individual will make up the dataset, with two images taken five years apart to model changes in facial features. The accuracy and effectiveness of each model will be assessed in the study, and the findings will be compared to determine which algorithm performs better in real-world circumstances. The results of this study will give important new information on how facial recognition models function and how they might be used in practical situations.

3 Literature Review

3.1 Convolutional Neural Networks

Convolutional Neural Network (CNN) has completely changed the field of computer vision, including facial recognition. FaceNet, ArcFace, and OpenFace are just a few facial recognition models that have extensively utilized CNNs. CNNs are employed in these models to extract face feature information. The most crucial patterns and attributes for recognising people are taught to CNNs using massive datasets of facial photos. The learned features can then be used by the CNNs to produce high dimensional feature vectors for every facial image.

For instance, in FaceNet, CNN is trained to produce a 128-dimensional vector that depicts the distinctive traits of each person's face. Then, two facial photos are compared using this feature vector to ascertain whether they are members of the same person. Similar to this, ArcFace and OpenFace use CNNs to extract features from facial photographs, which are then normalized and modified to increase the models' accuracy. To learn a distance metric between the feature vectors, these models also employ a metric learning methodology. Then, two facial photos are compared to see if they belong to the same individual using this distance measure. CNNs play a key role in facial recognition algorithms like FaceNet, ArcFace, and OpenFace. They make it possible for these models to recognise faces accurately and reliably by extracting and learning the elements that are most crucial for recognising people from facial photos.

3.2 FaceNet

Google researchers created FaceNet, a facial recognition model, in 2015. It is a deep learning-based approach that creates high-dimensional feature vectors by extracting features from facial photos using a Convolutional Neural Network (CNN).

FaceNet's capacity to acquire accurate feature representations of faces that are resilient to changes in lighting, position, and expression is one of its main features. This is accomplished by training the model on a sizable face picture dataset made up of millions of photographs of thousands of different people. The model is trained using a triplet loss function, which seeks to maximize the distance between feature vectors in photographs of distinct people and reduce the distance between feature vectors in images of the same person.

The Siamese network architecture used by the FaceNet model consists of two identical CNNs that employ the same set of weights. The anchor image is processed by one CNN, and the positive and negative images are processed by the other CNN. The triplet loss function is then used to compare the output of the CNNs, which promotes the model to learn distinguishing features for facial recognition.

The use of the cosine similarity score, a distance metric, to compare feature vectors is one of FaceNet's significant contributions. This distance measure allows the model to correctly recognise people from facial photographs and has been demonstrated to be useful in comparing high-dimensional feature vectors.

FaceNet has achieved cutting-edge performance on several benchmark datasets for facial recognition, including the MegaFace dataset, the YouTube Faces dataset, and the Labeled Faces in the Wild (LFW) dataset. It has also been utilized in several real-world applications, such as social media and security systems.

As a result, FaceNet is a very successful face recognition model that has proven to be remarkably accurate and resilient at recognising people from facial photos. One of the most popular models for facial recognition, it excels in performance thanks to the Siamese network architecture it employs and the cosine similarity score it uses to compare feature vectors.

3.3 ArcFace

ArcFace is a facial recognition model created by researchers at the National University of Singapore and released in 2018. It is a deep learning-based approach that creates high-dimensional feature vectors by extracting features from facial photos using a convolutional neural network (CNN). A softmax loss function with an additive angular margin is used to train the model, which pushes it to learn highly discriminative feature representations of faces.

ArcFace's capacity to produce highly discriminative feature vectors appropriate for face recognition tasks is among its main benefits. Using an angular margin loss function, which raises the angle between feature vectors of distinct persons while decreasing the angle between feature vectors of the same individual, is accomplished. As a result, the model can successfully identify people from facial photos thanks to feature vectors that are highly discriminative.

ArcFace utilizes a ResNet structure with either 100 or 200 layers, a well-known CNN design for image classification tasks. The model generates feature vectors from facial images through an embedding layer and subsequently categorizes these vectors into distinct individuals using a densely linked layer. On a variety of benchmark datasets dedicated to face recognition, including the LFW dataset, MegaFace dataset, and the Asian Face collection, ArcFace has consistently delivered exemplary outcomes. This model has been effectively integrated into platforms such as social media networks and various security infrastructures.

As a result, ArcFace is a highly successful facial recognition model that creates highly discriminative feature vectors for face identification tasks using a new loss function. It is one of the most popular facial recognition models thanks to its cutting edge performance on a variety of benchmark datasets and practical applications.

3.4 OpenFace

Researchers at Carnegie Mellon University developed the open-source facial recognition system known as OpenFace. It has been around since 2015 and uses deep learning methods. High-dimensional feature vectors called facial embeddings, which can be used for face recognition tasks, are created by the model.

To understand how to create facial embeddings, a deep neural network architecture that powers OpenFace was trained on a sizable dataset of facial photographs. To extract characteristics from facial images and create embeddings, the model combines fully connected and convolutional layers.

OpenFace's capability to accommodate variations in facial appearance, such as changes in lighting, position, and facial emotion, is one of its main benefits. This is accomplished by employing a variety of strategies for data augmentation during training, which teaches the model to produce embeddings that are resilient to these fluctuations.

On various benchmark datasets for face recognition, such as the MegaFace dataset, the YouTube Faces dataset, and the Labeled Faces in the Wild (LFW) dataset, OpenFace has demonstrated great accuracy. Security systems, social media sites, and virtual reality applications are just a few of the real-world uses for the model. OpenFace can be used for computer vision applications other than face identification, like facial landmark detection and facial emotion recognition. The concept is easily incorporated into other programs and is accessible as an open-source software library.

As a result, OpenFace is a strong facial recognition model that can produce superior facial embeddings that are resistant to fluctuations in facial appearance. It is a well-liked option for facial recognition applications due to its high accuracy on benchmark datasets and adaptability in different computer vision tasks.

3.5 RetinaFace

Researchers at the Chinese University of Hong Kong initially unveiled RetinaFace in 2019, a cutting-edge facial identification and alignment model. It is intended to find and align faces in difficult real-world circumstances and is based on deep learning techniques.

RetinaFace was created utilizing a multi-task learning architecture that carries out face landmark detection, facial posture estimation, and facial detection all at once. This enables the model to recognise and align faces in pictures with different poses, lighting conditions, and occlusions.

The model makes use of a version of the well-known RetinaNet object detection architecture that is tailored for spotting small things, like faces. To increase the precision of detection and alignment, it additionally makes use of a pyramid feature network.

RetinaFace's great accuracy in identifying and matching faces in difficult real world circumstances is one of its main features. The WiderFace and AFW datasets, two benchmark datasets, show that the model performs at the cutting edge. It has also been utilized in several real-world applications, including social media platforms, video surveillance systems, and security systems.

RetinaFace is a software library that is accessible as open-source and is simple to include in other computer vision programmes. The model is also very effective and has a quick processing speed that enables real-time applications.

In conclusion, RetinaFace is a facial alignment and detection model that can recognise and align faces in difficult real-world situations. It is a well-liked option for facial detection and alignment applications because of its cutting-edge performance on benchmark datasets and economy in processing speed.

4 Design Methodology

The research involved the collection of images featuring prominent individuals from various sources, followed by the selection of two images based on a timestamp of five years. This method was adopted to assess the robustness of each model in accommodating the changes that individuals undergo over their lifespan. The images were systematically organized into their respective directories and subjected to a cosine similarity test using three models, namely FaceNet, ArcFace, and OpenFace, in combination with RetinaFace, a backend detector designed for detecting faces within an image. Using a convolutional neural network, RetinaFace initially preprocesses the input image to create feature maps (CNN). The subsequent convolutional layers that are used to forecast the bounding boxes of the faces in the image as well as their corresponding facial landmarks use these feature maps as input. Using a multi-task learning method that is built into the model's design, RetinaFace extracts facial landmarks. The model is trained to predict the coordinates of numerous important facial landmarks, including the corners of the eyes, nose, and mouth, in addition to the bounding boxes of the faces in an image. The extraction of face landmarks in RetinaFace involves several essential phases. To create feature maps, the input image is first processed by a convolutional neural network (CNN). The subsequent convolutional layers that are used to forecast the coordinates of the facial landmarks use these feature maps as input. The model can then use these offsets to calculate the final coordinates of each landmark with respect to the bounding box coordinates once it has predicted the offsets for each facial landmark. The facial landmarks in the image can then be found using these coordinates. Convolutional neural networks and regression layers are combined to extract face landmarks in RetinaFace, and they work together to anticipate the precise coordinates of important facial characteristics inside an image. This makes it possible to detect multiple faces in noisy images.

4.1 Face Extraction by RetinaFace

An RGB image is used as the source for RetinaFace, which is then processed by a backbone network to extract feature maps at various scales. The bounding boxes and confidence ratings for faces at each scale are predicted by a set of detection heads using the feature maps that have been input into them.

RetinaFace employs a focus loss function that, during training, assigns greater weight to hard cases (i.e., faces that are challenging to recognize) in order to increase the accuracy of face detection. This enables the network to concentrate on learning to recognise faces in difficult lighting or occlusion conditions. RetinaFace locates the facial landmarks, such as the eyes, nose, and mouth, within each identified face using a different branch of the neural network once the faces have been identified. This is accomplished by regressing the coordinates of the facial landmarks concerning the identified face's bounding box.

Lastly, RetinaFace aligns the faces and crops them out of the original image using the faces and facial landmarks that it has identified. Applying an affine transformation based on the positions of the facial landmarks to the face image accomplishes this. Following that, the obtained face images can be further processed or examined for a variety of purposes, including face recognition.

4.2 Vectorization by FaceNet, ArcFace and OpenFace

FaceNet uses a convolutional neural network to extract a 128-dimensional feature vector from a facial image. In the feature space, feature vectors corresponding to the same individual tend to cluster closely, whereas those of distinct individuals are positioned more distantly. This distinct positioning arises because the network is trained using a triplet loss function. As a result, FaceNet excels at recognizing faces in photographs, even with variations in lighting, poses, and facial expressions.

ArcFace is a variation of FaceNet that enhances the discriminative power of the feature vectors by using an additional angle margin loss. By increasing the cosine similarity between them, this loss function pushes feature vectors of the same individual to be closer together than feature vectors of different individuals. Because of this, ArcFace is quite good at recognising faces even with few training data.

A 128-dimensional feature vector is extracted from a facial image by OpenFace using a deep neural network. Similar to FaceNet, the network is trained using a triplet loss function. However, to increase the precision of feature extraction, OpenFace additionally uses facial landmark detection and alignment. As a result, even when the poses and expressions in the photographs vary, OpenFace is quite effective at identifying faces.

FaceNet and ArcFace both do exceptionally well in recognising faces, but ArcFace excels in situations where there is little training data available. Contrarily, OpenFace uses facial landmark detection and alignment to improve its accuracy for face recognition in difficult lighting circumstances.

After RetinaFace detected the faces in each photograph, the images underwent vectorization. Subsequently, the resultant vectors were fed into each model. Each model calculated the gap between the vector representations of the two images, producing a similarity score and a threshold value for evaluative reasons. This approach is depicted in Fig. 1.

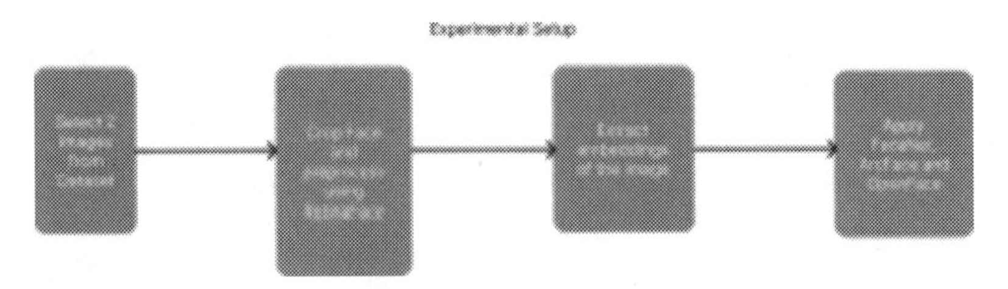

Fig. 1. .

4.3 Results

By comparing the similarity scores and identifying the true positives and true negatives, the photos were verified. The mean of the similarity scores for photographs belonging to the same individual were calculated and compared to the mean of the similarity scores for images belonging to different individuals to appropriately assess the results. The resulting data was examined, and a standard distribution curve was drawn to show the true positive and true negative rates. The density of the true positives and true negatives were represented in Figs. 2, 3 and 4, providing a clear depiction of the results.

The correlation between Euclidean distance and density for the input dataset is shown graphically in Fig. 2. According to the analysis, true positive images follow a symmetrical distribution, but true negative images show a skewed distribution. These results show that the FaceNet model can distinguish between photographs of the same person accurately if the Euclidean distance is less than or equal to 11, and between images of different people if the distance is larger than or equal to 11. These findings shed important light on the FaceNet model's effectiveness and its precision in authenticating images.

Figure 3 illustrates the graphical depiction of the relationship between Euclidean distance and density for the input dataset. True positive photos display a symmetrical distribution, while true negative images show a skewed distribution, according to the findings. These results imply that with an accuracy rate of less than or equal to 4 Euclidean distance, the ArcFace model can identify photographs of the same person with reasonable accuracy. Similar to this, the accuracy rate for photos of various people is greater than or equal to 4 Euclidean distances. The model's accuracy is lower than that of the FaceNet model, as shown by the density of true negatives, which is noticeably larger than that of genuine positives. Nonetheless, these findings shed important light on the efficiency and accuracy of the ArcFace model in image authentication as well as its potential for a range of uses.

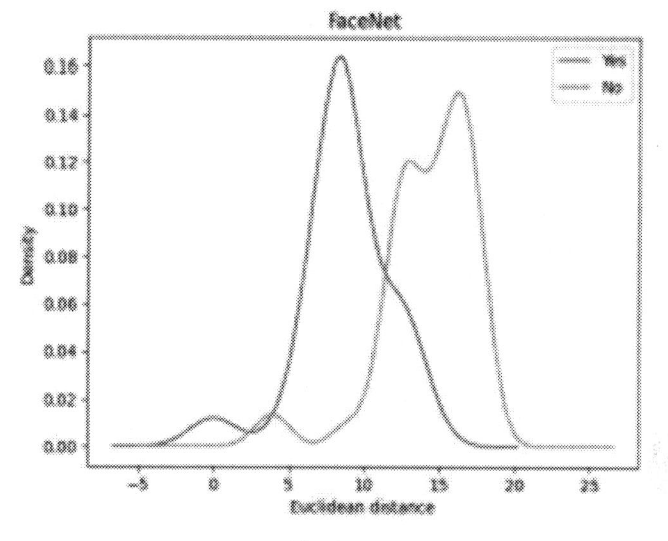

Fig. 2. .

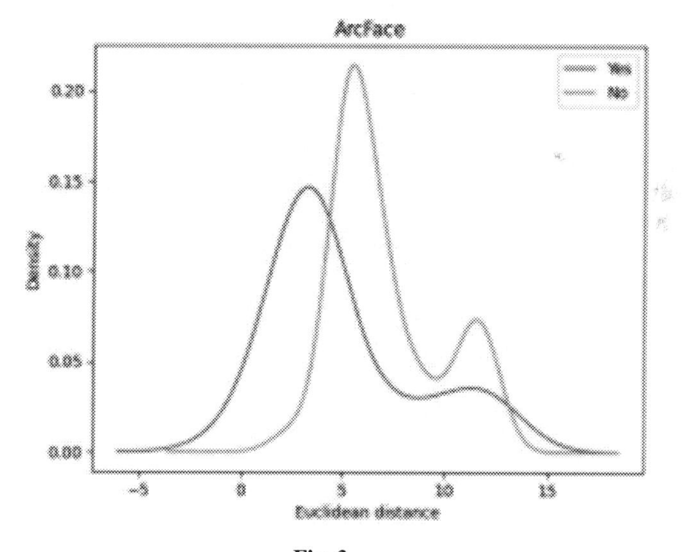

Fig. 3. .

The association between Euclidean distance and density for the input dataset is shown graphically in Fig. 4. True positive images have a symmetrical distribution, according to the research, while true negative images have a skewed distribution. These results indicate that the OpenFace model can recognise individuals between images with an accuracy rate of less than or equal to 0.9 Euclidean distance. Similar to this, the accuracy rate for photos of various people is greater than or equal to 0.9 Euclidean distances. The accuracy of the OpenFace model is worse than that of the FaceNet and ArcFace models, as seen by the fact that the density of genuine negatives is noticeably larger than that of true positives.

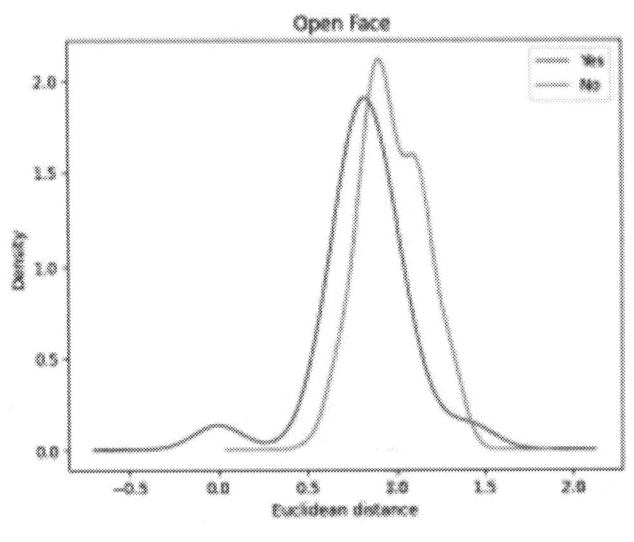

Fig. 4. .

The OpenFace model's effectiveness and accuracy in image authentication, as well as its potential for a wide range of applications, are nevertheless crucial insights provided by these studies.

Based on the above results, we can conclude that FaceNet's triplet loss function with online triplet mining outperforms the competition in face recognition tests. This is due to the triplet loss function's guarantee that embeddings for the same individual are more closely spaced apart than embeddings for various individuals, which is essential for precise facial recognition. Moreover, the FaceNet online triplet mining technique chooses instructive triplets on the fly during training, increasing the training process' accuracy and efficiency.

4.4 Loss Functions

Loss functions are used to train neural network models in face recognition algorithms like FaceNet, ArcFace, and OpenFace. The training aims to minimise the difference between the projected output of the model and the actual output, which is calculated using the loss function. We will talk about the loss functions each of these algorithms uses and how they operate in this comparison.

1. FaceNet Loss Function: FaceNet employs a neural network model and refines it using the triplet loss function. The triplet loss function is an adaptation of the contrastive loss function, which has garnered attention in the realm of image recognition endeavors. Within the framework of the triplet loss function for facial recognition, three core

components are integrated: an anchor image, a congruent image, and a disparate image. The foundational premise of this function is to ensure that the metric space between the anchor and the congruent image remains consistently tighter compared to the distance between the anchor and the disparate image. The distance between the anchor and positive images should be as short as possible while the distance between the anchor and negative images should be as long as possible according to the triplet loss function. The distance metric employed is commonly the cosine distance or Euclidean distance between the neural network-generated embeddings for each image.

Mathematically articulated, the formulation for triplet loss is given by: $L = \max(0, f(a) - f(p)^2 - f(a) - f(n)^2 + \alpha)$ where f(a), f(p), and f(n) denote the embeddings generated by the neural network for the anchor, congruent, and disparate images in that order. These embeddings are typically measured using metrics such as Euclidean or cosine distances. If the spatial relationship between the anchor and congruent images exceeds that of the anchor and disparate images by at least the margin (α), the loss evaluates to 0. Otherwise, when the differential is below this margin, the loss deviates from zero, prompting adjustments within the network to minimize this discrepancy.

Since the triplet loss function enables the neural network to learn embeddings that are distant for images of various persons and close for photographs of the same person, it is effective. Based on the neural network's embeddings, this facilitates the comparison and recognition of various faces. The triplet loss function does have certain drawbacks, though, including the potential for high computing costs, particularly when working with huge datasets. Researchers have created several triplet loss function changes and enhancements to solve this drawback, such as the use of hard or semi-hard negative samples, which can increase the face recognition system's effectiveness and accuracy.

2. ArcFace Loss Function: The additive angular margin loss, often known as the ArcFace loss, is a modified version of the softmax loss function that is used by ArcFace. By adding an angle margin between the embeddings of various classes, the ArcFace loss function is intended to improve the embeddings' ability to distinguish between distinct classes. The loss function can be defined mathematically as follows:

$$L = -\frac{1}{N} \sum_{y=i} \log \frac{e^{s(\cos(\theta_i)+m)}}{e^{s(\cos(\theta_i)+m)} + \sum_{j \neq i} e^{s(\cos(\theta_j))}},$$

where N is the total number of samples in the dataset, y is the ground truth label, i is the index of the current sample, j is the index of the other samples, s is the scaling factor, m is the additive margin, and is the angle between the current sample and the weight vector corresponding to its ground truth class.

The cosine similarity between the embedding vectors and the weight vectors of their respective classes is optimized using the additive angular margin loss function. To make the decision boundary more discriminative, a margin is added to the cosine similarity score for the proper class. To make the margin more resilient to intraclass fluctuation, it is applied in the cosine space rather than the Euclidean space. In face recognition tasks, it has been demonstrated that the additive angular margin loss function performs better than other well-known loss functions such as Softmax loss and triplet loss. It works especially well in jobs where there are many classes and there is a lot of intra-class variation, like in extensive facial recognition applications. The additive angular margin loss function is an effective method for enhancing the discriminative ability of embedding vectors in face recognition tasks. It adds an angular margin to the Softmax loss function to generate a decision boundary that is more resilient to intra class variance and increases the accuracy of the face recognition system.

3. OpenFace Loss Function: To train its neural network model, OpenFace likewise uses the triplet loss function, but it does so in a different way known as triplet loss with online triplet mining. The embeddings are more discriminative and there is less chance of false positives thanks to this method, which chooses hard negative examples from a batch of images. The loss function can be mathematically defined as $L = \max(0, f(a) - f(p)^2 - f(a) - f(n)^2 + \text{margin})$, where f(a), f(p), and f(n) are the embeddings of the anchor, positive, and negative images, respectively, represents the L2 norm, and margin is a hyperparameter that determines the minimum margin between the positive and negative distances. Online triplet mining includes choosing triplets when they are needed during training rather than pre-defining them. This prevents the use of all feasible triplets, which can be computationally expensive, and instead limits the use of the most relevant triplets for training. The online triplet mining technique involves the following steps:
   (a) Randomly choose an anchor representation from the training set.
   (b) Identify a congruent image corresponding to the same individual as the anchor.
   (c) Opt for a disparate image associated with an individual distinct from the anchor's.
   (d) Compute the embedding vectors for the anchor, congruent, and disparate images via the neural model.
   (e) Evaluate the triplet loss based on these derived embeddings.
   (f) Refine the neural model's parameters employing the backpropagation technique.

The success of the online triplet mining technique depends on the selection of informative triplets during training. Using the toughest negative triplet—i.e., the triplet that results in the greatest loss value—is a frequent strategy. This guarantees that the network concentrates on mastering the toughest samples.

Using semi-hard triplet mining is an alternative strategy that chooses triplets where the negative sample is farther distant from the anchor sample than the positive sample, but is still within a margin, is another method. As a result, the network can learn a more generalized representation without becoming stuck in a local minimum.

It has been demonstrated that using online triplet mining can significantly enhance the triplet loss function's face recognition performance. In large-scale face recognition applications where there are many training samples and the training procedure can be computationally expensive, it is especially helpful. The online triplet mining technique boosts the effectiveness and precision of the face recognition system by automatically choosing informative triplets while training.

5 Conclusion

In conclusion, our study assessed the effectiveness of three well-known facial recognition models - FaceNet, ArcFace, and OpenFace - in real-time applications and on sizable datasets. On large datasets, FaceNet performs better than ArcFace and OpenFace in terms of accuracy, speed, and efficiency. In contrast to ArcFace's additive angular margin loss function and OpenFace's combination of multiple loss functions, our analysis showed that FaceNet's triplet loss function with online triplet mining is particularly effective in learning embeddings that are more discriminative for face recognition tasks. FaceNet's architecture is additionally tailored for face recognition tasks using methods like joint optimization of the embedding and classification models and data augmentation methods like random cropping and horizontal flipping.

Our findings have significant ramifications for the design and implementation of facial recognition systems in practical settings, especially when processing big datasets in real-time. We recommend the use of FaceNet for such applications, as it has demonstrated superior performance in our comparative study.

Future research should investigate the possibility of mixing various embedding or loss functions to boost the effectiveness of facial recognition models even more. Also, to strengthen these models' robustness and dependability in real-world circumstances, it would be beneficial to look into how different elements like as lighting, facial expressions, and occlusions affect how well they work.

References

1. Baltrusaitis, T. et al.: OpenFace: an open-source facial behavior analysis toolkit. In: 2016 IEEE Winter Conference on Applications of Computer Vision (WACV) (2016)
2. Deng, J. et al.: Arcface: additive angular margin loss for deep face recognition. In: 2019 IEEE/CVF Conference on Computer Vision and Pattern Recognition (CVPR) (2019)
3. Jain, E., Mishra, S.: Face recognition attendance system based on real-time video. Int. J. Comput. Appl. **184**(6), 12–18 (2022)
4. Pattnaik, P., Mohanty, K.K.: AI-based techniques for real-time face recognition-based attendance system- A comparative study. In: 2020 4th International Conference on Electronics, Communication and Aerospace Technology (ICECA) (2020)
5. Schroff, F., et al.: FaceNet: a unified embedding for face recognition and clustering. In: 2015 IEEE Conference on Computer Vision and Pattern Recognition (CVPR) (2015)
6. Yang, H., Han, X.: Face recognition attendance system based on real-time video processing. IEEE Access **8**, 159143–159150 (2020)
7. Boutros, F., Damer, N., Kirchbuchner, F., Kuijper, A.: Elasticface: Elastic margin loss for deep face recognition. 2022 IEEE/CVF Conference on Computer Vision and Pattern Recognition Workshops (CVPRW) (2022). https://doi.org/10.1109/cvprw56347.2022.00164
8. Chi, C., Zhang, S., Xing, J., et al.: Selective refinement network for high performance face detection. In: Proceedings of the AAAI Conference on Artificial Intelligence, vol. 33, pp. 8231–8238 (2019). https://doi.org/10.1609/aaai.v33i01.33018231
9. Deng, J., Guo, J., Ververas, E., et al.: Retinaface: single-shot multi-level face localisation in the wild. In: 2020 IEEE/CVF Conference on Computer Vision and Pattern Recognition (CVPR) (2020). https://doi.org/10.1109/cvpr42600.2020.00525

10. Dirin, A., Delbiaggio, N., Kauttonen, J.: Comparisons of facial recognition algorithms through a case study application. Int. J. Interact. Mob. Technol. (iJIM) **14**, 121 (2020). https://doi.org/10.3991/ijim.v14i14.14997
11. Elmahmudi, A., Ugail, H.: Experiments on deep face recognition using partial faces. In: 2018 International Conference on Cyberworlds (CW) (2018). https://doi.org/10.1109/cw.2018.00071
12. Forsyth, D.: Object detection with discriminatively trained part-based models. Computer **47**, 6–7 (2014). https://doi.org/10.1109/mc.2014.42
13. Gidaris, S., Komodakis, N.: Object detection via a multiregion and semantic segmentation-aware CNN model. In: 2015 IEEE International Conference on Computer Vision (ICCV) (2015). https://doi.org/10.1109/iccv.2015.135
14. Majumdar, P., Agarwal, A., Singh, R., Vatsa, M.: Evading face recognition via partial tampering of faces. In: 2019 IEEE/CVF Conference on Computer Vision and Pattern Recognition Workshops (CVPRW) (2019). https://doi.org/10.1109/cvprw.2019.00008
15. Menezes, H.F., Ferreira, A.S., Pereira, E.T., Gomes, H.M.: Bias and fairness in face detection. In: 2021 34th SIBGRAPI Conference on Graphics, Patterns and Images (SIBGRAPI) (2021). https://doi.org/10.1109/sibgrapi54419.2021.00041
16. R AK, Solayappan VA, T SS, K RP (2021) Masked deep face recognition using ArcFace and ensemble learning. 2021 IEEE 2nd International Conference on Technology, Engineering, Management for Societal impact using Marketing, Entrepreneurship and Talent (TEMSMET). doi: https://doi.org/10.1109/temsmet53515.2021.9768777
17. Wang, M., Deng, W.: Deep face recognition: a survey. Neurocomputing **429**, 215–244 (2021). https://doi.org/10.1016/j.neucom.2020.10.081
18. Zhang, D., Hu, J., Li, F., et al.: Small object detection via precise region based fully convolutional networks. Comput. Mater. Continua **69**, 1503–1517 (2021). https://doi.org/10.32604/cmc.2021.017089
19. Zhang, S., Chi, C., Lei, Z., Li, S.Z.: Refineface: Refinement neural network for high performance face detection. IEEE Trans. Pattern Anal. Mach. Intell. **43**, 4008–4020 (2021). https://doi.org/10.1109/tpami.2020.2997456
20. Zhu, J., Li, D., Han, T., et al.: Progressface: scale-aware progressive learning for face detection. In: Vedaldi, A., Bischof, H., Brox, T., Frahm, JM. (eds.) Computer Vision – ECCV 2020. ECCV 2020 LNCS, vol. 12351, pp. 344–360. Springer, Cham (2020). https://doi.org/10.1007/978-3-030-58539-6-21

Hash Edward Curve Signcryption for Secure Big Data Transmission

S. Sangeetha and P. Suresh Babu(✉)

Department of Computer Science, Bharathidasan College of Arts and Science, Erode, India
sangee.siva2011@gmail.com, ptsuresh77@gmail.com

Abstract. Medical health service data is kept in huge dataset and shared through different devices. However, privacy and security of data sharing area significant concern, since data needs to be accessed from various locations in the distributed system. Therefore, a novel method called Hash Edward Curve Signcryption (HECS) is introduced for healthcare data communication in a secure manner. In the HECS method, related medical health service data is collected from the huge healthcare dataset. The proposed HECS method includes two major processes. First, Theil-Sen Robust Linear Regression is carried out to classifying the data and categorizes the data into number of classes. After the classification, the Hash Edward signcryption cryptographic technique is employed to perform the secured data transmission. In the designed cryptosystem, Pseudo ephemeral agreement key pair generation is performed for each session of data transmission. Then the HECS method performs the signcryption process for converting the data point into the encrypted data and generates the digital signature. Then in unsigncryption, the signature verification is carried out to check the user authenticity. This, in turn, secured data transmission is carried out with higher confidentiality. The assessment of the HESC method is carried with accuracy on classification and integrity of data. The quantitatively analysed results confirmed that the HECS method increases security than the orthodox methodology by means of improved data confidentiality rate and integrity rate.

Keywords: Big Data · secured data transmission · Theil-Sen Robust Linear Regression · Hash Edward Digital Signature Generation

1 Introduction

Medical data transmission between varied healthcare entities directs to achieve better patient health quality. However, the transmission of a large amount of healthcare data is a major challenging issue in terms of security and privacy, especially for large-scale healthcare systems. Hence security plays an essential role relating to healthcare data, there should be protection of data access from unauthorized users.

A BCSE model proposed in [1] on Big Data and identifying the attackers during the communication. But, it did not improve the performance stability where a large number of data samples were considered. Privacy-Preserving MRS Scheme was developed in [2]

S. Aurelia et al. (Eds.): ICCSST 2023, CCIS 1973, pp. 235–247, 2024.
https://doi.org/10.1007/978-3-031-50993-3_19

for safeguarding the patient data privacy by using the ElGamal Blind Signature method. However, it failed to obtain higher data confidentiality as well as integrity.

An IoT-Blockchain integration architecture was introduced in [3] for improving safety in medical health service data applications. However, the performance of overhead during the secure transmission was not minimized. The IoT-cloud-enabled healthcare data system was developed in [4] based on a searchable encryption method to preserve patient privacy. But the designed system failed to improve the security of data with latest encryption technique.

A system for securing medical health data was introduced in [5] to ensure the secure sharing of authenticated medical health data while safeguarding the privacy of patients' information. Though, it was unsuccessful in delivering enhanced data privacy through more effective and efficient data encryption algorithms. A novel encryption method was introduced in [6] for transmitting patient medical care data packets. But it was unable to offer advanced and effective algorithms that could enhance data privacy.

From the analyses of existing work, contributions of a novel HECS development is given below.

- To enhance the data to be transmitted securely, a new HECS method is developed based on Theil-Sen Robust Linear Regression and Pseudo ephemeral Kupyna hash Edward signcryption Cryptographic technique.
- After that, the Pseudo ephemeral Kupyna hash Edward signcryption technique is employed for improving the confidentiality of data transmission by performing the key generation, signcryption, and unsigncryption. The Kupyna hash function is applied to an Edward curve digital signature scheme to improve the data integrity. After that, signature verification is performed for avoiding unauthorized data access resulting in improved data confidentiality.
- Finally, comprehensive experimental evaluation is conducted to calculate efficiency of HECS methodology compared with related studies and various performance indicators.

The subsequent sections of the paper are organized in the following manner. Section 2 presents a comprehensive review of the relevant literature. Section 3 offers a description of the proposed HECS methods, accompanied by a well-structured architecture diagram. Section 4 outlines the experimentation process along with the dataset explanation. Section 5 outlines the performance of the proposed method by comparing with the metrics of the method already existed are compared to using various metrics. Finally, Sect. 6 provides a concluding summary of the paper.

2 Related Works

In [7], a system utilizing blockchain technology was introduced to enhance the security of medical IoT devices by means of Lamport Merkle Digital Signature. The designed scheme minimizes the computation time and overhead but a similar data classification was not performed to further minimize the time. A framework utilizing artificial intelligence and blockchain technology was created in [8] to enhance the management of medical records, ensuring both intelligence and security. However, the framework failed to incorporate the cryptographic hash algorithm, which is essential for enhancing data integrity. A medical data sharing method was developed in [9] based on consortium blockchain for medical data distribution and privacy protection. But the higher security of medical data sharing was not improved. Hyperelliptic curve cryptography was developed in [10] for privacy preservation and estimation of patient health data transmission. But, the data integrity was not improved.

An Efficient Ensemble structural design was developed in [11] for the privacy and safety of e-medical datasets. However, an improved cryptology technique was not used with better encryption. A hybrid data encryption approach was developed in [12] by combining multi-objective adaptive genetic algorithms for secure healthcare data transmission. But the higher integrity of medical data transmission was not improved.

3 Methodology

In network environment, transmitting data can be a huge challenge. It is crucial to prioritize secure transmission of patient health data collected from IoT device in order to reduce healthcare costs and enhance the quality of life for patients. To create a trustworthy and secure big data environment, it is necessary to identify possible attacks from attackers. Unfortunately, compromise of patient medical data occurs frequently by these attackers. Driven by these concerns, the focus of this paper is to develop a highly secure method called HECS for the transmission of medical data.

The proposed HECS methods performs secure data transmission with the help of different processes namely data classification and signcryption. Initially, the large numbers of patient data $D_i = D_1, D_2, D_3 \ldots D_n$ are collected from the dataset. After the data collection, the classification is performed using Theil-Sen Robust Linear Regression. The proposed machine learning technique employs Theil-Sen Robust Linear Regression for the analysis of patient medical data received from a server. This method utilizes median-based estimation for its computations. Then Prevosti's distance method is used for analysing the patient medical data and median of particular classes. After that, the damped least square method is applied for finding the minimum distance to group the data into particular classes. Thus, the input medical data is categorized into distinct classifications.

After the classification, the HECS performs transmission of data from sender to receiver using Pseudoephemeral kupyna HashEdward Signcryption cryptographic technique in secure manner. By applying the signcryption cryptographic technique, Pseudo ephemeral agreement key pairs are generated for each session of data transmission to avoid unauthorized access. A pseudo number generator is used for the key establishment process which is a deterministic random bit generator for generating random numbers. After the generation of key, the process of signcryption is carried out to produce cipher text and signature using the Kupyna hash Edward curve digital signature algorithm. On the receiver side, unsigncryption is carried out to validate the signature and retrieve the correct data. Thus the proposed cryptographic techniques help to transmit the data securely. The subsequent sections provide a concise explanation of the two processes involved in the HECS methods.

3.1 Theil-Sen Robust Linear Regression

First, the proposed HECS method performs the data analysis and is classified into various classes to reduce the data transmission complication. Therefore, the proposed method analyzes the data by Theil-Sen Robust Linear Regression. This machine learning technique measures the correlation between two variables using Theil-Sen Robust Linear Regression. The Theil-Sen Robust Linear Regression analysis in which finds the line that most closely fits the data according to a particular mathematical criterion i.e. generalized median-based estimator. The generalized median in a discrete set of sample data points in a Euclidean space is to minimize the sum of distance. The proposed HECS method uses Prevosti's distance to categorize the patient medical data points into distinct classes.

First, numbers of sample healthcare data $D_i = D_1, D_2, D_3 \ldots D_n$ are collected from the dataset.Then Prevosti's distance between the median and the data points is computed as given below,

$$\vartheta = \frac{1}{r}\sum\nolimits_{i=1}^{n}\sum\nolimits_{j=1}^{k}\left|D_i - \beta_j\right|^2 \tag{1}$$

where ϑ indicates a Prevosti's distance which is measured based on the median 'β_j' and data 'D_i', r denotes the number of data points. A damped least-squares method is used to address the non-linear least squares problems by finding the minimum distance.

$$Y = arg\ min\ \vartheta \tag{2}$$

where 'Y' denotes an output of the damped least-squares method, *arg min* symbolize an argument minimum function that aims to minimize the sum of distances denoted as 'ϑ'. Consequently, the regression function examines comparable input sample data and categorizes it into distinct classes.

The process of Theil-Sen Robust Linear Regression initially calculates the number of classes and the median. Then the Prevosti's distance between the median and data points is estimated. Then the damped least-squares technique is employed to find the minimum distance. Based on the regression results, similar data are correctly classified with higher accuracy.

3.2 Pseudoephemeral Kupyna HashEdward-Signcryption-Based Secure Data Transmission

After the data classification, secure data transmission is performed using Pseudoephemeral Kupyna Hash Edward-Signcryption technique. Signcryption is a public key cryptographic algorithm in which both digital signature and encryption functions are performed simultaneously. Encryption and digital signature are essential methods in cryptography that ensure the confidentiality and integrity of data during transmission. Traditionally, two separate steps such as signature-then-encryption process are required to achieve security objectives. However, the Edward-Curve Signcryption combines both digital signature and message encryption in a single logical step, providing an efficient and secure cryptographic solution. The proposed HECS technique uses the Edward curve digital signature algorithm to perform fast public-key digital signatures. Its main advantages are to obtain higher performance and secure implementations than the other signature generation algorithms. During the signature generation, Kupyna cryptographic compression function is create the hashes by applying Davies–Meyer hash function.

A signcryption method comprises the following process as shown below.

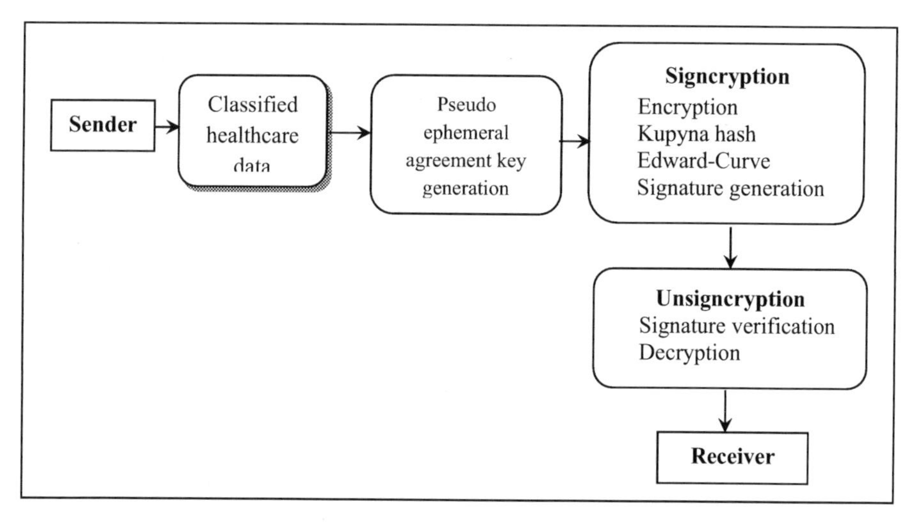

Fig. 1. Block Diagram Pseudo Ephemeral Kupyna Edward Signcryption

Figure 1 depicts the Pseudo Ephemeral Kupyna Edward Signcryption model which improves the security of healthcare data transmission with higher confidentiality and integrity. To enhance secure communication between sender and recipient, the suggested cryptographic approach comprises the generation of pseudo ephemeral agreement keys, signcryption, and unsigncryption. The subsequent subsections provide a concise explanation of these procedures.

3.2.1 Pseudo Ephemeral Agreement Key Generation

Generation of a pair of keys in cryptographic technique is an important process to encrypt or decrypt the medical health care data. The pseudo ephemeral agreement key is generated for every session in this proposed HESC method. Once the session completed the generated key will be disabled and new key will be generated for the next session. This technique improves the confidentiality level and avoids access by attackers.

By applying the Edward-curve digital signature technique, Let us consider the signing Pseudo ephemeral agreement private key is randomly chosen 'φ_p'. With the private key 'φ_{pr}', the signing session ephemeral agreement public key is generated as given below,

$$\varphi_{pb} = \varphi_{pr}.Q \tag{3}$$

where, φ_{pb} denotes a signing session ephemeral agreement public key, φ_{pr} the private key, Q denotes a base posit or generator point of the Edward curve. As a result, a pair of private and public keys is generated. The public key being distributed for subsequent processing where as private key being kept confidential.

3.2.2 Signcryption

The two operations namely encryption as well as signature generation are performed with the proposed Signcryption after generating the public key as signing key and private key as verification key. Encryption is the process of converting healthcare data i.e. plain text into cipher text form.

As shown in Fig. 2, the block diagram, signcryption technique converts the patient medical data to cipher text by the process of encryption. Let us examine the classified Patient medical data $CD_i = D_1, D_2, D_3 \ldots D_k$. Therefore, the public key encryption is performed on the sender side with the Pseudo ephemeral agreement public key of the receiver.

$$\rho_c \leftarrow En\big(CD, \varphi_{pb}(R)\big) \tag{4}$$

where 'ρ_c' denotes a ciphertext of patient data, '*En*' indicates encryption, $\varphi_{pb}(R)$ symbolize recipient public key for a session. Simultaneously, the sender performs the signature generation process with its pseudo ephemeral agreement private key. The signature generation is carried out using the Edward-curve algorithm.

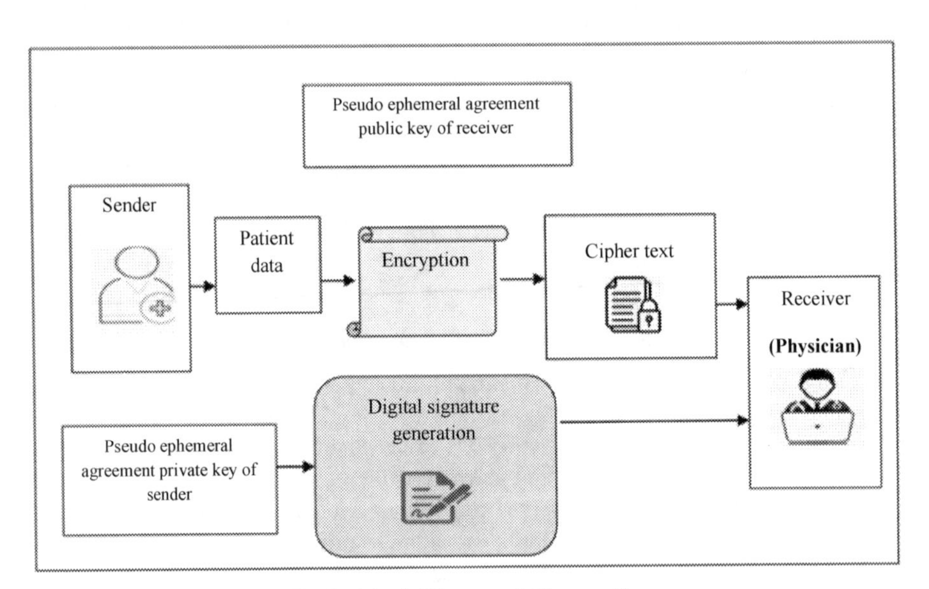

Fig. 2. Block Diagram of Signcryption

The sender first generates hash value of their private key by using the Kupyna hash function. The Meyer hash function is used to produce hashes by using Kupyna hash compression function. Subsequently, the Davies–Meyer hash function is provided with private key of sender as input initially.

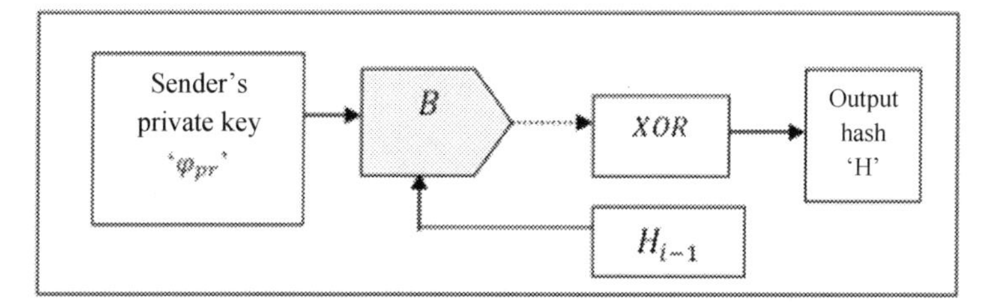

Fig. 3. Davis Mayer Single-Block-Length Compression

Figure 3 above depicts the Davies–Meyer hash function, wherein the input 'φ_{pr}' is received and the hashes generate previously (H_{i-1}) is initially assigned as '0'. In Fig. 3, the symbol 'B' denotes a block cipher and the input (φ_{pr}) functions as the key for the block cipher, undergoing XOR operation with the hashes generate previously. The subsequent procedure helps to produce final hashes

$$H = [B_{\varphi_{pr}}(H_{i-1}) \oplus H_{i-1}] \tag{5}$$

where, H *represents* a hash value. Finally, the outcome is taken from the compression function.

After that, generate the 'f' from the upper 32 bytes of hash and the input data

$$f = H'(H(32:)||CD) \tag{6}$$

where, || denotes a concatenation. Then the sender matches the 'f' onto the curve as given below,

$$V = f.Q \tag{7}$$

Next, the sender computes 'u' as given below,

$$u = f + \left[H\left(V||\varphi_{pb}||CD\right)\right].\varphi_{pr} \tag{8}$$

Therefore, the signature generated by the sender is (V, u). The value of V and u is 32 bytes long. Totally the signature is 64 bytes long. The generated signature and encrypted data are transmitted to the receiver (i.e. physician).

3.2.3 Unsigncryption

The final procedure of the suggested method is unsigncryption. When the receiver gets the cipher text and signature from the sender, they first verify the signature. A valid signature is used to consider the data was created by an authorized sender and it is not modified by any intruders (i.e. attacks). The receiver first performs the signature verification with the sender's public key, before the data decryption.

First receiver generates 'G' using V, φ_{pb} and CD as given below,

$$G = H\left(V||\varphi_{pb}||CD\right) \tag{9}$$

Then, the receiver also creates the two verification values,

$$s_1 = u.Q \tag{10}$$

$$s_2 = V + \varphi_{pb}.G \tag{11}$$

Then verify $s_1 == s_2$.

When the values s_1, *and* s_2 are matched, it is considered as valid and the receiver acquires the ciphertext. Conversely, if the signature is deemed invalid, the receiver does not receive the original data. These procedures strengthen the security of transmitting patient data. If the signature is said to be valid and then the decryption process is performed as given below,

$$D \leftarrow Dn\left[\varphi_{pr}(Receiver), \rho_c\right] \tag{12}$$

From (12), D denotes original patient data, Dn indicates the decryption, $\varphi_{pr}(Receiver)$ denotes a private key of the receiver, ρ_c denotes a ciphertext. This approach improves the safety of medical healthcare data while transmitting through networks by mitigating the risk of attacks. The algorithmic procedure for secure data transmission is outlined below:

// Algorithm 1: Pseudo Ephemeral Kupyna Hash Edward Signcryption Based Secure Data Transmission
Input: Dataset, Classified data $CD_1, CD_2, CD_3, \ldots. CD_k$ **Output:** Improve the secure data transmission

Begin
Step 1: Collect the number of classified data $CD_1, CD_2, CD_3, \ldots. CD_k$,
Step 2: for each session
Step 3: enerate the Pseudo ephemeral agreement private and public key 'φ_{pr}' and 'φ_{pb}'
Step 4: end for
// Signcryption
Step 5: For each patient data CD
Step 6: Perform encryption with the receiver public key $\rho_c \leftarrow En(CD, \varphi_{pb}\ (R))$
Step 7: Generate the digital signature (V, u) using sender's private key
Step 8: Send ciphertext (ρ_c) and digital signature$(V, u)$to the receiver
Step 9: End for
// Verification of signature and decryption
Step 10: The cipher text and digital signature is received by receiver
Step 11: Verifies the signature
Step 12: If ($s_1 == s_2$)then
Step 13: Signature is valid
Step 14: The receiver uses their private key to decrypt the data
Step 15: else
Step 16: Invalid signature
Step 17: decoding not carried out
Step 18: end if
Step 19: Data transmission done safely
End

Algorithm 1 illustrates the procedure for ensuring safety transmission of healthcare data through the Pseudo ephemeral Kupyna hash Edward signcryption. The input consists of classified data. Initially, public and private keys are generated for each session using Pseudo ephemeral agreement. Subsequently, encryption of data by the sender is done using the public key of recipient using the and generates a digital signature using their own private key. Then sender transmit the data and digital signatures to the recipient. Digital signature verification is carried out in the recipient's end. The signature is deemed valid only if both the signature matches and the receiver obtain the original data. Conversely, if the signatures do not match, the signature is considered invalid, resulting in the receiver not receiving the original data. This process significantly enhances the authentication and secrecy of data.

4 Assessment Settings

An evaluation of the novel HECS method and conventional method BCSE [1], and PMRSS [2] in Java is carried out in this section. The evaluation is conducted using the WUSTL EHMS 2020 Dataset for (IoMT) Cyber security study, which can be accessed at https://www.cse.wustl.edu/~jain/ehms/index.html. The WUSTL-EHMS-2020 medical dataset was developed using EHMS testbed. Network flow metrics and patients' biometrics are collected from this system. The patients' health level is obtained from sensors attached to their bodies and the data is transmitted to the database set. However, the data before arriving to the server it is intruded by the intruders. Hence, the HECS method assumes accountable to locate and catch various form of attacks during the distribution of the patients' biometric data, thus preventing the integrity and confidentiality of the data. The set of data contains a total of fourty four attribute and 16,000 occurrences. Out of these, thirty-five attributes represent internet flow problems, 8 attributes represent medical biometric and 1 attribute represents the label (0 or 1). Label 1 represents the there is an attacker in occurrence, label 0 represents other occurrences.

5 Performance Comparison

Comprehensive evaluation of HECS method and established methods, namely BCSE [1] and PMRSS [2], is presented in this section. Two, namely Accuracy on classification and Integrity Rate of Data, are utilized to evaluate the performance of these methods. The performance of each metric is thoroughly examined through the use of a table and graphical representation.

Accuracy on Classification: It refers as the proportion of medical data that is accurately categorized into the respective classes. Mathematically, the accuracy is calculated using the formula

$$Acc = \left[\frac{ncc}{n}\right] * 100 \tag{13}$$

The term Acc represents the accuracy on classification, while ncc represents the count of data that have been correctly classified, n signifies the total medical data. The accuracy on classification is expressed in %.

Integrity Rate of Data: It refers the percentage of data remains unaltered or unaffected by any unauthorized intrusions or attacks. This rate is calculated using the following formula:

$$I_{rate} = \left[\frac{NNI}{n}\right] * 100 \quad (14)$$

where, I_{rate} represents the integrity rate of data, *NNI* indicates the data items that have not been changed by any external sources and 'n' represents the total number of data items. The resulting integrity rate of data is expressed in percentage.

Table 1. Accuracy on classification

Data Range	Accuracy on classification (%)		
	HECS	BCSE	PMRSS
1600	96.25	86.25	90.62
3200	96.87	85.93	90.78
4800	97.50	86.45	91.87
6400	97.65	86.09	92.03
8000	96.25	88.12	90.62
9600	97.39	88.02	91.77
11200	96.87	86.6	89.10
12800	97.26	87.89	91.79
14400	97.91	88.19	91.66
16000	96.87	87.5	91.25

The performance analysis of the accuracy on classification for the proposed HECS method, along with two existing methods, namely BCSE [1] and PMRSS [2], is presented in Table 1. The experiments were conducted using patient data ranging from 1600 to 16000 from the dataset. Among the three methods, the HECS method demonstrates a higher performance in terms of classification than the conventional methods. Statistical estimation is used to prove the classification correctness. To illustrate, when considering a dataset of 1600 data points, the accuracy achieved by applying the HECS method is 96.25%, whereas the existing methods [1] and [2] achieve accuracies of 86.25% and 90.62% respectively. Similar experiments were conducted nine times for each method, resulting in varying outcomes depending on the number of data points. Ultimately, the HESC method efficacy is correlated with conventional method. Table 1 overall performance results indicate that the classification accuracy of HECS approach increase by 11% than in BCSE [1] and 7% than in PMRSS [2].

As illustrated in Fig. 4, a performance comparison on accuracy of classification was conducted using three distinct methods, namely HECS and BCSE [1], and PMRSS [2]. The horizontal axis of the graph represents the data ranges, whereas classification

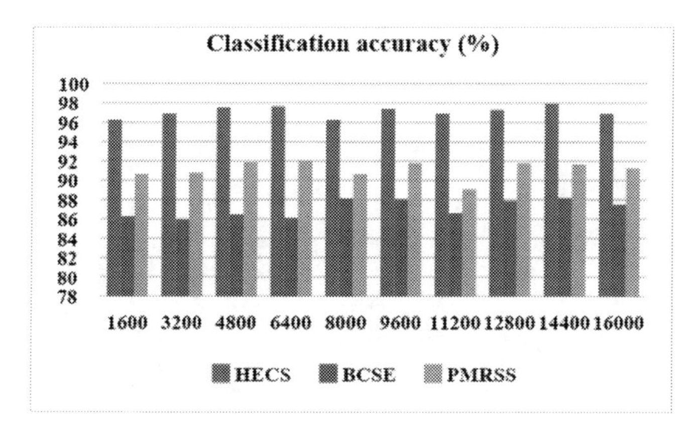

Fig. 4. Accuracy on Classification Comparisons

accuracy is represented in vertical axis. The finding specifies that the suggested HECS method outperformed existing methodology. This can be attributed to the utilization of Theil-Sen Robust Linear Regression for classifying the data into distinct categories. The classification technique scrutinized the medical data and median of the different classes based on Prevosti's distance. Subsequently, the damped least square method is employed to determine the minimum distance and classify the data into a specific category.

Table 2. Integrity Rate of Data

Data Range	Integrity rate of Data (%)		
	HECS	BCSE	PMRSS
1600	95.31	85.31	89.06
3200	95.62	85.62	87.5
4800	95.83	85.66	89.58
6400	96.87	85.62	89.06
8000	95.37	86.87	89.37
9600	96.87	87.5	90.83
11200	96.42	86.16	87.94
12800	96.87	86.95	89.53
14400	97.43	87.91	90
16000	96.37	86.12	90.03

Table 2 and Fig. 5 exhibit the outcomes of integrity rate of data achieved through three distinct methods, namely HECS and BCSE [1], PMRSS [2], by using the data gathered from the classified database. In order to conduct the experiment, data ranging from 1600 to 16000 were considered. Compared to the existing performance results, the HECS method demonstrated that the integrity rate of data attained is superior that

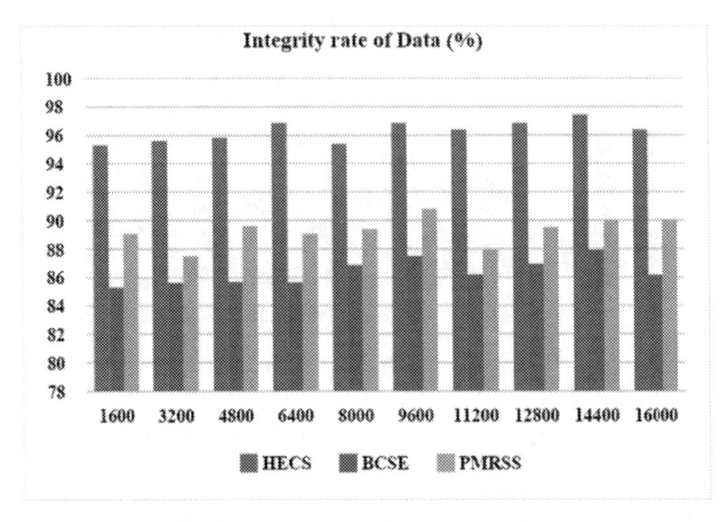

Fig. 5. Integrity Rate of Data Comparisons

the conventional methods. For the purpose of this experiment, 1600 data points were selected. The data integrity rate using HECS was found to be 95.31%. Similarly, the data integrity rates for techniques [1] and [2] were observed to be 85.31% and 89.06% respectively. Furthermore, different results were obtained for each method and subsequently compared. The comparison results indicate that the HECS method enhances the rate of data integrity by 11% in comparison to [1] and an 8% improvement than in [2]. This can be attributed to the fact that HECS performs signature verification by employing the Edward curve verification algorithm. Additionally, the Kupyna cryptographic compression function is utilized to generate the hashes for medical data provided. During the signature generation process, the Kupyna cryptographic hash function is applied in conjunction with the Davies-Meyer compression function to generate the hashes. Consequently, the data remains unaltered by any unauthorized individuals, thereby enhancing the data integrity rate.

6 Conclusion

In this research paper, to deal with huge healthcare data security a new security technique called HECS is implemented which based on Theil-Sen Robust Linear Regression and Pseudo ephemeral Kupyna Hash Edward Signcryption Cryptographic technique. HECS methods collect a variety of data and perform the data analysis using Theil-Sen Robust Linear Regression. The regression function classifies similar types of data into different classes. Finally, Pseudo ephemeral Kupyna hash Edward signcryption is employed in HECS for secure data transmission with the help of key generation, signcryption, and unsigncryption. In this way, the HECS method is used for achieving higher data integrity during data transmission. The experimental assessments are conducted in relation to various datasets and the outcomes are compared. The quantitatively analyzed numerical results indicate that the proposed HECS methodology exhibits superior performance in terms of accuracy on classification and integrity rate of data when compared to other existing methodologies.

References

1. Li, F., Yu, X., Ge, R., Wang, Y., Cui, Y., Zhou, H.: BCSE: blockchain-based trusted service evaluation model over big data. Big Data Min. Anal. **5**(1), 1–14 (2022). https://doi.org/10.26599/BDMA.2020.9020028
2. Sun, Y., Liu, J., Yu, K., Alazab, M., Lin, K.: *PMRSS*: privacy-preserving medical record searching scheme for intelligent diagnosis in IoT healthcare. IEEE Trans. Ind. Inf. **18**(3), 1981–1990 (2022). https://doi.org/10.1109/TII.2021.3070544
3. Bataineh, M.R., Mardini, W., Khamayseh, Y.M., Yassein, M.M.B.: Novel and secure blockchain framework for health applications in IoT. IEEE Access **10**, 14914–14926 (2022). https://doi.org/10.1109/ACCES.2022.3147795
4. Wang, K., Chen, C.-M., Tie, Z., Shojafar, M., Kumar, S., Kumari, S.: Forward privacy preservation in IoT-enabled healthcare systems. IEEE Trans. Ind. Inf. **18**(3), 1991–1999 (2022). https://doi.org/10.1109/TII.2021.3064691
5. Wang, B., Li, Z.: Healthchain: a privacy protection system for medical data based on blockchain. Future Internet **13**(10), 1–16 (2021). https://doi.org/10.3390/fi13100247
6. Besher, K.M., Subah, Z., Ali, M.Z.: IoT sensor initiated healthcare data security. IEEE Sens. J. **21**(10), 11977–11982 (2021). https://doi.org/10.1109/JSEN.2020.3013634
7. Alzubi, J.A.: Blockchain-based Lamport Merkle digital signature: authentication tool in IoT healthcare. Comput. Commun.. Commun. **170**, 200–208 (2021). https://doi.org/10.1016/j.comcom.2021.02.002
8. Chamola, V., Goyal, A., Sharma, P., Hassija, V., Binh, H.T.T., Saxena, V.: Artificial intelligence-assisted blockchain-based framework for smart and secure EMR management. Neural Comput. Appl.Comput. Appl. **35**, 22959–22969 (2022). https://doi.org/10.1007/s00521-022-07087-7
9. Zhang, D., Wang, S., Zhang, Y., Zhang, Q., Zhang, Y.: A secure and privacy-preserving medical data sharing via consortium blockchain. Secur. Commun. Netw. **2022**, 1–15 (2022). https://doi.org/10.1155/2022/2759787
10. Prasanalakshmi, K.M., Karthik Srinivasan, S., Shridevi, S.S., Yu-Chen, H.: Improved authentication and computation of medical data transmission in the secure IoT using hyperelliptic curve cryptography. J. Supercomput.Supercomput. **78**, 361–378 (2022). https://doi.org/10.1007/s11227-021-03861-x
11. Kasım, O.: An efficient ensemble architecture for privacy and security of electronic medical records. Int. Arab J. Inf. Technol. **19**(2), 272–280 (2022). https://doi.org/10.34028/iajit/19/2/14
12. Denis, R., Madhubala, P.: Hybrid data encryption model integrating multi-objective adaptive genetic algorithm for secure medical data communication over cloud-based healthcare systems. Multimedia Tools Appl. **80**, 21165–21202 (2021). https://doi.org/10.1007/s11042-021-10723-4

An Energy Efficient, Spontaneous, Multi-path Data Routing Algorithm with Private Key Creation for Heterogeneous Network

K. E. Hemapriya[1](✉) and S. Saraswathi[2]

[1] Sri Krishna Arts and Science College, Coimbatore, India
hemapriyake@skasc.ac.in

[2] Nehru Arts and Science College, Coimbatore, India

Abstract. In the recent years many investigation studies are carried out in heterogeneous sensor networks owing to its countless applications in all fields like health care, defense, environment surveillance, industrial based applications etc. In general, heterogeneous network have many fruitful advantages as precise sensing of parameters, manipulation of sensed data and passing sensed dataset to a receiver point in an effective manner. This type of network may be incorporated in a large-scale environment in an unstructured network with energy limitations. As known that sensor has a defined amount of power, memory and coverage area. For realistic implementation of nodal point in the investigation premises, it should be defined in an effective manner both in physical and coverage metrics. Hence it is unavoidable to propose a methodology for the effective usage of power conservation to achieve an optimistic performance. Several research studies were carried out in former years based on Fuzzy technique, Neural Networks, genetic-algorithm and many more improved approaches. This paper proposes a Population based (Ant Colony Optimization-ACO) investigative approach for a multi-path data forwarding to heterogeneous network as spontaneous energy proficient multi-path data routing (SEPMDR). In the initial stage, the proposed approach discovered its efficient neighboring sensor nodal before forwarding data packets with improvised data security factor. The privacy in the data transmission is implemented by using security key generation for each data transmission carried out in the heterogeneous network. The stability of the proposed technique in terms various network parameters with previously defined research strategies is evaluated and simulation results are conferred with Network Simulator Tool.

Keywords: WSN · Heterogeneous Network · Data routing · Power optimization

1 Introduction

Heterogeneous sensor network is a network with pre-defined count of nodal points. The sensor nodes are a handy module with a pre-defined power storage that depends upon coverage capable of each sensor module [1]. It is enabled to take decision on the mobility over the sensed area to communicate to other nodal points for performing data hopping.

S. Aurelia et al. (Eds.): ICCSST 2023, CCIS 1973, pp. 248–263, 2024.
https://doi.org/10.1007/978-3-031-50993-3_20

The hopping will be carried out to the surrounding nodes for successful data packet transmission to nearest Base Station. Each and every nodal point is used to sense various analytical parameters like atmospheric pressure, air humidity tracking and surveillance.

The sensor nodes in a network structure may either be an organized type or unorganized type [2]. An outlined schematic representational distribution of sensor nodal points in the sensing area connecting it with a base station is shown in Fig. 1.

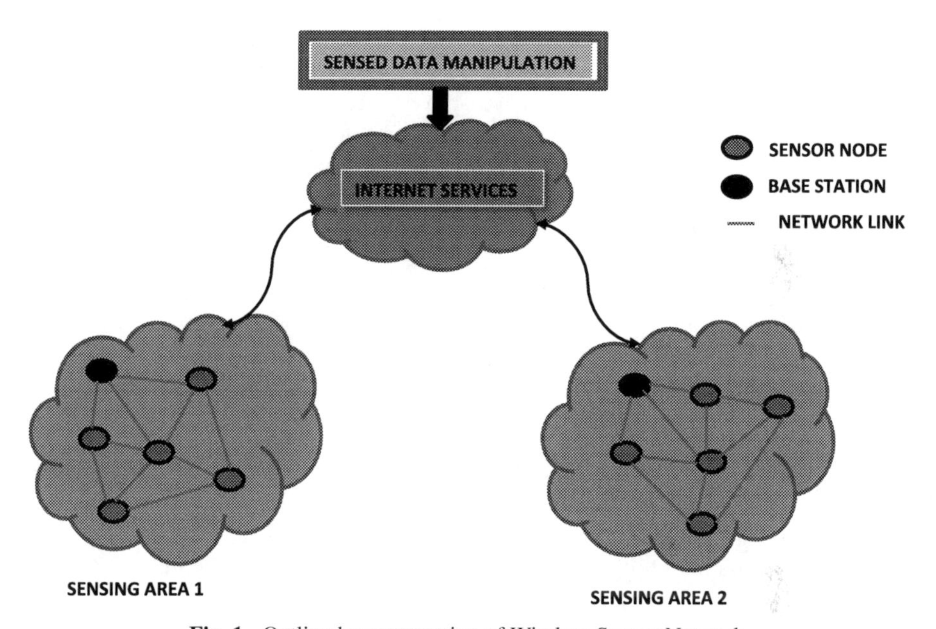

Fig. 1. Outlined representation of Wireless Sensor Network

For an organized type of network arrangement, the creation of sensor nodes link will be in a defined format. It will be able to sense the data in a controlled manner with prescribed routing paths between the neighboring nodes. In this type of network structure, occurrence of error in the data transmissions is minimum. On other hand, unorganized type sensor nodes will be deployed in large scale realistic application. The occurrence probability of error will be high due to the varying network characteristics of the network structure.

For avoiding the error occurrence in an unorganized type, sensor nodes will be equipped with self-identification of surrounding nodes in the network for establishing a confirmed network link. This will make an uninterrupted data transmission between source node and base station. If the link gets disconnected in between the data transmission, the process of establishing confirmed network link will be initiated again.

1.1 Difficulties in Formation of a Heterogeneous Network

The challenges faced in the organization of heterogeneous sensor network are discussed as [3],

- ***Fault tolerance:*** Nodes in a network may enter into inactive or sleep mode at any time of data transmission due to decreased power backup. The capability of a network to withstand erroneous situation and continuous data transmission as before is termed as fault tolerance. In a heterogeneous type of network, formation of a stable fault tolerance will be harder as its network characteristics will vary in accordance to network load.
- ***Topology used:*** In general, a heterogeneous network will have varying topology for each of data transmission. Hence, an optimized topology structure must be followed in the network formation.
- ***Scalability:*** A sensor network will have a pre-defined nodal point depending on their position and coverage. For increasing coverage inside a network, it may be required to raise the count of sensor nodes. This is known as scalability. But it will increase the sensor density of a network structure.
- ***Transmission channel:*** The path through which a data hopping is given is known as the transmission channel. For a heterogeneous type, the path will be unstable as its network load is a changeable metrics. Hence it is difficult to decide on an optimized transmission channel.

In a sensor network, nodes maybe distribute all over the sensing area. It helps data transmission to be carried out in a multi-hop manner. This effects sensor nodes near to the base station and may lose its power in the process of data forwarding to the base station [3–5]. It makes use of optimized data routing strategy for multi-hop data transfer. The above discussed issues make degradation in performance and conventional data routing mechanism.

By focusing on above discussed challenges for enhancing lifespan of sensor nodes in a heterogeneous network, it is possible to highlight the factors to be considered on the development of proposed strategy. They are listed below:

1. Selection of optimized routing path for transmission of data packets from sender nodal point to receiver nodal point.
2. As a heterogeneous network is operational in all time spans, maintenance of power backup in the nodes during stable condition is a challenging aspect.
3. Optimization of other basic network metrics is also a difficult factor.

To resolve the above stated issues, data forwarding frequency based methodology will be a suitable methodology for maintaining efficiency of the heterogeneous network structure. This methodology is based on Intelligent Swarm based colony optimization. In this methodology, the data route, in which data packets are passed maximum number of times, will have high frequency of data forwarding. In turn, it implies the corresponding path as a shortest for all data routing process.

In this research study a novel routing methodology is proposed based on colony optimization mechanism for a heterogeneous network. The heterogeneous network characteristics have varying aspect depending on its network load. The proposed strategy (SEPMDR) will sustain given network load and make the basic network metrics stable. The remaining part of the paper is structured as Sect. 2 which supports literature survey of the proposed system followed by Sect. 3 entitled the basic perception of intelligent swarm based optimization mechanism. The Sect. 4 explains about proposed system with simulation results depicted in Sect. 5. The Sect. 6 provides conclusion and highlights of the proposed methodology.

2 Literature Survey

In this segment, an inclusive survey of all investigation work supporting the proposed system is studied elaborately with their limitations on their corresponding work. It is stated as follows.

The research strategy described in [6] explains remarkable capabilities of a small creature (ant) to successfully complete intricate task through a chemical substance, Pheromone. This will make others to find the path towards their nest. Likewise the path which has the maximum probability of data forwarding was chosen as the optimal path for forgoing data routing cycles. If a path is selected on their frequency, nodes lying in the corresponding path will get drained shortly.

Hence, the aspect of choosing a path on data passing frequency may not be an optimal solution. In [7] a basic network parameter table was maintained in all the nodal points of a network. If any data transmission has to occur, initially an ant must identify the optimal path for data transmission and update the node ID in all network tables from source to destination. This procedure may happen for all data transmission. This updating process will make delay in data transmission. Hence this updating must be handled by another optimized mechanism.

A power aware mechanism in [8] explains about power usage of all nodal point in a heterogeneous network for both control packets forwarding and usual data packet forwarding. The mechanism stated attains a stable maintenance of power usage at every node in a network and decreases by 52% than their usual energy consumption.

In [9] a specific node will be assigned as an administrative node for all the basic operations in the data manipulation. Before starting any data transmission, administrative node will check for an optimal route for transmission. Then, it will initiate data transmission between source and base station. This strategy gives a sustainable performance in network efficiency. But the processing time span to decide on data route may add unwanted network delay.

A cluster based methodology is described in [10]. It tells about dividing of entire network into smaller regions termed as *cluster* for making a feasible data transmission between the nodes in a cluster structure. It is effective for data transmission within a cluster, but there may be a delay in transmission between two neighboring clusters. Hence, this mechanism must be improved for inter cluster data transmission.

In [11] an effective way of data routing methodology, LMRSPM (Local Monitoring, Route Scheduling, Planning Manager) using Ant centered data collection among the sensor nodes. Here in forward direction a moving nodal point could be used to traverse through the entire network on optimized data routing process. In this, each node will have a basic network metrics table, having details of every data hopping cycle. If it is a new data hopping process, it will get updated in the table. Once it reaches the destination, it will be redirected in opposite direction towards the source for the confirmation of customized data path. The performance of this strategy shows a remarkable improvement in data transmission with low data error rate.

An efficient routing methodology described in [12, 13] has two stages of data routing optimization. First stage is data route identification over the sensor nodes in a network. An ant centered module traverse over the network and intimate nodes regarding the data transmission. Sensor nodes will update details regarding data transmission and starts data transmission successfully. If there is data route failure second stage will take care of route discovery which makes the data transmission to be in an effective manner.

An efficient protocol named as, MCSBCH (Modern Cluster Supervisor Based Cluster Head) is explained by [14] in which basic network metrics will be manipulated in realistic and current status for route discovery. The updating of data route in all sensor nodes which lies in the data route is done by forward directed Ant-structured modules. At that time, the current power backup of each node will be analyzed, and then the route will be finalized. This will avoid the circumstances of node failure due to power draining issues. This method of route discovery shows an improved performance while comparing it to conventional route discovery mechanism like S-LEACH, AODV.

2.1 Contribution of Proposed Research Mechanism

From the above high-lightened survey, it could be understood that there are numerous approaches for the improvement of energy conservation, data route discovery, route reconstruction for data transmission in a heterogeneous network structure. But still, further energy optimization maybe attained by considering current network overhead, data packet load and overall throughput.

3 Intelligent Swarm Adapted Colony Based Optimization Methodology

In the swarm based strategy, an ant-sized creature runs in their habitation area in search of its foodstuff in a random manner. While it can return to its living location, a chemical-based substance will disposed out from the small sized creature. For other ant-sized creatures, the chemical substance will be a reference path between their nest and foodstuff. The more chemical substance that is disposed, the higher is the use of that corresponding route. This implies that it is the shortest path. Hence for transferring the foodstuff it will use the shortest path. Using the same characteristics, shortest path will be identified for making a successful data transmission in heterogeneous sensor network. A drafted way of representation of Swarm optimization is depicted in the Fig. 2.

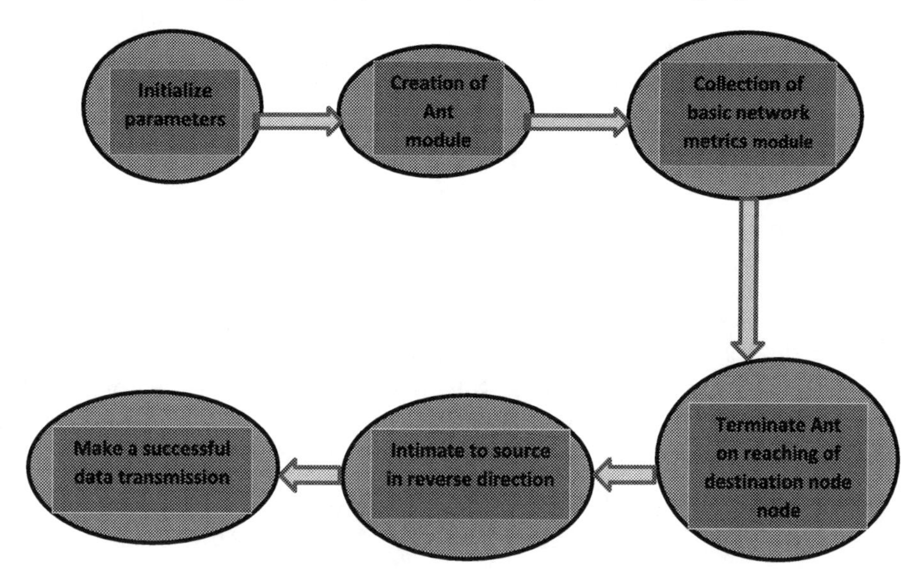

Fig. 2. Flow Mechanism of Swarm Optimization Mechanism

4 Spontaneous Energy Proficient Multi-path Data Routing (SEPMDR)

In this segment, the formation of basic network metrics and operational flow of proposed system will be discussed.

4.1 Formation of Basic Network Metrics

Here, the proposed system's network QoS (Quality of service) metrics like power assigned, network throughout, network delay etc., will be formulated in the steps given below.

***Step 1*:** Assume a source node as N_{src} which wants to transmit data to a destination node N_{des}. This may be manipulated by analyzing all basic network metrics. For making an optimal data route all other nodes will be visited and basic *network routing node table* will be structured as R_{table} with hopping distance and other QoS parameter details for every node.

Step 2: For initialization, the packet transmission source node will pass over all the nodes and forms the details of network hopping distance as 1-hop, 2-hop separately in R_{table}. This will be done by forward directed-Ant termed as Ant_{frwd}.

Step 3: The selection of suitable data route is decided by analyzing data passing frequency (high evaporation of pheromone stage). The path having higher probability will be elected as optimal data route for data transmission.

Step 4: In the meantime, when Ant_{frwd} reaches destination node, it will be reversed by Backward Directed-Ant termed as $\text{Ant}_{\text{bckwd}}$. This is for conforming data route from N_{src} to N_{des}.

Step 5: Finally, the data packet transmission is carried over elected data path without any network interruption.
Step 6: For each of the data transmission a private key will be generated. This is to increase security of the data transmission.

The flow mechanism of the proposed methodology is explained in the Fig. 3a. In the generation of private key, as shown in Fig. 3b, each data transmission will generate a key for all the active nodes in the network structure. It also gets updated in the key manager module. This is useful for reconstruction of data route in case of node failure. Finally, a performance analysis is taken on each of the data transmission.

The performance consists of manipulation of various network metrics and has an updated list of active nodes after completion of data transmission. This will help to eradicate inactive nodes in the forthcoming data transmission.

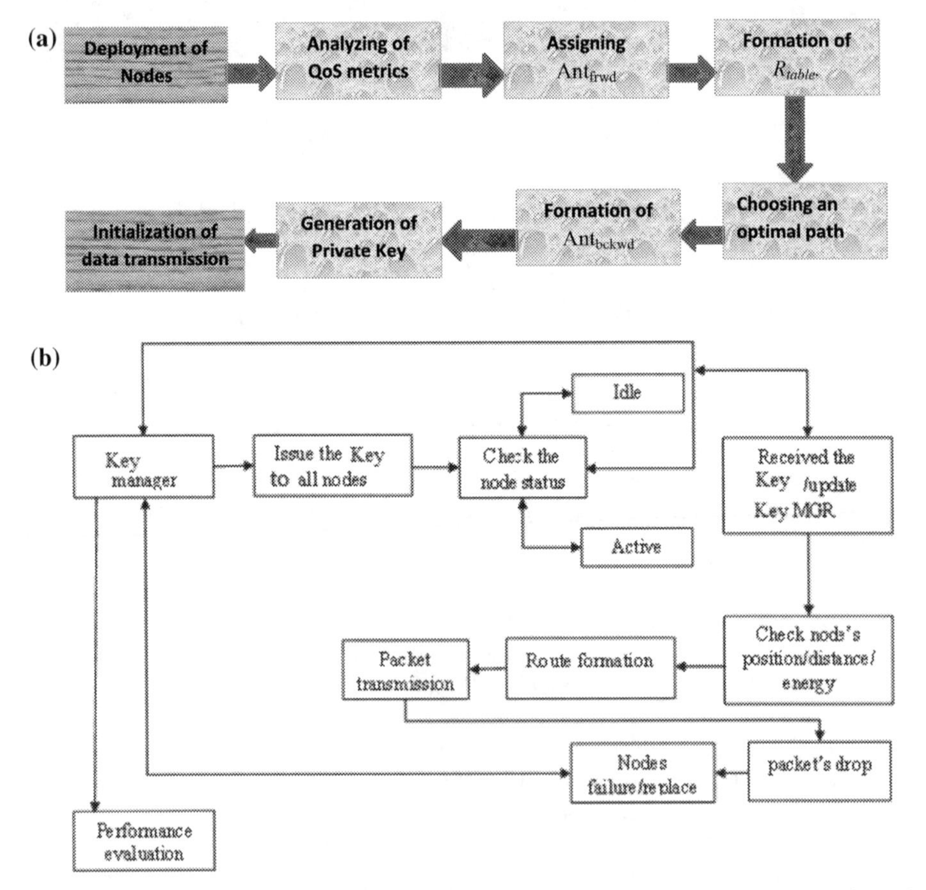

Fig. 3. a. Flow Mechanism of Proposed Methodology SEPMDR b. Private Key generation of Proposed Methodology SEPMDR c. Selection of optimal Data route of Proposed Methodology (Color figure online)

(c)

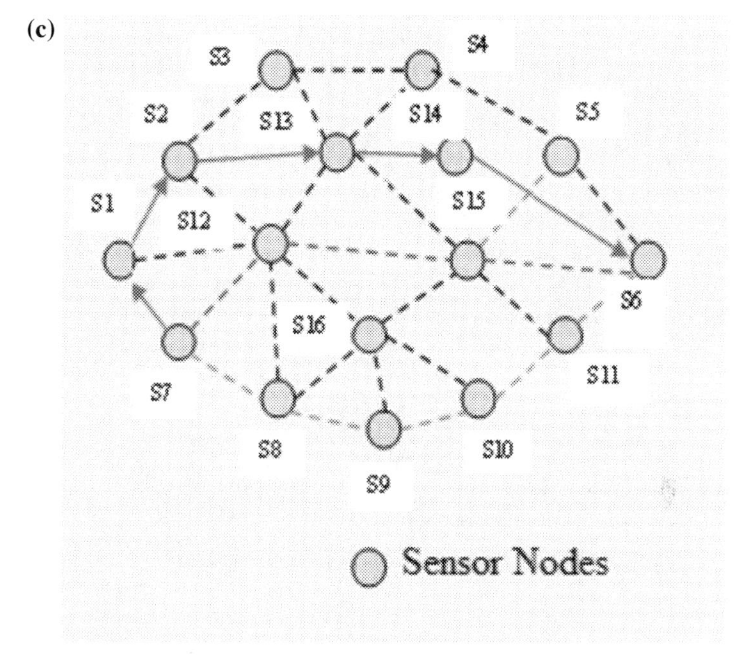

Fig. 3. *(continued)*

In Fig. 3(c), the red line (busy path) indicates that data transmission is still occurring between S7 and S6, the green line indicates that the path is present (S7 to S6), and the blue line indicates that the path is not being reused for transmission (S7 to S6). Some sensor nodes are idle and have no communication requirements.

4.2 Algorithm 1 for SEPMDR

Step 1: Nodes *Sn* will be deployed in the sensing Area.
Step 2: Initial *Start-up Message* will be sent.
Step 3: A stable network link will be established by replying to the *Start-up Message,* with initial power as p(i).
Step 4: Node will send data termed as N_{src}.
Step 5: An Ant_{frwd} is designated for calculating hopping distance between adjacent nodes using the relation., $S_n(d) = \sqrt{(x_i - x_j)^2 + (y_i - y_j)^2}$ sensor nodes distance between neighboring nodes. $(x_i, x_{j,} y_i, y_{j-XY\ Coordiantes})$.
Step 6: Network Hopping distance of all adjacent nodes are structured in the R_{table}.
Step 7: The path having the highest data frequency is elected as optimal route for data transmission.
Step 8: For conforming link between source node and receiving node an $\text{Ant}_{\text{bckwd}}$ will be redirected towards the source node.
Step 9: Initiate data transmission between source N_{src} and destination node N_{des}. For each data transmission, a private key will generated for increasing the security in data transmission.

4.3 Algorithm 2 for Private Key Creation

Step 1: Deploy the key to all nodes.
Step 2: Check the node status.
Step 3: If Node is Idle, Key creation will be restarted from Key Manager.
Step 4: If Node is Active, it will be updated in Key Manager.
Step 5: Verification of Node's Position/Distance/Energy for Route Formation.
Step 6: Packet Transmission Started.
Step 7: If there is any Drop/Failure in Data Packet Transmission appropriate node will be replaced by Key manager (Key mgr).
Step 8: Performance evaluation of deployed nodes and there by a Private Key is created (Table 1).

Table 1. Summary of notations

Symbol	Description
Sn	Number of sensor nodes in the network
$Nsrc$	Source node
N_{des}	Destination node
$S_n(d)$	Node distance
R_{table}	Network nodes table
Ant_{frwd}	Forward directed-Ant termed as $\mathrm{Ant}_{\mathrm{frwd}}$
Ant_{bckwd}	Backward Directed-Ant termed as $\mathrm{Ant}_{\mathrm{bckwd}}$
p(i)	Initial energy

4.4 Operational Phases of SEPMDR

The proposed SEPMDR has the following stages of operational phases for electing an optimal data route for data transmission.

4.4.1 Establishing Network Link

At the first stage of network formation, nodes are distributed in all sensing areas of application. Next, the nodes have to form a link between all of the adjacent nodal points. This is to confirm a network path between all of the nodes. It is fulfilled by passing a *startup message* to all of its adjacent nodes. On replying to this message, each node has to pass their basic network metrics to other nodes in the network structure.

This type of control messages may be sent periodically to a heterogeneous network. As heterogeneous network characteristics will vary according to the network load.

4.4.2 Passing of Data Packet

Assume that a node designated as N_{src} wants to send data packet to destination node given as N_{des}. The node has to send startup message to all of its adjacent nodes by Ant_{frwd}. On getting the reply from all other nodes, the active node list gets updated in the network table. The path having maximum probability will be picked as data route and transmission will be carried over the route. A private key will be generated for the transmission.

4.4.3 Sustainable Maintenance of Data Route

The selection of data route is fulfilled by the path having maximum probability. In the meantime, Ant_{bckwd} will traverse in reverse direction from destination nodal point to source point. This is to confirm a stable link between source and destination nodal points. Also the path is followed by the private key generated for each data transmission and passed over to all the nodes in the network. While a message is received, each node will check the key. If the key gets matched, then it will hop to the next stage or it will be dropped by the corresponding node.

Using this technique, it is possible to check in the intermediate stage of network path for any data congestion that occurs. If the congestion persists, it will be able to re-transmit data packet from the point where the congestion occurred.

5 Performance Evaluation and Its Results

Table 2. Basic Simulation Parameter

Parameters	Values
Operating System	Ubuntu
Data Routing mechanism	Basic Ant based Strategy
Simulator	NS(all-in-one)-2
Antenna Type	Omni Antenna
Model used for Simulation	Power Aware Model
Starting power status of Nodes	1000 J
MAC Type	IEEE 802.11
Network Dimension	3000 m × 1000 m
Type of network	Wireless mode

For performance evaluation of the proposed system, all sensor nodes are randomly placed in the sensing area and its positions are noted down with reference to XY Coordinates. Depending on the coverage area of each node the position may vary. In particular, any of the application area to be precisely sensed more number of sensor nodes will be positioned in the location. The randomly positioned sensor nodal points are depicted in the Fig. 4 with its reference in all three axes X, Y, Z planes. The basic simulation parameters for the proposed parameters are given in Table 2.

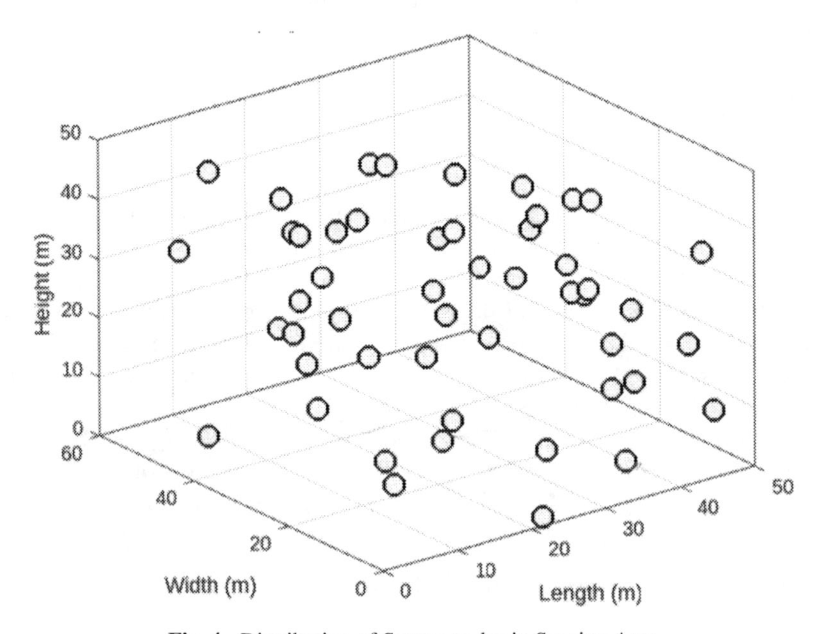

Fig. 4. Distribution of Sensor nodes in Sensing Area.

5.1 Power Conservation of Nodes

In a heterogeneous sensor network, the network performance will vary in accordance to the network on that particular time. Hence it is important to conserve the energy backup of all live nodes in a sensor network to maintain stability. The proposed mechanism achieves a sustainable power saving strategy for a heterogeneous sensor network up to 20%, when compared to other methodologies as shown in the Fig. 5 and it is tabulated in Table 3.

Table 3. Energy Consumption of nodes

Simulation Time (ms)	SEPMDR (Joules)	MCSBCH (Joules)	LMRSPM (Joules)
1	177	182	216
2	163	213	213
3	152	212	219
4	143	187	207
5	135	186.2	217
6	121	184	221

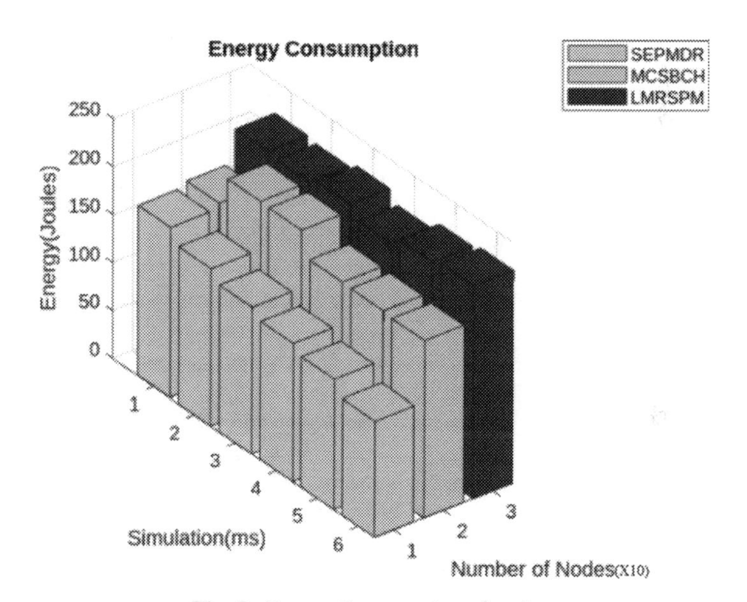

Fig. 5. Energy Consumption of nodes.

5.2 Comparison of Packet Delivery Ratio (PDR) of the Proposed System

For a data transmission to be carried over in a heterogeneous network, packet delivery ratio must be high when compare to other mechanisms. In a heterogeneous network, load is floating metrics, so PDR must be maintained in same level. This implies that the performance of a methodology should be stable under any circumstances in a network structure. The outcomes of the proposed system are given in Fig. 6 and the achieved PDR is insisted in Table 4.

Table 4. Packet Delivery Ratio of various Research Studies

Simulation Time (ms)	MCSBCH	LMRSPM	SEPMDR
1	87.76	85.41	95.64
2	85.50	83.01	96.69
3	82.70	80.13	97.60
4	79.87	77.23	98.84
5	77.73	75.08	99.31
6	74.97	83.06	99.84

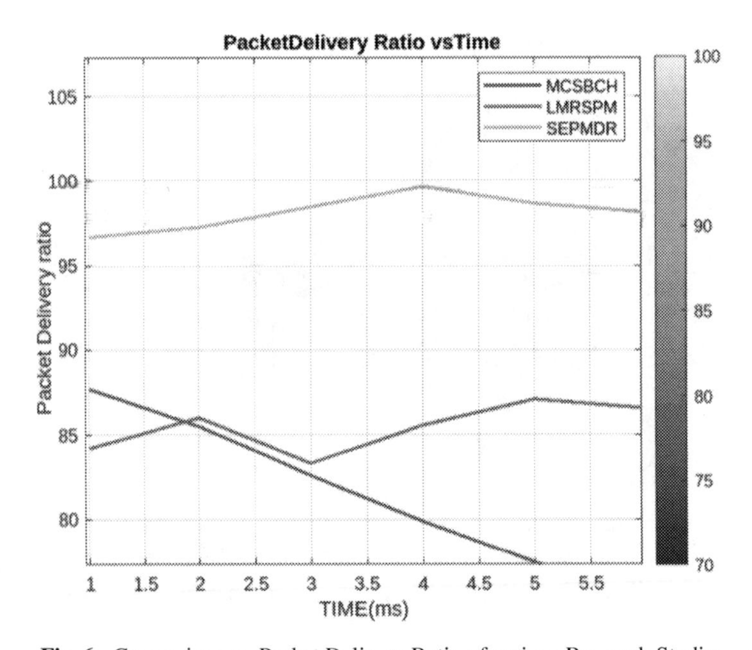

Fig. 6. Comparison on Packet Delivery Ratio of various Research Studies

5.3 Evaluation of Routing Overhead of the Different Methodologies

Generally routing overhead is a metrics which gives an account of the amount of network control packets needed for data route reconstruction, optimal data route selection, analyzing of network performance in a periodic manner. The mechanism which is having a minimum number of control packets will have low routing overhead. This implies that the corresponding mechanism is working in an efficient manner. Based on this factor, the proposed system accomplishes a reasonable minimum routing overhead when compared to other methodologies, as shown in Fig. 7 and a comprehensive view of required routing overhead is tabulated in Table 5.

Table 5. Routing Overhead of various mechanism

Simulation Time (ms)	SEPMDR (bytes)	MCSBCH (bytes)	LMRSPM (bytes)
1	186.3	196.2	217.6
2	210.4	223.1	210.4
3	204.2	214.3	218.5
4	184.1	186.4	223.8
5	179.4	193.7	205.3
6	174.9	173.6	222.8

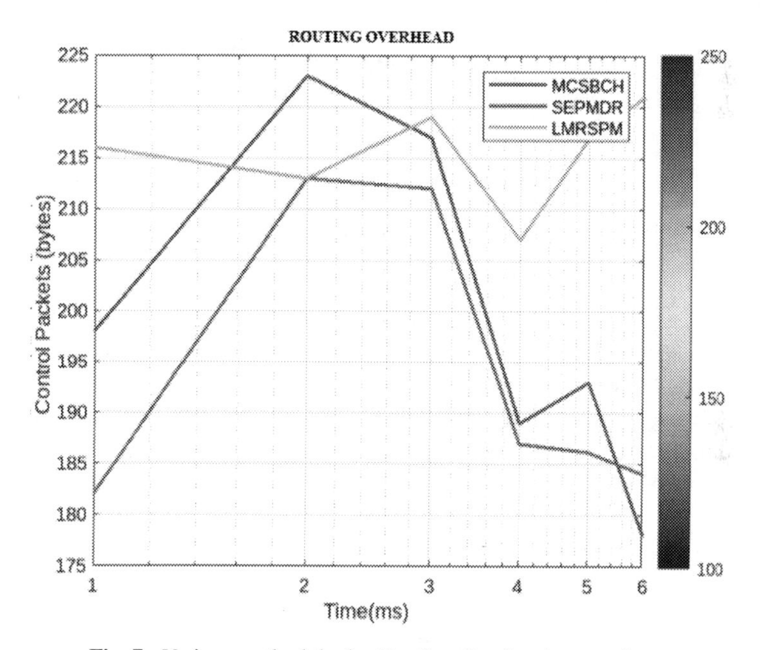

Fig. 7. Various methodologies Routing Overhead comparison

5.4 Comparison of Network Throughput

It is a metrics that tells about how successfully a data transmission is carried out between a source point and destination point. For optimizing this factor, it is required to make a trade between the node's power conservation and the amount of data transmission finished. Hence, the methodology possesses a maximum throughput it maybe said that the mechanism works efficiently in data transmission and maintain a sustainable performance in its entire network situation. From the diagrammatical representation on Fig. 8, it is shown that the proposed methodology has a higher throughput in comparison to various methodologies.

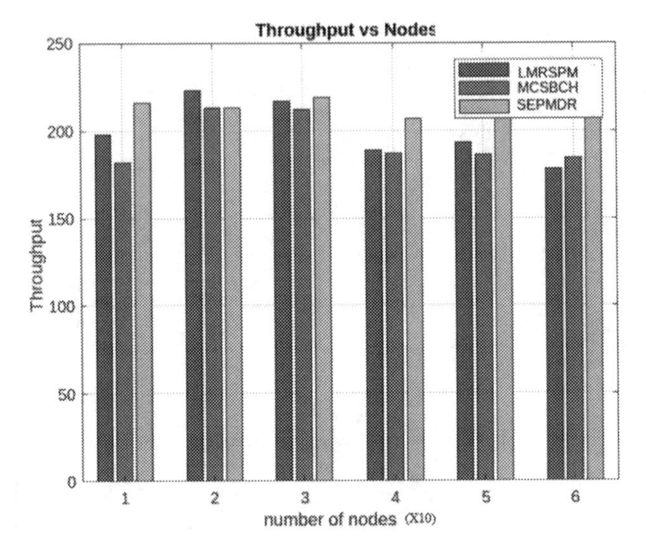

Fig. 8. Comparison of network-throughput

6 Conclusion and Future Scope

Today with the developments in data handling strategies, wireless mode of data communication is an unavoidable technique. This is due to their data handling and manipulation capability of nodal points in a network structure. The path must possess maximum immunity against various factors affecting network performance in terms of basic metrics like network throughput, routing overhead, etc.

In spite of implementing an effective methodology for making an optimal data transmission between source and destination points, the data security becomes another important factor on affecting data privacy of a network. Hence, on structuring an efficient data routing mechanism for heterogeneous network it is required to keep an eye on both optimal data transmission that takes place in multipath with maximized data privacy. So by giving importance to these issues, the proposed methodology in the research is SEPMDR. The proposed system has a capability of making the finest data transmission in a multipath manner. It also increased data privacy by generating a separate private key for each and every data transmission carried over in a heterogeneous network. In future, the performance of the proposed system can be increased by implementing an external intrusion detection strategy for making a sustainable heterogeneous network against any type of fraudulent attacks from the external world.

References

1. Deepika, D., Anand, N.: Complete scenario of routing protocols, security leaks and attacks in MANETs. J. Proc. IJARCSEE **3**(10) (2013)
2. ul Hassan, M., Al Awady, A., Mahmood, K., Ali, S., Saeed, M.K.: Node relocation techniques for wireless sensor networks: a short survey. Int. J. Adv. Comput. Sci. Appl. IJACSA **10**(11), 323–329 (2019). https://doi.org/10.14569/ijacsa.2019.0101145

3. Nayyar, A., Singh, R.: IEEMARP-a novel energy efficient multipath routing protocol based on ant Colony optimization (ACO) for dynamic sensor networks. Multimed. Tools Appl. **79**, 352221–435252 (2020). https://doi.org/10.1007/s11042-019-7627-z
4. Saeed, M.K., ul Hassan, M., Mahmood, K., et al.: Efficient solution for connectivity restoration (ESCR) in wireless sensor and actor-networks. Wirel. Pers. Commun.. Pers. Commun. **117**(3), 2115–2134 (2020). https://doi.org/10.1007/s11277-020-07962-3
5. Kaur, M., Sarangal, M., Nayyar, A.: Simulation of jelly fish periodic attack in mobile ad hoc networks. Int. J. Comput. Trends Technol. (IJCTT) **15**(1), 20–22 (2014)
6. Hema Priya, K.E., Saraswathi, S.: A survey on cluster stranded routing protocol practices with strategic supervision for endangered data transmission in manet. Des. Eng. (2021)
7. Dorigo, M., Stützle, T.: Ant colony optimization: overview and recent advances. In: Gendreau, M., Potvin, J.Y. (eds.) Handbook of Metaheuristics, vol. 272, pp. 311–351. Springer, Cham (2019). https://doi.org/10.1007/978-3-319-91086-4_10
8. Swaminathan, K., Ravindran, V., Ram Prakash, P., Satheesh, R.: A perceptive node transposition and network reformation in wireless sensor network. In: Iyer, B., Crick, T., Peng, S.L. (eds.) ICCET 2022, vol. 303, pp. 623–634. Springer, Singapore (2022). https://doi.org/10.1007/978-981-19-2719-5_59
9. Mirjalili, S., Dong, J.S., Lewis, A.: Ant colony optimizer: theory, literature review, and application in AUV path planning. In: Mirjalili, S., Dong, J.S., Lewis, A. (eds.) Nature-inspired optimizers. SCI, vol. 811, pp. 7–21. Springer, Cham (2020). https://doi.org/10.1007/978-3-030-12127-3_2
10. Swaminathan, K., Ravindran, V., Ponraj, R., Satheesh, R.: A smart energy optimization and collision avoidance routing strategy for IoT systems in the WSN domain. In: Iyer, B., Crick, T., Peng, S.L. (eds.) ICCET 2022, vol. 303, pp. 655–663. Springer, Singapore (2022). https://doi.org/10.1007/978-981-19-2719-5_62
11. Kumar, S., Nayyar, A., Kumari, R.: Arrhenius artificial bee Colony algorithm. In: Bhattacharyya, S., Hassanien, A., Gupta, D., Khanna, A., Pan, I. (eds.) International conference on innovative computing and communications, vol. 56, pp. 187–195. Springer, Singapore (2019). https://doi.org/10.1007/978-981-13-2354-6_21
12. Swaminathan, K., Vennila, C., Prabakar, T.N.: Novel routing structure with new local monitoring, route scheduling, and planning manager in ecological wireless sensor network. J. Environ. Prot. Ecol. **22**(6), 2614–2621 (2021)
13. Sun, Y., Dong, W., Chen, Y.: An improved routing algorithm based on ant colony optimization in wireless sensor networks. IEEE Commun. Lett.Commun. Lett. **21**(6), 1317–1320 (2017)
14. Mondal, S., Ghosh, S., Dutta, P.: Energy efficient data gathering in wireless sensor networks using rough fuzzy C-means and ACO. In: Bhattacharyya, S., Sen, S., Dutta, M., Biswas, P., Chattopadhyay, H. (eds.) Industry interactive innovations in science, engineering and technology. LNNS, vol. 11, pp. 163–172. Springer, Singapore (2018). https://doi.org/10.1007/978-981-10-3953-9_16
15. Ravindran, V., Vennila, C.: Energy consumption in cluster communication using MCSBCH approach in WSN. J. Intell. Fuzzy Syst. **43**(1), 1669–1679 (2022)

A Hybrid Model for Epileptic Seizure Prediction Using EEG Data

P. S. Tejashwini[1(✉)], L. Sahana[1], J. Thriveni[1], and K. R. Venugopal[2]

[1] Department of CSE, University of Visvesvaraya College of Engineering, Bengaluru 560001, India
tejashwinirnk@gmail.com
[2] Bangalore University, Bengaluru, Karnataka, India

Abstract. More than 65 million people's quality of life is affected by a neurological brain condition epilepsy. When a seizure can be anticipated, therapeutic action can be employed to stop it from happening. The data analysis of epileptic seizures makes use of EEG impulses. In the interim, amplitude integrated Electroencephalography (aEEG) has shown promise in the identification of epileptic episodes. This article describes a method for automatically identifying epileptic episodes in EEG readings. Three steps make up the suggested methodology: preprocessing, feature selection, and classification. The current study proposes a deep learning-based seizure prediction system that includes preprocessing scalp EEG signals, extracting key characteristics implementing convolutional neural networks (CNN), and classifying them with the as distance of vector machines. The put forward framework demonstrates its powerful capability in the automatic the identification of seizures, as evidenced by its competitive hypothetical results on EEG datasets analyzed to state-of-the-art approach with a success rate of 98.57%
.

Keywords: Convolutional Neural Network · Electroencephalography · Support Vector Machine

1 Introduction

Epilepsy is a neurological brain disorder condition that affects more than 1% of the global population and is distinguished by frequent seizures [1]. Although medication and surgical treatments can alleviate symptoms in over 30% of cases, patients continue to experience seizures despite these interventions. To prevent seizures, it is crucial to predict them before they occur. Recording the nerve activity in the brain through EEG impulses can assist in this endeavor. These electrical signals can be obtained through scalp EEG or intracranial EEG (iEEG) recordings, depending on the location of the electrodes [2]. EEG recordings can detect abrupt changes in brain nerve activity linked with neurological disorders.

The plot in Fig. 1 depicts continuous EEG signal recordings from three channels over the course of an hour, revealing distinct states: pre-ictal, ictal, and post-ictal. The preictal

S. Aurelia et al. (Eds.): ICCSST 2023, CCIS 1973, pp. 264–274, 2024.
https://doi.org/10.1007/978-3-031-50993-3_21

state, which occurs 30 mins before a seizure, is particularly informative as it provides insight into the onset of a seizure [3]. Due to the distinct differences in amplitude and frequency between preictal and inter ictal states, accurate categorization of these signals can result in the epileptic seizure's prediction. Preventing seizures with medicine can be accomplished by spotting the launching of the preictal state as soon as possible using techniques shown in Fig 2 and 3. [4]. EEG signals can be recorded using headsets which can then be converted to digital format and saved for processing at sampling rates ranging from 200 Hz to 5000 Hz. A neurologist then uses specialised software to annotate these impulses in order to pinpoint the launching and conclusion of seizures. The period of time between 30 and 90 mins prior to the beginning of a seizure is known as the preictal condition.

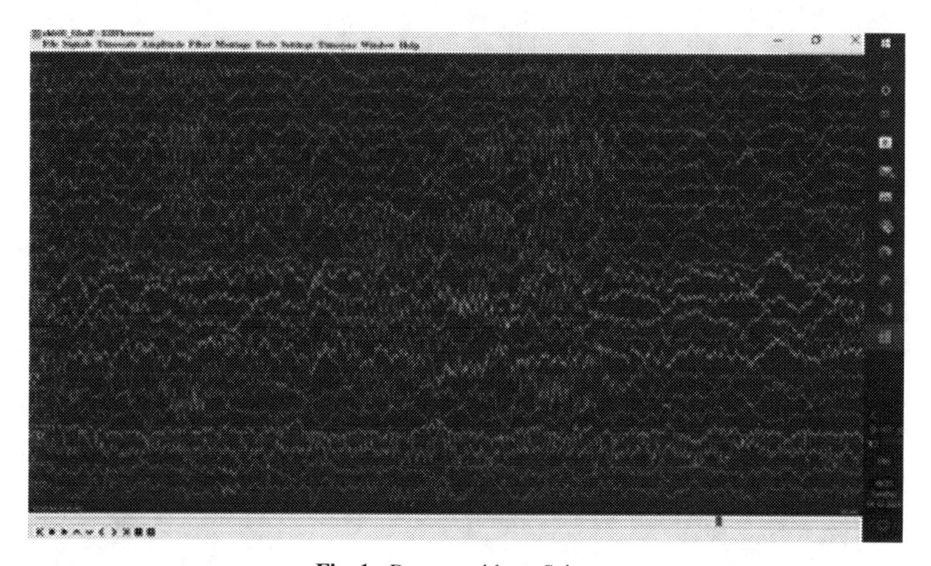

Fig. 1. Dataset without Seizure.

Many researchers have proposed employing deep learning and machine learning to anticipate seizures. The techniques include feature extraction, preprocessing, and categorization. Preprocessing is the first step, which includes reducing noise in EEG impluses and improving the Signal to Noise (SNR) ratio. Preprocessing EEG data can be done by applying a variety of methods. One frequently used technique is to filter the signals in the time domain using bandpass Butterworth [5] and notch filters [6]. Applying spatial filters, such as the common spatial pattern filter or the optimised spatial pattern filter is another strategy in [7] that can increase the signal-to-noise ratio. Another helpful preprocessing technique is empirical mode decomposition in [8], which retains low-frequency components while extracting intrinsic mode functions to improve signal-to-noise ratio. Additionally, EEG signals can be preprocessed using the Fourier transform and wavelet transform to make them appropriate for input into CNN [9]. Direct characteristics extraction from raw data is possible with CNNs, and the derived features

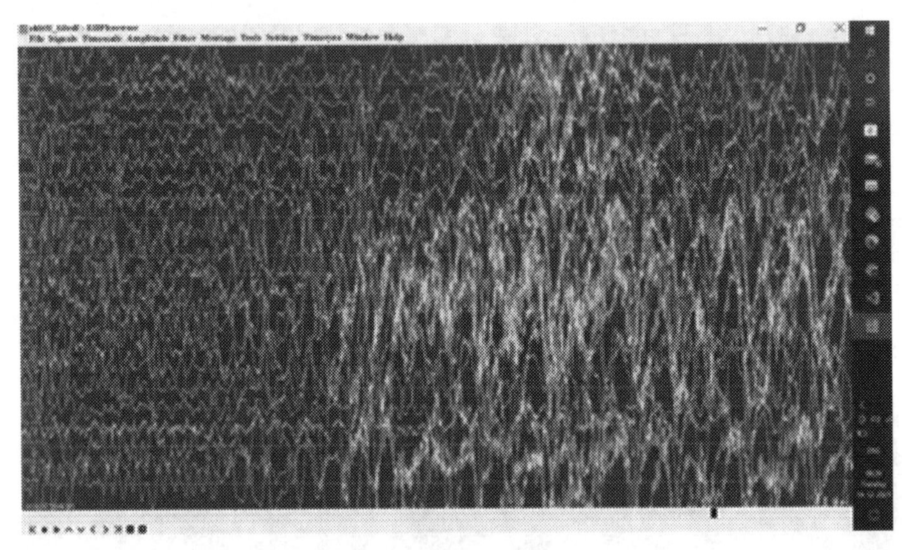

Fig. 2. Dataset with and without Seizure.

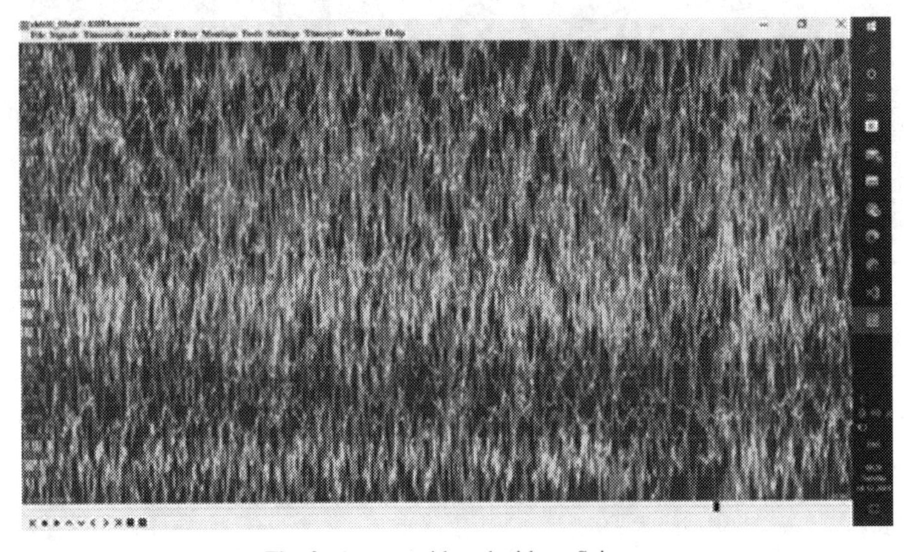

Fig. 3. Dataset with and without Seizure.

are optimized deploying the class information included with the data. After the characteristics are extracted, classification is carried out employing deep learning or machine learning classifiers. Support vector machines (SVM) [10], random forest [11], k-nearest neighbors [12], Naive Bayes [13], and multilayer perceptron [14] are a few examples of machine learning classifiers that have been applied for classification. For classification, deep learning classifiers such CNNs [15], recurrent neural networks [16], and long short-term memory [17] networks have also been utilized. Overall, seizure prediction deploying deep learning algorithms has yielded positive results, and it is anticipated that these strategy will continue to advance with additional study and improvement.

1.1 Organization

This paper introduces a seizure prediction approach that utilizes deep learning strategy. The Sect. 2 of this paper delves into the current state of seizure prediction methods that utilize scalp EEG data. In the Sect. 3, we ex plain the methodology proposed in this paper. The Sect. 4 discusses the outcomes obtained during the implementation and in the Sect. 5, we conclude the proposed approach and offer suggestions for future work.

2 Related Work

Table 1. Comparison of intracranial EEG signals-based seizure prediction techniques.

Method	Pre-Processing	Features	Classifiers	Sensitivity	Specificity
Mehla *et al.* [18]	Fourier intrinsic band functions	CNN	Kruskal–Wallis	95.37%	92%
Wu *et al.* [19]	Morphological filter	Frequency and Time domain	Random Forest	90.23%	89.47%
Rout *et al.* [20]	Decomposition using VMD	HT	EMRVFLN	95.74%	91.176%
Malekzadeh *et al.* [21]	Butterworth band pass filter	DT-CWT	KNN, SVM,CNN	96.73%	93.36%
Malekzadeh *et al.* [22]	Band pass filter	TQWT	CNN–RNN	92.71%	90.13%

Mehla et al. [18] put forward a plan which split into three primary elements. The dissection of the EEG impluses into a fixed amount of FIBFs is carried out in the first section. The characteristics are derived from FIBFs in the second stage and important features are chosen by applying the Kruskal-Wallis test. The significant features are given to the support vector machine classifier in the final stage. The suggested approach offers superior outcomes in contrast to cutting-edge techniques by using 10-fold cross-validation (Table 1).

Wu et al. [19] developed a unique approach that utilizes cEEG-based and aEEG-based seizure identification method. The cEEG-based seizure identification method splits the cEEG signals into 5 s epochs with a 4 s offset, and then retrieves multi-domain properties from each epoch. Then seizure detection is carried out by random forest classification. After translating the cEEG signals into aEEG signals, the morphological filter is employed by the aEEG-based seizure identification algorithm to accomplish spike detection and assess whether there are seizures.

Rout et al. [20] in this paper analyzed the EEG information to detect and categorise epileptic seizures. Variational mode decomposition (VMD), the Hilbert transform (HT), and the suggested error-minimized random vector functional link network (EMRVFLN) are used participants.

Malekzadeh et al. [21] indicates a novel method to recognize epileptic seizures in EEG data by integrating deep learning (DL) simulation with non - linear behavior based on fractal dimension (FD). The data sets were prepossessed using a Butterworth band pass filter during the preprocessing stage. The EEG signals were fragmented into several sub-bands using the dual-tree complex wavelet transform (DT-CWT). The classification stage was accomplished using the KNN, SVM and convolutional autoencoder (CNN-AE) techniques.

Malekzadeh et al. [22] proposed strategy for feature extraction, preprocessing, and classification. Then, eliminated artifacts from the EEG data sets using a band-pass filter. EEG signal decomposition is accomplished using the Tunable Q Wavelet Transform (TQWT). The suggested CNN-RNN model has been fed the retrieved characteristics as input, and promising results have been reported. To show the efficacy of the suggested CNN-RNN classification process, the K-fold cross-validation with k = 10 is applied in the classification stage.

3 Proposed Methodology

In the put forward method, a strategy for predicting the commencement of the preictal state was provided a short minutes earlier the seizure starts. The pro posed strategy's flowchart is shown in Figure 4. CHBMIT scalp EEG data set was adopted which includes 24 subjects and signals that were collected using 23 electrodes and digitized at 256 Hz. Using the "edf-read" function, these signals are first transformed into mat files. EEG data are subjected to a Butterworth bandpass filter to eliminate baseline noise [23] and power line interference [24]. CNN were utilized to extract characteristics [25]. Because they are extracted employing class information taken into consideration, these traits offer lower inter class variance. Following feature extraction, SVM was utilized to replace all connected layers. While CNN was utilised to obtain the features, SVM [26] was employed to differentiate between interictal and preictal segments. The subsections that follow provide a short explanation of STFT, CNN and SVM.

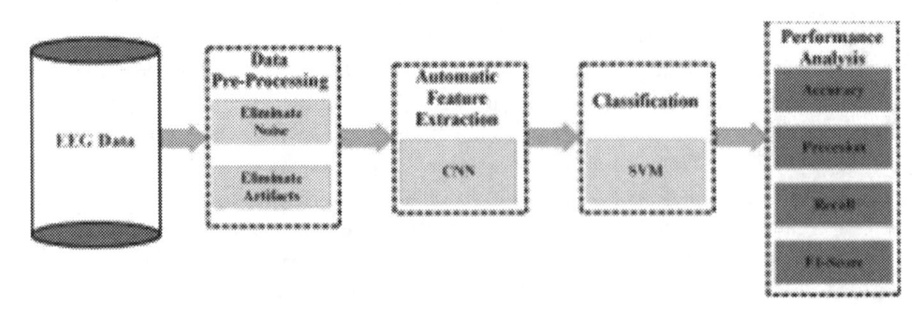

Fig. 4. Flow Diagram of Epileptic Seizure Prediction

3.1 Preprocessing of EEG signals

The Short Time Fourier Transform (STFT) is a signal processing technique for tracking a transmission's frequency content through time. The STFT divides a signal into smaller segments and each segment is then given a Fourier Trans form. The outcome is the creation of a time-frequency representation of the signal, which enables the tracking of the amplitude and phase of each frequency component across time [27].

3.2 Feature Extraction

The proposed framework utilizes preprocessed EEG impluses to extract features using a proprietary three-layer CNN architecture. This type of neural network is commonly used for automating the feature selection and classification of both data and images. EEG impulses are dissected into 30 epoch with batch size 16 and multi-domain characteristics are selected from each epoch. The architecture encompasses multiple convolutional layers, pooling layers and standard artificial neural network layers for classification, with the activation functions relu and softmax used after the convolutional layers.

CNN model takes an input image with dimensions of 256 $\times$ 256 pixels. The kernel size of 3 $\times$ 3 refers to the size of the sliding window that is deployed to apply the filter. Each filter will take 3*3 pixel section of the input image and perform a mathematical operation on those pixels. The output of this operation will be a single pixel in the feature map.

3.3 Classification

SVM were employed to classify between interictal and preictal states after extracting features from CNN. There are two categories of SVMs: linear and non linear SVM.SVM locate support vectors and create a decision boundary using slope and intercept if our data is linearly separable and refer to these as linear SVM. If the data is not sequentially separable, SVM may not be able to categorize it using a linear border. To address this issue, SVM maps the data into a higher-dimensional scope, making it easier to distinguish the data. This is accomplished using a kernel technique. To categorise interictal state and preictal state in our work linear SVM was employed (Fig. 5).

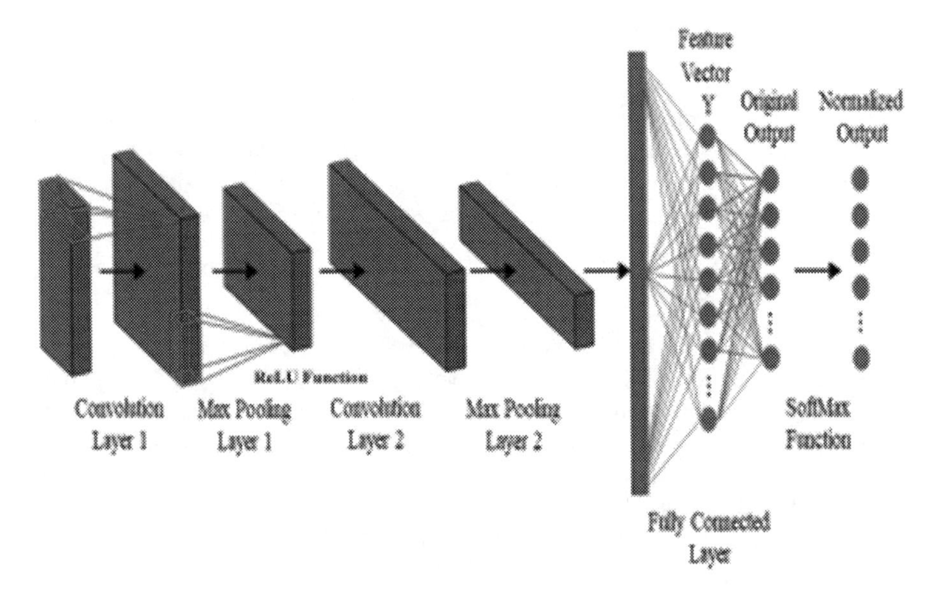

Fig. 5. Feature Selection using CNN

4 Performance Analysis

4.1 Dataset

The Children's Hospital Boston and the Massachusetts Institute of Technology collaborated to establish the CHB-MIT dataset, a collection of EEG recordings that is openly accessible. The dataset includes 23 paediatric epilepsy patients' EEG recordings, each of which was made for roughly 23 h. The dataset information was produced to support algorithm development and evaluation for autonomous seizure identification and forecasting. The dataset contains a variety of EEG recordings, including scalp, subdural, and stereo recordings that were made with various electrodes and tools. From 256 Hz to 400 Hz, the data was sampled at various frequencies. The patients in the dataset ranged in age from 1 month to 19 years and had a variety of epilepsy types, such as generalised epilepsy, temporal lobe epilepsy, and frontal lobe epilepsy. Both the waking and sleeping states, as well as the interictal and ictal periods, were recorded for each subject.

4.2 Implementation Details

The ultimate result of this CNN layer will be a $256 \times 256 \times 32$ tensor, where the third dimension (32) indicates the number of feature maps produced by this layer. This output will be added as input data to the next layer of the CNN architecture (Table 2).

4.3 Performance Analysis

The Fig.7 depicts the analysis between the suggested method and the current state-of-the-art approach. The efficiency of the recommended classification approach is assessed

Table 2. Proposed CNN model summary.

Parameters	Values
Image Size	256*256
Batch Size	16
Epoch	30
kernel size	3*3
Filter size	32
Activation Function	Relu and Softmax

Table 3. Performance Metrix

Parameters	Score (%)
Accuracy	98.57%
Precision	98.97%
F1-Score	98.59%
Sensitivity	99.20%
Specificity	99.12%

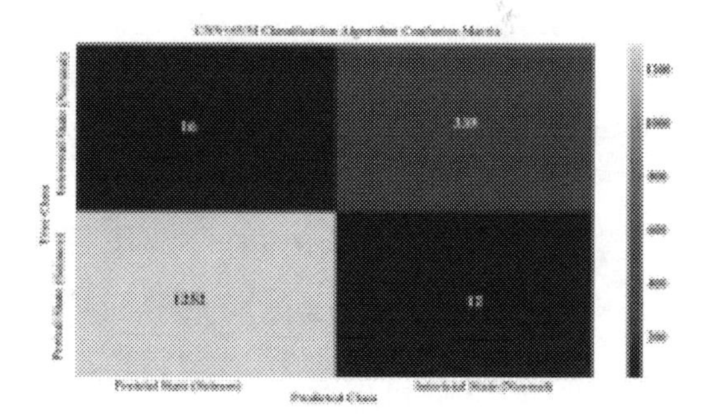

Fig. 6. Confusion Matrix.

using five performance measure such as accuracy, precision, F1 Score, sensitivity and specificity. The five performance measure can be assessed utilizing confusion matrix. A confusion matrix evaluates the performance of CNNs for classification tasks. The four possible conclusion of a binary classification task are represented by the four cells of the confusion matrix (Fig. 6).

True positives (TP): The model accurately predicts 1252 positive classes. False positives (FP): The model inaccurately predicts 12 positive classes. False negatives (FN): The model inaccurately predicts 16 negative classes. True negatives (TN): The model accurately predicts the 335 negative classes. The success rate of a binary classification model depends on both sensitivity and specificity. The formula for accuracy is:

$$\text{accuracy} = (TP + TN)/\text{total number of cases}$$

Sensitivity and specificity affect the quantity of true positives and true negatives, respectively.

Sensitivity is the quantity of real true cases that are accurately identified as positive by the model. The formula for sensitivity is:

$$\text{sensitivity} = TP/(TP + FN)$$

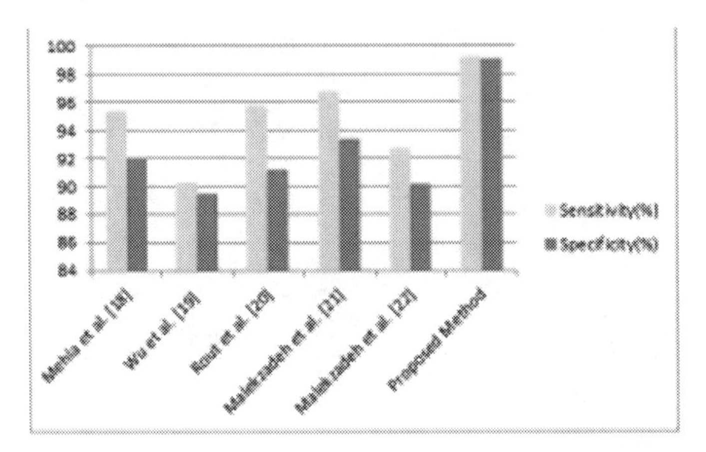

Fig. 7. Comparison of Proposed method with the state of the art method.

Specificity is the quantity of real false cases that are accurately identified as negative by the model. The formula for specificity is:

$$\text{specificity} = \text{TN} / (\text{TN} + \text{FP})$$

If a model has high sensitivity, it means that it correctly identifies a high proportion of positive cases. This results in more true positives and fewer false negatives, which increases the accuracy of the model.

A model properly detects a significant share of negative instances when it has high specificity. As a result, the model is more accurate because there are fewer false positives and more true negatives.

Figure 8 outlines the count of accurate and inaccurate predictions made by the model on a set of test data.

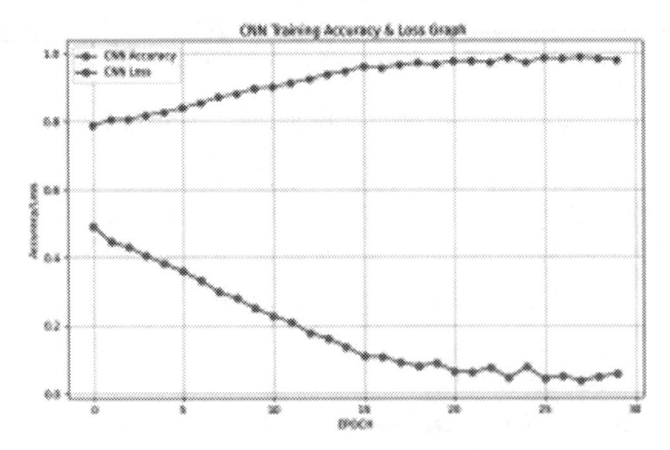

Fig. 8. CNN training Accuracy and Loss Graph

5 Conclusion

We have built a deep learning-based method for identifying epileptic seizures that has demonstrated encouraging outcomes. Deep learning algorithms can identify trends and accurately forecast seizures by studying EEG signals. For epilepsy patients, this may result in early intervention and better quality of life. By reducing the risks, this successful seizure prediction can improve the quality of life for those with epilepsy. Our methodology has a greater level of sensitivity and specificity than other approaches since it uses machine learning classifiers and CNN-based feature extraction. The focus of future study might be on improving preprocessing to increase signal-to-noise ratio and lowering the statics of parameters used for deep learning approaches for feature selection and/or classification. The need for extensive and varied training and testing datasets, as well as the problem of generalization to new patients, are still issues that need to be resolved. To create more solid and dependable models that may be applied in clinical settings, more research is required.

Overall, deep learning-based epileptic seizure prediction has the potential to greatly enhance epilepsy management and therapy, opening the door for future customized medicine techniques. In accordance with previous modernizing techniques, our system provides patient-specific seizure prediction.

References

1. Nasseri, M., et al.: Semi-supervised training data selection improves seizure forecasting in canines with epilepsy. Biomed. Signal Process. Control **57**, 101743 (2020)
2. Le Van Quyen, M., et al.: Anticipation of epileptic seizures from standard EEG recordings. Lancet **357**(9251), 183–188 (2001)
3. Robinson, P., Rennie, C., Rowe, D.: Dynamics of large-scale brain activity in normal arousal states and epileptic seizures. Phys. Rev. E **65**(4), 041924 (2002)
4. Hazarika, N., Chen, J.Z., Tsoi, A.C., Sergejew, A.: Classification of EEG signals using the wavelet transform. Signal Process. **59**(1), 61–72 (1997)
5. Rasekhi, J., Mollaei, M.R.K., Bandarabadi, M., Teixeira, C.A., Dourado, A.: Preprocessing effects of 22 linear univariate features on the performance of seizure prediction methods. J. Neurosci. Methods **217**(1–2), 9–16 (2013)
6. Bandarabadi, M., Teixeira, C.A., Rasekhi, J., Dourado, A.: Epileptic seizure prediction using relative spectral power features. Clin. Neurophysiol. **126**(2), 237–248 (2015)
7. Ang, K.K., Chin, Z.Y., Wang, C., Guan, C., Zhang, H.: Filter bank common spatial pattern algorithm on BCI competition IV datasets 2a and 2b. Front. Neurosci. **6**, 39 (2012)
8. Usman, S.M., Khalid, S., Akhtar, R., Bortolotto, Z., Bashir, Z., Qiu, H.: Using scalp EEG and intracranial EEG signals for predicting epileptic seizures: review of available methodologies. Seizure **71**, 258–269 (2019)
9. Antoniades, A., Spyrou, L., Took, C.C., Sanei, S.: Deep learning for epileptic intracranial EEG data. In: 2016 IEEE 26th International Workshop on Machine Learning for Signal Processing (MLSP), pp. 1–6. IEEE (2016)
10. Williamson, J.R., Bliss, D.W., Browne, D.W., Narayanan, J.T.: Seizure prediction using eeg spatiotemporal correlation structure. Epilepsy behavior **25**(2), 230–238 (2012)
11. Van Diessen, E., Otte, W.M., Braun, K.P., Stam, C.J., Jansen, F.E.: Improved diagnosis in children with partial epilepsy using a multivariable prediction model based on eeg network characteristics. PLoS ONE **8**(4), e59764 (2013)

12. Chaovalitwongse, W.A., Fan, Y.-J., Sachdeo, R.C.: On the time series k-nearest neighbor classification of abnormal brain activity. IEEE Trans. Syst. Man Cybernet. Part A: Syst. Humans **37**(6), 1005–1016 (2007)
13. Sharmila, A., Geethanjali, P.: Dwt based detection of epileptic seizure from EEG signals using Naive Bayes and k-NN classifiers. IEEE Access **4**, 7716–7727 (2016)
14. Orhan, U., Hekim, M., Ozer, M.: EEG signals classification using the k-means clustering and a multilayer perceptron neural network model. Expert Syst. Appl. **38**(10), 13475–13481 (2011)
15. Acharya, U.R., Oh, S.L., Hagiwara, Y., Tan, J.H., Adeli, H.: Deep convolutional neural network for the automated detection and diagnosis of seizure using eeg signals. Comput. Biol. Med. **100**, 270–278 (2018)
16. Tsiouris, K.M., Pezoulas, V.C., Zervakis, M., Konitsiotis, S., Koutsouris, D.D., Fotiadis, D.I.: A long short- term memory deep learning network for the prediction of epileptic seizures using EEG signals. Comput. Biol. Med. **99**, 24–37 (2018)
17. Petrosian, A., Prokhorov, D., Homan, R., Dasheiff, R., Wunsch, D., II.: Recurrent neural network based prediction of epileptic seizures in intra-and extracranial eeg. Neurocomputing **30**(1–4), 201–218 (2000)
18. Mehla, V.K., Singhal, A., Singh, P., Pachori, R.B.: An efficient method for identification of epileptic seizures from eeg signals using Fourier analysis. Phys. Eng. Sci. Med. **44**, 443–456 (2021)
19. Wu, D., et al.: Automatic epileptic seizures joint detection algorithm based on improved multi-domain feature of cEEG and spike feature of aEEG. IEEE Access **7**, 41551–41564 (2019)
20. Rout, S.K., Biswal, P.K.: An efficient error - minimized random vector functional link network for epileptic seizure classification using VMD. Biomed. Signal Process. Control **57**, 101787 (2020)
21. Malekzadeh, A., Zare, A., Yaghoobi, M., Alizadehsani, R.: Automatic diagnosis of epileptic seizures in EEG signals using fractal dimension features and convolutional autoencoder method. Big Data Cogn. Comput. **5**(4), 78 (2021)
22. Malekzadeh, A., Zare, A., Yaghoobi, M., Kobravi, H.-R., Alizadehsani, R.: Epileptic seizures detection in EEG signals using fusion handcrafted and deep learning features. Sensors **21**(22), 7710 (2021)
23. Daud, S.S., Sudirman, R.: Butterworth bandpass and stationary wavelet transform filter comparison for electroencephalography signal. In: 2015 6th International Conference on Intelligent Systems, Modelling and Simulation, pp. 123–126. IEEE (2015)
24. Ahmed, M., Farooq, A., Farooq, F., Rashid, N., Zeb, A.: Power line interference cancellation from EEG signals using RLS algorithm. In: 2019 International Conference on Robotics and Automation in Industry (ICRAI), pp. 1–5. IEEE (2019)
25. Jana, R., Mukherjee, I.: Deep learning based efficient epileptic seizure prediction with EEG channel optimization. Biomed. Signal Process. Control **68**, 102767 (2021)
26. Stojanović, O., Kuhlmann, L., Pipa, G.: Predicting epileptic seizures using nonnegative matrix factorization. PloS One **15**(2), e0228025 (2020)
27. Usman, S.M., Khalid, S., Bashir, S.: A deep learning based ensemble learning method for epileptic seizure prediction. Comput. Biol. Med. **136**, 104710 (2021)

Adapting to Noise in Forensic Speaker Verification Using GMM-UBM I-Vector Method in High-Noise Backgrounds

K. V. Aljinu Khadar(✉), R. K. Sunil Kumar, and N. S. Sreekanth

Department of Information Technology, Kannur University, Kannur, India
{aljinu,sunilkumarrk,nssreekanth}@kannuruniv.ac.in

Abstract. The performance of the GMM-UBM-I vector in a forensic speaker verification system has been examined in the context of noisy speech samples. This analysis utilised both Mel-frequency cepstral coefficients (MFCC) and MFCCs generated from auto-correlated speech signals. The noisy signal's auto correlation coefficients are concentrated around the lower lag, whereas the autocorrelation coefficients near the higher lag are very small. Thus, in addition to retain the periodic nature, autocorrelation-based MFCC is also robust for analyzing speech signals in intense background noise. The performance of MFCC and auto-correlated MFCC depends heavily on the quality of the sample. It works best with data that is free of noise, but it suffers when used on real-world examples, ie, with noisy data. The experiment on speaker verification for forensic purposes involved the addition of White Gaussian Noise, Red Noise, and Pink Noise, with a Signal-to-Noise Ratio (SNR) range spanning from −20 dB to + 20 dB. The performance of both methods was affected drastically in call cases but autocorrelation-based MFCC gave better results than MFCC. Thus, autocorrelation-based MFCC is a valuable method for robust feature extraction when compared with MFCC for speaker verification purposes in intense background noise. The verification accuracy in our method is improved even in very high noise levels (−20 dB) than the reported research work.

Keywords: MFCC · Gaussian Mixture Model · I-Vector · Forensic Speaker Verification

1 Introduction

Speaker recognition refers to the automated technique of determining the identity of the speaker by analysing the information included in the speech signal. Voice print recognition and speech biometric recognition are synonyms for speaker recognition [1]. Speaker recognition is a method of biometric authentication that employs voice features inside the voice waveform to automatically identify the speaker's physiological and behavioral characteristics [2]. Closed-set and open-set recognition are two types of speaker recognition systems that are commonly used [3]. The concept of a closed set pertains to scenarios where the unidentified speaker is expected to belong to a predetermined group

S. Aurelia et al. (Eds.): ICCSST 2023, CCIS 1973, pp. 275–287, 2024.
https://doi.org/10.1007/978-3-031-50993-3_22

of acknowledged speakers. Conversely, the notion of an open set pertains to scenarios where the unidentified speaker may originate from unregistered individuals. Moreover, in practical implementation, speaker identification systems can be categorised into two distinct types: text-dependent recognition and text-independent recognition.

In a forensic examination, a person's voice can be quite significant. Speech is the most natural and common way for humans to express information and the speech signal can carry a wide range of information. The method of determining if an individual's voice is the source of a questioned voice recording is known as forensic speaker recognition (FSR). It's the science that tries to figure out whether or not the recorded voice belongs to the suspect. FSR is founded on the idea that each person's voice is as unique and similar as the fingerprints or DNA of an individual. Speech production is influenced by shape, size, and other similar physical parameters of vocal apparatus i.e., vocal tract, and the cavity of the mouth, and it also has a high correlation with pronunciation abilities, regional accent, and other factors [4].

Speaker profiling is a methodology that entails the extraction of personal details on an unidentified perpetrator from incriminating speech material. These details encompass various factors, including but not limited to gender, age, dialect, features of respiration, phonation, articulation, and speaking style [5]. The growing popularity of Information and Communication Technologies (ICT) and the wide spread accessibility of mobile phones have led to their misuse, rendering them a potent instrument for engaging in criminal activities such as abduction, extortion, blackmail, the dissemination of obscene content, anonymous harassment, ransom demands, terrorist communication, match-fixing, and similar offenses. Criminals are progressively capitalising on these communication channels, operating under the assumption that their identities will remain concealed if they maintain a mask. Fortunately, this assertion is no longer valid. A person's voice can be successfully recognized and investigation agencies can zero down to a person or group of people who committed the crime [6].

The objective of forensic speech comparison is to aid a court of law in ascertaining if a provided sample of speech is said by a specific individual who is under the surveillance of the investigating agency. Often, the records of the questioned and known speakers don't match up when it comes to the intrinsic or extrinsic conditions of the speakers, or both. Speaker-intrinsic variability can arise from a multitude of reasons, such as alterations in speaking style, variations in speech duration, fluctuations in channel characteristics, and other related influences. The presence of background noise in a noisy setting necessitates speakers to modify their speech patterns. The modification is influenced by the diverse amounts and types of ambient noise, including office noise, ventilation system noise, traffic noise, crowd noise, and any noise that may be caught during the recording procedure [7]. Speaker recognition accuracy has been shown to be severely harmed by environmental noise mismatch between training and testing datasets [8, 9]. To increase speaker verification accuracy in noisy environments, more research is needed. This article presents a novel approach for adapting speaker verification to accommodate varying levels of background noise. This work is carried out in Malayalam, which is a low-recourse and official language of Kerala and is a Dravidian language spoken in India.

2 Data Acquisition

In this case, we used a variety of mobile phone devices and PCs to record a continuous speech data set. In order to ensure the wide variance in the data set like the channel variation, dialectical variations, age variations, and speaking style variations in the speech data, the samples were collected from various sources. The duration of the speech data is not constant. A total of 100 speakers (50 males and 50 females) read the same set of Malayalam sentences in the data set. There are ten different sentences in each group. The sample of 100 individuals was partitioned into five distinct groups, with each group consisting of 20 participants. Within each group, there was an equal distribution of gender, with 10 male and 10 female participants. We first tested 20 speakers (10 male and 10 female) from each of the four groups reading the same Malayalam sentences of a duration of 6–8 s.

3 Feature Extraction

Speech parameterization involves converting a speech signal into a set of feature vectors that emphasize speaker-specific characteristics. The goal of this transformation is to create a new representation that is more compact and suitable for speaker modeling. Various speech parameterizations, such as Mel-Frequency Cepstral Coefficients (MFCC) [10], Linear Predictive Coefficients (LPC) [11], and Linear Predictive Cepstral Coefficients (LPCC) [12], are frequently utilised in modern speaker recognition methodologies. In the field of speaker recognition, the Perceptual Linear Prediction (PLP) [13] based feature extraction methodology and Mel Frequency Cepstral Coefficients (MFCC) are widely recognised as the most efficient, cost-effective, and universally-accepted techniques for extracting features. The utilisation of Mel Frequency Cepstral Coefficients (MFCC) was implemented as a means of extracting characteristics for the objective of speaker verification in this research.

4 Mel-Frequency Cepstral Coefficients (MFCC)

For feature extraction, MFCCs are the most popular method when it comes to voice and speaker recognition tasks. Davis and Mermelstein [14] proposed the MFFCs in the 1980s. Humans' auditory perception is non-uniform which is linear at low frequency (1000 Hz) and becomes nonlinear after 1000 Hz. Thus, the recorded speech signal must be nonlinearly mapped into the perceived scale. Mel scale maps the measured frequency exactly like human auditory perception. From a computational perspective, a group of overlapped triangular band passes filters that simulate the characteristics of the Mel scale is used and is shown in the figure. In MFCC, these filter banks are applied on the speech spectrum to get the Mel scale power spectrum. The output of the log filter bank is then obtained using the log operator. Finally, 20 MFCCs are generated using discrete cosine transforms (DCT). The 13th parameter is the log energy computed from each speech segment. In order to recover the dynamic information of the speech that is lost during frame-by frame analysis, temporal derivatives, or ΔMFCCs and $\Delta\Delta$MFCCs, were recorded as the first and second derivatives of the MFCC. Thus, each speech segment's final feature vector has 20 MFCCs, 20 ΔMFCCs, and 20 $\Delta\Delta$MFCCs (Fig. 1).

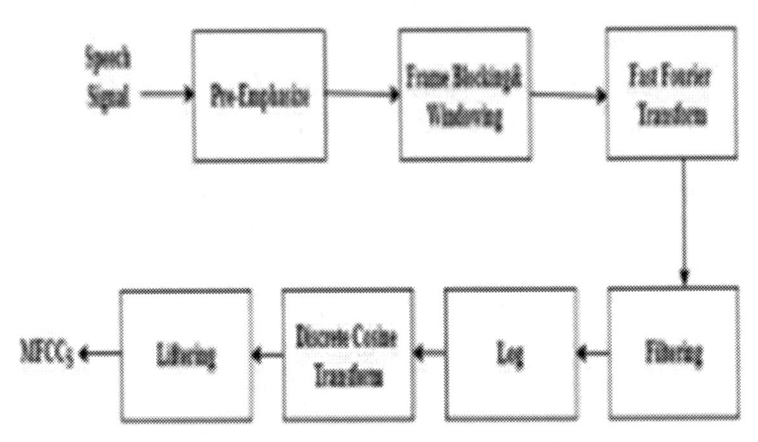

Fig. 1. Feature Extraction Using MFCC

5 Speaker Verification System Using GMM-UBM I Vector Frame Work

A typical Automatic Speaker Recognition algorithm, encompassing both identification and verification, is structured in stages, comprising a front-end module responsible for extracting relevant features, and a back-end module dedicated to computing the similarity of speaker features. In order to achieve the objective of speech recognition, a diverse range of algorithms are utilised during both the initial processing and subsequent phases [15]. Figure 2, shows a typical stage-by-stage speaker verification design. The front-end module converts speaker utterances that are initially in a low-dimensional format into high-dimensional feature vectors. The backend analyzed the input result and used threshold values to compare feature similarity.

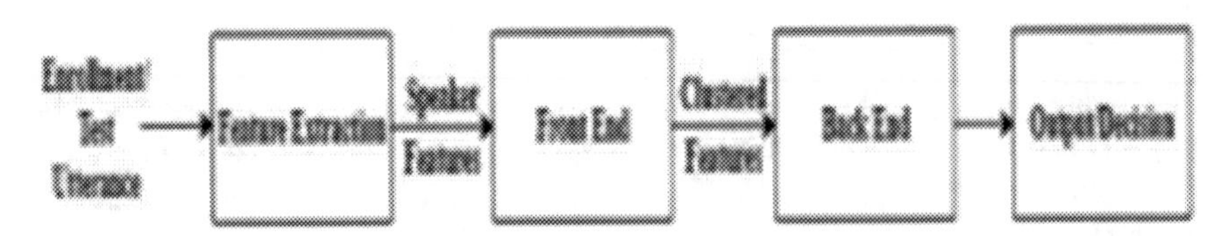

Fig. 2. Speaker Verification Systems

The front end converts the speech signal into a more usable format, typically a set of feature vectors. In this study, the Gaussian Mixture Model (GMM)-Universal Background Model (UBM) and I Vector Method algorithms were employed in the front end, while the Probabilistic Linear Discriminant Analysis (PLDA) and Cosine Distance Scoring (CSS) algorithm were utilised in the back end. Speaker recognition systems employ three primary methodologies: template matching, probabilistic models, and artificial neural networks. The Gaussian mixture model and universal background model (GMM-UBM) are often employed techniques in the probabilistic approach for assessing acoustic characteristics [16]. Intersession variability is among the most challenging aspects of GMM-UBM systems. The suggested methodology, referred to as Joint Factor Analysis (JFA), seeks to tackle the issue of variability by explicitly modeling inter-speaker variability and channel or session variability as separate variables [17, 18]. Dehak, Najim, et al. (2010) made a significant finding regarding the channel factors within the Joint Factor Analysis (JFA) framework. They observed that these channel factors not only encompassed information pertaining to the acoustic channel but also provided valuable insights about the individual speakers. As a result, the researchers introduced an innovative methodology that entailed merging the channel and speaker spaces to form a cohesive total variability space [19]. Following the preliminary procedures, which encompassed the application of backend methodologies like as linear discriminant analysis (LDA) and within-class covariance normalisation (WCCN) to mitigate intersession variability, a scoring technique referred to as the cosine similarity score was utilised.

Matejka et al., 2011 [20] proposed using a probabilistic LDA (PLDA) model in stead of cosine similarity scoring. N. Dehak, 2011 proposed an I vector system in GMM super vector space that does not distinguish between speaker and session effects [21]. The objective is to establish a comprehensive variability space that encompasses both speaker and session variability concurrently. In recent years, the utilisation of vector-based approaches has exhibited exceptional performance in the field of speaker verification. Despite the fact that I vector were initially developed for speaker verification, they have since been applied to a variety of problems. In a substantial proportion of instances involving forensic speaker recognition, the voice sample may have originated from a distinct channel environment, or the duration of the speech sample obtained from a forensic case exhibits substantial variation. The GMM-UBM I vector method is highly adaptable to channel variations and speech utterance duration, which makes it ideal for forensic speaker verification systems.

In real forensic speaker verification cases the speech data may derived from different channels also the duration of the speech data is highly variable. So this framework highly suitable for forensic cases. In this paper, we used the GMM UBM-I vector framework to investigate speaker verification. Our goal is to demonstrate how this framework can be adapted to deal with high levels of background noise. We used the I-vector to verify speakers in a dataset of 1000 speeches from 100 speakers (10 speeches per speaker). There are 50 men and 50 women among the speakers. The verification procedure consists of three steps. The development phase is the step in which we provide a background speaker set. We will provide a known speaker data set in the second phase, which is known as the enrolment phase, and we will need a verification phase for every verification system. During this phase, we will send a test data with the claimed id. The procedure used here is depicted in the flow chart below (Fig. 3).

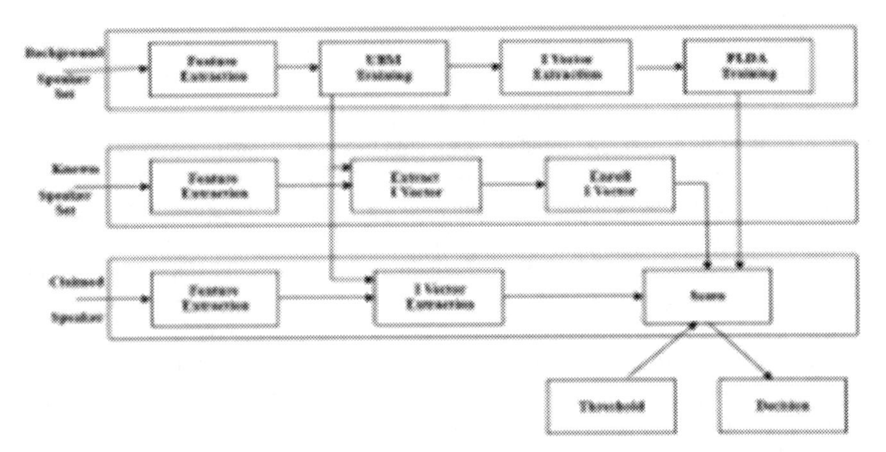

Fig. 3. Speaker Verification Using I Vector

Speaker verification system is implemented using I-vector based system. The number of mixtures was established as eight. During the process of expectation maximization (EM) and maximum a posteriori (MAP) estimation, the total number of iterations performed was 10. Additionally, the dimensionality of the I-vector was specified to be 100. The verification experiment is conducted by adding three different noises such as Red, Pink, and white noise with SNR varying from 20 dB to −20 dB. The experimental findings are presented in Table 1. The equal error rate (EER %) for 8 mixtures/13 MFCC features is found to be 0.214%.

Figure 4 shows verification accuracy of the system with three different noises. The verification accuracy for clean speech is 95% and it decreases drastically by adding noise when MFCC alone as a speech parameter. Since most of the forensic case speech data may be derived from a noisy real environment. So a novel feature extraction method based autocorrelation based MFCC is used in the next session to get better recognition accuracy even in highly noisy environments.

Table 1. Performance of FSR for the feature MFCC with SNR of differed noise type

Method	Noise Type	Accuracy (%)					Clean Speech
		20 dB	10 dB	0 dB	−10 dB	−20 dB	
MFCC	White	89	86	80	62	49	95
	Pink	89	84	79	63	53	
	Red	94	90	88	80	64	

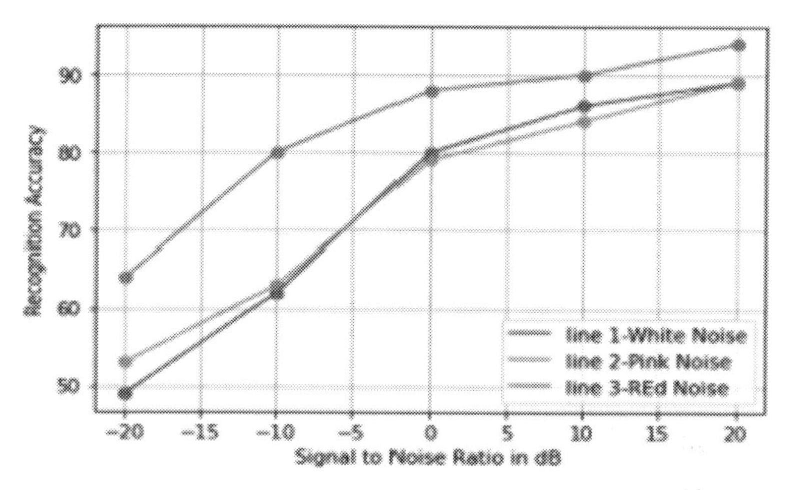

Fig. 4. Performance of FSR for the feature MFCC with SNR of differed noise type

6 Modified Feature Extraction for Noise Adapting

Speech has a set of very distinctive traits that capture human physiological features, language, background noise, etc. So, converting these traits to machine readable format is an important task in speech processing termed feature extraction. The method of feature extraction must have an acceptable accuracy even when there is ambient noise. We employed Mel Frequency Cepstral Coefficients (MFCC) as a characteristic in this work. The performance of a speech recognition system depends significantly on different operations in the pre-processing stage. Amplitude Normalization, Pre-emphasis Filtering, Framing, and Windowing are the commonly used pre-processing steps employed in this work. Many works are reported in autocorrelation-based Mel Frequency Cepstral Coefficients calculation in the fields of speech recognition in the noisy environment [22–26].

For feature extraction, we adopted the approach described by Amita Dev et al. 2010 [22]. The key step in acoustic speech feature extraction processes is trans forming the pre-processed time-domain signal to the autocorrelation domain to study the performance of the speaker verification system in intense background noise in the GMM-UBM –I vector framework. Because this work is being done on noisy speech signals, it is worth examining the power of autocorrelation to sup press the noise. This study employs the autocorrelation method to extract MFCC features in a robust manner; particularly in intense background noise for forensic speaker verification purposes.

MFCC is the widely utilized acoustic feature in speech recognition systems. MFCC performs well for speech recognition in a clean environment, but its performance deteriorates in noisy speech processing. Background noise has a significant impact on power spectrum estimates in MFCC, lowering the recognition rate. Since the autocorrelation method is robust in noisy conditions, it is used as an initial step in the MFCC algorithm, which enhances the efficiency of MFCC features in noisy speech. We compared the accuracy of the forensic speaker verification system with MFCC and autocorrelation-based

MFCC features. The steps involved in the auto correlation-based MFCC feature are as follows. Initially, the relative higher-order autocorrelation coefficients are extracted. Next, the fast Fourier transform (FFT) is employed to evaluate the amplitude spectrum of the speech signal, which is subsequently segmented based on frequency. Finally, the magnitude spectrum that has been separated is converted into coefficients resembling Mel-frequency cepstral coefficients (MFCCs). The recovered features are referred to as Mel-frequency cepstral coefficients (MFCCs) derived from a Differentiated Relative Higher-Order Autocorrelation Sequence Spectrum [22]. The speaker verification experiment was again conducted using an autocorrelated MFCC feature using the same GMM UBM-I vector method. Table 2 shows the performance of the FSR using signal corrupted by adding different noise namely white, pink, and red noise with SNR varying from − 20 dB to 20 dB. The figure shows the block diagram of the autocorrelated MFCC feature extraction method [27] (Figs. 5, 6, 7 and 8).

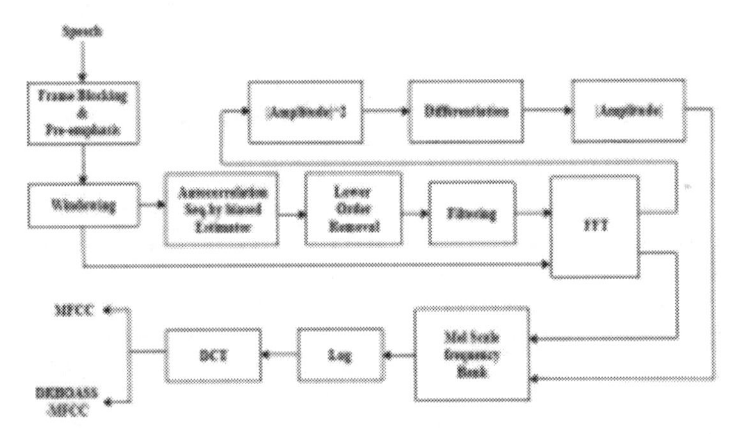

Fig. 5. Revised Feature Extraction Method

Table 2. Performance of FSR for the feature Auto correlated - MFCC with SNR of different noise type

Method	Noise Type	Accuracy (%)					Clean Speech
		20 dB	10 dB	0 dB	−10 dB	−20 dB	
Auto correlated MFCC	White	95	90	83	72	57	95
	Pink	94	88	82	71	64	
	Red	94	91	89	87	74	

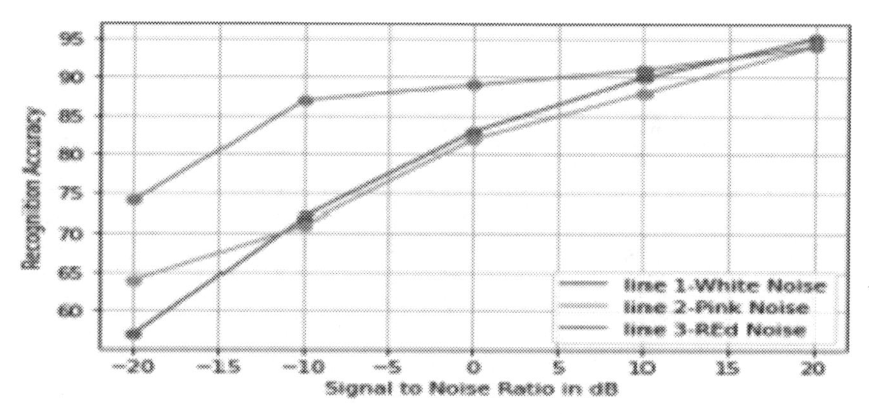

Fig. 6. Performance of FSR for the feature Auto correlated MFCC with different noise

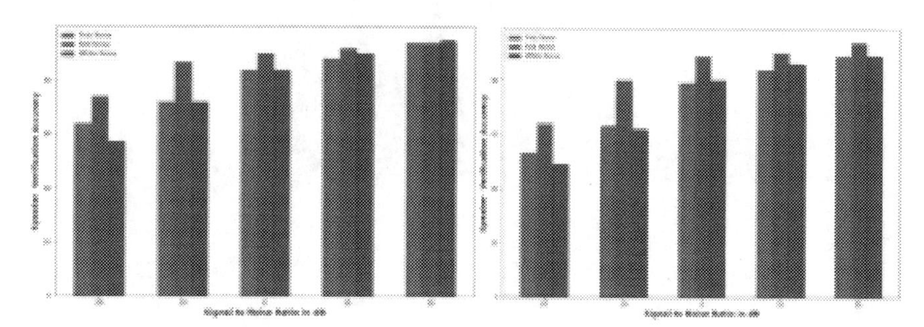

Fig. 7. Comparison of Speaker verification accuracy using MFCC and Auto correlated – MFCC as features for Pink, Red, and White Noise

The graph illustrates FSR's performance on the feature autocorrelation-based MFCC with SNRs of various noise types We performed the verification experiment by trying to introduce three different noises, namely White Gaussian Noise, Red Noise, and Pink Noise, with SNR ranging from −20 dB to +20 dB. Both methods are more affected by pink noise added speech signal than white Gaussian noise, but less so by red noise. This is due to pink noise heavily distracting the structure of the autocorrelated spectrum [27]. In the context of speaker verification in a noisy environment, the Auto-correlation-based Mel-frequency cepstral coefficients (MFCC) method demonstrates robustness when compared to the conventional MFCC approach. Our method demonstrates enhanced verification accuracy even under high noise levels (−20 dB) as compared to the study conducted by Ju-ho Kim et al. on Extended U-Net for Speaker Verification in Noisy Environments [28] (Table 3).

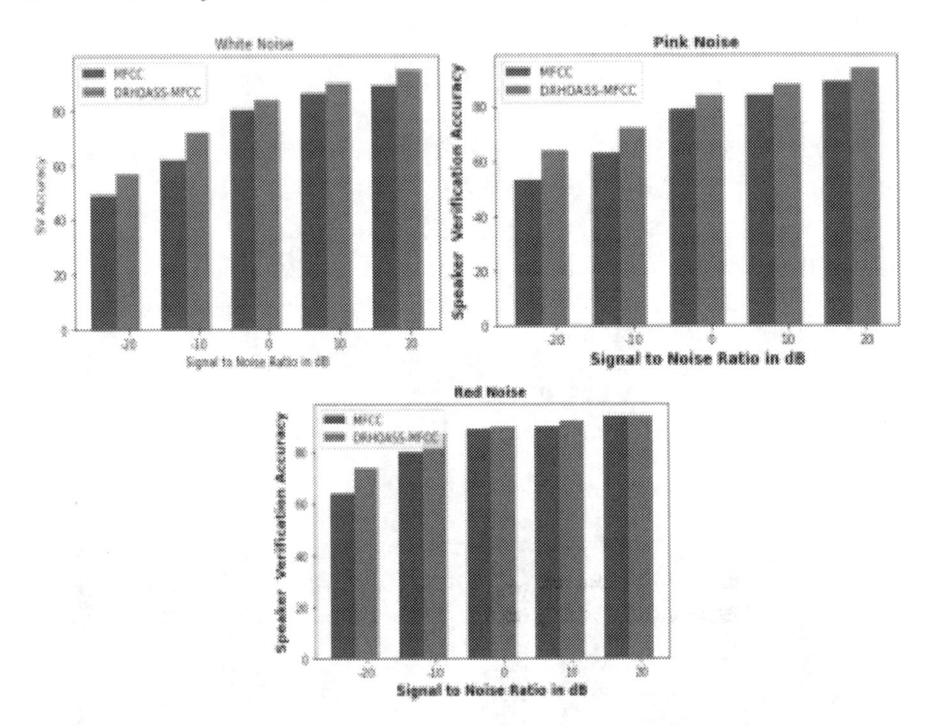

Fig. 8. Comparison of Speaker verification accuracy based on different noise types

Table 3. Performance Analysis of FSR- Percentage improvement using Auto Correlated Feature using Different Noise

Noise SNR in dB	Verification Accuracy in % for MFCC feature	Verification Accuracy in % For Autocorrelation based MFCC feature	Percentage of Improvement (%)
White Noise			
−20	49	57	8
−10	62	72	10
0	80	84	4
10	86	90	4
20	89	95	6
Pink Noise			
−20	53	64	11
−10	63	72	9
0	79	84	5

(*continued*)

Table 3. (*continued*)

Noise SNR in dB	Verification Accuracy in % for MFCC feature	Verification Accuracy in % For Autocorrelation based MFCC feature	Percentage of Improvement (%)
10	84	88	4
20	89	94	5
Red Noise			
−20	64	74	10
−10	80	87	7
0	89	90	1
10	90	92	2
20	94	94	0

7 Conclusion

MFCC-based autocorrelation based with noisy speech samples. It produces excel lent results with noise-free samples, but its performance suffers when applied to noisy data. White Gaussian Noise, Red Noise, and Pink Noise were introduced into the system during the verification experiment, with Signal-to-Noise Ratio (SNR) values ranging from −20 dB to +20 dB. The noise had a significant impact on the performance of both methods in all cases, but autocorrelation-based MFCC significantly outperforms MFCC. Autocorrelation-based MFCC is a useful method for robust feature extraction for forensic speaker verification when compared to MFCC for speaker verification using the GMM-UBM-I vector framework. The verification accuracy in our method is improved even in very high noise levels (−20 dB) than the reported research work.

References

1. Furui, S.: Speaker-independent and speaker-adaptive recognition techniques. Adv. Speech Signal Process. 597–622 (1992)
2. Chiu, T.-L., Liou, H.-C., Yeh, Y.: A study of web-based oral activities enhanced by automatic speech recognition for EFL college learning. Comput. Assisted Lang. Learn. **20**(3), 209–233 (2007)
3. Kabir, M.M., et al.: A survey of speaker recognition: fundamental theories, recognition methods and opportunities. IEEE Access **9**, 79236–79263 (2021)
4. Ajili, M., et al.: Phonological content impact on wrongful convictions in forensic voice comparison context. In: 2017 IEEE International Conference on Acoustics, Speech and Signal Processing (ICASSP). IEEE (2017)
5. Amino, K., Osanai, T., Kamada, T., Makinae, H., Arai, T.: Historical and procedural overview of forensic speaker recognition as a science. In: Neustein, A., Patil, H. (eds.) Forensic Speaker Recognition, pp. 3–20. Springer, New York (2012). https://doi.org/10.1007/978-1-4614-0263-3_1

6. Tull, R.G., Rutledge, J.C.: Cold speech for automatic speaker recognition. In: Acoustical Society of America 131st Meeting Lay Language Papers (1996)
7. Benzeghiba, M., et al.: Impact of variabilities on speech recognition. In: Proceeding of the SPECOM (2006)
8. Mandasari, M., McLaren, M., van Leeuwen, D.A.: The effect of noise on modern automatic speaker recognition systems. In: 2012 IEEE International Conference on Acoustics, Speech and Signal Processing (ICASSP). IEEE (2012)
9. Hasan, T., et al.: CRSS systems for 2012 NIST speaker recognition evaluation. In: 2013 IEEE International Conference on Acoustics, Speech and Signal Processing. IEEE (2013)
10. Logan, B.: Mel frequency cepstral coefficients for music modeling. ISMIR **270**(1) (2000)
11. Atal, B.S.: The history of linear prediction. IEEE Signal Process. Mag. **23**(2), 154–161 (2006)
12. Hariharan, M., Chee, L.S., Yaacob, S.: Analysis of infant cry through weighted linear prediction cepstral coefficients and probabilistic neural network. J. Med. Syst. **36**, 1309–1315 (2012)
13. Hermansky, H.: Perceptual linear predictive (PLP) analysis of speech. J. Acoustical Soc. Am. **87**(4), 1738–1752 (1990)
14. Naing, H.M.S., et al.: Filterbank analysis of MFCC feature extraction in robust children speech recognition. In: 2019 International Symposium on Multimedia and Communication Technology (ISMAC). IEEE (2019)
15. Bai, Z., Zhang, X.-L.: Speaker recognition based on deep learning: an overview. Neural Netw. **140**, 65–99 (2021)
16. Reynolds, D.A., Quatieri, T.F., Dunn, R.B.: Speaker verification using adapted Gaussian mixture models. Digit. Signal Process. **10**(1–3), 19–41 (2000)
17. Kenny, P., et al.: Joint factor analysis versus eigenchannels in speaker recognition. IEEE Trans. Audio Speech Lang. Process. **15**(4), 1435–1447 (2007)
18. Kenny, P., et al.: A study of interspeaker variability in speaker verification. IEEE Trans. Audio Speech Lang. Process. **16**(5), 980–988 (2008)
19. Dehak, N., et al.: Cosine similarity scoring without score normalization techniques. Odyssey (2010)
20. Matějka, P., et al.: Full-covariance UBM and heavy-tailed PLDA in I-vector speaker verification. In: 2011 IEEE International Conference on Acoustics, Speech and Signal Processing (ICASSP). IEEE (2011)
21. Dehak, N., et al.: Front-end factor analysis for speaker verification. IEEE Trans. Audio Speech Lang. Process. **19**(4), 788–798 (2010)
22. Dev, A., Bansal, P.: Robust features for noisy speech recognition using MFCC computation from magnitude spectrum of higher order autocorrelation coefficients. Int. J. Comput. Appl. **10**(8), 36–38 (2010)
23. Farahani, G., Ahadi, S.M.: Robust features for noisy speech recognition based on filtering and spectral peaks in autocorrelation domain. In: 2005 13th European Signal Processing Conference, pp. 1–4. IEEE (2005)
24. Shan, Z., Yang, Y.: Scores selection for emotional speaker recognition. In: Tistarelli, M., Nixon, M.S. (eds.) ICB 2009. LNCS, vol. 5558, pp. 494–502. Springer, Heidelberg (2009). https://doi.org/10.1007/978-3-642-01793-3_51
25. Lau, L.: This is a sample template for authors. J. Digit. Forensics Secur. Law **9**(2), 1–2 (2014)
26. Farahani, G.: Autocorrelation-based noise subtraction method with smoothing, overestimation, energy, and cepstral mean and variance normalization for noisy speech recognition. EURASIP J. Audio Speech Music Process. **2017**(1), 1–16 (2017). https://doi.org/10.1186/s13636-017-0110-8

27. Bibish Kumar, K.T., Sunil Kumar, R.K.: Viseme identification and analysis for recognition of Malayalam speech intense background noise. Ph.D. thesis (2021)
28. Kim, J., et al.: Extended U-net for speaker verification in noisy environments. arXiv:2206.13044v1 (2022)

Classification of Code-Mixed Tamil Text Using Deep Learning Algorithms

R. Theninpan and P. Valarmathi(✉)

School of Computer Science and Engineering, Vellore Institute of Technology, Chennai 600127, Tamil Nadu, India

theninpan.r2019@vitstudent.ac.in, valarmathi.sudhakar@vit.ac.in

Abstract. Natural Language Processing (NLP) is a vast subject with applications in many fields in today's modern world. The goal of NLP is to achieve human like language processing for a variety of activities or applications. The internet is full of textual data in many different languages. Although a large number of internet comments found in public spaces are often positive, a significant portion are toxic in nature. We first need to separate the good from the bad before classifying the different levels of toxicity. This will lessen any unintentional prejudice towards certain individuals or entity and lessen negativity on social media. Our primary goal is to identify, categorize, and analyze the toxicity that now plagues social media platforms. This study focuses on classifying Code Mixed Tamil text using deep learning algorithms. Tamil as a language has many obstacles to be overcome in this NLP task. Since Tamil's grammar structure, specific features are unique and complex, it is actually hard to make a model that can consistently perform for any data from the language of Tamil. The agglutinative nature of Tamil is a major problem when it comes to tasks like classification since the context gets twisted when the single word is split into corresponding morphemes. Since there are many studies conducted on Code Mixed text of other languages with deep learning algorithms, this paper aims to find the effectiveness of XLNet and Bi-LSTM on Code-Mixed Tamil.

Keywords: Artificial Intelligence · Deep Learning · Code-Mixed Text · Natural Language Processing · Text Classification · Toxic Classification

1 Introduction

Text classification is one of the NLP tasks that gets the most attention from the community. Any new model or approach that is invented which can be used in different NLP tasks is never not used in text classification. Text classification is a crucial part of the internet as most of the technologies that people use nowadays have some kind of classification task mixed in it to increase the user friendliness of the technology. The Social medias that have different people wo hails from different parts of the world needs this text classification to regulate that online public space. Text classification can be used in n number of ways to better the experience of users in social media by suggesting what

S. Aurelia et al. (Eds.): ICCSST 2023, CCIS 1973, pp. 288–298, 2024.
https://doi.org/10.1007/978-3-031-50993-3_23

they like by analyzing what they say in the public space and protect people from any negativity by blocking or censoring comments and posts that tends to hurt a particular user beliefs or feelings. Multi-label text classification is a type of task that is very useful in e-Learning, healthcare and e-Commerce.

The research studies conducted in this classification tasks are larger in number compared to other tasks. Most of the studies on text classification are carried out for the English language and some of the other major European languages. But people on the internet do not communicate only in English. Very often people tend to use their own language whether completely with their own script and dialects or using English scripts to type sentences in their languages by using similar sounding letters. Multilingual speakers frequently use the communication technique known as "Code-Mixing", which involves combining words and phrases from two distinct languages in a single text or speech utterance. Due to its coexistence with noisy, monolingual text, identifying Code Mixed text can be difficult. Numerous Code-Mixing measures have been widely applied over time to determine and validate the quality of Code-Mixed text.

The online community is not only comprised of English-Speaking users and there are many groups and sub communities in the public space that is dedicatedly run for other language users. The automated speech regulation algorithms in these online spaces tend to overlook the negative comments and hate speech on social media that is typed in Code-Mixed text as transliterating and applying the normal algorithms used for English could not accurately classify those texts as they tend to lose the meaning of certain words that is negative and could not sense the intent of the comment if the context also gets lost in the translation. Text Classification is usually done with neural networks and deep learning to get the best outcome. Research works done in Tamil shows that Bi LSTM works the best with CodeMixed Tamil text and even when using ensemble models and transfer learning methods like ULMFiT, the mixture of Bi-LSTM performs better than any state-of-the-art technology or standalone algorithms [1, 2, 8]. The consistency of Bi-LSTM through different research works shows that we cannot overlook the potential of it whenever we try new experiment on Code-Mixed Tamil text. In this study we aim to use XLNet and Bi-LSTM to classify the texts and compare the results with other algorithms that exists.

2 Related Work

In the recent years, the research conducted for Indian regional languages have been many. While major languages in the north like Hindi, Urdu have a lot of research work done on it, Dravidian languages like Tamil, Malayalam, Telugu, Kannada have very little research done on them that too in the last few years. The userbase of social media and e-Commerce consumers are now at a level where manual detection and removal of negative information and sentimental analysis on a particular consumer product is difficult without the help of AI. It is even more so for a language like Tamil. The studies that have been conducted in the past few years have used various different approach to this classification tasks. Sentiment analysis is a major sub field of study in text classification, according to S. Anbukkarasi et al. can be conducted in lexicon based or machine learning based approach [8].

S. Anbukkarasi et al. [8] collected Tamil (not Code-Mixed) tweets from the internet and fed them into a type of RNN algorithm to classify them into positive, negative or neutral. They have used a modified version of LSTM algorithm to carry out the work, called Bi-Directional LSTM which is used in most of the studies that shows better results than existing models. They have preprocessed the Tamil text into combined characters instead of Unicode characters to avoid the loss of meaning. This approach is more yielding in terms of results. Whereas in the study conducted to find the best algorithm that is suitable for classifying Code-Mixed Tamil text based on the sentiment Bidirectional LSTM is used along with ULMFIT to get a better result. After the usual preprocessing which involves removal of unnecessary characters and recurrent entities, Feature extraction is carried out using mBert and ULMFIT along with word2Vec method. This feature extracted data is fed into the Bi-LSTM model which is fine tuned due to ULMFIT to get the sentiment of the data [1]. They have compared the results of Bi-LSTM combined with ULMFIT to other algorithms like SVM, naive bayes, GRU and LSTM. Out of them Bi-LSTM performs better for Code-Mixed Tamil texts. M. Subramanian et al. had a huge dataset with more than 2,80,000 Code-Mixed Tamil texts to conduct their study on. The authors have focused on finding the efficiency of using transformer models on Tamil text. They have compared BERT model with other machine learning algorithms/models like Naïve bayes multinomial classifier, KNN, logistic regression etc where BERT outperforms all of them with an accuracy of 82.71% [3].

Named entity recognition is a type of classification problem. In a study conducted in the year 2022 [2] named entity recognition is attempted on Tamil text using deep learning algorithms. All the deep learning algorithms used for this experiment has shown significantly higher accuracy levels since the objective of this experiment is to classify entities already identified into predetermined categories. But this study provided an insight into how to handle the text data properly to optimize the classification process. The study conducted to classify Tamil news article [4] into appropriate topics using deep learning follows the same preprocessing techniques as any classification tasks like removing null space special characters and duplicates. Four different models are built for the experiment to compare and contrast the results obtained. RNN seems to perform better than other models used in every evaluation metrics. Compared to the work of C. S. K. Selvi et al. the Topic categorization of Tamil news articles using pre trained word2Vec embedding with CNN by S. Ramraj et al. [9] compare CNN combined with pretrained word embedding with other models that uses TFIDF vectorization to see which model performs better. Like expected CNN with Pretrained embedding performed better than the rest of the models.

S. Thavareesan et al. tackled one of the major obstacles when it comes to Tamil text classification which is the difference in morphology of Tamil compared to English. The grammatical nuances of Tamil make it difficult but at the same time necessary for a method to identify them. They also propose a novel approach to find the alternative words for the unseen words using fastText and Word2Vec embeddings. Multi-label emotion classification experiment study strives to solve the problem of identifying multiple emotion from a single text data which has Code-Mixed text in it. I. Ameer et al. have worked to label 12 different emotions based on the user's mental state that reflects on the text data collected from them. They have used classical machine learning techniques and deep learning models. By going a step further, they have also tried transfer learning models like BERT and XLNet. The best model turned out to be classical machine learning methods with word unigram for Code-Mixed text of English and Urdu [5]. In another experiment conducted for Code-Mixed Hindi and English texts to classify them based on predetermined categories, the approach R. Priyadharshini et al. handled for classifying native script is to transliterate the native script to roman script and use embedding models to proceed with the entity recognition. This transliterating approach is used in many other studies that involves non-English languages but have similarities to the English language. This approach works better only for languages that have very close similarities to English. Since Hindi is an Indo-European language it shows better results in that particular study [11]. Many of the studies conducted on Indian languages used Bi-LSTM which seems to perform better than many state-of-the art techniques that already exists for these tasks. The study performed by S. Thara and P. Poornachandran compares various different transformer models to identify the language of the text corpus data. In the case of Malayalam language, the authors had to acquire data and then transliterate the data to the correct roman script before they were able to proceed with the identification process. The research conducted involves different types of BERT models like camemBERT, DistilBERT, XLMRoBERTa, Electra etc. This study involves the identification and categorization of Malayalam English text data to create a large dataset for word-Level language identification.

3 Methodology

The architecture diagram in Fig. 1 gives an overview of the experiment conducted in this study. The Code-Mixed Tamil text data used for this experiment is a dataset used by Chakravarthi, B. R. et al. [13], which has roughly 20,000 comments in Malayalam and English, 7,000 comments in Kannada-English, and 44,000 comments in Tamil English. This dataset seems fit for the task we are undertaking which is offensive text classification. The data is first cleaned and preprocessed to fit the model that will be used down the line. The preprocessing removes all the unnecessary, useless part of the text data by removing all the punctuation, empty spaces and special characters.

Once the preprocessing is done the data is used to train the XLNet model and Bi-LSTM model. The prediction results that is obtained after predicting the test data using the trained model is evaluated using different performance metrics. Finally, the results are tabulated and visualized to get a better understanding of the models' performances.

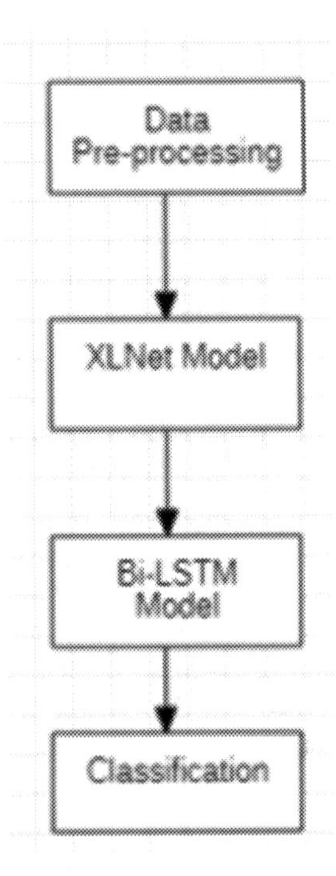

Fig. 1. Architecture diagram of the proposed method

3.1 Code-Mixed Text Data

The mingling of two or more languages or linguistic subtypes in speech is known as Code-Mixing. The terms "Code-Mixing" and "Code-Switching" are sometimes used interchangeably by academics, particularly when discussing syntax, morphology, and other formal features of language. Others rely on more precise definitions of Code Mixing; however, these precise meanings may vary across linguistics, education theory, communications, and other subfields. Code-Mixed data is what that is prevalently used in social medias like YouTube, twitter, Instagram etc. When trying to classify texts that are purely mono-linguistic the approach that is used for it can be similar to English text classification tasks but for Code-Mixed text needs different approach. Especially Tamil Code-Mixed text can be hard to identify and understand by the machine. Tamil language requires the previous word's meaning to understand the intent of the current word. The machine cannot classify a single text sentence using one of the word present in it. Most of the base line models or deep learning algorithms face problems doing this. Bi-LSTM steps into solve this problem. The dataset used here has more than 60,000 YouTube comments made up of multiple Code-Mixed language text. The collection consists of 44,000 Code-Mixed Tamil texts with the rest from other Dravidian languages. The dataset is already collected and split into test, train and validation set. Sample data is given in Table 1.

Table 1. Code-Mixed Tamil data

id	text	category
Tam_1	[illegible]...	Positive
Tam_2	Teruk an irukku ... mokke movie .. waste of time	Negative
Tam_3	manitha samuthaayam amaipil irunthu intha pada...	Positive
Tam_4	JJ mam we miss u	Positive
Tam_5	Subtitle me trailer dekhne wale like karo	not-Tamil
...	...	...
Tam_4398	Ithukum dislike potta kammanatti koovaingalam...	Negative
Tam_4399	Suyama Sinthikiravan than super Hero Seama da...	Mixed_feelings
Tam_4400	Super thalaiva ... Nee mass dha eppavume	Positive
Tam_4401	[illegible]...	unknown_state
Tam_4402	Semma thalaiva alu athikama akirukum enimale e...	Positive

3.2 Data Preprocessing

Without data preprocessing any NLP task would not yield good results. A good preprocessing is necessary for classification models, especially text data from social medias. Data from social media can contain more than just Tamil-English texts. It can contain special characters like '!', '@', dates, emoticons and numbers etc. During preprocessing step, the special characters, stop words and null spaces are removed to get a clean text data of just words. Stop words in Tamil are many and necessary to be removed to get just the contextual words individually. In Fig. 2 some of the Tamil stop words are given. These words provide no meaning and are not of any use when trying to classify texts based on whether they are offensive are not.

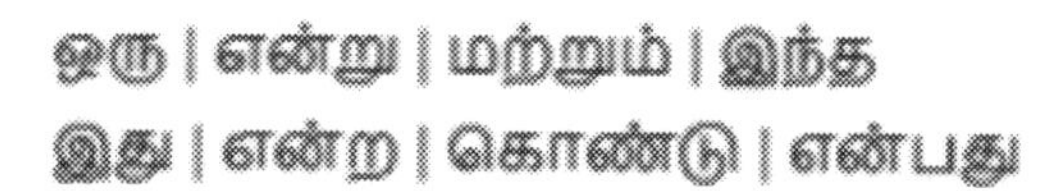

Fig. 2. Tamil Stop Words

After removing stop words, special characters and null spaces, it is also important to remove repeating/recurrent characters and words. People tend to comment words like "Semmaaaaa" where the extra 'a' are not needed and they can also type the words multiple times to show their like or dislike towards something. Finally, after the removal of emojis and numbers the data preprocessing is done.

3.3 Bidirectional - Long Short-Term Memory

Making any neural network have the sequence information in both directions—backwards (future to past) or forward—is known as bidirectional long-short term memory (Bi-LSTM) (past to future). A bidirectional LSTM differs from a conventional LSTM in that our input flows in two directions. We may make input flow in one way, either backwards or forward, using the standard LSTM. To maintain both past and future information, Bi-directional input can be made to flow in both ways.

3.4 XLNet

Just like Bi-LSTM, XLNet is also used for its ability to learn bidirectional contexts by maximizing the expected likelihood over all permutations of the input sequence factorization order [14]. Pretraining methods based on denoising autoencoding, such as BERT, perform better than those based on autoregressive language modelling because they can model bidirectional contexts. BERT neglects the dependency between the masked positions and experiences a pretrain-finetune discrepancy since it relies on masking the input to corrupt it. With these advantages and disadvantages in mind, they [14] suggest XLNet, a generalised autoregressive pretraining method that (1) makes it possible to learn bidirectional contexts by maximising the expected likelihood over all variations of the factorization order and (2), thanks to its autoregressive formulation, outperforms the drawbacks of BERT. Furthermore, Transformer-XL, a cutting-Edge autoregressive model, is incorporated into pretraining by XLNet.

4 Experiments and Results

After the data was cleaned by removing all the stop words, special characters, null space etc., the clean texts are fed into the Models one by one. We compare the performance of XLNet + Bi-LSTM with Normal LSTM and Bi-LSTM. The model loss vs accuracy graph in Fig. 3 shows the performance of LSTM for this Code-Mixed data. As we train with more and more data the loss goes down and the accuracy increases only slightly. Whereas, in Bi-LSTM alone the accuracy starts off high keeps increasing and the loss decreases drastically as evident in Fig. 4. The LSTM results in an accuracy of 74% while Bi-LSTM beats by a large margin with 86.8%. The F1 scores are also different for LSTM and Bi-LSTM, with the former getting 74 while the latter gives 81. But both these models perform exceptionally better than the proposed XLNet and Bi-LSTM model.

While these standalone models performed exceptionally well, the proposed system of XLNet with Bi-LSTM seems to not work out as its accuracy and loss proved so in the Fig. 5(a) and (b).

The comparative analysis between the models is recorded as a table below. Table 2 shows the differences that we have seen so far between the models clearly and one look at that would tell us what the superior model among them would be for now.

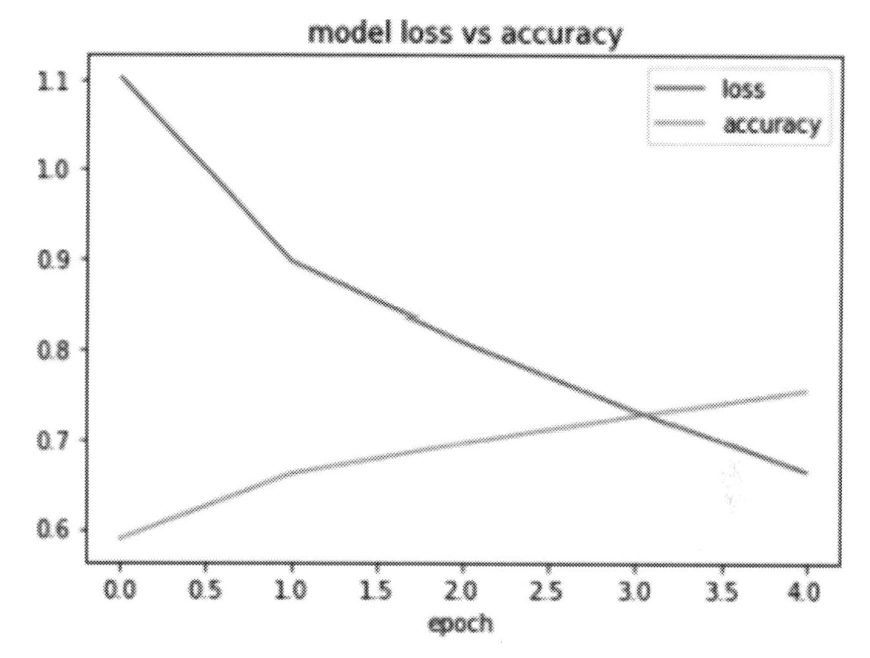

Fig. 3. Loss vs Accuracy for LSTM

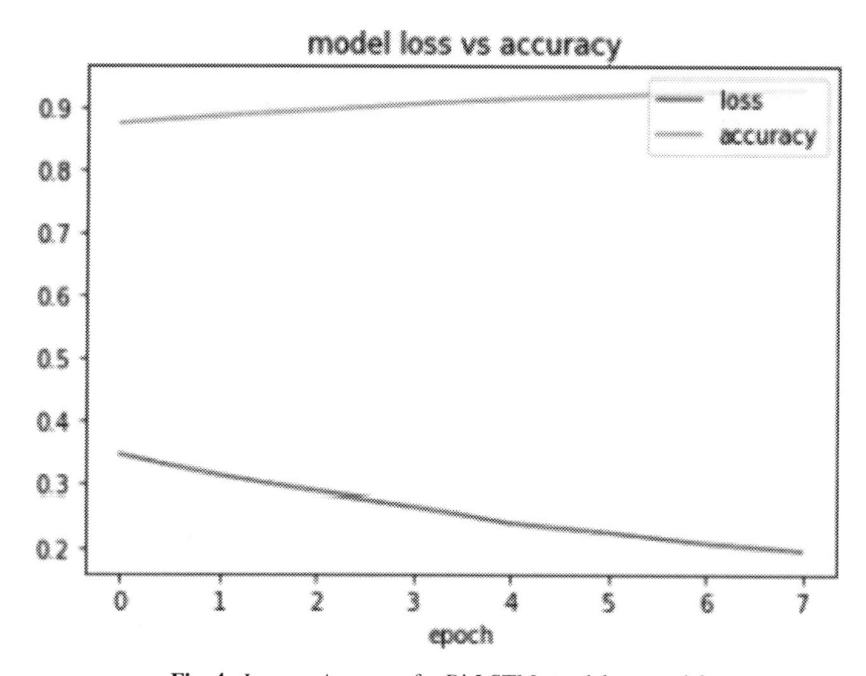

Fig. 4. Loss vs Accuracy for Bi-LSTM standalone model

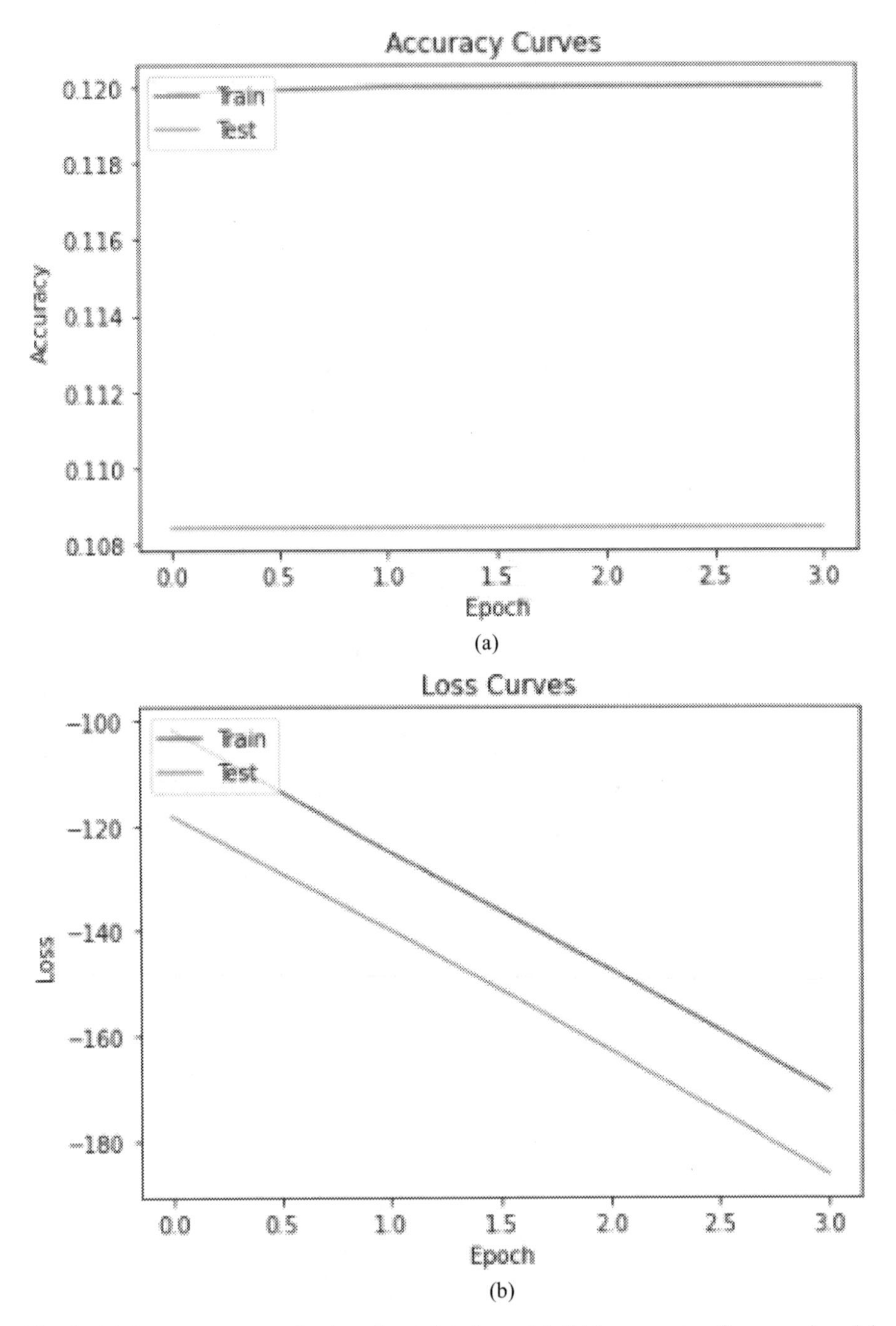

Fig. 5. (a) Accuracy curves of train and test data in model. (b) Loss curves of proposed model

Table 2. Comparative Table of models experimented with

Models	Accuracy (in %)	F1 Score
LSTM	77	74
Bi-LSTM	86.8	81
XLNet + Bi-LSTM	12	6

5 Conclusion

Classification of Code-Mixed Tamil text is carried out in this study using XLNet with Bi-LSTM. We conclude from the results that standalone Bi-LSTM performs better than LSTM and XLNet + Bi-LSTM. The Bi-LSTM model gives an accuracy of 86.8% and an F1 score of 81 while LSTM performs with an accuracy of 77% and an F1 score of 74. While the stand alone showed better performance the proposed system of XLNet and Bi-LSTM didn't work as expected. It gave an accuracy of 12% that too with heavy loss. The study fails to confirm whether XLNet can accurately classify Code-Mixed Tamil text with the help of Bi-LSTM. Maybe the fitting of XLNet was wrong or the mixture of Bi-LSTM with XLNet produced heavy loss that the learning process was hindered because of it. There can be other factors that come with the data. Since Tamil is not a European language, like any other models XLNet would also struggle with it. On top of it we are using code-mixed data which brings in more complexity to the experiment. Handling the data well could be a solution to this experiment with more Tamil data and pre-trained transformer models would help solve the classification problem of Code-Mixed Tamil text data.

References

1. Nithya, K., Sathyapriya, S., Sulochana, M., Thaarini, S., Dhivyaa, C.R.: Deep learning based analysis on code-mixed Tamil text for sentiment classification with pre-trained ULMFiT. In: 2022 6th International Conference on Computing Methodologies and Communication (ICCMC), pp. 1112–1116 (2022). https://doi.org/10.1109/ICCMC53470.2022.9754163
2. Anbukkarasi, S., Varadhaganapathy, S., Jeevapriya, S., Kaaviyaa, A., Lawvanyapriya, T., Monisha, S.: Named entity recognition for Tamil text using deep learning. In: 2022 International Conference on Computer Communication and Informatics (ICCCI), pp. 1–5 (2022). https://doi.org/10.1109/ICCCI54379.2022.9740745
3. Subramanian, M., Adhithiya, G.J., Gowthamkrishnan, S., Deepti, R.: Detecting offensive Tamil texts using machine learning and multilingual transformer models. In: 2022 International Conference on Smart Technologies and Systems for Next Generation Computing (ICSTSN), pp. 1–6 (2022). https://doi.org/10.1109/ICSTSN53084.2022.9761335
4. Selvi, C.K., Induja, N., Lekshmi, S.L., Nagammai, S.: Topic categorization of Tamil news articles. In: 2022 International Conference on Computer Communication and Informatics (ICCCI), pp. 1–6 (2022). https://doi.org/10.1109/ICCCI54379.2022.9741061
5. Ameer, I., Sidorov, G., Gómez-Adorno, H., Nawab, R.M.A.: Multi-label emotion classification on code-mixed text: data and methods. IEEE Access **10**, 8779–8789 (2022). https://doi.org/10.1109/ACCESS.2022.3143819

6. Sabri, N., Edalat, A., Bahrak, B.: Sentiment analysis of Persian-English code-mixed texts. In: 2021 26th International Computer Conference, Computer Society of Iran (CSICC), pp. 1–4 (2021)
7. Thara, S., Poornachandran, P.: Transformer based language identification for Malayalam English code-mixed text. IEEE Access **9**, 118837–118850 (2021). https://doi.org/10.1109/ACCESS.2021.3104106
8. Anbukkarasi, S., Varadhaganapathy, S.: Analyzing sentiment in Tamil tweets using deep neural network. In: 2020 Fourth International Conference on Computing Methodologies and Communication (ICCMC), pp. 449–453 (2020). https://doi.org/10.1109/ICCMC48092.2020.ICCMC-00084
9. Ramraj, S., Arthi, R., Murugan, S., Julie, M.S.: Topic categorization of Tamil news articles using PreTrained Word2Vec embeddings with convolutional neural network. In: 2020 International Conference on Computational Intelligence for Smart Power System and Sustainable Energy (CISPSSE), pp. 1–4 (2020). https://doi.org/10.1109/CISPSSE49931.2020.9212248
10. Thavareesan, S., Mahesan, S.: Word embedding-based Part of Speech tagging in Tamil texts. In: 2020 IEEE 15th International Conference on Industrial and Information Systems (ICIIS), pp. 478–482 (2020). https://doi.org/10.1109/ICIIS51140.2020.9342640
11. Priyadharshini, R., Chakravarthi, B.R., Vegupatti, M., McCrae, J.P.: Named entity recognition for code-mixed Indian corpus using meta embedding. In: 2020 6th International Conference on Advanced Computing and Communication Systems (ICACCS), pp. 68–72 (2020). https://doi.org/10.1109/ICACCS48705.2020.9074379
12. Yadav, K., Lamba, A., Gupta, D., Gupta, A., Karmakar, P., Saini, S.: Bi-LSTM and ensemble based bilingual sentiment analysis for a code-mixed Hindi-English social media text. In: 2020 IEEE 17th India Council International Conference (INDICON), pp. 1–6 (2020). https://doi.org/10.1109/INDICON49873.2020.9342241
13. Chakravarthi, B.R., et al.: DravidianCodeMix: sentiment analysis and offensive language identification dataset for Dravidian languages in Code-Mixed text. Lang. Resour. Eval. (2022). https://doi.org/10.1007/s10579-022-09583-7
14. Yang, Z., Dai, Z., Yang, Y., Carbonell, J., Salakhutdinov, R., Le, Q.V.: XLNet: generalized autoregressive pretraining for language understanding (2019). https://doi.org/10.48550/ARXIV.1906.08237

The Road to Reducing Vehicle CO_2 Emissions: A Comprehensive Data Analysis

S. Madhurima, Joseph Mathew Mannooparambil, and Kukatlapalli Pradeep Kumar(✉)

Computer Science and Engineering, Christ University, Bangalore, Karnataka, India
Kukatlapalli.kumar@christuniversity.in

Abstract. In recent years, the influence of carbon dioxide (CO_2) releases on the environment have become a major concern. Vehicles are one of the major sources of CO_2 emissions, and their contribution to climate change cannot be ignored. This research paper aims to investigate the CO_2 emissions of vehicles and compare them with different types of engines, fuel types, and vehicle models.

The study was carried out by gathering information about the CO_2 emissions of vehicles from the official open data website of the Canadian government. Data from a 7-year period are included in the dataset, which is a compiled version. There is a total of 220 cases and 9 variables. The data is analyzed using statistical methods and tests to identify the significant differences in CO_2 emissions among different Car Models. The results indicate that vehicles with diesel engines emit higher levels of CO_2 compared to those with gasoline engines. Electric vehicles, on the other hand, have zero CO_2 emissions, making them the most environmentally friendly option. Furthermore, the study found that the CO_2 emissions of vehicles vary depending on the type of fuel used. The study also reveals that the CO_2 emissions of vehicles depend on the model and age of the vehicle. Newer models tend to emit lower levels of CO_2 compared to older models. In conclusion, this study provides valuable insights into the CO_2 emissions of Cars and highlights the need to adopt cleaner and more sustainable transportation options.

Keywords: Data Analysis · CO_2 Emissions · Road Transport · Pollution

1 Introduction

Carbon dioxide emissions from vehicular sources are a significant contributor to greenhouse gases in the atmosphere. These discharges originate from the scorching of fossil gasses such as gasoline and diesel and subsidize to climate alteration. Vehicles are accountable for about one-third of all carbon dioxide releases in the United States. SUVs and trucks release roughly twice as much carbon dioxide annually as the usual passenger vehicle, which emits around 4.6 metric tons. There are number of ways to reduce carbon dioxide emissions from vehicles. One is to improve fuel efficiency. This can be done by designing more efficient engines, using lighter materials in construction, and improving aerodynamics. Another way to reduce emissions is to switch to alternative fuels such as electricity or biofuels. Increasing the use of public transportation, walking,

S. Aurelia et al. (Eds.): ICCSST 2023, CCIS 1973, pp. 299–309, 2024.
https://doi.org/10.1007/978-3-031-50993-3_24

and biking are also effective ways to reduce vehicular emissions. We can greatly reduce carbon dioxide emissions and aid in the fight against climate change by cutting down on the amount of vehicle miles we drive. Upcoming sections provides the information on literature review, results and conversation tailed by conclusion.

2 Literature Background

2.1 Assessment of Carbon Dioxide Emissions of Internal Combustion Engine Vehicle and Battery e-Vehicles

In this sub section, the CO_2 emissions of battery electric vehicles (BEV) are compared with those of gasoline and diesel internal combustion engine vehicles (ICV). As reference regions for vehicle operation, the US, the EU, Japan, China, and Australia were chosen. The fastest-growing contributor to global warming emissions is the transportation sector. Despite this, present laws only apply to a vehicle's functioning stage, i.e., its tailpipe releases from the tank to the wheels. There are currently no regulations that take a vehicle's other life cycles into account [1–3]. The variation in fuel and electric consumption of cars by region may have an impact on LCA. The study displays the total amount of driving, from 0 km to 200,000 km. The scope of this study is life cycle of vehicles. The analysis excepted removal and reprocessing of left-over resources in the vehicle production and in this article, the CO_2 emissions of battery electric vehicles (BEV) are compared with those of gasoline and diesel internal combustion engine vehicles (ICV). The firmest rising foundation of GHG emissions is the shipping segment [4]. Despite this, present laws only apply to a vehicle's functioning stage, i.e., its tailpipe emissions from the tank to the wheels. The other life cycle of a vehicle is not currently taken into account by regulations. A life-cycle assessment (LCA), which takes into account CO_2 emissions of vehicles during their operational and end-of-life phases, can be used as an objective method to quantify GHG emissions over the course of a vehicle's lifetime.

The variation in fuel and electric consumption of cars by area could have an impact on LCA. Lifetime driving distance is needed for The LCA study of automobiles. ICVs and BEVs' lifetime driving distances were listed at 150,00 km for the EU and 100,000 km for Japan. The scope of this study is life cycle of vehicles. The study did not include waste materials recycling and disposal during car production and maintenance phase [5].

2.2 Electric Vehicles CO_2 Emissions

Electric vehicles (EVs) are often touted as a more environmentally friendly alternative to traditional internal combustion engine (ICE) vehicles because they produce zero tailpipe emissions. However, the overall environmental impact of an electric vehicle, including its carbon dioxide (CO_2) emissions, depends on several factors:

a. **Electricity Source**: The CO_2 emissions associated with an electric vehicle largely depend on the source of the electricity used to charge it. If the electricity comes from renewable sources like wind, solar, or hydropower, the emissions are very low. In regions where the electricity grid relies heavily on coal or other fossil fuels, the emissions associated with charging an EV can be higher.

b. **Energy Efficiency**: EVs are generally more energy-efficient than ICE vehicles. They convert a higher percentage of the energy from their fuel source (electricity) into actual vehicle movement, which can reduce overall emissions.
c. **Battery Life and Recycling**: The environmental impact of EVs is influenced by the lifespan of their batteries and the ability to recycle or repurpose them at the end of their life. Recycling can reduce the demand for new raw materials and minimize environmental impact.

While assessing the CO_2 emissions of electric vehicles, it's crucial to consider the entire life cycle, including manufacturing, electricity source, operation, and end-of-life management. Studies have shown that, in many regions with a relatively clean electricity grid, EVs produce fewer emissions over their lifetime compared to equivalent ICE vehicles. However, the environmental benefits can vary significantly depending on local factors. A major advantage of PEVs over ICEVs is rapid systemic change in the energy sector Lower upstream emissions result directly from lower carbon intensity. We don't take low-carbon or carbon-free fuels into account in the study; instead, we pay attention to indirect emissions from electric vehicles (PEVs) and changes related to the energy shift in power generation [6].

Every year when a PEV is in service, its annual GHG emissions decrease as the generation mix gets better. Without taking into account biofuels, this is significantly different from ICEV. When PEV disposal is taken into account, this effect might even intensify. The life-cycle carbon dioxide emissions from plug-in electric vehicles are impacted by changes to the electricity mix and battery capacity. So, the conclusion is Peer-to-peer (PEV) electric vehicles (or plug-in hybrids) are an important step in reducing Transport-related carbon emissions are a concern, but they shouldn't be put off [7].

2.3 Study of CO_2 Emissions in Different Temperatures and Road Loads

As humans, we have long been concerned about the impact of vehicle emissions on our environment. Although there have been numerous research on vehicle CO_2 emissions, it is still unclear how ambient temperature affects CO_2 emissions. The study was conducted to examine the relationship between CO_2 emissions and ambient temperature quantitatively. This study involved testing of the worldwide harmonized Light cars Test Cycle (WLTC) at four distinct values (−10 °C, 0 °C, 23 °C, and 40 °C), two small-duty fuel cars were verified [8].

It revealed that amplified interior confrontation, rather than gasoline enrichment, was the primary cause of increased CO_2 emissions in cold ambient temperatures. According to the study, CO_2 emissions from small engines were more responsive to changes in load while those larger engines were more susceptible to temperature variations. Also, air conditioning increased distance-specific CO_2 emissions significantly for both vehicles tested.

The study also identified there is a significant relationship between transient CO_2 emission, coolant temperature, and cumulative CO_2 emission, with an R2 greater than 0.98. This finding suggests that the engine's state of warm-up might be described using cumulated CO_2 emission/coolant temperature, which would enhance transient CO_2

modeling. This helps us better understand how temperature affects vehicle CO_2 emissions, and its results could be used to inform microscopic CO_2 models and CO_2-related guidelines [9].

2.4 Policies in Road CO_2 Emissions

To reduce emissions, policymakers implement a variety of policy approaches. Current methods of policy evaluation often focus on the effects of individual policies, but this can limit the ability to identify effective combinations of policies. Due to this uncertainty, a new approach is used that uses statistical analysis to detect structural breaks in CO_2 emissions, which could be caused by any number of previously unknown policies. By identifying these "causes of effects", we can make systematic inferences about the effectiveness of policy combinations. This study demonstrates eleven effective policy changes that decreased emissions by 8% to 26% [10]. The most effective strategy combines gasoline or carbon taxes with incentives for green vehicles, demonstrating that CO_2 emissions can be decreased to that of the EU's zero-emission targets. Overall, these findings suggest that policy mixes can be effective in reducing emissions and provide policymakers with evidence-based guidance for designing effective policy mixes in the future [11].

2.5 CO_2 Emissions in Commercial Passenger Cars

The global market for passenger vehicles is experiencing significant switch to hybrid and fully electric vehicles from internal combustion engines and other vehicles. The desire to lessen greenhouse gas emissions and mitigate the climatic effect of the transportation industry is what is driving this change. This research shows comprehensive comparisons between the over-all life-cycle glasshouse gas discharges from various commercially accessible passenger vehicles with various drivetrains and fuel sources to highlight the importance of this change. This research is conducted using simple models to calculate the cycle GHG releases of the vehicles in expressions of figures for Carbon dioxide equivalents based on the control weight and maximum motor command of the vehicle. The study analyses 790 different vehicle variants and quantifies based on the most recent assessed emission coefficient values, the production, utilization, and recycling emissions individually [12]. Overall, the study highlights the urgent need to transition from traditional combustion engine vehicles to vehicles powered by cleaner and more sustainable energy sources. It makes the policymakers, manufacturers, and consumers in making informed decisions to lower emissions of greenhouse gases and mitigate climate impact of the transportation sector [13, 14].

3 Result and Discussion

This section provides a comprehensive data analysis considered from a particular data set explaining the parameters carbon dioxide emission. Visualization such as box plot analysis, histogram, and stem and leaf plot were observed and described in this regard. Statistical parameters are also described with appropriate values and tabular forms.

3.1 Calculation of Basic Parameters

The Table 1 shows the calculation of basic parameters like mean, median, mode, standard deviation and variance of CO_2 Emissions (g/km).

Table 1. Table depicting central tendency measures.

CO_2 Emissions(g/km)	
Valid	220
N	
Missing	1
Mean	246.61
Median	239.00
Mode	184
Std. Deviation	65.381
Variance	4274.731

3.2 Data Visualization

Figure 1 is a histogram visualization which shows the frequency of Fuel Consumption City (L/100 km) of given dataset. The highest frequency lies in the range 10 to 15 L/100 km.

The bar chart shows the count of different numbers of cylinders in groups. The highest count is for 4 cylinders with count value of about 100 and the lowest count is for 10 cylinders with count value of about 2. This is illustrated in Fig. 2.

The above pie chart depicted in Fig. 3 shows the graphical representation of different categories of cylinders. From this pie chart we can know that 4 cylinders are represented in huge quantity shown in green and 10 cylinders are represented in small quantity shown in yellow. The stem and leaf plot shows the representation of Fuel Consumption City (L/100 km) in tabular form. The plot is divided into stem and leaf, for example 58 can be represented as 5 in stem and 8 in leaf which represented in Fig. 4. The boxplot in Fig. 5 shows the representation of Fuel Consumption. The boxplot contains few outliers, and the median is found to about 11.9 L/100 km.

3.3 Statistical Test and Significance Value

ANOVA Test:

Table 2 shows the ANOVA test between CO_2 emissions and Fuel Consumption. The significance value is 0.000 which is less than 0.05, therefore there is significant difference between CO_2 emissions and Fuel Consumption.

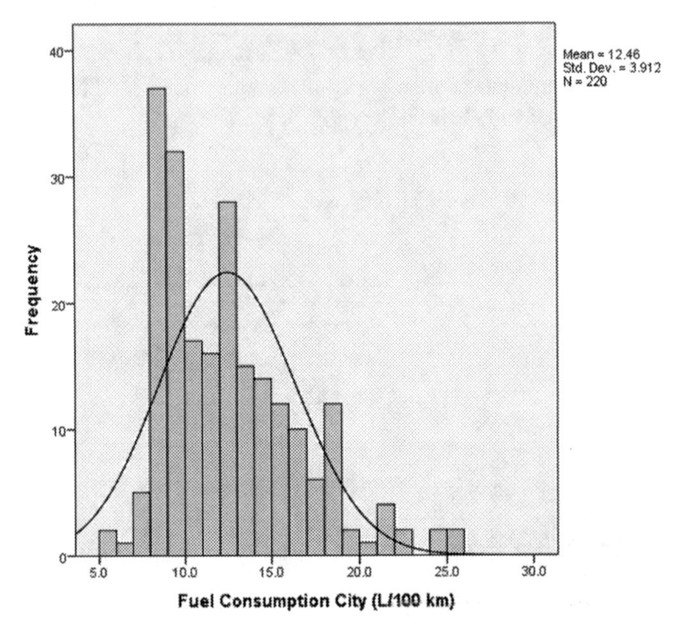

Fig. 1. Output of variable plotted as a histogram.

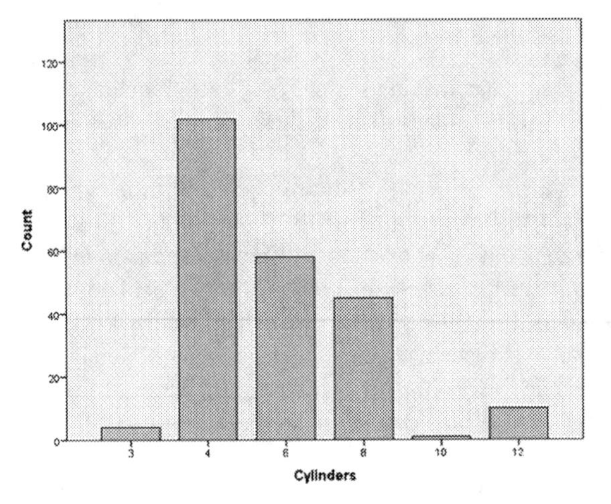

Fig. 2. Output of variable as bar chart

Variables taken:

i CO_2 emissions (g/km) = Dependent list
ii Fuel consumption city (L/100km) = Factor

Independent Sample T Test:
This test between engine size as the test variable and registration as grouping variable is

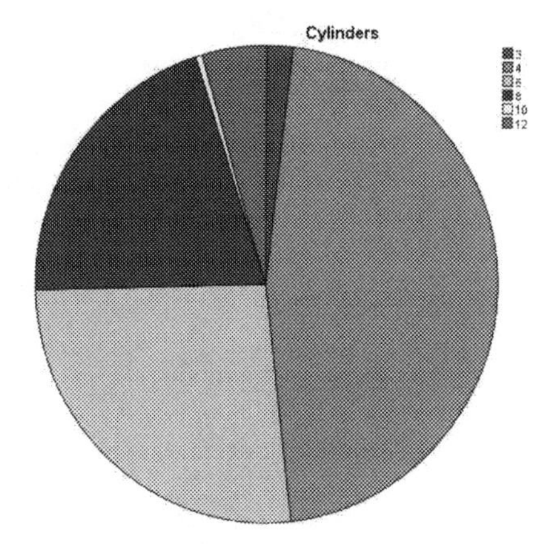

Fig. 3. Output of variable as pie chart

```
Fuel Consumption City (L/100 km) Stem-and-Leaf Plot

 Frequency    Stem &  Leaf

     2.00        5 .  58
     1.00        6 .  0
     5.00        7 .  01679
    37.00        8 .  0000012222444445555555556667888899999
    32.00        9 .  0111122222233344444445666667788999
    17.00       10 .  00001114455668899
    16.00       11 .  0111223334588888
    28.00       12 .  01111223344466666677777889999
    15.00       13 .  012333445677899
    14.00       14 .  01223334666778
    12.00       15 .  022222367777
    10.00       16 .  0000123678
     6.00       17 .  223455
    12.00       18 .  000011122259
     2.00       19 .  09
     1.00       20 .  0
     4.00       21 .  5555
     2.00       22 .  12
     4.00 Extremes    (>=24.0)

 Stem width:        1.0
 Each leaf:       1 case(s)
```

Fig. 4. Output of variable as stem and leaf plot

demonstrated in Table 3. It has a significance value of about 0.841 which is greater than 0.05, therefore Engine size and registration varied significantly.

Variables taken:

Engine size(g/km) = Test variable
Registration = Grouping variable

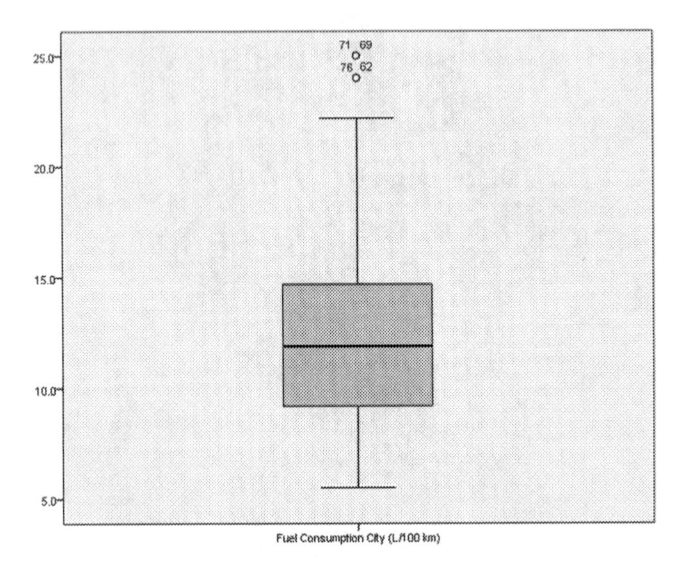

Fig. 5. Output of variable plotted as Boxplot.

Table 2. Output of ANOVA test

CO_2 Emissions(g/km)					
	Sum of Squares	df	Mean Square	F	Sig.
Between Groups	921873.959	96	9602.854	82.643	.000
Within Groups	14292.200	123	116.197		
Total	936166.159	219			

3.4 Correlation and Regression Analysis

Pearson Correlation Analysis:

The relationship between cylinders and registration is seen in Table 4. Here the correlation value is shown to be 0.007 which is between 0 and 1, therefore it has a weak positive correlation.

Considering straight line equation in order to find the value of y when x is given (assume x = 120).

Equation – $y = bx + c$
Here $x = 12$
$b = 0.56$
$c = -1.24$
$y = 0.56 \times 120 + -1.24$
$= 67.2 - 1.24 = 65.96$

The correlation coefficient obtained $r = 0.928$(obtained from SPSS).

Table 3. Output for independent sample t test

	registration	N	Mean	Std. Deviation	Std. Error Mean
Engine Size(L)	not register	57	3.18	1.488	.197
	register	162	3.13	1.479	.116

Independent Samples Test

		Levene's Test for Equality of Variances		t-test for Equality of Means						
		F	Sig	t	df	Sig. (2-tailed)	Mean Difference	Std. Error Difference	95% Confidence Interval of the Difference	
									Lower	Upper
Engine Size(L)	Equal variances assumed	.127	.722	.201	217	.841	.046	.228	−.404	.495
	Equal variances not assumed			.200	97.604	.842	.046	.229	−.408	.500

Table 4. Output of Pearson correlation analysis

Pearson Correlations

		Cylinders	registration
Cylinders	Pearson Correlation	1	.007
	Sig. (2-tailed)		.922
	N	220	219
registration	Pearson Correlation	.007	1
	Sig. (2-tailed)	.922	
	N	219	219

Regression Analysis:
(See Table 5).

Table 5. Analysis of Regression

Coefficients					
Model	Unstandardized Coefficients		Standardized Coefficients	t	Sig.
	B	Std. Error	Beta		
(Constant)	−1.240	.385		−3.224	.001
CO_2 Emissions(g/km)	.056	.002	.928	36.834	.000

a. Dependent Variable: Fuel Consumption City (L/100 km)

4 Conclusion

The literature review is done for the Comprehensive data analysis of the CO_2 Emission dataset and various operations are performed like basic calculation on statistical parameters, visualization of data, significance test, correlation, and regression analysis. The data is analyzed to identify the significant differences in CO_2 emissions among different Car Models. The results conclude that Electric cars are becoming more and more popular, and they produce no emissions at all. Other alternative fuels such as biodiesel and natural gas can also help to reduce emissions. Also, newer models tend to emit lower levels of CO_2 compared to older models. Reducing vehicle CO_2 emissions is crucial to combat climate change and creating a sustainable future. There are several measures that individuals can take to reduce their vehicle's emissions, including choosing fuel-efficient vehicles, driving less or carpooling, using public transportation, biking or walking, maintaining their vehicles properly, and adopting eco-driving habits.

However, reducing vehicle emissions also requires collective action such as including government policies that promote cleaner transportation options and incentivize the adoption of electric and hybrid vehicles. It also requires investment in clean energy infrastructure, such as charging stations for electric vehicles and renewable energy sources. Ultimately, it is everyone's responsibility to contribute to reducing vehicle CO_2 emissions and promoting sustainable transportation options. By taking individual actions and advocating for collective action, we have the power to make the earth cleaner and healthier for both current and future generations.

References

1. Khan, M., Kar, N.C.: Hybrid electric vehicles for sustainable transportation: a Canadian perspective. World Electr. Veh. J. **3**(3), 551–562 (2009)
2. Acar, C., Dincer, I.: The potential role of hydrogen as a sustainable transportation fuel to combat global warming. Int. J. Hydrogen Energy **45**(5), 3396–3406 (2020)
3. Denny, D., Kumar, K.P.: Secure authenticated communication via digital signature and clear list in VANETs. ECS Trans. **107**(1), 20065 (2022)
4. Kawamoto, R., et al.: Estimation of CO_2 emissions of internal combustion engine vehicle and battery electric vehicle using LCA. Sustainability **11**(9), 2690 (2019)
5. Lombardi, L., et al.: Comparative environmental assessment of conventional, electric, hybrid, and fuel cell powertrains based on LCA. Int. J. Life Cycle Assess. **22**, 1989–2006 (2017)

6. Martz, A., Plotz, P., Jochem, P.: Global perspective on CO_2 emissions of electric vehicles. Environ. Res. Lett. **16**(5), 054043 (2021)
7. Xu, B., et al.: Have electric vehicles effectively addressed CO_2 emissions? Analysis of eight leading countries using quantile-on-quantile regression approach. Sustain. Prod. Consum. **27**, 1205–1214 (2021)
8. Wang, Y., et al.: Quantitative study of vehicle CO_2 emission at various temperatures and road loads. Fuel **320**, 123911 (2022)
9. Giechaskiel, B., Komnos, D., Fontaras, G.: Impacts of extreme ambient temperatures and road gradient on energy consumption and CO_2 emissions of a euro 6d-temp gasoline vehicle. Energies **14**(19), 6195 (2021)
10. Koch, N., et al.: Attributing agnostically detected large reductions in road CO_2 emissions to policy mixes. Nat. Energy **7**(9), 844–853 (2022)
11. Solaymani, S.: CO_2 emissions patterns in 7 top carbon emitter economies: the case of transport sector. Energy **168**, 989–1001 (2019)
12. Buberger, J., et al.: Total CO_2-equivalent life-cycle emissions from commercially available passenger cars. Renew. Sustain. Energy Rev. **159**, 112158 (2022)
13. Seboldt, D., et al.: Hydrogen engines for future passenger cars and light commercial vehicles. MTZ Worldwide **82**(2), 42–47 (2021)
14. Kumar, K., et al.: A distinctive approach on safe driving through interactive and behavioral analysis. Int. J. Comput. Sci. Eng. **7**, 710–714 (2019). https://doi.org/10.26438/ijcse/v7i2.710714

A Deep Learning Based Bio Fertilizer Recommendation Model Based on Chlorophyll Content for Paddy Leaves

M. Nirmala Devi(✉), M. Siva Kumar, B. Subbulakshmi, T. Uma Maheswari, Karpagam, and M. Vasanth Kumar

Department of Computer Science and Engineering, Thiagarajar College of Engineering, Madurai, Tamil Nadu, India
nirmaladevi2004@gmail.com, {mskcse,bscse}@tce.edu

Abstract. Rice is the main agricultural product in India, with 90% of the population consuming it as a staple food. Nutrient deficiencies in rice leaves during growth period causes imbalances leading to reduce in crop yield. Growth of the paddy plants is highly related to the chlorophyll content present in it. Chlorophyll content in a leaf is a key indicator of greenness of a leaf and identifies the nutrient deficiencies in plants. Chlorophyll in plants contains nitrogen which is responsible for the photosynthesis process. In this research, chlorophyll and nitrogen contents are measured for the paddy leaves. Here, Support Vector Machine (SVM) Regression and Convolutional Neural Network (CNN) models are used to measure the chlorophyll and nitrogen content in plants. In this research, the color was the main parameter used to quantify chlorophyll and nitrogen contents in plants and RGB color model is used. Bio fertilizers is an influencing factor in yield progression and physiological processes as Bio fertilizers supply necessary nutrients to plants and enhance chlorophyll content in leaves. The chlorophyll and nitrogen concentrations were measured and then based on the measured nitrogen concentration, the appropriate bio fertilizer is recommended by the SVM classification model to enhance the nitrogen content in the plants. Here the CNN (99.98%) algorithm works better than the SVM algorithm in prediction of Chlorophyll and nitrogen contents. Hence, the Convolutional Neural Network model is built to predict the chlorophyll and nitrogen contents for the paddy leaves based on color and recommends the appropriate bio fertilizer to improve plant growth.

Keywords: Paddy Leaves · Leaf color · Leaf Chlorophyll · Nitrogen · Nutrient deficiencies · Bio fertilizer Recommendation · Convolutional Neural Network · Support Vector Machine Classification · Regression models

1 Introduction

Rice is the main agricultural product in India, with 90% of the population consuming it as a staple food. Due to this the rice production plays an important role in the society. Plant growth is highly dependent on the Chlorophyll Content. Chlorophyll is an essential

S. Aurelia et al. (Eds.): ICCSST 2023, CCIS 1973, pp. 310–321, 2024.
https://doi.org/10.1007/978-3-031-50993-3_25

component in the process of photosynthesis, which produces oxygen and sustains plant life. Traditionally, chlorophyll content is measured using the wet chemical extraction method done in a laboratory, which is time-consuming and costly. However, improvement in recent technology has led to the development of non-destructive plant evaluation methods.

Recently, Deep Learning methods such as convolutional neural network (CNN) methods have been used in agricultural research to conduct plant evaluations in a non-destructive approach. Non-destructive approach is a method of measurement of the chlorophyll and nitrogen content in plants without cutting of the leaf from the plants. ANN requires feature extraction with the human intervention, while CNN automates the feature extraction process from digital image data. In our study, color was the main parameter considered to quantify chlorophyll content in plants.

Chlorophyll and Nitrogen Contents in the plants can be measured based on the leaf color. Here Image processing tool such as open cv is used to extract the extract color of the leaf. The RGB color model is used to extract color of the leaf. With the RGB values of a leaf, the chlorophyll content can be measured.

Nitrogen is a vital nutrient required by plants for various metabolic processes, including the synthesis of chlorophyll, amino acids, and nucleotides. Plants need a sufficient supply of nitrogen to produce healthy leaves, stems, and roots. However, in many soils, the nitrogen levels may be inadequate, leading to reduced crop yield and poor plant health. Bio fertilizers have significance in yield development and physiological processes, as they supply plants with necessary nutrients and enhance chlorophyll content in leaves. Bio fertilizers are the living organisms which play a crucial role in enhancing nitrogen content in plants, which is essential for their growth and productivity.

The recommendation of appropriate bio fertilizers based on the chlorophyll content in leaves can significantly enhance the efficiency of nitrogen intake by plants. A deficiency in chlorophyll can indicate that a plant is not receiving enough nutrients, including nitrogen. By accurately measuring the chlorophyll content in leaves, bio fertilizer recommendation models can determine the level of nitrogen deficiency in the soil and recommend the appropriate bio fertilizer to enhance the nitrogen content in the soil. The use of bio fertilizers has several advantages over traditional chemical fertilizers. Chemical fertilizers contain high levels of nitrogen, which is a hazard to the environment if not used properly.

In Summary, the recommendation of appropriate bio fertilizers is an important step in enhancing the nitrogen content in plants. With the help of deep learning models that accurately measure the chlorophyll content in leaves, bio fertilizer recommendation models can determine the level of nitrogen deficiency in the soil and recommend the appropriate bio fertilizer to enhance the nitrogen content in the soil.

Our research aims to develop an optimal model to quantify chlorophyll and nitrogen content in leaves and recommend appropriate bio fertilizers to enhance chlorophyll content. By using the CNN algorithm and considering color as the main parameter, digital images of plants can be learned properly and their correlation with chlorophyll content and nitrogen levels in leaves can be determined.

2 Related Works

Recently, Deep learning techniques have been used to quantify chlorophyll content in plant leaves using image processing. M. I. Khoshrou et al. (2021) discussed about a deep learning for quantifying the chlorophyll content of tomato plant leaves through image processing. This method uses a convolutional de-noising auto-encoder to reduce background noise and a deep auto-encoder network to extract important parts of the plant leaf image, which is then fed into a neural network using support vector regression to estimate chlorophyll content [1].

Paul, Subir et al. (2020) explained about a convolutional auto-encoder (CAE) with image data to estimate the canopy mean chlorophyll content of pear trees. Anomalies in the trees were detected using multidimensional scaling on the auto-encoder characteristics before fitting nonlinear regression models. SVM was examined as nonlinear regression models for predicting chlorophyll content [2].

J. Gao et al. (2022) discussed that the amount of chlorophyll in plant leaves is a crucial sign of how well a crop is growing. This work employs near-ground multispectral data to nondestructively estimate the chlorophyll content of maize. The feature wavebands of the two datasets in the experiment are initially chosen from the 11 available wavebands. These studies can offer helpful guidelines for near-earth multispectral data-based tracing of the chlorophyll content in maize. They can also serve as guidelines for the creation of Knowledge based agricultural tracking systems for other crops, which were tested exclusively on maize and produced reliable results in this study. Sahar Y et al. (2021) have shown that Using Life cycle assessment (LCA), discovered that the main environmental hotspots for organic rice production were compost production, CH-4 emissions from the affected fields, nitrogen related emissions, and the mechanisation of the paddy land practices [3–5].

El-Hendawy et al. (2022) projected that Plant chlorophyll (Chl) is one of the crucial components in the monitoring of plant stress and indicates the ability of plants for photosynthetic activity, but measuring it in the lab is typically time- and money-inefficient and only uses a tiny portion of the leaf. The Chl content was calculated in the following units: SPAD value, concentration per plant, and content per area. In order to estimate the three units of Chl content were examined. According to the findings, Chl area and Chl plant fared better for the majority of indices within each SRI type, but Chl SPAD performed badly. The PLSR models, based on the four SRI types singly or in combination, still had trouble estimating chloride SPAD by showing a high correlation with chloride plant and then chloride area. The most useful indices were retrieved by the SMLR models from each SRI form, and they accounted for 73–79%. All features of Chl content all had an impact on how well the various SMLR predictive models performed in forecasting Chl content. In conclusion, this work shows that spectral reflectance data may be used to accurately [6].

Suruban, C et al. (2020) discussed that In Samoa, a variety of organic resources high in nitrogen (N) are locally accessible. For vegetable production, none of them were tested. The OAs' N mineralization patterns and application rates varied significantly from one another. Composted mucuna had the highest levels of N mineralization had the maximum biomass yield and N uptake [7].

Abbas Soleymanifard et al. (2022) analyzed that Six different genotypes of safflower were grown under rained circumstances, and the effects of nitrogen (N) fertilization was investigated. In two crop years, 2015–2016 and 2016–2017, a factorial experiment based randomised designs was carried out [8].

Sakshi Agrawal et al. (2022) have shown that the economy of most developing countries are based on agriculture, and the majority of their inhabitants depend on it to survive. In the era of climate change, global agricultural systems are facing difficult and unprecedented difficulties. The world's population is increasing each year, which calls for higher agricultural output. In response, there has been a push to use cutting-edge technologies like nanotechnology. The current situation in emerging regions that fall under the purview of agriculture and create environmental remediation solutions has sparked a great deal of interest in nanotechnology with plant systems. Plant-mediated nanoparticle (NP) synthesis is cost-effective, environmentally benign, and provides long-term product safety [9].

Doris Ying et al. (2022) presented the study that a novel method for predicting the chlorophyll content of microalgae using color models and linear regression was given. Software called SPSS was used to do the analysis. Acetone was suggested as a viable extraction solvent by the model that best fit the experimental data. The red-green-blue model in picture analysis provided a greater correlation than the cyan-magenta-yellow-black model. In summary, utilizing prediction models to estimate chlorophyll content is quicker, more effective, and less expensive [10].

Abhishek et al. (2023) discussed that all crops have a foothold in the earth's soil. The most crucial measure of soil health utilized for crop selection, mechanical manipulation, irrigation control, and fertilizer management is soil texture. The storage and movement of air and water in the soil, as well as root growth, the availability of plant nutrients, and the activity of various microorganisms, are all influenced by the texture of the soil. The fertility, quality, and health of the soil are all impacted by these characteristics taken together. The traditional approach to analyzing soil texture is labor-intensive, time-consuming, and difficult. With the aid of machine learning (ML), it is now possible to swiftly and in real-time evaluate the physical, chemical, and biological characteristics of soil. This method is environmentally friendly because it doesn't use any potentially harmful chemicals. Machine learning is able to anticipate nonlinear properties and learn complex features. Convolutional Neural Networks (CNN) use convolutional layers to quickly complete tasks like object detection, image synthesis, and image classification by automatically learning features from the input data. After picture sub setting, data preprocessing, and image augmentation are finished, soil texture photographs are provided as the input dataset. This results in a soil texture prediction model based on CNN with an affordable accuracy of 87.50% [11].

In Summary, in this research work using CNN, SVM Regression and Classification model the nitrogen content in paddy leaves are detected using the chlorophyll content. It helps in appropriate recommendation of Bio fertilizers.

3 Methodology

(See Fig. 1).

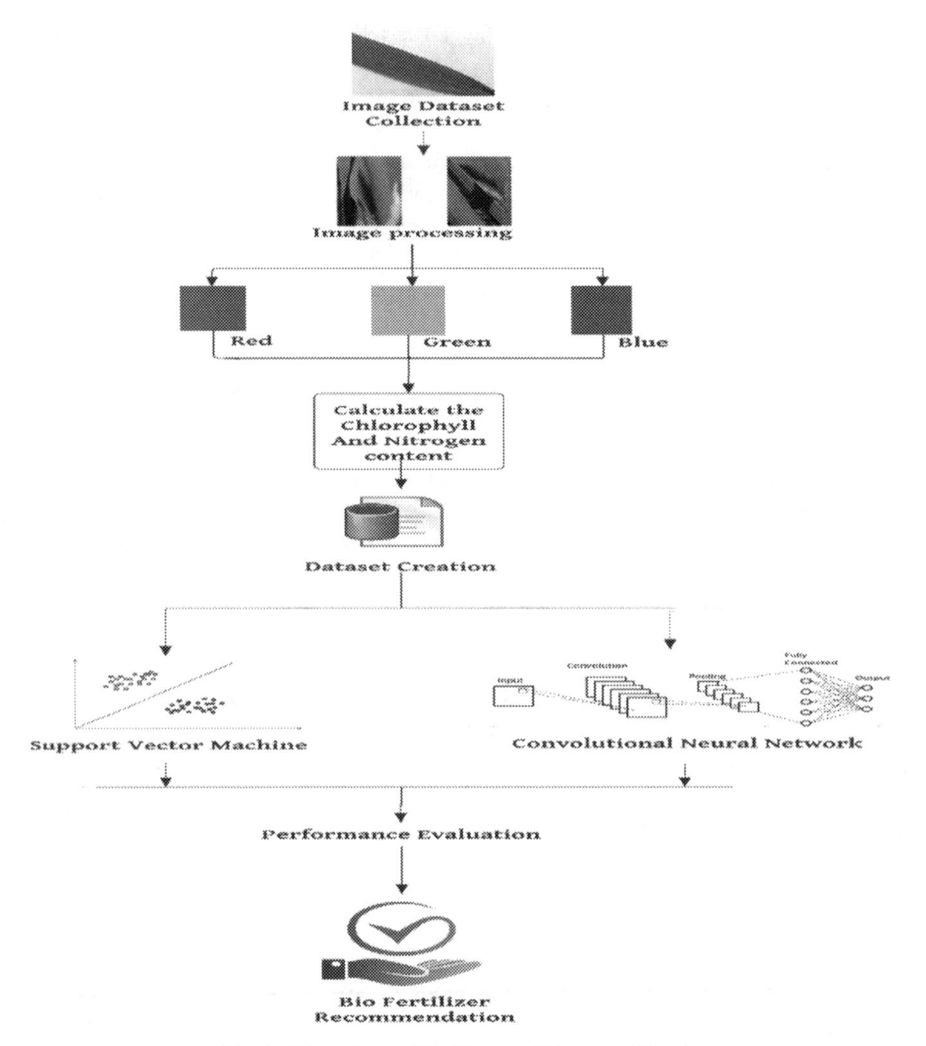

Fig. 1. Flow chart of the Proposed System of Design

3.1 Dataset Collection

In this research, the dataset required is Paddy leaves Image Dataset. The two datasets Rice Leaf Diseases and Thermal images - diseased & healthy leaves - paddy are combined for this work. The datasets can be collected from the Kaggle website which is publicly accessible to everyone. This combined dataset contains 3356 .jpg images of healthy and diseases affected paddy leaves with white paper as a background. The images are grouped into 4 classes such as healthy and based on the type of disease they are classified into hispa, brown spot, leaf smut classes. There are nearly 700 images in each class.

3.2 Image Preprocessing

After the image dataset is collected, then the image processing needs to be done in order to extract the exact color of the leaf image. In our research, Open CV is used for Image processing. Generally, image is stored internally in the pixel format. Firstly, the image is transformed in to the balk and white image to extract the accurate intensity of the color in each pixel of the image. Gaussian blur is used to handle the outliers from the image and image is segmented as into Fore Ground (FG) and Back Ground (BG) using the Otsu's method of Image thresholding algorithm. Then the image is converted to the RGB and HSV color model. To extract the exact RGB and HSV values.

3.3 Calculation of Chlorophyll Content of a Leaf Image

Growth status of a plant is highly dependent on the Chlorophyll Content. Chlorophyll content in a leaf is a key indicator of the greenness of a leaf and identifies the nutrient deficiencies in plants. Hence chlorophyll influences in the crop production and yield, it is needed to calculate the chlorophyll content in plants. The procedure to calculate the chlorophyll content in leaves based on the extracted RGB values is mentioned below:

- Firstly, check the B component in the RGB values. If it is greater than 90, then it is white paper
- Find the leaf image in the image and make its background as black.
- Using the RGB values, the chlorophyll content in leaves can be measured using the formula mentioned below:

$$\text{Chlorophyll content} = G - (R/2) - (B/2) \tag{1}$$

3.4 Calculation of Nitrogen Content of a Leaf Image

Nitrogen is a vital nutrient required by plants for various metabolic processes, including the synthesis of chlorophyll, amino acids etc,. Plants need a sufficient supply of nitrogen to produce healthy leaves, stems, and roots. However, in many soils, the nitrogen levels may be inadequate, leading to reduced crop yield and poor plant health. Hence it is necessary to measure the nitrogen content in plants in order to supply the required nutrients to the soil to enhance the plant growth. The procedure to calculate the nitrogen content in leaves based on the extracted RGB values is mentioned below:

- Firstly, check the B component in the RGB values. If it is greater than 90, then it is white paper
- Find the leaf image in the image and make its background as black.
- Find the average of each component and divide it by 255 to get average in the range of 0 to 1.
- Find max and min among these average values
- Find the HSB values using the below mentioned algorithm:
  - if max value is equal to R, then the average is equal to Hue, $H = ((G - B)/(\text{maximum} - \text{minimum})) * 60$

- if max value is G, then average is equal to Hue, H = (((B − R)/(maximum − minimum)) + 2) * 60
- if maximum value is equal to B, then the average is equal to Hue, H = (((R − G)/(maximum − minimum)) + 4) * 60
- saturation can be calculated using the formula: (maximum − minimum)/ maximum
- Brightness of the leaf is equal to the maximum value.

- Using the extracted HSB values, the Nitrogen content in leaves can be measured using the formula (2) mentioned below:

$$Nitrogen = ((H - 60)/60 + (1 - S) + (1 - B))/3 \tag{2}$$

3.5 Dataset Creation

All the measured RGB, HSB, chlorophyll and Nitrogen values are appended in a CSV file and then this dataset is used for the further process of training the model.

3.6 Data Preprocessing

Data preprocessing is the needy step in any of the machine learning and data science applications because of major role played by the data in their application developments. Data preprocessing is the method which is used for removing all the null values, irrelevant data, handling missing data, scaling and normalizing the data, handling categorical data and reducing noise.

3.7 Prediction Model Development

Here the model is trained for prediction of chlorophyll and nitrogen content using the two algorithms such as Support Vector Machine and Convolution Neural Network.

Support Vector Machine Regression

SVM Regression, a supervised learning algorithm which will predict numerical continuous values. Unlike SVM classification, which is used for classifying discrete values, SVM regression is used for predicting continuous values. It finds a separator that best fits the data points while also minimizing the margin violations. The SVM regression algorithm tries to minimize these target violations while also minimizing the weight between the predicted regression line and the data points. SVM regression is a powerful algorithm to predict continuous valued outcomes., especially when dealing with complex data sets with non-linear relationships between the feed features and the target variable.

Here our model is trained using the SVM regression algorithm. Firstly, the dataset is divided into modelling and validation dataset with 80% as model training data and 20% as validation data. Then the model training data is fed as the input to the SVM Regression model and the validation data is used to test the model. The SVM model returned the chlorophyll content and nitrogen content with the accuracy of 99.96% and 99.86%.

Convolutional Neural Network

CNN, a deep learning neural network which is a perfect model for image and video analysis. CNNs consist of several layers, including (i) convolutional layers, (ii) pooling layers, and (iii) fully connected layers. The convolutional layers use a set of filters on the input image to extract features and patterns. The pooling layers deduct the dimensionality of the feature maps by dividing them as samples and retaining the most relevant information. The fully connected layers classify the extracted features into different categories, such as objects or scenes. CNNs are capable of learning and identifying complex patterns and structures in images and videos, such as edges, corners, and textures, making them an effective tool for various computer vision applications, including object detection, facial recognition, and image segmentation. They have shown remarkable performance in many image classification tasks, such as the ImageNet challenge, where they outperformed human-level accuracy. The popularity and success of CNNs are due to their ability to automatically learn features and patterns from raw image data, reducing the need for manual feature engineering.

Firstly, the dataset is divided into model-training and model-testing dataset of 80:20 ratio. Then the training dataset is fed as the input to the CNN model and then the model undergoes several operations such as Convolution it applies filters to extract the features from the image. In pooling operation, max pooling is used in which it creates the feature maps to the max value in order to deduct the attributes of the image and then it is converted to the 1D convolution. Later it is passed as output to the output layer and the testing data is used to test the model. The CNN model returned the chlorophyll content and nitrogen content with the accuracy of 99.98% and 99.96%.

3.8 Performance Evaluation

In this research, the two models SVM and CNN model have been compared and evaluated for training the data. The results show that CNN model has more accuracy of 99.96% compared to SVM model of accuracy of 98.86%.

3.9 Bio Fertilizer Recommendation Model

The recommendation of appropriate bio fertilizers is an important step in enhancing the nitrogen content in plants. Bio fertilizers provide a natural and eco-friendly alternative to chemical fertilizers and can improve soil health and plant growth. With the help of deep learning models that accurately measure the chlorophyll content in leaves, bio fertilizer recommendation models can determine the level of nitrogen deficiency in the soil and recommend the appropriate bio fertilizer to enhance the nitrogen content in the soil.

The following Table 1 gives the summary of recommendation of various bio fertilizers based on the Nitrogen content.

In this research, SVM Classification model is used to recommend the appropriate bio fertilizer based on the measured nitrogen content range as mentioned in the Table 1. The recommendation model is trained using the trained dataset. Then the proposed methodology is evaluated using the test dataset and accuracy of the model is measured.

Table 1. Recommendation of various bio fertilizers based on Nitrogen content for paddy leaves.

Bio fertilizer	Nitrogen content in paddy leaf
Azolla	Low
Blue green algae	Moderate
Azosprillium	Low to moderate
Rhizobium	Moderate to High

4 Results and Discussions

4.1 Chlorophyll and Nitrogen Content Prediction Model

Here the Support vector machine Regression and Convolutional Neural Network (CNN) models are used for prediction of the chlorophyll content and the models returned an accuracy of 99.96% and 99.98% as shown in Fig. 2. CNN model has more accuracy than SVM model. Hence the CNN works well for even low sample size data. Hence CNN model has been selected among SVM in our research.

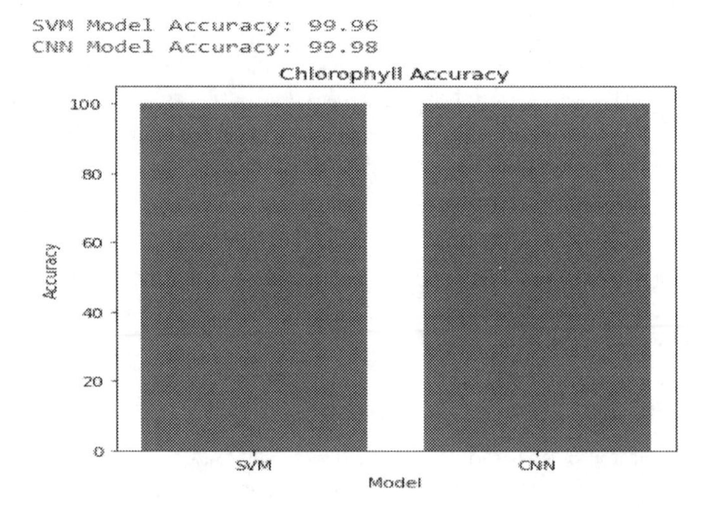

Fig. 2. Accuracy of Chlorophyll content measured for SVM and CNN models

Nitrogen Content is predicted using both SVM Regression and the CNN algorithms. Here also the CNN algorithm performed better than the SVM with the accuracy of 99.98% as shown in Fig. 3. Hence the CNN algorithm works well for Nitrogen prediction in leaves. Therefore, CNN algorithm is used for Nitrogen measurement in paddy leaves.

CNN and SVM are the best machine learning models are applied for regression activities. However, CNNs have been found to work better than SVMs for predicting chlorophyll and nitrogen content. The reason for this is that CNNs are particularly

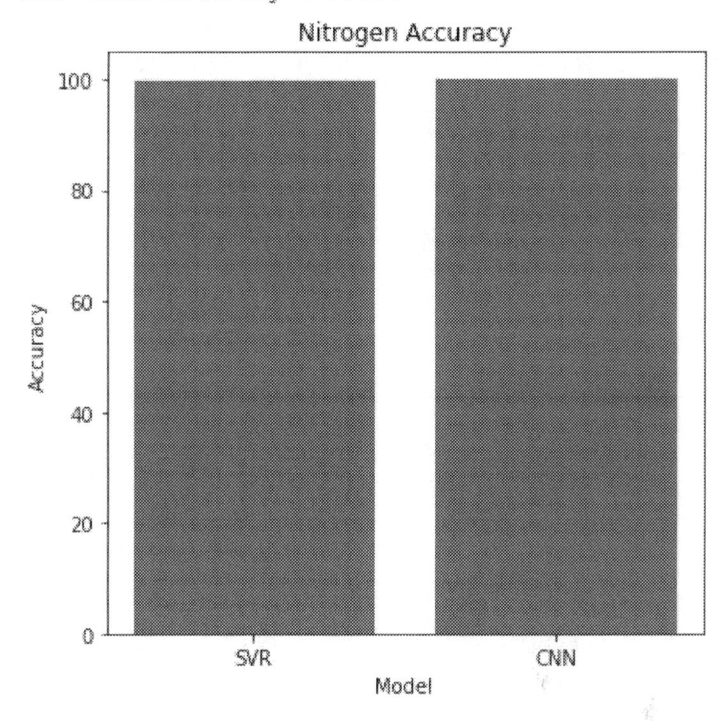

Fig. 3. Accuracy of Nitrogen content measured for SVM and CNN models.

effective at processing images, and chlorophyll and nitrogen content prediction is often based on captured images. CNNs are implemented to identify patterns and features within the image data by applying appropriate filters to the feeding data, which are then combined to form higher-level representations of the data. This makes them well-suited to extracting features from images that are important for predicting chlorophyll and nitrogen content.

In contrast, SVMs are more suited to tasks where the data has dividing line between classes or groups. SVMs use a kernel function to convert the input data into Larger-dimensional space, where the separation between classes is more apparent. However, in regression tasks where the data is continuous and does not have clear class boundaries, SVMs can be less effective than CNNs. Furthermore, CNNs can understand the features automatically from the input data, whereas SVMs require manual feature engineering, which can be laborious and less accurate.

Overall, CNNs are better suited to chlorophyll and nitrogen content prediction tasks because they are better at processing image data and can automatically learn features from the input data, resulting in more accurate predictions and SVMs are better suited for classification tasks because data has vivid classification between categories.

4.2 Bio Fertilizer Recommendation Model

Bio fertilizers is an essential element in improving soil health and plant growth. With the help of deep learning models that accurately measure the chlorophyll content in leaves, bio fertilizer recommendation models can determine the level of nitrogen deficiency in the soil and recommend the appropriate bio fertilizer to enhance the nitrogen content in the soil. Here the SVM classification model has returned high accuracy of 99.29% as shown in Fig. 4.

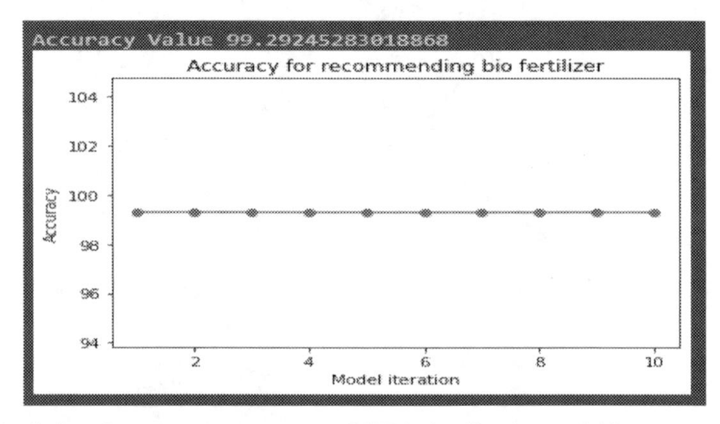

Fig. 4. Analytics of accuracy measurement of SVM classification model in recommending the biofertilizer

5 Conclusion

In this research, a CNN model has been developed to predict the chlorophyll and nitrogen contents in plant leaves using digital images. It showed that the CNN was powerful for regression problems and also a CNN architecture could be well-trained even in the presence of low sample size data. Moreover, since the pigment content in plants was strongly correlated with its visual color, our experiment showed that the filters on model, which were generally known to be superior in recognizing objects using shape features, were also superior in recognizing color features. The user simply needs to input the leaf image of the plant to the model. Then, the model would display a prediction of the chlorophyll pigment and nitrogen contents in plants in real-time. In this research, the CNN (99.98%) model works better than the SVM (98.86%) model for prediction of chlorophyll and nitrogen content in plants. Hence the CNN model is used for regression tasks. After identifying the Chlorophyll and nitrogen contents present in the leaves through the leaf color, nitrogen deficiencies are determined. Based on nitrogen deficiency in a leaf the appropriate bio fertilizer will be recommended by SVM classification model to enhance the nitrogen contents in the soil. Thus, a deep learning model is built to recommend a bio fertilizer based on the chlorophyll content present in the leaves.

References

1. Khoshrou, M.I., Zarafshan, P., Dehghani, M., Chegini, G., Arabhosseini, A., Zakeri, B.: Deep learning prediction of chlorophyll content in tomato leaves. In: 2021 9th RSI International Conference on Robotics and Mechatronics (ICRoM), Tehran, Iran, Republic of Islamic, pp. 580–585 (2021). https://doi.org/10.1109/ICRoM54204.2021.9663468
2. Paul, S., Poliyapram, V., İmamoğlu, N., Uto, K., Nakamura, R., Kumar, D.N.: Canopy averaged chlorophyll content prediction of pear trees using convolutional auto-encoder on hyperspectral data. IEEE J. Sel. Top. Appl. Earth Obs. Remote Sens. **13**, 1426–1437 (2020). https://doi.org/10.1109/JSTARS.2020.2983000
3. Gao, J., Zhang, Z.: Study on deep learning model for online estimation of chlorophyll content based on near ground multispectral feature bands. IEEE Access **10**, 132183–132192 (2022). https://doi.org/10.1109/ACCESS.2022.3230355
4. El-Hendawy, S., et al.: Combining hyperspectral reflectance indices and multivariate analysis to estimate different units of chlorophyll content of spring wheat under salinity conditions. Plants **11**(3), 456 (2022). https://doi.org/10.3390/plants11030456
5. Youseftabar, S., Sharifabad, H.H., Heravan, I.M.: Rice cropping system in semi-mechanized and traditional planting patterns in Northern Iran. J. Plant Nutr. **44**(7), 970–987 (2021). https://doi.org/10.1080/01904167.2020.1845380
6. Singh, A., Thakur, A., Sharma, S., Gill, P.P.S., Kalia, A.: Bio-inoculants enhance growth, nutrient uptake, and buddability of citrus plants under protected nursery conditions. Commun. Soil Sci. Plant Anal. **49**(20), 2571–2586 (2018). https://doi.org/10.1080/00103624.2018.1526946
7. Suruban, C., Kader, M.A., Solaiman, Z.M.: Influence of various composted organic amendments and their rates of application on nitrogen mineralization and soil productivity using Chinese cabbage (Brassica rapa L. var. Chinensis) as an indicator crop. Agriculture **12**, 201 (2022). https://doi.org/10.3390/agriculture12020201
8. Soleymanifard, A., Mojaddam, M., Lack, S., Alavifazel, M.: Effect of *Azotobacter Chroococcum* and nitrogen fertilization on some morphophysiological traits, grain yield, and nitrogen use efficiency of safflower genotypes in rainfed conditions. Commun. Soil Sci. Plant Anal. **53**(6), 773–792 (2022). https://doi.org/10.1080/00103624.2022.2028815
9. Agrawal, S., Kumar, V., Kumar, S., Shahi, S.K.: Plant development and crop protection using phytonanotechnology: a new window for sustainable agriculture. Chemosphere **299**, 134465 (2022). https://doi.org/10.1016/j.chemosphere.2022.134465. https://www.sciencedirect.com/science/article/pii/S0045653522009584. ISSN 0045-6535
10. Tang, D.Y.Y., et al.: Application of regression and artificial neural network analysis of Red-Green-Blue image components in prediction of chlorophyll content in microalgae. Bioresource Technol. **370**, 128503 (2023). https://doi.org/10.1016/j.biortech.2022.128503. https://www.sciencedirect.com/science/article/pii/S0960852422018363. ISSN 0960-8524
11. Abishek, J., Kannan, P., Devi, M.N., Prabhaharan, J., Sampathkumar, T., Kalpana, M.: Soil texture prediction using machine learning approach for sustainable soil health management. IJPSS **35**(19), 1416–1426 (2023). https://doi.org/10.9734/ijpss/2023/v35i193685

A Comparison of Multinomial Naïve Bayes and Bidirectional LSTM for Emotion Detection

S. K. Lakshitha[1], V. Naga Pranava Shashank[1], Richa[2(✉)], and Shivani Gupta[1]

[1] Vellore Institute of Technology, Chennai, Tamilnadu, India
[2] BIT Mesra, Ranchi, Jharkhand, India
richasingh.bv@gmail.com

Abstract. Emotion detection is an area of sentiment analysis that focuses on the extraction and evaluation of feelings. Many deep learning and machine learning researchers have found emotional content in text. People's lives all across the world were significantly impacted by the COVID-19 pandemic. The social networking site twitter were very helpful in documenting people's emotion and views. In literature there are two methods that has been used extensively for the emotion identification; one is Multinomial Naive Bayes model, and another is Bidirectional Long Short-Term Memory (LSTM). In this research, we have identified the efficient approach among the two mentioned approaches. We have compared the multinomial naive bayes model and bidirectional LSTM for identifying emotion. To identify the emotion in text, the tweet from twitter is utilized; that contains emotions in the form of text. The results show that bidirectional Long-Short Term Memory approach outperforms as compared to the multinomial naïve bayes approach (MNB). The bidirectional LSTM approach enhances the emotion detection over the MNB approach. The time taken for the training of the tweets differs for the two approaches. Due to its longer training period, the multinomial naive bayes technique may not be suitable for use with huge datasets.

Keywords: Emotion Detection · Multinomial Naïve Bayes · Bidirectional LSTM · COVID-19 · Tweets

1 Introduction

The general state of mental and physical health was significantly impacted by the COVID-19 pandemic, [1, 2]. Twitter and other social networking platforms are excellent tools for capturing the user emotion and opinions. People have turned to social media during these challenging times to express their concerns, viewpoints, and observations on the global pandemic [4, 5]. In this paper, we have collected the tweets between August 2020 to September 2020 and analyzed the tweets to locate the change in emotions of the users. The emotions of happiness, grief, rage, pleasure, hatred, fear, etc. are taken into account. Despite the fact that the word "feelings" lacks a regular structure, cognitive science places a high priority on studying emotions [16]. Natural language processing (NLP) is b used for recognizing emotions in text. The use of word embeddings is one of the most important applications of NLP. Learning-based systems perform better since they take into account the semantic and syntax properties of the text [15].

S. Aurelia et al. (Eds.): ICCSST 2023, CCIS 1973, pp. 322–332, 2024.
https://doi.org/10.1007/978-3-031-50993-3_26

There are two methods that is found as widely used approaches among the available emotion detection models. These two methods are bidirectional LSTM method and multinomial naive bayes models. In this Paper, bidirectional LSTM models and multinomial naive bayes models are compared to get the better approach among the two approaches in terms of training time and accuracy. An LSTM normally encodes sequence data context in a single direction i.e. from left to right. While encoding the sequence of data, it is crucial for some applications to consider both the left and right contexts of the input. A bi-directional LSTM uses two distinct LSTMs with different hyperparameters such as number of epochs, activation function, decay rate etc.to encode the sequence data in both directions. The final representation is then created by combining the concealed states of the two networks. The model performs well when applied to tasks involving sentiment categorization and sequence prediction.

The Multivariate Bernoulli Naive Bayes model (BNB) states that when a word appears in a document, the value of the attribute corresponding to that word is either recorded as one or zero. Multinomial Naive Bayes (MNB), a development of Bernoulli Naive Bayes (BNB), was released. MNB considers word frequency and information while assuming that the document is a collection of words. Structured modelling tasks, such as classes with several documents and labels, are easy to incorporate because of its probabilistic nature. The literature review that lists various machine learning and deep learning algorithms for emotion detection is discussed in Section II of the study, which is organized as a whole. The approach, which covers data preparation and analysis for text emotion detection, is covered in Section III. The results and discussion are followed by the next section. We identify future directions in Section V, where we draw a conclusion to our paper.

2 Literature Review

For the purpose of identifying human activity, the authors [6] have compared the Unidirectional and Bidirectional LSTM. To assess the accuracy of each model, they analyse two datasets. When using bidirectional LSTM, the results showed a negligible increase in training accuracy for the two datasets. For the UniMiB SHAR and WISDM (Wireless Sensor Data Mining) datasets, respectively, Bidirectional LSTM achieved an accuracy of 99% and 98.11%. For the UniMiB SHAR and WISDM datasets, respectively, an accuracy of 98.87% and 97.68% was attained using Unidirectional LSTM. [7] have combined an efficient text pre-processing technique with bidirectional LSTM. Three datasets were used in the experiment: two datasets of tweets from Twitter and the ISEAR dataset. Based on their ability to accurately classify data, the models are assessed. The classification accuracy they used as a gauge of evaluation is calculated as the number of precise predictions divided by the overall number of instances in the test dataset. The performance of the model was then found to be significantly impacted by a suitable text pre-processing. [3] has discussed a number of machine learning models for text classification. They assert that when deep learning models are compared, a bi-channel CNN model outperformed other deep learning models, and when deep learning algorithms are compared to conventional machine learning algorithms, Logistic Regression performed better than other deep learning models.

The authors of [8] describe an improved word representation that, by including sentiment information into the standard TF-IDF approach, provides weighted vectors. The sentiment analysis of RNN, CNN, LSTM, and NB are contrasted with the suggested sentiment analysis approach. They noticed that the BiLSTM model can better extract the text representation of the comments by taking the context information into full consideration. The strategy they suggested was 91.54% accurate. Accuracy rates for RNN, CNN, LSTM, and Naive Bayesian were 87.18%, 85.10%, 88.46%, and 86.02%, respectively. The BiLSTM-based sentiment analysis of comments took a lot of effort to train the model.

In order to examine the audits and help the clint in deciding whether consumers are satisfied, the writers in [9] have concentrated on looking into the survey dataset. The data is subjected to the Bi-LSTM algorithm, and a ranking is provided. Additionally, they contrasted various algorithms, including CNN, RNN, and Bi-LSTM. When Enhanced Bi-LSTM's accuracy was compared to that of other algorithms, it was higher at 97.32%. In the Bi-LSTM model which is enhanced, the Bi-LSTM model serves as the model's input layer, and the LSTM model with a system of attention and the SoftMax serves as the model's output layer. [11] has suggested using Multinomial Nave Bayes to identify emotions in tweets. For feature extraction, they employed two models: unigram and unigram with parts of speech (POS) tags. For the study of emotions, they utilised a classifier called Naive Bayes. The accuracy of the unigram model and the unigram with POS tag model was 68.25% and 70.25%, respectively. They noticed that the outcomes would be more pleasing if two or more algorithms were integrated and additional data was learned. [10] have created a framework concept focused on the TF-IDF module and MNB algorithm. They noted that the MNB method is a quick, simple, and contemporary text classification system. The model's accuracy was determined to be 90% or such. [13] designed a word embedding method that was created using unsupervised learning and coupled with n-grams features and word emotion polarity score features in order to create a sentiment feature from a collection tweets. For twitter sentiment analysis, they make use of five kinds of Twitter data. They suggested that a deep convolutional neural network using word vectors that have already been trained performs well when analyzing sentiment on Twitter.

A hybrid approach of the greedy search and MNB algorithms has been proposed by [14]. They worked with data from IMDB. The trials were carried out using WEKA, an open-source machine learning platform. 10 fold cross validation was used to test each method. They used a greedy search strategy along with naive bayes to reach an accuracy of 84.10%.

3 Proposed Methodology

Using statistics and natural language processing, emotion analysis examines the feelings expressed in a text. Based on different emotions mentioned in the text, such as fear, rage, happiness, sadness, love, inspiring, or neutral, to the text will be executed. There are several places where users' emotions can be seen, such as tweets, comments, and reviews.

3.1 Bidirectional LSTM

Figure 1 [17] shows the Bi-LSTM organization. It refers to any neural network that has the capacity to retrieve sequence data in both the forward (from the present to the future) and reverse (from the past to the future) orientations. In a bidirectional LSTM, as opposed to a conventional LSTM, our comprehension is reciprocal. We can enable input to flow either backwards or forwards using a conventional LSTM. When there is bidirectional input, information can be moved in both directions. This ensures to maintain both the present and the future.

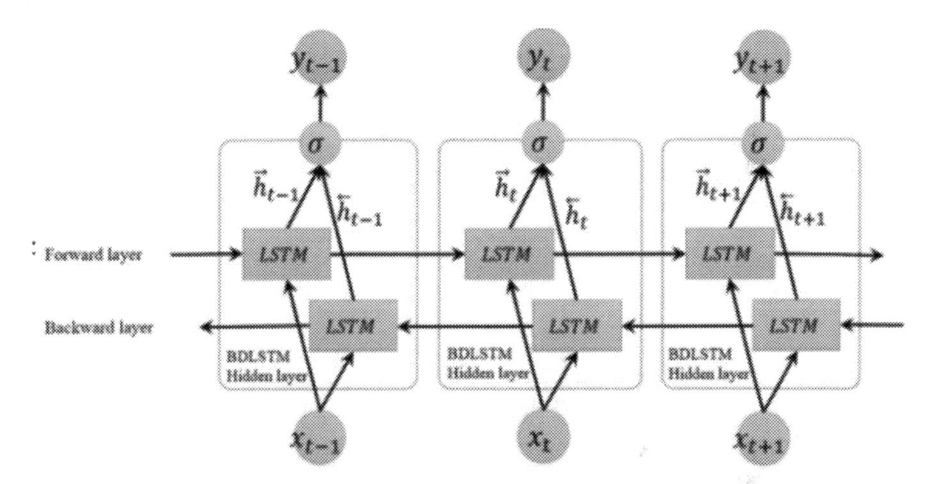

Fig. 1. Bidirectional LSTM Layer Structure

During the preprocessing of the data, the null elements has been removed and a tokenizer has been applied. Further, padding has been used to make each sentence an equal amount of length. Because the model can only comprehend numbers, vectorization has performed on the data by starting with a predefined glove vector of 50d.

After preprocessing each emotion is one-shot encoded in our model after preprocessing, after which it is trained and assessed. A loss function called cross entropy has the formula

$$E = -y \cdot \log(Y^{\wedge}) \quad (1)$$

where Y^ is defined as the softmaxj(logits), E is defined as the error, y is the label, and logits are the weighted sum. Since softmax contains an exponential component, cross-entropy is a good complement to softmax. A cost function that contains a natural log element will result in a convex cost function. Adam, a derivative of the Adaptive Moment Estimation method, is the optimisation algorithm that is being employed. Adam uses a gradient average that decays exponentially with time. In order to provide an adaptable learning rate, it additionally makes use of the exponentially decaying average of previous squared gradients.

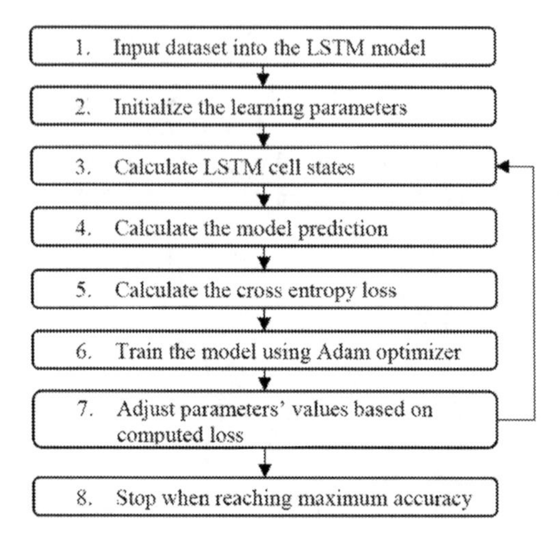

Fig. 2. Tweets Emotion Analysis Steps using Bidirectional LSTM

The model's training procedure is shown in Fig. 2. Initializing the learning parameters comes first after feeding the LSTM model with raw data. The model prediction is then computed after computing the LSTM cell states. We train the model using the Adam optimizer after determining the cross-entropy loss and then modify the training parameters, as necessary. This procedure is repeated until the model reaches its highest level of accuracy.

3.2 Multinomial Naïve Bayes

The MNB, which is a member of the family of Bayes theorem-based probability analyzers, relies on the unbiased premise that the values of some features are invariably different from those of others. This strategy of employing supervised learning approach makes a solid hypothesis that all characteristics in the training data are interconnected, or the Bayes assumption. In terms of CPU usage, time, and memory, naive Bayesian is efficient. It can even perform well even with relatively less training sets and lower intensive computing. MNB is used when the data recommendations have a multinomial distribution [12].

Identification of features and labels is the initial step in classification. The classification outcome we seek is label. The classification process has two parts. Figure 3 depicts a training phase and a testing phase. The classifier model is trained using the provided training data set during the training phase, and its performance is assessed during the testing phase. Performance can be evaluated in terms of recall, accuracy, and precision.

The MNB classifier's features are the words in a sentence, and its values are the frequency with which a word appears in a sentence. The NRCLexicon (an MIT-approved project that predicts sentiment and emotion of a given text) dataset was pre-processed, and the emotion of the sentence was acquired. The labelled dataset will be used to train the multinomial naive bayes classifier.

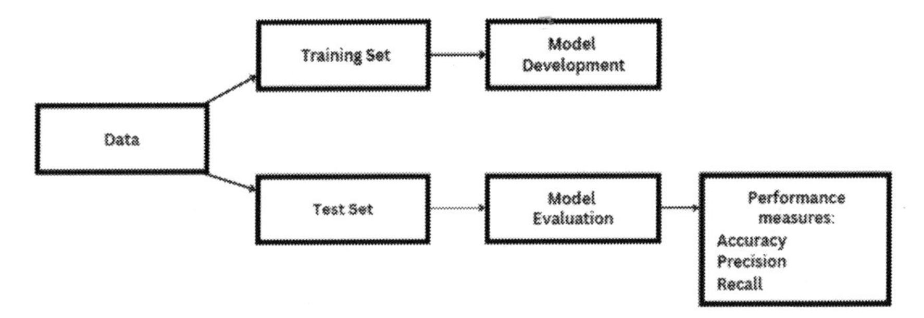

Fig. 3. Tweets Emotion Analysis Steps using Multinomial Naïve Bayes

The probability of obtaining a class C utilising MNB

$$\text{P(Class} = \text{c} \mid \text{Features)} = \frac{P(Features \mid Class = c) \cdot P(Class = c)}{P(Features)} \propto$$

$$\text{P(Features} \mid \text{Class} = \text{c)} \cdot \text{P(Class} = \text{c)}$$

where, c is the class of the possible outcomes.

For instance, terms like good, loved, pleasant, etc. were employed to indicate the likelihood of having favourable emotions. Consequently, the formula can be changed as

P (Y = +|X = good, liked, bad, unpleasant, nice, pleasant, rude)$\propto$

P (X = good, liked, bad, unpleasant, nice, pleasant, rude | Y). P (Y = +)

A matrix is created internally once the probability of each word utilised in the classifier is computed. Since multiplication is used in the probability computation, we are unable to have terms with a probability of 0. This means that laplace smoothing can be employed, where the probability of discovering a feature given a class 'c' can be changed by computing - the number of times the feature occurs in c + 1/(total number of features in class 'c' + number of features).

$$\text{P(Feature} \mid \text{class} = \text{c) is calculated as } \frac{Number\ of\ times\ the\ feature\ occurs\ in\ c + 1}{Total\ number\ of\ features\ in\ class\ 'c' + number\ of\ features}$$

The words in each sentence are processed as features, the probability is calculated, and the emotion predicted with the highest probability value is returned when all preprocessing have been completed and a model has been developed.

4 Results and Discussion

This section includes the results and discussion for the comparison of the two approaches. The section first describes the dataset and then it further focuses on the analysis.

4.1 Datasets

In this study, we incorporated the data of user tweets from an open-source website (https://www.kaggle.com), during the COVID-19 lockdown period (August-September

2020) for experimental purpose. The data set consisting of more than 2,50,000 tweets were extracted from kaggle and it contains the raw tweets related to keywords including but not restricted to COVID-19, coronavirus, lockdown, etc. The dataset consists of extracted tweets from the twitter platform for analysis. The raw tweets were unable to produce accurate results in emotion recognition, as we realized when we dug deeper into the cleaning of data to find important features. The biggest problems were stop words in tweets, hashtags (#tags), @mentions, and web links (URLs). We utilised regular expressions to eliminate the URLs, @mentions, and #tags from the tweets. The stop words were managed by the NLTK library inside of the Textblob.

4.2 Experimental Results and Analysis

We contrast the capabilities of the Naive Bayes machine learning model and the Bidirectional LSTM deep learning model. Before training the models, we preprocess the text data to ensure accuracy. We can see that the Bi-LSTM based model is outperforming the rival model in terms of performance. When trained on a small sample set, Naive Bayes is unable to attain significant accuracy. This is brought on by insufficient data, overfitting, unrepresentative data, and the assumption of conditional independence. Bidirectional LSTM-based models exhibit steady improvement as the dataset length grows, as seen in Fig. 4.

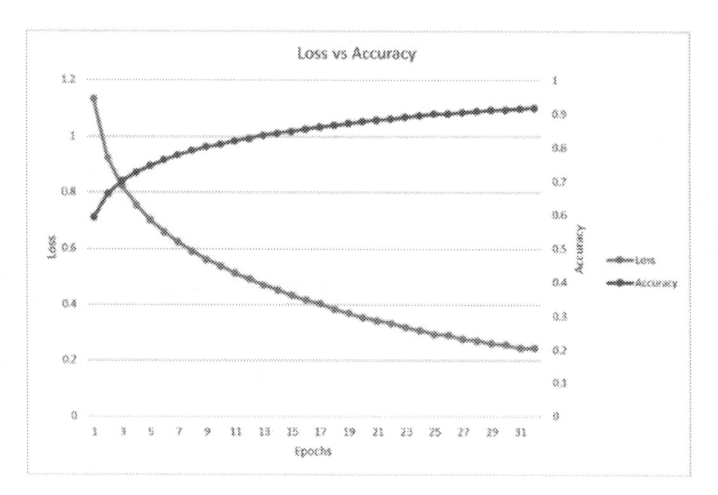

Fig. 4. Loss function and accuracy of BiLSTM model training and verification

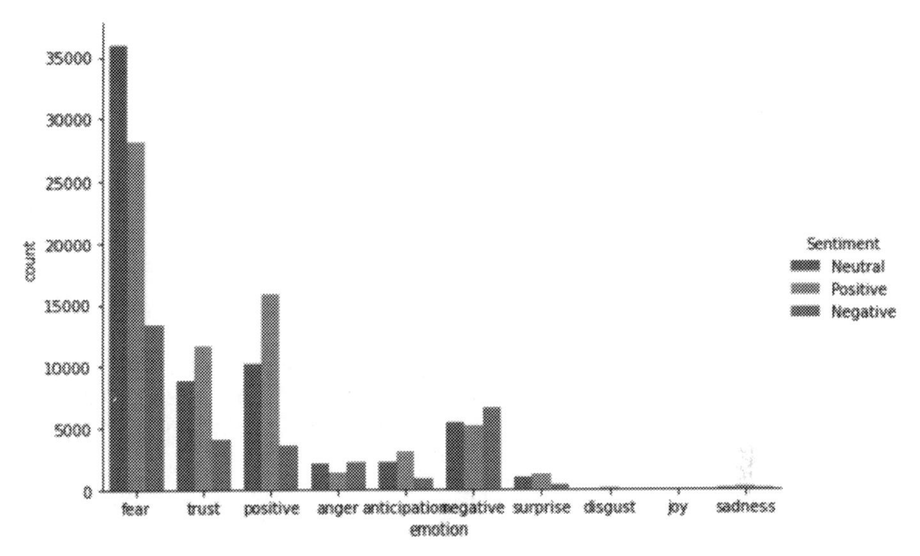

Fig. 5. Comparison of different emotions in the dataset

In Fig. 5 we can observe people were in fear during the months of August-September 2020. Very few people are having extremely negatives thoughts about COVID-19. The graph depicts how many times the emotion is classified as positive, negative or neutral sentiment. The emotions which we have taken are anger, anticipation, disgust, fear, joy, negative, positive, sadness, surprise and trust.

Table 1. COVID-19 training accuracy

Parameter	COVID-19(August-September 2020)
Epochs	32
Iterations	547
Batch size	128
Learning rate	1
Dropout	0.5
Model	Accuracy
Multinomial Naïve Bayes	76.61%
Bidirectional LSTM	91.82%

Table 1 shows the parameter values used in the training phase. A single cycle of training the machine learning model is referred to as an epoch. Here the number of epochs is 32 and the number of iterations is 547. The number of samples processed before the model is updated is known as the batch size. Batch size controls how well the error gradient is estimated when training neural networks. The batch size here is 128.

One of the hyper-parameters that describes how our network's weights are adjusted in relation to the loss gradient descent is learning rate. The learning rate is 1 and the dropout is 0.5. Also, the accuracy for both multinomial naïve bayes and bidirectional LSTM is 76.61% and 91.82% respectively. The difference in accuracy was less than 15.21%.

Since Bidirectional LSTM can learn complicated relationships between many features and how they affect the final prediction, it performs better than Multinomial Nave Bayes, which is based on the premise that features are conditionally independent. By keeping track of previous inputs, LSTMs are intended to capture long-term dependencies in sequential data. Since the meaning of a sentence might depend on its full context, they are therefore well suited for tasks like text classification or sentiment analysis. The order and context of the data are not taken into account by Naive Bayes, on the contrary hand. LSTMs are more noise-resistant than Naive Bayes because they can be trained to filter out noisy data and concentrate on the important information. Naive Bayes, in contrast, is susceptible to noise in the data and sensitive to irrelevant information.

5 Conclusion and Future Direction

The comparison of the bidirectional LSTM and multinomial naive bayes models for emotion detection has been studied in this study. The two methods were discussed using the COVID-19 twitter dataset. The ability to observe emotion in text is made possible by text emotion recognition, which classifies information using deep learning and NLP. The ability of the recommended ways to directly learn features from raw data has been demonstrated in the results, making them very feasible techniques for emotion recognition. The results showed that the bidirectional LSTM increased the training accuracy for this dataset. With the bidirectional LSTM, we were able to achieve accuracy of up to 91.82%, and with the multinomial naive bayes, accuracy of up to 76.61%.

The extent of the dataset employed in the study project was constrained by technological limitations. Future studies will concentrate on applying the models to real-time datasets gathered from various sources and enhancing accuracy. The improvement of emotion recognition, modelling the intensity of the feelings, allowing many emotion classes to be active at once, and investigating alternative emotion class models will be further points of emphasis. Additionally, extensive research could be done to identify alternative algorithms that can use contextual data and sophisticated word embedding techniques, use various evaluation metrics like precision, recall, and confusion matrix, and provide more accurate emotion detection. Future research in this area of study appears promising because more than a million bytes of data are produced daily.

References

1. Rajput, N., Ahuja, B., Rathi, V.: Word frequency and sentiment analysis of Twitter messages during Coronavirus pandemic (2020). https://doi.org/10.48550/arXiv.2004.03925
2. Mansoor, M., Gurumurthy, K., Anantharam, U., Prasad, V.: Global sentiment analysis of COVID-19 tweets over time (2020)

3. Suneera, C., Prakash, J.: Performance analysis of machine learning and deep learning models for text classification, pp. 1–6 (2020). https://doi.org/10.1109/INDICON49873.2020.9342208
4. Tareq, A., Hewahi, N.: Sentiment analysis of tweets during COVID-19 pandemic using BLSTM. In: 2021 International Conference on Data Analytics for Business and Industry (ICDABI), Sakheer, Bahrain, pp. 245–249 (2021). https://doi.org/10.1109/ICDABI53623.2021.9655932
5. Vijay, T., Chawla, A., Dhanka, B., Karmakar, P.: Sentiment analysis on COVID-19 Twitter data. In: 2020 5th IEEE International Conference on Recent Advances and Innovations in Engineering (ICRAIE), Jaipur, India, pp. 1–7 (2020). https://doi.org/10.1109/ICRAIE51050.2020.9358301
6. Alawneh, L., Mohsen, B., Al-Zinati, M., Shatnawi, A., Al-Ayyoub, M.: A comparison of unidirectional and bidirectional LSTM networks for human activity recognition. In: 2020 IEEE International Conference on Pervasive Computing and Communications Workshops (PerCom Workshops), Austin, TX, USA, pp. 1–6 (2020). https://doi.org/10.1109/PerComWorkshops48775.2020.9156264
7. Sumanathilaka, T.G.D.K., Selvarai, V., Raj, U., Raiu, V.P., Prakash, J.: Emotion detection using bi-directional LSTM with an effective text pre-processing method. In: 2021 12th International Conference on Computing Communication and Networking Technologies (ICCCNT), Kharagpur, India, pp. 1–4 (2021). https://doi.org/10.1109/ICCCNT51525.2021.9579844
8. Xu, G., Meng, Y., Qiu, X., Yu, Z., Wu, X.: Sentiment analysis of comment texts based on BiLSTM. IEEE Access **7**, 51522–51532 (2019). https://doi.org/10.1109/ACCESS.2019.2909919
9. Sushmitha, M., Suresh, K., Vandana, K.: To predict customer sentimental behavior by using enhanced Bi-LSTM technique. In: 2022 7th International Conference on Communication and Electronics Systems (ICCES), Coimbatore, India, pp. 969–975 (2022). https://doi.org/10.1109/ICCES54183.2022.9835947
10. Abbas, M., Ali, K., Memon, S., Jamali, A., Memon, S., Ahmed, A.: Multinomial Naive Bayes classification model for sentiment analysis (2019). https://doi.org/10.13140/RG.2.2.30021.40169
11. Sharupa, N.A., Rahman, M., Alvi, N., Raihan, M., Islam, A., Raihan, T.: Emotion detection of twitter post using multinomial Naive Bayes. In: 2020 11th International Conference on Computing, Communication and Networking Technologies (ICCCNT), Kharagpur, India, pp. 1–6 (2020). https://doi.org/10.1109/ICCCNT49239.2020.9225432
12. Helmi Setyawan, M.Y., Awangga, R.M., Efendi, S.R.: Comparison of multinomial Naive Bayes algorithm and logistic regression for intent classification in chatbot. In: 2018 International Conference on Applied Engineering (ICAE), Batam, Indonesia, pp. 1–5 (2018). https://doi.org/10.1109/INCAE.2018.8579372
13. Jianqiang, Z., Xiaolin, G., Xuejun, Z.: Deep convolution neural networks for Twitter sentiment analysis. IEEE Access **6**, 23253–23260 (2018). https://doi.org/10.1109/ACCESS.2017.2776930
14. Chirawichitchai, N.: Sentiment classification by a hybrid method of greedy search and multinomial Naïve Bayes algorithm. In: 2013 Eleventh International Conference on ICT and Knowledge Engineering, Bangkok, Thailand, pp. 1–4 (2013). https://doi.org/10.1109/ICTKE.2013.6756285
15. Guo, J.: Deep learning approach to text analysis for human emotion detection from big data. J. Intell. Syst. **31**, 113–126 (2022). https://doi.org/10.1515/jisys-2022-0001

16. Hasan, M., Rundensteiner, E., Agu, E.: Automatic emotion detection in text streams by analyzing Twitter data. Int. J. Data Sci. Anal. **7**, 35–51 (2019). https://doi.org/10.1007/s41060-018-0096-z
17. https://www.gabormelli.com/RKB/Bidirectional_LSTM_%28BiLSTM%29_Training_Task. Bidirectional LSTM (BiLSTM) Training Task - GM-RKB, 2 May 2023

Hybrid Region of Interest Based Near-Lossless Codec for Brain Tumour Images Using Convolutional Autoencoder

Muthalaguraja Venugopal(✉) and Kalavathi Palanisamy

Department of Computer Science and Applications, The Gandhigram Rural Institute (Deemed to be University), Gandhigram, India
muthalagurajav@gmail.com, pkalavathi.gri@gmail.com

Abstract. One of the most significant industries producing digital images worldwide is radiology. The advancements in radiological equipment have ensured the production of high-definition digital medical images. Irrespective of the growth in storage and network facilities, these high-definition images still face the problem of high storage space requirements and high transmission costs. To handle aforementioned problems and pave way for faster, hassle-free transmission and effective storage of such medical images for telemedicine services, we need enhanced image compression techniques. Medical images tend to have data with both high and low clinical importance for downstream analysis and treatment. Compression algorithms must be developed in order to handle both high and low clinically important data at the same time to improve the compression standard of medical images. We propose a Convolutional autoencoder technique for Region of Interest based hybrid near-lossless medical image compression aided by "You Only Look Once" (YOLO) deep learning algorithm. This work aims to achieve ROI-based near-lossless compression with notable compression ratio and medical image quality. To achieve this ROI-based near-lossless compression, we employed a combination of YOLO object detection algorithm, Convolutional autoencoder, and Haar wavelet transform with SPIHT encoding on grayscale Magnetic Resonance brain tumour images. The proposed approach was evaluated against several existing standard compression methods. Results inferred that our proposed method assured the near-lossless image compression scenario by maintaining the quality of medical images after decompression and comparatively reduced the storage and transmission cost by ensuring an effective compression ratio.

Keywords: Medical Image compression · Convolutional Autoencoder · Wavelet Transform · SPIHT · Region of Interest · YOLO · Near-Lossless

1 Introduction

The digital image processing domain has shown tremendous development in the recent past. With the advancement of cutting-edge technologies, digital image processing has reached a new peak in image quality aspect and the production of image data has

S. Aurelia et al. (Eds.): ICCSST 2023, CCIS 1973, pp. 333–350, 2024.
https://doi.org/10.1007/978-3-031-50993-3_27

increased worldwide. Alongside the digital imaging domain, storage and network facilities have also emerged significantly, but when considering the growth of the digital imaging sector and multiple terabytes of image data produced every day via different sources, it becomes essential to handle the image data effectively. The enormous amount of image data produced globally clearly creates a real-time problem across various domains in storage and transmission aspects. Among various sources of image data, radiology produces high-definition medical images with the help of modern radiological equipment. In radiology starting from X-Ray to today's modern hybrid radiological imaging, various medical imaging modalities have been drastically developed [1]. As a result of advancements in medical imaging modalities, production of high-quality medical images has already started transforming diagnosis and treatment methods, particularly in telemedicine services. The need to store such medical images for future reference and the transfer of such images over a network for telemedicine became inevitable. Since the data obtained from the medical imaging modalities are of high quality and voluminous, it undoubtedly leads to higher transmission and storage cost. To handle such huge data efficiently, compression becomes a vital tool [2, 3]. Employing suitable compression techniques on medical images will allow us to effectively handle the storage and transmission problems in real-time. Numerous compression methods like Arithmetic coding [4], Huffman [5], Run Length Coding (RLC) [6], Joint Photographic Experts Group (JPEG) [7] are available for both general-purpose images and medical images. Among these, wavelet-based methods are known for their exceptional efficiency in maintaining image quality while delivering a notable Compression Ratio (CR) [8, 9].

In general, data compression may be bisected into lossy and lossless methods. Lossy compression algorithms may be applied on areas where loss of certain data during compression process may not affect the end result. In areas where data loss is not acceptable lossless methods may be applied. There is also a third dimension of compression technique which widely known as the Near-Lossless compression approach. This method tries to balance the benefits of the lossy and lossless compression approaches. It losses data but only a negligible amount, like the name suggests it can be understood as nearly lossless compression technique but not completely lossless. Now considering the image data, it accepts lossy, lossless and near-lossless compression techniques according to the nature of the image and the area where they are used after compression process. Lossy technique provides high Compression Ratio (CR) but with compromised image quality while Lossless provides low CR with equal image quality to the original image. Near-Lossless provides high CR like a lossy compression technique and high image quality nearly equivalent to lossless compression [10–12].

With regard to medical images, there are different imaging modalities available in radiology like X-Ray, Computed Tomography (CT), Magnetic Resonance Imaging (MRI), Positron Emission Tomography (PET), Ultra Sound (US) and many others. These images contain data of clinical importance for diagnosis and treatment. So, the lossy compression on medical image may seem to be unacceptable but it is not completely true. To ensure the applicability of suitable compression algorithms for medical images we studied the nature of such images. As a result, medical image was fragmented into multiple regions with data of different level of importance for down stream analysis [13]. It is clear that medical images have both data with high clinical importance and data with low or negligible clinical importance. The regions of a medical image can be basically split into Background and Region of Interest (ROI). Further the ROI can be subdivided into Secondary Region of Interest (S-ROI) and Primary Region of Interest (P-ROI). The background data of the medical image can be neglected during compression as it may be rebuilt during the decompression procedure.

The S-ROI region has data with low clinical importance but cannot be neglected and the P-ROI region contains data with high clinical importance which is critical and cannot be lost during the compression process. Now to decide the type of compression algorithm that could be applied on a medical image taking into account of the three regions, the background region and S-ROI region can be compressed using lossy compression technique as the data loss in those areas will not affect the purpose of medical image in downstream analysis. P-ROI region can be compressed by either near-lossless or lossless compression approach [14]. Compression process can be approached in two ways namely homogeneous and hybrid. Homogeneous refers to the usage of single compression algorithm over the entire data and the hybrid approach refers to the use of more than one compression algorithm on the data.

In recent past, data compression algorithms have shown a notable transition from traditional approaches to learning based approaches. Deep learning methods such as Autoencoder (AE), Generative Adversarial Network (GAN) are now being popularly used for image compression tasks. Even in some cases of transform-based learning approaches like Discrete Cosine Transform (DCT) and Discrete Wavelet Transform (DWT) deep learning is being used to achieve certain goals. Deep learning-based compression methods are mostly lossy in nature as they tend to drop some data in the process of producing latent representation. Among various deep learning approaches the autoencoder based methods are mostly preferred in the recent literature due to its compression performance. One major factor to be noted when employing a learning based algorithm is that they are data dependent unlike traditional algorithms which are mostly data independent.

We propose a hybrid Near-Lossless ROI based compression approach named "HRNL Codec" for 2D MRI brain tumour images. Though different modalities can be used to analyze a particular disease, unique medical conditions may require specific type of medical imaging technique for accountable diagnosis and treatment of that disease. MRI scans are widely used for the study of brain pathology. It produces 2D and 3D images of the brain using a strong magnetic field and radio frequency current pulsed through a patient. We focus on compression of 2D MRI brain tumour images due to its global usage in localized treatment and also in telemedicine services. We have opted for a hybrid compression approach in this work to gain the advantage in compressing the different regions of the images using different compression techniques. This hybrid approach paves way to achieve Near-Lossless image compression scenario.

2 Related Works

In the literature, several image compression techniques were developed by re searchers to solve the image compression problem in different aspects. Some of the notable works are based on DWT, ROI, hybrid and learning based methods which are as follows. DWT based methods takes an important place in the image compression domain and various DWT based methods were over time. Chen and Tseng [15] developed a method in which correlation analysis is done to select the best wavelet basis function and then the adaptive prediction model was used. In order to provide more precise predictions, this approach used several prediction equations at various wavelet sub-bands to address the multicollinearity issue in wavelet-based compression. As compared to alternative approaches already in use, the experimental results showed better compression for US, CT and MRI images. A hybrid method was proposed by Ammah and Owusu [16] using wavelet transform and vector quantization and Huff man coding to compress the medical images while maintaining the perceptual quality of the medical images. The results of this proposed method outperformed other recent techniques. Another hybrid approach for lossless image compression was proposed by Starosolski [17] using DWT and prediction. This approach applies the prediction technique on the wavelet subbands to perform better. This method was able to achieve a significant compression ratio improvement when compared to JPEG2000 but with a increased time complexity. A wavelet-based deep auto encoder decoder was developed by Mishra [18] for image compression. This method adopts wavelet decomposition as a preprocessing step and then uses deep autoencoder to encode the image for individual subbands. It exhibited a notable compression performance in terms of image quality aspects when compared to existing methods. Many ROI based methods were also proposed in the recent past. The method proposed by Ansari and Anand [19] is a context-based compression technique based on SPIHT in which coefficients corresponding to a spatial location are organized in a tree form and are then encoded. It performed well when compared to other standard approaches for ultrasound medical images. Lossless ROI based method proposed by Bairagi and Sapkal [20] for Digital Imaging and Communications in Medicine (DICOM) images which mainly focus on removing the noisy background using Integer wavelet transform and distortion limiting technique. In the method developed by Zuo et al. [21], the medical image is separated as ROI and NROI using region-based contour method and ROI is com pressed using the JPEG-LS algorithm and lossy wavelet-based compression is applied to the NROI region. Compared with the standard lossless and lossy methods, this proposed method performed well in terms of CR and similarity indexes. Devadoss and Sankaragomathi [22] proposed a near-lossless method for compression by partitioning the ROI into smaller blocks and then using Burrows-Wheeler Transform and Move to Front Transform for ROI compression and hybrid fractal compression technique for the NROI region. This method performed well in terms of PSNR compared with JPEG and produced a superior CR and PSNR than SPIHT and DWT. In [23] Haojie Liu et al. proposed a lossy compression method with Deep Convolutional Neural Networks (CNN) which outperformed BPG, WebP, JPEG and JPEG2000 in Multi Scale-SSIM. A Discrete Cosine Transform (DCT) like CNN based transform was proposed by Dong Liu et al. [24]. It achieved higher compression efficiency than DCT. Zhengxue Cheng et al. [25] developed an autoencoder architecture which per form compression by decomposing the image by various sampling operations. It out performs state of the art methods providing higher coding efficiency. Pengfei Guo et al. [26] proposed an image compression framework using CNN for retinal image and it delivers better CR, as well as achieves a remarkable Multi Scale-SSIM compared to existing methods. Krishnaraj et al.

[27] modelled a DWT based deep learning model to achieve better reconstruction after compression. This model gave better CR and reconstructed image quality than Super-resolution CNN, JPEG and JPEG 2000. David Alexandre et al. [28] suggested a image compression method with autoencoder that makes use of residual blocks to convert the input image into feature maps. By several criteria, this approach outperformed JPEG and BPG, however it provided results that were inferior to BPG. The lossy image compression method proposed by Jun-Hyuk Kim et al. [29] was based on an asymmetric autoencoder and used a decoder pruning approach. The proposed method was assessed using PSNR and bits per pixel measurements. The outcomes of the experimental evaluation proved the effectiveness of this method's decoder pruning procedures and encoder.

The literature evidently shows that DWT, ROI and learning based methods had shown their efficiency individually and also in a hybrid approach on image compression tasks. The above-mentioned methods address the image compression problem in various aspects which faces different challenges as follows.

- In region-based techniques, to compress the ROI and remaining regions separately, automatic and accurate ROI detection is important. This is a notable challenge faced by region based techniques.
- As suggested in the literature, using hybrid compression techniques for images yields better compression performance. In this, employing an efficient lossy compression algorithm is another challenge observed in the literature.
- In learning-based methods, development of compression technique with minimal data loss and maximum compression performance becomes a challenge especially for medical image compression.

Consequently, to overcome these issues, we proposed "HRNL Codec", a hybrid region-based approach by combining the effectiveness of autoencoder and wavelet based encoding technique to attain near lossless compression standard. The proposed method uses the following techniques to overcome the above said challenges.

- The proposed "HRNL Codec" uses a learning based CNN model "YOLO" to per form automatic and accurate ROI detection. This helps to overcome the above said challenge in the region-based techniques.
- To address the challenge in the hybrid compression technique the proposed method uses autoencoder technique in S-ROI for lossy compression and wavelet-based technique in the P-ROI region.
- The "HRNL Codec" uses convolutional autoencoder for image compression with minimal data loss as convolutional layers works better with image data. This helps overcome the challenge faced in the learning based methods.

The proposed method uses the combination of above said techniques to resolve the challenges found in the literature to attain near lossless compression standard for 2D grayscale medical images.

3 Materials and Methods

A ROI based hybrid near lossless compression approach for 2D grayscale MRI brain tumour images using the Convolutional Autoencoder (CAE) and SPIHT encoding with haar wavelet decomposition which is aided by YOLO object detection algorithm for ROI detection is proposed in this work. Figure 1 depicts the overall architecture, while the following sections detail each of the HRNL Codec's modules.

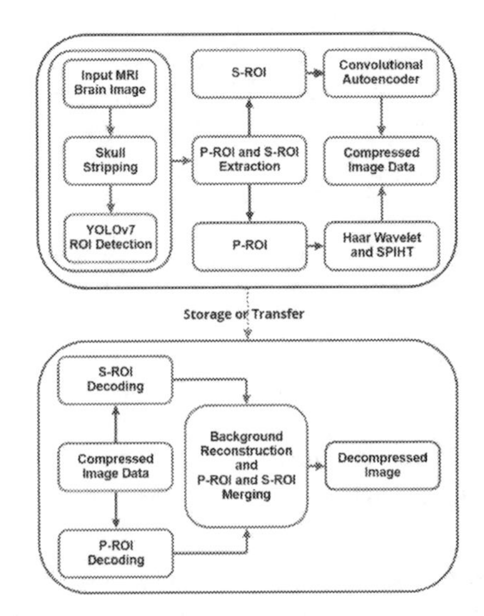

Fig. 1. Architecture of HRNL Codec (Proposed)

3.1 Preprocessing

The goal of any compression algorithm is to generate minimum possible representation of the input image after compression. To achieve this, one of the main tasks is to eradicate the redundant or needless data in image. This needless data in the image will add only a low value to the important information present in the image. In medical images, redundant data are present in different regions depending on the considered medical image type. The skull in the brain scans is a familiar redundancy in the medical images. The skull region adds less important to the brain scans in many occasions. This skull region plays a negligible part in diagnosis and treatment of brain tumour [30, 31]. During the compression of brain scans removing the skull region and considering only the actual brain region will improve the CR and tumour detection accuracy. In this work, skull stripping is done as preprocessing step before performing ROI detection. Numerous skull stripping methods are available in the past literature. The method proposed by [32] was used for skull stripping in this preprocessing step. After preprocessing the skull stripped image is taken for ROI detection and further processing. Fig. 2 represents a sample skull stripped image.

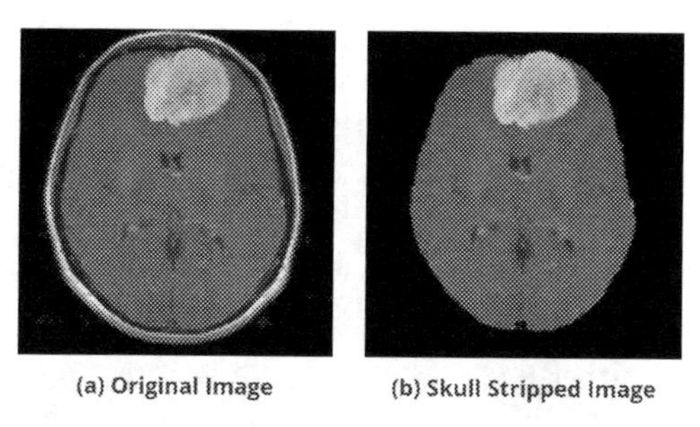

Fig. 2. Brain portion extraction in MRI image

3.2 Region of Interest

The purpose of medical imaging is to provide sufficient information about the disorder for further analysis and treatment. In clinical assessment using medical images, ROI detection is one of the most critical tasks. Accuracy in detecting the ROI is adequate for further downstream analysis and treatment. Various ROI detection methods have been proposed in the past and they have played an import role in the medical sector [14, 33–35]. Thresholding, Edge detection, Active Contour Models, Watershed, Morphological operations, Template matching and Region growing are some of the popular traditional techniques for ROI detection. Recently, deep learning techniques like CNN, R-CNN, U-Net, FCN, DBN and GAN are being used popularly for ROI detection. All these traditional and learning based methods has their uniqueness in the ROI detection. This work used a learning-based algorithm, version 7 of "You Only Look Once" (YOLO) for ROI detection [36]. YOLO is a popular object detection deep learning model which was created for detecting objects in images and known for its speed and accuracy in object detection. This CNN based model divides the image into a grid and predicts bounding box, probability and confidence score for individual cells in the grid. The advantage of using this model is that it performs object detection and classification in one forward pass through the network while other models achieve this with a minimum of two passes through the network. YOLO achieves this efficient object detection by using multiple convolutional, feature fusion and upsampling layers. An important feature of YOLOv7 is that it predicts the coordinates directly and does not depend on predefined anchor boxes to generate bounding box which lowers the computational complexity. Another important reason for employing YOLOv7 for ROI detection in this work is its capability to accurately spot and classify several objects in real-time. This feature will add advantage to the proposed hybrid method to identify the Background, S-ROI and P-ROI from input image efficiently. In the proposed HRNL Codec, the YOLOv7 is trained using 5000 MRI brain images with and without tumour region and then tested with 250 different MRI brain scans of the BR35H Brain tumour detection dataset [37] and collected dataset. It performed very well, providing 98% average accuracy in S-ROI and P-ROI detection. We have utilized the YOLO model to automate the ROI detection and classification process as given in Fig. 3 and made use of the coordinates of the detected regions to segment them and apply different compression algorithms.

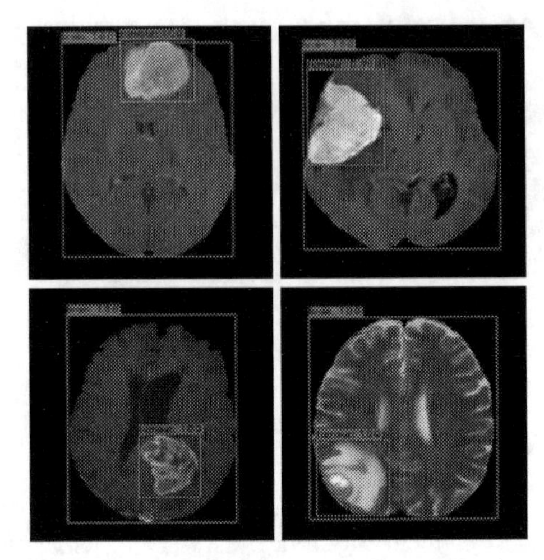

Fig. 3. P-ROI and S-ROI detection using YOLOv7 [Small box indicating tumour region is P-ROI; Large box indicating the brain region represent S-ROI; Remaining is the background]

3.3 Background Elimination

In most cases background of the medical image will add zero importance for diagnosis and treatment of any disorder. The background of the image will be composed of zero intensity pixels and take up additional unwanted space in the image. As the background possess very less important data it can be removed from the medical image during compression process. Anyway, we need the background of the image after decompression of the image to locate the ROI region accurately. The background of the image can be reconstructed at the decompression stage as it contains only the zero intensity pixels. As we get the coordinates of the ROI during the ROI detection stage using those coordinates, we can generate back the background of the image at the decompression end.

3.4 Wavelet Decomposition and SPIHT Encoding

Wavelet decomposition is a powerful mathematical tool used for effective time frequency analysis in image processing. It is known for its ability to break down an image into different frequency components using which significant information can be obtained for analysis. Wavelets are preferred for image compression for their ability to eliminate the redundancy in image by splitting original image into different sub bands with negligible data loss. The wavelet transform function is used for decomposition of image into different sub bands. The transform function convolves the given image using wavelet functions. Different frequency range is addressed by different sub bands after decomposition process. Sub band holds smallest detail to overall structure of image according to its frequency. There are various wavelets available for decomposition like db2, Haar, biorthogonal and any others. Haar wavelet decomposition technique, which uses Haar wavelets as basis functions to decompose image, is used in this work due to its simplicity and efficiency in image compression task [38–40]. Haar obtains this decomposed sub bands just by using simple averaging and differencing functions on the original image.

Basic Haar wavelet is a mathematical step function of positive and negative steps of one amplitude each. Haar represents an image by using the scaled and translated set of basis functions of the wavelet. Haar wavelet decomposition divides the image into four equal sub bands of different frequencies namely an approximation with low frequency information and three detail sub bands with high frequency information. The decomposition process can be carried out on the approximation sub band for different levels. For each level four new sub bands will be created in the approximation sub band which paves way for multi resolution analysis. After decomposition of the image SPIHT is used for encoding. SPIHT is a wavelet-based algorithm which provides balance between CR and image quality. We have used only a single level decomposition to reduce the chance of losing information since we are using medical images. After applying wavelet decomposition, the SPIHT algorithm is used and the data is encoded [41, 42]. Figure 4 shows the sample P-ROI region and a single level haar wavelet decomposition of an ROI region.

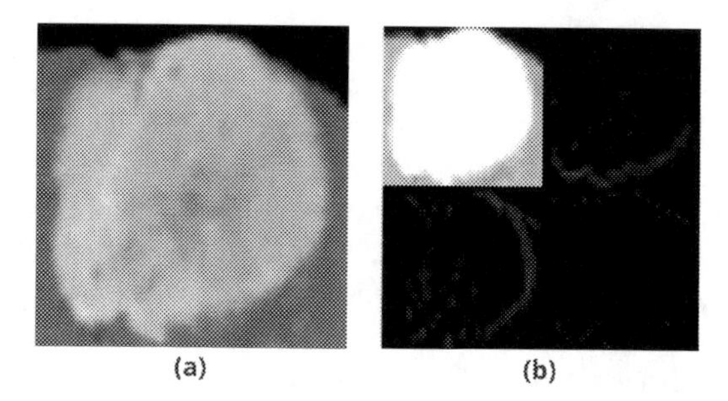

Fig. 4. a) P-ROI b) Single Level decomposition of P-ROI

3.5 Convolutional Autoencoder

Autoencoders are basically artificial neural networks which can be used for unsupervised learning tasks like compression. They have the ability to learn to encode data into lower dimension and decode them back using the latent data automatically. A complete autoencoder has two segments, an encoder to compress the data and a de coder to retrieve it back. Both encoder and decoder part are neural networks which are trained to reduce the reconstruction error between the initial and reconstructed data using backpropagation. Autoencoder is a lossy compression method which suffer loss of data during the dimensionality reduction process [43]. Training an autoencoder is one of the major challenges as are heavily data dependent. They perform better at tasks for which they have trained for. Defining the number of layers and hyperparameter tuning are critical tasks in training an autoencoder which decides the functional efficiency of the autoencoder. Among different autoencoders we have employed CAE in this work due to its ability to handle image data. CAE's are made up of convolution and deconvolution layers. With the help of the convolution operation its takes ad vantage of the spatial data of the image and applies pooling operations to decrease the dimension of image to produce a latent space representation. This latent is then mapped to original dimension of the image by using deconvolution layers in the decoder part. The number of layers in the autoencoder may vary based on serval conditions depending on the application [44].

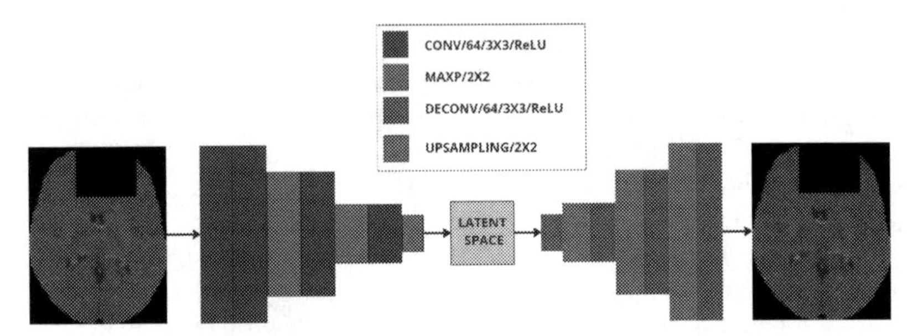

Fig. 5. Convolutional Autoencoder in the proposed HRNL Codec for S-ROI

The CAE in the proposed method was trained using 5000 2D grayscale brain scans of BR35H brain tumour detection dataset from Kaggle [37] and other collected dataset reshaped to 256 × 256. Out of these 5000 images, for 4500 were used for training, 250 each for validation and testing the CAE in the proposed model. Mean Squared Error loss function were used for the CAE training and the CAE was trained with early stopping criteria to handle the overfitting problem for 150 epochs. The hyperparameters of the proposed CAE given in Fig. 5 were tuned by kerasTuner and trail & error methods. The hyperparameters which are used in the CAE in the proposed method for compression of S-ROI are given in Table 1.

Table 1. Hyperparameters used in the proposed lossy CAE for S-ROI compression.

Hyperparameters	Specifics
Number of Conv/Deconv Layers	4/4
No. of Filters	64
Size of Kernel	3 × 3
Padding	Same
Optimizer and Learning rate	Adam and 0.0001
Batch size	8
Activation function	ReLU
Regularization techniques	Early stopping, Batch Normalization

3.6 Proposed Methodology

A 2D grayscale MRI brain tumour image of size 256 × 256 is taken as the input for the HRNL Codec and processed as per the following steps for hybrid Near-Lossless compression.

1. **Skull stripping:** A 2D grayscale MRI brain tumour image is taken as input and it is skull stripped using the method proposed by [32] to eliminate the additional data in the image.
2. **ROI detection and Extraction:** The P-ROI and S-ROI are detected and located with the aid of YOLO v7. The location coordinates are obtained for both ROIs. The P-ROI and S-ROI regions are extracted using the location coordinates obtained during ROI detection stage and the background data present in the S-ROI part is eliminated.
3. **Encoding:** A hybrid compression approach in which P-ROI is decomposed using the Haar wavelet decomposition followed by SPHIT encoding and S-ROI is encoded using the lossy Convolutional autoencoder's encoder module. The coordinate details of the ROIs are merged with the encoded data of ROIs and stored or transmitted
4. **Decoding:** The encoded P-ROI data can be decoded by applying inverse encoding operation and applying inverse wavelet transform operations. The S-ROI region is decoded by using the Convolutional autoencoder's decoder module.
5. **Reconstruction:** The zero-intensity background is recreated for entire image size and by using the coordinate details of the ROIs taken from the decoded data, the decoded P-ROI and S-ROI are reconstructed to their exact locations on the zero intensity background image to achieve the complete decompressed image data.

4 Evaluation Metrics

4.1 Peak Signal to Noise Ratio (PSNR)

PSNR is used to measure the quality of the restored images. Higher PSNR represents higher quality of the restored images which is required in the medical images. It is calculated using below Eq. (1)

$$PSNR = 10\log_{10}\left(MAX_I^2/MSE\right) \quad (1)$$

where, MAX_I is the maximum fluctuation in an image. MSE is the Mean Squared Error calculated using the below Eq. (2)

$$MSE = 1/mn\sum_{a=1}^{m}\sum_{b=1}^{n}\left[I(a,b) - I'(a,b)\right]^2 \quad (2)$$

where, I(a,b) is the original image, I′ (a,b) is the obtained decompressed image and m, n are the dimensions of the image.

4.2 Structural Similarity Index Metric (SSIM)

SSIM is a function of luminance comparison, contrast and structural comparison term. The value of SSIM lies between 0 to 1. Values near 1 indicate better results. It is calculated using the equation given in Eq. (3)

$$SSIM\,(a,b) = [l(a,b)]^{\alpha} \cdot [c(a,b)]^{\beta} \cdot [s(a,b)]^{\gamma} \tag{3}$$

where,

$$l(a,b) = \frac{2\mu_a\mu_b + c_1}{\mu_a^2 + \mu_b^2 + c_1} \tag{4}$$

$$c(a,b) = \frac{2\sigma_a\sigma_b + c_2}{\sigma_a^2 + \sigma_b^2 + c_2} \tag{5}$$

$$s(a,b) = \frac{2\sigma_{ab} + c_3}{\sigma_a\sigma_b + c_3} \tag{6}$$

where, μ_a, μ_b, σ_a, σ_b, and σ_{ab} are the local means, standard deviations, and cross covariance for images a, b.

4.3 Compression Ratio (CR)

If s1 represents the total bits of original image and s2 represents the total bits of the compressed image, then CR is the ratio between the s1 and s2 and is given by Eq. (7)

$$CR = \frac{s1}{s2} \tag{7}$$

4.4 Bits Per Pixel (BPP)

BPP denotes total bits required to represent a single sample and is given by (8)

$$BPP = \frac{Compressed\ file\ size}{Number\ of\ pixels} \tag{8}$$

5 Results and Discussions

The YOLO model and the convolutional autoencoder model used in the proposed HRNL Codec was trained using a 16 GB RAM and 4 GB GPU enabled workstation with 5000 images each for 150 epochs with early stopping. The total training time was approximately 10 hours. The proposed HRNL Codec was also compared with the standard methods like JPEG, JPEG2000, EZW and SPIHT in terms of different metrics. These methods were used for comparison of the proposed method's results as these are most widely used standard image compression techniques in various image processing domains. Moreover, most of the methods in the literature have compared their results with the above said standard methods due to their standard and wide usage. Hence, we have used these standard methods as benchmarks for comparing our method's results. The comparisons are depicted as charts and discussed in this section. The results of the proposed HRNL Codec for the selected samples are given in the Table 2 in terms of different metrics for quantitative and qualitative analysis. The selected images from our dataset are presented in Fig. 6. These sample images have been selected by considering the non-uniformity of the tumour position, shape and size to demonstrate the reliability of the HRNL Codec. The first column of Fig. 6 is the original image. The skull stripped images are in the second column of Fig. 6. YOLOv7 predictions are given in Fig. 6(c). The extracted S-ROI and P-ROI regions for the selected image are presented in Fig. 6(d) and Fig. 6(e), respectively. The last column in Fig. 6 is the decompressed image obtained by the proposed HRNL Codec method.

Table 2. Results of the proposed HRNL Codec for selected image samples.

Image No.	PSNR	SSIM	BPP	CR
1	37.51	0.9364	0.18	44.76
2	43.45	0.9663	0.60	13.15
3	41.23	0.9452	0.54	14.77
4	38.43	0.9125	0.52	15.36
5	36.84	0.9275	0.16	51.52
6	40.52	0.9645	0.56	14.23
7	39.37	0.9032	0.52	15.17
AVG	**39.62**	**0.9365**	**0.44**	**24.13**

The graph in Fig. 7 clearly shows that in terms of CR, the proposed HRNL Codec performs better than the prevailing standard methods for the selected samples. The proposed HRNL Codec is also compared with the prevailing methods in BPP and the result is presented in Fig. 8. It can be observed that our HRNL Codec method has obtained low BPP values than the compared methods. The comparison is also made in terms of PSNR and that resultant graph is given in the Fig. 9. This graph clearly shows the PSNR improvement over the compared approaches. It is evident that the HRNL Codec approach has produced an average of 24.13, 39.62 and 0.44 in CR, PSNR and BPP, respectively, producing a Near-Lossless compression standard, which, when compared to current techniques, is superior. As per the obtained results, our HRNL Codec has performed better, by providing a good compression ratio with a notable visual quality. This shows performance improvement of proposed method than the existing methods. The improvement in visual quality and reduced com pressed data size is crucial for any image compression technique to use it in practical scenarios like telemedicine services.

From Fig. 7, 8 and 9 we can say that our HRNL Codec has compressed the image size to a greater extent than the prevailing methods. It is to be noted that our HRNL Codec method has maintained better image visual quality in PSNR than other methods, which is of utmost essential for medical images in telemedicine services.

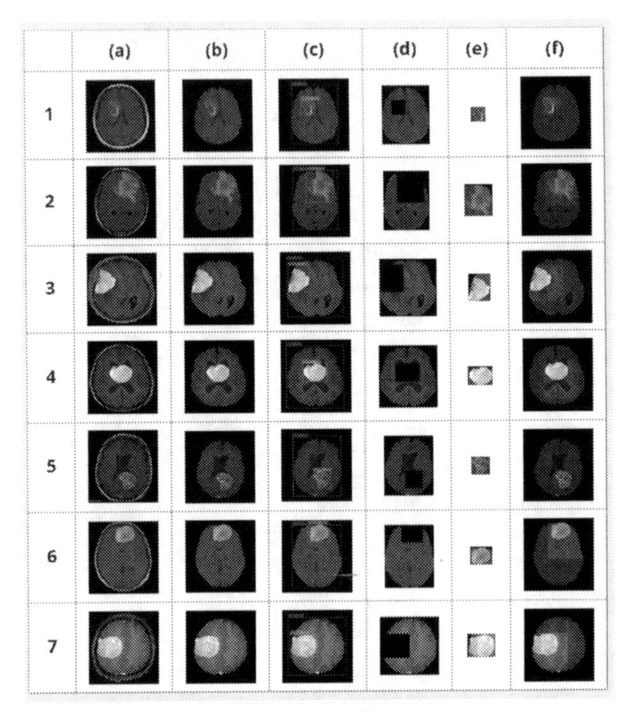

Fig. 6. Results of the proposed HRNL Codec (a) Input images (b) Skull Stripped images (c) YOLOv7 predicted images (d) Extracted S-ROI (e) Extracted P-ROI (f) Reconstructed images

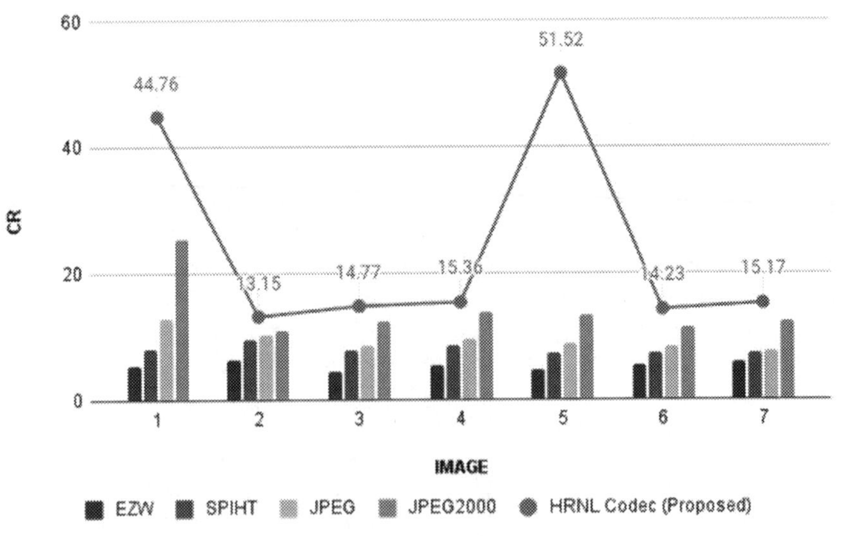

Fig. 7. HRNL Codec (Proposed) compared with existing methods in CR

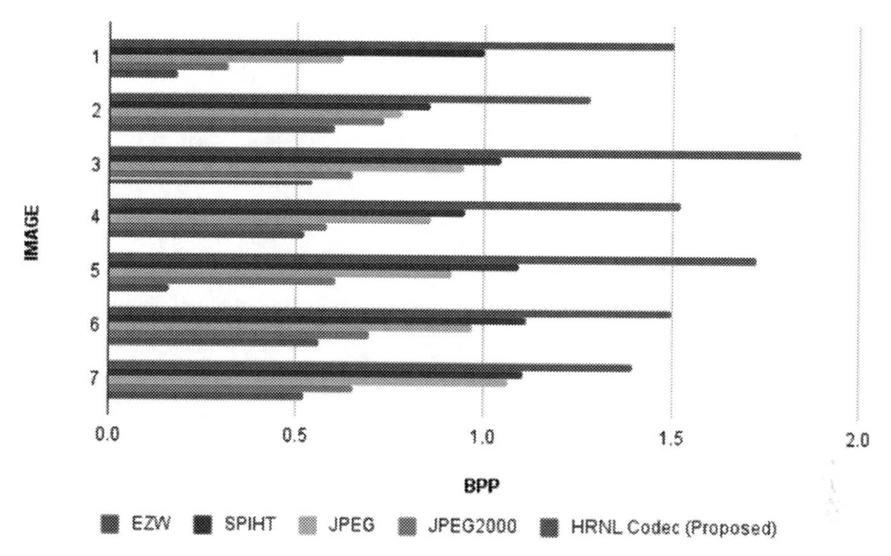

Fig. 8. HRNL Codec (Proposed) compared with existing methods in BPP

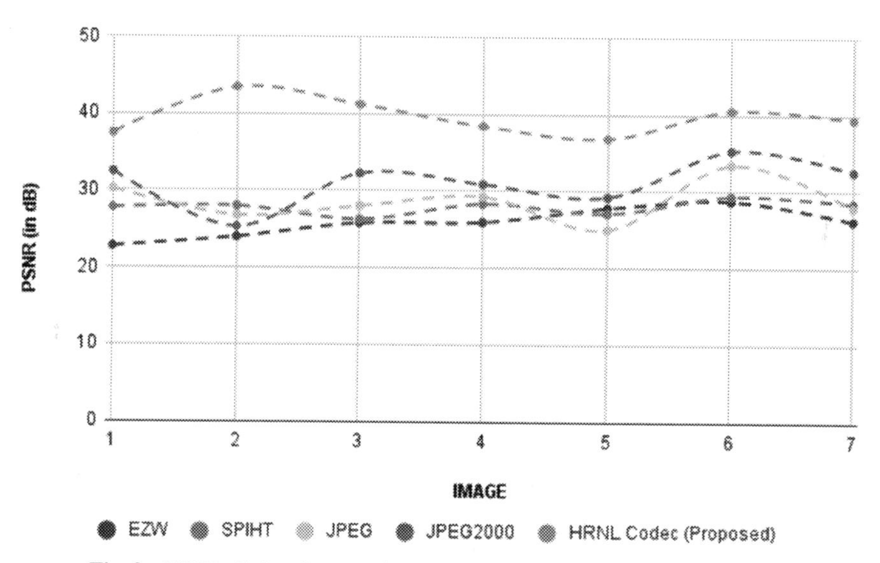

Fig. 9. HRNL Codec (Proposed) compared with existing methods in PSNR

6 Conclusions

This work aimed to achieve ROI-based near-lossless compression by improving the CR and maintaining the quality of medical image. Results of proposed HRNL Codec assured the near-lossless compression scenario by maintaining the decompressed medical image quality and comparatively reduced the storage and transmission cost by ensuring an effective compression ratio when compared with other existing standard methods. In comparison with the existing methods, our HRNL Codec method performed better in metrics like CR, PSNR and BPP. Making note on limitations of our HRNL Codec the computation complexity of our method is a bit high as two deep learning techniques were employed. However, with the growth in the computational facilities such as use of GPU will aid in practical application of this method. Secondly, S-ROI region compressed using convolutional autoencoder may drop some low clinically important data during compression which must also be addressed effectively. As a future scope the method may be enhanced to reduce the computational complexity and a separate enhancement module may be developed for the S ROI region to compensate the dropped data in that region.

Acknowledgement. This work was supported by Council of Scientific and Industrial Research (CSIR), File Number: 25(0304)/19/EMR-II, Human Resource Development Group, Government of India.

References

1. Miller, C.G., et al.: Medical Imaging in Clinical Trials. Netherlands. Springer, London (2014). https://doi.org/10.1007/978-1-84882-710-3
2. Ansari, M.A., Anand, R.S.: Recent trends in image compression and its application in telemedicine and teleconsuktation. In: National Systems Conference, No. July, pp. 59–64 (2008)
3. Zukoski, M.J., Boult, T., Iyriboz, T.: A novel approach to medical image compression. Int. J. Bioinform. Res. Appl. Bioinform. Res. Appl. **2**(1), 89–103 (2006)
4. Puech, W.: Efficient adaptive arithmetic coding based on updated probability distribution for lossless image compression. J. Electron. Imaging **19**(2), 023014 (2010)
5. Vaish, A., Kumar, M.: A new Image compression technique using principal component analysis and Huffman coding. In: Proceedings of 2014 3rd International Conference on Parallel, Distributed and Grid Computing, PDGC 2014, pp. 301–305 (2015)
6. Tu, C., Liang, J., Tran, T.D.: Adaptive runlength coding. IEEE Signal Process. Lett. **10**(3), 61–64 (2003)
7. Raid, A.M., et al.: JPEG image compression using discrete cosine transform - a survey. Int. J. Comput. Sci. Eng. Surv. **5**(2), 39–47 (2014)
8. Walker, J.S., Nguyen, T.Q.: Wavelet-Based Image Compression. Wavelet-Based Image Compression. The Transform and Data Compression Handbook (2001)
9. Boopathiraja, S., Kalavathi, P., Dhanalakshmi, C.: Significance of image compression and its upshots - a survey. Int. J. Sci. Res. Comput. Sci. Eng. Inf. Technol. 1203–1208 (2019)
10. Boopathiraja, S., Palanisamy, K., Surya Prasath, V.B.: On a hybrid lossless compression technique for three-dimensional medical images. J. Appl. Clin. Med. Phys.Clin. Med. Phys. **22**(8), 191–203 (2021)
11. Boopathiraja, S., Kalavathi, P.: A near lossless three-dimensional medical image compression technique using 3D-discrete wavelet transform. Int. J. Biomed. Eng. Technol. **35**, 191–206 (2019)
12. Sharma, U., Sood, M., Puthooran, E.: A block adaptive near-lossless compression algorithm for medical image sequences and diagnostic quality assessment. J. Digit. Imaging **33**(2), 516–530 (2020)
13. Kasute, S.D., Kolhekar, M.: ROI based medical image compression. Int. J. Sc. Res. Netw. Secur. Commun. **5**(1) (2017). ISSN 2321-3256

14. Kaur, M., Wasson, V.: ROI based medical image compression for telemedicine application. Procedia Comput. Sci. **70**, 579–585 (2015)
15. Chen, Y.T., Tseng, D.C.: Wavelet-based medical image compression with adaptive prediction. Comput. Med. Imaging Graph. **31**(1), 1–8 (2007)
16. Ammah, P.N.T., Owusu, E.: Robust medical image compression based on wavelet transform and vector quantization. Inform. Med. Unlocked **15**, 100183 (2019). https://doi.org/10.1016/j.imu.2019.100183. ISSN 2352-9148
17. Starosolski, R.: Hybrid adaptive lossless image compression based on discrete wavelet transform. Entropy **22**(7), 751 (2020). https://doi.org/10.3390/e22070751
18. Mishra, D., Singh, S.K., Singh, R.K.: Wavelet-based deep auto encoder-decoder (WDAED)-based image compression. IEEE Trans. Circuits Syst. Video Technol. **31**(4), 1452–1462 (2021). https://doi.org/10.1109/TCSVT.2020.3010627
19. Ansari, M.A., Anand, R.S.: Context based medical image compression for ultrasound images with contextual set partitioning in hierarchical trees algorithm. Adv. Eng. Softw.Softw. **40**(7), 487–496 (2009)
20. Bairagi, V.K., Sapkal, A.M.: ROI-based DICOM image compression for telemedicine. Sadhana – Acad. Proc. Eng. Sci. **38**(1), 123–131 (2013)
21. Zuo, Z., Lan, X., Deng, L., Yao, S., Wang, X.: An improved medical image compression technique with lossless region of interest. Optik **126**(21), 2825–2831 (2015)
22. Devadoss, C.P., Sankaragomathi, B.: Near lossless medical image compression using block BWT–MTF and hybrid fractal compression techniques. Cluster Comput. **22**(s5), 12929–12937 (2019)
23. Liu, H., et al.: Deep image compression via end-to-end learning. In: Proceedings of the IEEE Conference on Computer Vision and Pattern Recognition Workshops (2018)
24. Liu, D., Ma, H., Xiong, Z., Wu, F.: CNN-based DCT-like transform for image compression. In: Schoeffmann, K., Chalidabhongse, T.H., Ngo, C.W., Aramvith, S., O'Connor, N.E., Ho, Y.-S., Gabbouj, M., Elgammal, A. (eds.) MMM 2018. LNCS, vol. 10705, pp. 61–72. Springer, Cham (2018). https://doi.org/10.1007/978-3-319-73600-6_6
25. Cheng, Z., et al.: Energy compaction-based image compression using convolution al autoencoder. IEEE Trans. Multimed. **22**(4), 860–873 (2019)
26. Guo, P., Li, D., Li, X.: Deep OCT image compression with convolutional neural networks. Biomed. Opt. Express **11**(7), 3543–3554 (2020)
27. Krishnaraj, N., et al.: Deep learning model for real-time image compression in Internet of Underwater Things (IoUT). J. Real-Time Image Proc. **17**(6), 2097–2111 (2020)
28. Alexandre, D., Chang, C.-P., Peng, W.-H., Hang, H.-M.: An autoencoder-based learned image compressor: description of challenge proposal by NCTU, pp. 2539–2542 (2019). http://arxiv.org/abs/1902.07385
29. Kim, J.H., Choi, J.H., Chang, J., Lee, J.S.: Efficient deep learning-based lossy image compression via asymmetric autoencoder and pruning. In: IEEE International Conference on Acoustics, Speech and Signal Processing, ICASSP, vol. 2020-May, pp. 2063–2067 (2020). https://doi.org/10.1109/ICASSP40776.2020.9053102
30. Mohsin, S., Sajjad, S., Malik, Z., Abdullah, A.H.: Efficient way of skull stripping in MRI to detect brain Tumor by applying morphological operations, after detection of false background. Int. J. Inf. Educ. Technol. (2012)
31. Kalavathi, P., Prasath, V.B.S.: Methods on skull stripping of MRI head scan images—a review. J. Digit. Imaging **29**(3), 365–379 (2016)
32. Somasundaram, K., Kalavathi, P.: A hybrid method for automatic skull stripping of magnetic resonance images (MRI) of human head scans. In: 2010 2nd International Conference on Computing, Communication and Networking Technologies, ICCCNT 2010 (2010). https://doi.org/10.1109/ICCCNT.2010.5592550

33. Gokturk, S.B., Tomasi, C., Girod, B., Beaulieu, C.: Medical image compression based on region of interest, with application to colon CT images. Ann. Rep. Res. Reactor Inst. **3**, 2453–2456 (2001)
34. Varma, M.K., Bagadi, M.: ROI based image compression in baseline JPEG. Int. J. Eng. Res. Appl. **4**(9), 168–173 (2014)
35. Abu-hajar, A., Sankar, R.: Region of interest coding using partial-SPIHT. In: IEEE International Conference on Acoustics, Speech and Signal Processing, vol. 3, pp. 257–260 (2004)
36. Wang, C.-Y., et al.: YOLOv7: trainable bag-of-freebies sets new state-of-the-art for real-time object detectors (2022). https://doi.org/10.48550/arXiv.2207.02696
37. Hamada, A.: Br35H Brain Tumour Detection 2020. https://www.kaggle.com/datasets/ahmedhamada0/brain-tumor-detection
38. Talukder, K.H., Harada, K.: Haar Wavelet Based Approach for Image Compression and Quality Assessment of Compressed Image (2010). http://arxiv.org/abs/1010.4084
39. Kalavathi, P., Boopathiraja, S.: A wavelet based image compression with RLC encoder. Comput. Methods Commun. Tech. Inform. 289–292 (2017). ISBN 978-81-933316-1-3
40. Boopathiraja, S., Kalavathi, P.: A medical image compression technique using 2D-DWT with Run length encoding. Glob. J. Pure Appl. Math. **13**(5), 87–96 (2017)
41. Belyaev, A.A., Yevtushok, O.S., Ryaboshchuk, N.M.: Lossless image compression algorithm based on haar transform. In: 2021 IEEE Conference of Russian Young Researchers in Electrical and Electronic Engineering (ElConRus). IEEE (2021)
42. Kamargaonkar, C., Sharma, M.: Hybrid medical image compression method using SPIHT algorithm and Haar wavelet transform. In: 2016 International Conference on Electrical, Electronics, and Optimization Techniques (ICEEOT). IEEE (2016)
43. Theis, et al.: Lossy image compression with compressive autoencoders. https://doi.org/10.48550/arXiv.1703.00395
44. Cheng, et al.: Deep convolutional autoencoder-based lossy image compression. https://doi.org/10.48550/arXiv.1804.09535

An Empirical and Statistical Analysis of Classification Algorithms Used in Heart Attack Forecasting

Gifty Roy(✉), Reshma Rachel Cherish, and Boppuru Rudra Prathap

Department of Computer Science and Engineering, School of Engineering and Technology, CHRIST (Deemed to Be University), Bangalore, India
gifty.roy@mtech.christuniversity.in

Abstract. The risk of dying from a heart attack is high everywhere in the world. This is based on the fact that every forty seconds, someone dies from a myocardial infarction. In this paper, heart attack is predicted with the help of dataset sourced from UCI Machine Learning Repository. The dataset analyses 13 attributes of 303 patients. The categorization method of Data Mining helps predict if a person will have a heart attack based on how they live their lives. An empirical and statistical analysis of different classification methods like the Support Vector Machine (SVM) Algorithm, Random Forest (RF) Algorithm, K-Nearest Neighbour (KNN) Algorithm, Logistic Regression (LR) Algorithm, and Decision Tree (DT) Algorithm is used as classifiers for effective prediction of the disease. The research study showed classification accuracy of 90% using KNN Algorithm.

Keywords: Heart attack · Machine Learning · Datamining algorithms · Prediction

1 Introduction

Cardiovascular diseases (CVD) are on the rise internationally, with figures indicating that heart attacks and strokes contribute to 85% of global mortality rates. Furthermore, related data studies show that non-communicable diseases have resulted in approximately 17 million premature deaths among individuals under the age of 70 and is categorized under the domain of mortality caused by CVDs. The distinguishing feature of a CVD from a cardiac arrest is that a heart attack occurs suddenly, usually resulting in instant mortality, but CVDs build gradually over time and produce a life-or-death situation over time. Several studies have identified numerous behavioral, physical, and psychological risk factors that contribute to heart disease and related illnesses.

A heart attack usually happens when the flow of blood to the heart is obstructed in one or more coronary arteries. A buildup of cholesterol and other particles in the heart causes the obstruction. Data mining techniques are routinely used to forecast heart at tacks. Medical practitioners can use mining techniques to anticipate disease early on, allowing patients to receive timely care.

S. Aurelia et al. (Eds.): ICCSST 2023, CCIS 1973, pp. 351–362, 2024.
https://doi.org/10.1007/978-3-031-50993-3_28

A data mining-based heart disease prediction algorithm can assist medical practitioners in diagnosing heart disease status obtained from the patient's prior health records. Techniques for predicting heart attacks using machine learning algorithms, such as classification techniques, are also popular these days. Machine learning models are designed to train and forecast the occurrence of heart attack using Random Forest Algorithm, Support Vector Machine (SVM) Algorithm, K-Nearest Neighbor (KNN) Algorithm, Logistic Regression Algorithm, and Decision Tree Algorithm.

The need of the hour is for efficient, accurate, and early-stage identification of cardiac problems. Although the use of technology in numerous health issues has transformed the healthcare world's outlook on illness detection strategies, persistent work in creating methods for early intelligent detection of heart failure difficulties is still required.

2 Literature Review

The main focus of [1] was to predict heart disease using different data mining algorithms. Classification algorithms, association rule mining, and clustering were the mining techniques used to indicate the heart disease rate from the data set. MAFIA, C4.5 algorithm, and K-mean clustering are the different algorithms used. Various tasks in volve preprocessing the medical data set to remove the irrelevant data using K-Mean clustering. The Maximal Frequent Itemset Algorithm is then applied to the clustered data to find the most frequently occurring pattern. The C4.5 algorithm is applied to the most frequent way to derive a decision tree. A decision tree is then applied to find the occurrence of heart diseases in the person. Medical data sets containing different attributes are used.

The major challenge faced is selecting the appropriate attributes for the data set, which is enormous. 94% accuracy is obtained with the K-Mean-based Mafia and ID3 and C4.5 algorithms compared to the K-Mean method with MAFIA and K-Mean method with ID3 alone. The future scope included in this paper is to predict Heart disease on real-time datasets and the efficiency comparison of algorithms used here with other data mining algorithms.

The objective of [2] is to use various data classification techniques to predict heart attack symptoms and compare the accuracy of each to find the most efficient one. Techniques used involve the Tanagra tool to classify the preprocessed data, and unsupervised machine learning methods like K-Mean, K-Nearest Neighbor (Knn), and Entropy-based clustering are also done. Tanagra tool, a data mining software, is used to retrieve the data required for the experiment. Algorithms such as KNN, Mean, and Evaluation entropy are applied to the data set to compute the performance of each algorithm based on the test output accuracy and efficiency. The output of each training set is recorded based on which the accuracy is measured.

UCI Machine Learning website is the source of the dataset used in [2]. The dataset consists of 3000 records for 14 attributes. Data available is heterogenous, and certain social constraints also apply to them. The Entropy based Mean is found to be the mostaccurate one among other algorithms used. It gives 82.9% accuracy, and the time taken to obtain the result is also less than other methods.

Existing prediction systems available for heart disease prediction are not capable enough to make intelligent early-stage clinical decisions. The method proposed in [3] attempts to create a prototype that is better equipped to extract hidden patterns and relationships from a historical database. [3] mainly emphasizes that vast amounts of data are not mined enough to make accurate predictions, uncompromising on two parameters, accuracy and efficiency. Hadoop Framework with Hadoop Distributed File System and Naïve Bayes Classification Method are the algorithms used. Tasks per formed involves the Data set collection from internet sources. A data warehouse is maintained using the HDFS - Hbase; Apache's Mahout project was used to train the dataset; pattern identification is done using the Naïve Bayes technique. The Cleveland Heart Disease Database was used for this analysis. The patterns have been studied and analyzed using an experiment dataset, which has been praised for its scalability and high quality. However, [3] notes that using the evolutionary algorithm to improve prediction accuracy, and Big Data extraction was also noted as challenge. In [4], the authors conduct experiments and surveys to see if surveillance systems exhibit any form of self-control when exposed to unanticipated dangers. [4] There are five main indicators used in diagnosis. Data analysis methods like Naive Bayes, support vector machines, random forest, decision tree, and linear discriminant analysis have all been applied to medical records in the diagnosis of cardiovascular disorders. In addition to allowing doctors to foresee the conditions and risk factors associated with coronary heart disease, this diagnosis can also help them anticipate the illness's various episodes and how they might be related to different segments of the population. Data gathered through some of the aforementioned ways or from other sources form the basis of computer-aided diagnosis methodologies. In one way, a confusion matrix is compared against accuracy, specificity, sensitivity, and F1 score measures. We noticed that the KNeighbors classifier performed better in the ML technique after applying data preprocessing to the dataset's 13 features. Compile real-time data from the user using sensors and other monitoring devices. Dependency of people using smart healthcare applications has increased tremendously. Because there is a large amount of data related to diseases, physicians can benefit from using data mining techniques and procedures in various situations.

It is important to find a heart attack as soon as possible because waiting can cause serious harm to the heart muscle, called the myocardium, which can lead to illness and death. [5] Coronary Heart Disease (CHD), lung diseases that last a long time, cancer, and diabetes are all getting more common. They looked at 25 things that were linked to the symptoms of a myocardial infarction in [5]. Using data mining techniques, decision tree, and random forest algorithms, they achieved a success rate of 96%. Researchers made a pilot model by putting together the data gathered from patients in a clinical setting with Acute Myocardial Infarction who were admitted to more than one hospital (AMI).

A series of questions was proposed for use in a mobile phone application. The system uses wearable sensors to get live data sourced from the user and send the information to their physician. When a person has a heart condition, their cell phone automatically finds out where they are by using GPS, GSM Cell-id, or Wi-Fi. It then transmits audio and text messages to both their physician and other emergency contacts that have been incorporated into it. Because there aren't many external devices, it can be pricey for most people and hard to use for those who aren't as tech-savvy. Using swarm algorithms to choose features worked well. Some studies classify cardiovascular disease. Shao et al. [6] looked into how heart disease can be categorized using hybrid intelligent modeling systems. This study looks at 13 risk factors to predict heart disease. This study comes up with a new hybrid method for evaluating several risk factors. MAR, LR, and ANN are all parts of this hybrid structure (ANN). LR and MAR bring down the values of risk variables that are encoded. With the rest of the encoded factors, a neural network is trained. In simulations, the hybrid technique does better than the single-stage neural network [6]. The algorithms Naïve Bayes and Decision tree were tested to see how well they could predict heart disease. The Naive Bayes method was less accurate (82.35%) than the Decision Tree method (98.03%) [5]. Particle Swarm Optimization (PSO) was used along with particle techniques (Random Forest, AdaBoost, and Bagged Tree) to improve their ability to predict what would happen [7]. Heart log is made up of 14 attributes and 270 samples from the UCI database [7]. Depression following a myocardial infarction makes stress riskier, it seems. Researchers showed that persons with cardiovascular illness are more likely to be depressed. If you're depressed after a myocardial infarction, you're more likely to have another one within two years, according to research [8]. Many causes can cause heart disease, including stress, anxiety, and despair. These items increase catecholamines, which impact metabolic pro cesses, blood pressure, and heart rate [9]. Data mining has three basic forms, per Reference [11]. They're databases, AI, and statistics. Part of AI are self-learning machines. Even though each axis has a clear meaning, data mining is challenging to define in a single sentence. The answer apparently comes from Reference [10], which says, "Data Mining is the non-trivial process of extracting meaningful information from existing data." This is why this knowledge is reflected in the data as patterns that can solve specific environmental concerns. C.V. Aravinda [12] proposes a method to help medical personnel make accurate diagnosis and prognoses of cardiac disease while also giving the most effective treatment alternatives. Deep learning techniques such as neural ne works, random forests, and decision tree classifiers are used to predict heart illness using sensor data. An experiment suggests that cardiac illness can be predicted with 90% accuracy.

The chances of having congenital heart disease in newborn babies are detected using Weighted SVM, Weighted Random Forest and Logit by examining the various features of women during and before pregnancy [13]. The study was conducted using the imbalanced data obtained from the survey and the weighted Support Vector Machine out performed the other two followed by the Logit. The research [14] focused on finding the key features that cause heart attacks and the correlation between them. A new dataset with prominent features was built and used different classifiers such as Naïve Bayes, Random Forest, Function SMO and AdaBoostM1 to compute the results. All the classifiers provided the constant result irrespective of the variations in each method. AdaBoostM1 is considered to be the best among the 4 classifiers considered in [14].

Eleven Machine Learning classifiers were used to provide a prediction of the existence of coronary heart disease. The research focused [15] on collecting the dataset and classifying it into "normal" and "diseased" labels. The dataset was trained by performing data cleansing to remove missing and inconsistent data. Exploratory Data Analysis (EDA) was performed to check the correlation between different attributes in the dataset. The 11 algorithms were then applied, and the model was evaluated for accuracy. Random Forest showed the highest accuracy of 96% in [15].

This research mainly focused [16] on using an automated machine learning algorithm for cardiovascular disease (CVD) prediction targeting the data of a client bank's participants. The proposed ML-based "AutoPrognosis" model which consisted of techniques such as data imputation, feature processing, classification, and calibration algorithms improved the risk prediction in contrast to the conventional approaches.

3 Problem Statement

Cardiovascular Diseases (often referred to as CVD) is regarded as one of the primary factors of mortality around the world. Consequently, the obligation to anticipate life style patterns that could lead to heart attacks may, to some extent, keep a person aware of whether or not they are safe on the radar.

The classification strategy that is part of Data Mining helps with forecasting the likelihood of a person having a heart attack based on the lifestyle choices that they make. A number of different classification algorithms, including Random Forest Algorithm, Support Vector Machine (SVM) Algorithm, K-Nearest Neighbor (KNN) Algorithm, Logistic Regression Algorithm, and Decision Tree Algorithm, are subjected to a comparative and quantitative research study. This study aims to evaluate the degree to which each of these algorithms accurately predicts the occurrence of a heart attack.

4 Data Mining Techniques

Random Forest, a type of supervised machine learning algorithm that uses several classifiers to handle complicated problems and improve the model's overall performance. The model's performance is improved by averaging the results of many decision trees made from different parts of the dataset. More trees are added to make the model more accurate and keep it from being too accurate. By picking samples from the dataset, the training set decision tree is made. The outcome with the most votes will be the final prediction. This algorithm is better than other algorithms because it can accurately predict the outcomes of large datasets in less time while keeping a high level of accuracy. The Random Forest Algorithm works with many different kinds of data. The heart dataset has been cleaned up and given a number of features. Sample data is then divided into segments called "train" and "test," which are evaluated based on a set of criteria. The training data is then used with a Random Forest Classifier to classify the results of the test set. You can check how well the model works by comparing what was expected to what actually happened. So, we can find out how well the model works in general at predicting heart attacks.

Support Vector Machine, SVM algorithm is a machine learning technique under "supervised machine learning." Using this method, a decision boundary or hyperplane is made in a space with N dimensions. The line it makes between the points of data is clear. For the SVM algorithm, the best results come from small datasets. The heart dataset, which was made with a support vector machine, was used to check how well our model worked. A dataset is taken and an SVM model is used on it to try to predict heart attacks. All thirteen characteristics are seen as independent variables, or set X, which are used to make a prediction, or set Y. There are two different parts of the dataset: the training and testing dataset. The dataset is said to be scaled or standardized when the mean is taken out. When differences are kept to one unit or less, this is called standardization. When the data has different-sized attributes, rescaling them can be use fulfor a number of machine learning techniques. Standard Scaler, a Python function, is used to help with the scaling process. In an SVM classifier, the "kernel" parameter is treated as if it were data that could be broken up in a linear way. Kernel functions are used to solve SVM-constrained optimization so that the needed dot products can be made. Based on the test data for the combined 'X' variable, it is possible to guess what will happen with the dependent variable 'Y'.

K-Nearest Neighbor, KNN is a simple algorithm for machine learning that uses the supervised learning method. The K-NN classifier assumes that new data or cases are the same as those that were already known. The new instance is then put in the category that fits it best out of the ones that are already there. It stores all the data that can be accessed and groups new data pieces together based on how similar they are. The KNN algorithm quickly puts new data into the right category when it comes in. The da taset is only saved when KNN is being trained. When new information comes in, it is put into a category that makes sense for it. People compare the KNN algorithm to the most accurate models because it can make very accurate predictions. The first step is to count how many neighbors K there are. Find the Euclidean distances between each of the K places that are close to each other. Find the K neighbors who live closest to you by using the Euclidean distance. Before putting the new data points in the category with the most neighbors, we have to count how many data points there are in each category based on these k neighbors.

Logistic Regression: An example of a supervised learning algorithm is the logistic regression algorithm. The principle objective of this algorithm is to estimate the likeli hood of a binary outcome. The method is mostly about figuring out how likely an event, outcome, or observation is by putting it into one of two categories. These are the most important parts of the algorithm. This method creates a regression model that identifies data as belonging to the "1" category and predicts the likelihood of a given set of data. This approach can be called the method's "fundamental goal." When a dependent target variable is categorical, as in this case for a dataset predicting heart attacks, the output variable, denoted by the letter y, predicts whether or not a person with a certain set of characteristics is likely to have a heart attack.

Decision Tree is a non-parametric tree-like structure utilized in regression and classification problems. Decision trees enhance the predictability, clarity, and precision of forecasts. Because they can handle both classification and regression, the tools are effective at fitting non-linear relationships. Their divisions serve as examples of wise decision-making. It contains leaf, branch, and root nodes. The node at the tree's root. Decision nodes are reached via branching from the root node. Leaf nodes indicate all dataset outcomes. A decision tree is utilized to illustrate techniques with conditional control statements. Trees of decisions may be categorical or continuous. To aggregate target variables, a decision tree based on categorical variables is utilized. In trees with continuous variables, there exists a fixed objective variable. Start by examining the root node, S, which contains all the data. Utilize the Attribute Selection Measure to deter mine the optimal dataset attribute (ASM). Create optimal S subgroups. Create the best node in the decision tree. Create a decision tree from the subsets of the dataset. Continue until you can no longer differentiate the nodes, at which point you can designate the final node as the leaf node.

The Naïve Bayes Algorithm is a supervised machine learning technique that can be used to do classification tasks. Naïve Bayes uses the Bayes theorem to figure out how likely something is without any other information. When a dataset is given to Naive Bayes, it is turned into frequency tables, the likelihood of the given characteristics is calculated, and the Bayes Theorem is used to estimate the likely outcome. This method puts pieces of text into groups. Preparing the data for a heart attack prediction model makes it easier to put things into categories. The model is trained with 80% of the training dataset, and the other 20% is used to test how well it works. The data is then "normalized" to make sure that predictions are the same and accurate. After preprocessing, the training set fits the Naive Bayes model, and the prediction function predicts what will happen with the test set. By comparing the expected output to the actual data, you can tell if the model is right or wrong. Naive Bayes can make accurate predictions from large data sets.

5 Methodology

The dataset used for the prediction of Heart Attack is sourced from UCI Machine Learning Repository. The dataset considers 13 attributes for 303 patients. All the attributes considered in the dataset are lifestyle influences for the probability of the occurrence of a heart attack. The dataset considered does not contain any missing values and hence a majority of the pre-processing steps can be skipped. Heart attack prediction can be done by using various classification-based Machine Learning techniques. The classification techniques demonstrated in this comparative study include Random Forest, Support Vector Machine (SVM), K-Nearest Neighbor (KNN), Logistic Regression, Decision Tree classification, and Naïve Bayes Classification Algorithms. Table-1 Gives the attributes of dataset considered for the experimental purposes.

The attributes used in the dataset are represented in Table 1. Consider the heart dataset that is implemented using the six algorithms. The algorithms are applied separately to the data set.

A detailed description of each of the 13 attributes used in the dataset are explained here:

Index 0 indicates the age of the person, index 1 indicates the gender of the person, index 2 indicates the chest pain type, index 3 indicates the resting blood pressure value, index 4 indicates serum cholesterol, index 5 indicates the fasting blood sugar value, index 6 indicates the resting electrocardiographic value, index 7 indicates the maximum heart rate achieved, index 8 indicates the exercise induced angina value, index 9 indicates the old peak value, index 10 indicates the slope of the peak exercise ST segment, index 11 indicates the count of major vessels, index 12 indicates the type of defect, and index 13 indicates the target value. The detailed Methodology shown in Fig. 1. The basic step for all algorithms is to import the dataset into the Google colaboratory environment. The next step is to extract the independent and the dependent variables. The independent variables are considered as set X using which a prediction set Y i.e., the output is predicted. In

Table 1. Heat attack dataset attributes.

Index as per Dataset	Attributes Considered
0	AGE
1	SEX
2	CP
3	TRTBPS
4	CHOL
5	FBS
6	RESTECG
7	THALACHH
8	EXNG
9	OLDPEAK
10	SLP
11	CAA
12	THALL
13	OUTPUT

this case, the 13 independent variables are the determining factors to predict whether a person is likely to have a heart attack or not. The next step is to split the dataset into two sets namely the training and testing set. The training set contains past records. The testing set contains the data that is going to be applied to the algorithm model. During the test phase, the heart attack prediction dataset is used to estimate how well the model has been trained to identify the occurrence of heart attack. The next step is to scale the features, and normalize them to a standard scale. Rescaling data attributes with different scales is useful for many machine learning techniques. This is sometimes called "normalization," and the properties are rescaled to fall between 0 and 1. This is done with the Standard Scaler Python function. The next step is to fit the models for each of the 6 algorithms which are represented in the below Table 2.

Table 2. Model Fitting Parameters for 6 Algorithms.

Model	Algorithm Parameters
K-Nearest Neighbour	neighbours=5; metric=minkowski
Naive Bayes	lassifier=GaussianNB
SVM	Kernel=linear (linearly separable data)
Logistic Regression	random_state=42
Random Forest	criterion=entropy; N_estimators=14
Decision Tree	criterion=entropy

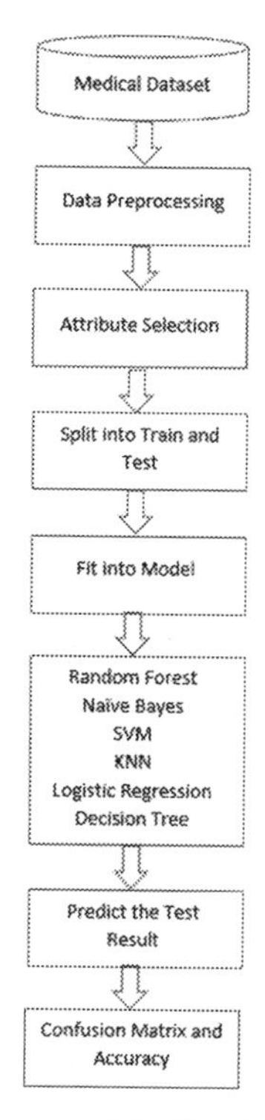

Fig. 1. Methodology used for Heart Attack Prediction.

At last the output for the dependent variable 'Y' is predicted based on the test data ('X_test'=20% of the dataset). The training dataset 'X' (X_train which constituted 80% of the heart dataset) was used to train the model. The actual and predicted values of the dependent output variable ('Y') are the categorization of the 'y-test' and 'y_pred' values 10

A confusion matrix (CM) is computed for the data set which represents a clear categorization of the number of samples in the dataset for which the actual vs predicted values are (0,0), (0,1), (1,0), and (1,1).

The confusion matrix obtained for each of the 6 algorithms for the Heart Attack prediction dataset is represented in the table format that follows. The interpretation of the matrix for each of the matrix positions namely TN = number of persons who do not have any abnormality in attribute values and will not have a chance of heart attack; FP = number of persons who can be misdiagnosed with a heart attack prediction diagnosis but are actually less likely to have an attack; FN = number of persons who can be misdiagnosed with a heart attack prediction diagnosis and are actually more likely to have an attack; TP = number of persons who have an abnormality in the attribute values and can be positively predicted to have a heart attack (Table 3).

Table 3. Model Fitting Parameters for 6 Algorithms.

Model	(0,0)	(0,1)	(1,0)	(1,1)
	TN	FP	FN	TP
K-Nearest Neighbour	27	2	4	28
Naive Bayes	26	3	5	27
SVM	25	4	4	28
Logistic Regression	25	4	5	27
Decision Tree	26	3	8	24
Random Forest	26	4	3	29

6 Results

The accuracy of each model was assessed. K-NN achieved the highest total accuracy of 90.16% followed by Random Forest with 88.52%. The Naive Bayes and SVM models scored the next-highest total accuracy of 86.88%, followed by Logistic Regression at 85.24%. The Decision Tree had the lowest total accuracy of 81.96%.

The accuracy metrics for six algorithms computed based on 13 individual attributes are represented in the figure below (Fig. 2, Fig. 3).

Method	age	sex	cp	trtbps	chol	fbs	restecg	thalachh	exng	oidpeak	slp	caa	thall
Random Forest	63.93443	55.73771	83.60656	50.81967	49.18033	52.45902	52.45902	65.57377	70.4918	70.4918	72.13115	72.13115	78.68853
Naive Bayes	62.29508	55.73771	83.60656	59.01639	52.45902	52.45902	52.45902	68.85246	70.4918	67.21312	72.13115	72.13115	80.32787
Logistic Regression	55.73771	55.73771	83.60656	59.01639	54.09836	52.45902	52.45902	72.13115	70.4918	68.85246	72.13115	72.13115	78.68853
SVM	62.29508	55.73771	68.85246	52.45902	52.45902	52.45902	52.45902	72.13115	70.4918	68.85246	72.13115	72.13115	78.68853
K-Nearest Neighbour	50.81967	55.73771	65.57377	45.90164	60.65574	55.73771	47.54098	72.13115	70.4918	65.57377	72.13115	72.13115	78.68853
Decision Tree	55.73771	55.73771	83.60656	49.18033	55.73771	52.45902	67.21312	63.93443	70.4918	68.85246	72.13115	72.13115	78.68853

Fig. 2. Accuracy of Model for Individual Attributes

On the basis of the computed confusion matrix, the accuracy of the heart attack pre diction for each of the 6 algorithms is represented in the below table format (Table 4).

Table 4. Accuracy Metrics for 6 Algorithms.

Model	Accuracy
K-Nearest Neighbour	90.163934
Naive Bayes	86.885246
SVM	86.885246
Logistic Regression	85.245902
Random Forest	88.52
Decision Tree	81.967213

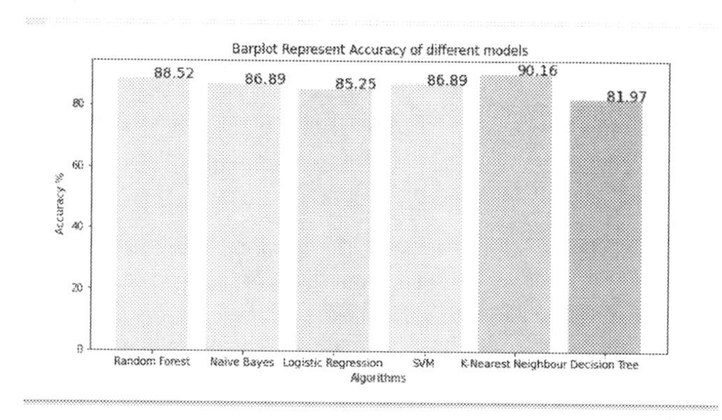

Fig. 3. Accuracy Depiction Chart for 6 Algorithms

7 Conclusion

Analysed the problem of summarizing different algorithms in data mining. The main focus was on using six different classification algorithms for the effective prediction of a heart attack. An existing dataset was taken for this prediction while focusing on parameters such as nil missing values and a good combination of lifestyle influencing attributes that contributed to the existence of cardiovascular ailments. Based on the computed accuracy rates of the algorithms used, KNN Algorithm with a neighbour set (k=5) gave the highest accuracy of pre diction. This can further be enhanced and expanded by combining multiple algorithm together to form a best hybrid algorithm. Although most lifestyle habits have been incorporated as attributes in the dataset used here, it can be further expanded by adding attributes such as smoking, drinking, unhealthy eating habits, obesity.

References

1. Banu, M.N., Gomathy, B.: Disease predicting system using data mining techniques. Int. J. Tech. Res. Appl. **1**(5), 41–45 (2013). e-ISSN: 2320-8163

2. Jaya Rama krishniah, V.V., Chandra Sekar, D.V., Ramchand, K.R.H.: predicting the heart attack symptoms using biomedical data mining techniques, **1**(3) (2012). ISSN – 2278–1080
3. Ghadge, P., Girme, V., Kokane, K., Deshmukh, P.: Intelligent heart attack prediction system using big data. Int. J. Recent Res. Math. Comput. Sci. Inf. Technol. **2**(2), 73–77 (2015)
4. Srinivas, K., Rao, G.R., Govardhan, A.: Analysis of coronary heart disease and prediction of heart attack in coal mining regions using data mining techniques. In: 2010 5th International Conference on Computer Science & Education, pp. 1344–1349 (2010). https://doi.org/10.1109/ICCSE.2010.5593711
5. Raihan, M., Mondal, S., More, A., Boni, P.K., Sagor, M.O.F.: Smartphone based heart attack risk prediction system with statistical analysis and data mining approaches. Adv. Sci. Technol. Eng. Syst. J. **2**(3), 1815–1822 (2017). https://doi.org/10.25046/aj0203221
6. Shao, Y.E., Hou, C.D., Chiu, C.C.: Hybrid intelligent modeling schemes for heart disease classification. Appl. Soft Comput.Comput. **14**, 47–52 (2014)
7. Yekkala, I., Dixit, S., Jabbar, M.A.: Prediction of heart disease using ensemble learning and particle swarm optimization. In: 2017 International Conference on Smart Technologies for Smart Nation (SmartTechCon), pp. 691–698. IEEE (2017)
8. Lépine, J.P., Briley, M.: The increasing burden of depression. Neuropsychiatric Dis Treat 7: 3 (2011)
9. Gielen, S., Schuler, G., Adams, V.: Cardiovascular effects of exercise training: molecular mechanisms. Circulation **122**(12), 1221–1238 (2010)
10. Marti, K.: Stochastic Optimization Methods, vol. 3. Springer, Berlin (2005)
11. Ahmet, I., Lakatta, E.G.: Wellbeing-of-The Right Heart Forecasts-The Fate-Of-The Left Heart During Heart Attack. J. Cardiac Fail. **26**(10), S29 (2020)
12. Aravinda, C.V., et al.: A deep learning approach for the prediction of heart attacks based on data analysis. In: Deep Learning for Medical Applications with Unique Data. Academic Press, pp. 1–18 (2022)
13. Luo, Y., Li, Z., Guo, H., Cao, H., Song, C., Guo, X., et al.: Predicting congenital heart defects: a comparison of three data mining methods. PLoS ONE **12**(5), e0177811 (2017)
14. Snigdha, A.R., Tasnim, S.N., Miah, K.R., Islam, T.: Early prediction of heart attack using machine learning algorithms. In: 2nd International Conference on Computing Advancements (ICCA 2022), March 10–12, 2022, Dhaka, Bangladesh. ACM, New York, NY, USA (2022)
15. Hassan, C.A.U., et al.: Effectively predicting the presence of coronary heart disease using machine learning classifiers. Sensors **22**(19), 7227 (2022)
16. Alaa, A.M., Bolton, T., Di Angelantonio, E., Rudd, J.H., Van der Schaar, M.: Cardiovascular disease risk prediction using automated machine learning: a prospective study of 423,604 UK Biobank participants. PLoS ONE **14**(5), e0213653 (2019)
17. Dataset considered is from UCI Machine Learning. https://archive.ics.uci.edu/ml/datasets/Heart+Disease

Healthcare Data Analysis and Secure Storage in Edge Cloud Module with Blockchain Federated Sparse Convolutional Network++

R. Krishnamoorthy(✉) and K. P. Kaliyamurthie

Bharath Institute of Higher Education and Research, Chennai, India
r.krishcse@gmail.com

Abstract. In current scenario, most difficult requirements are managing massive amount of multimedia data generated by Internet of Things (IoT) devices, which can only be managed with the cloud. The intelligent Edge Cloud computing technology operates in a distributed environment and emerges as a solution. This research aims to use Edge Cloud computing to reduce latency in e-healthcare. This study proposes a novel method for using machine learning to analyze healthcare data and storing it in an edge cloud module. The input is monitored healthcare data that is collected, processed, and analyzed with the help of a blockchain-federated sparse convolutional network++, which also improves network security. The malicious attacks are then identified and malicious data is stored using a centralized edge cloud computing module and authentication process. The experimental analysis is conducted on a variety of healthcare datasets and is based on a security analysis of the data in terms of data transmission rate, computation cost, communication overhead, random accuracy, mean average precision (map), and specificity. The proposed method achieved a data transmission rate of 65%, a computation cost of 51%, a communication overhead of 75%, random accuracy of 85%, mean average precision (map) of 63%, and specificity of 79% .

Keywords: Healthcare data analysis · Machine learning · Edge cloud module

1 Introduction

Healthcare IoT applications must rely on cloud computing as a platform. It is used to store, process, and analyze IoT device-generated healthcare data. However, there are limitations to cloud computing. A standard number of hops on way to target server, which causes propagation delay, is one of these factors. i.e. the time the packet spends traveling over a fiber or wire) [1]. The delay in service will also be exacerbated by response time of cloud server replication as well as data replication from one cloud data center to another. Another factor that impedes network response is an excessive workload on cloud data center. The body sensor network's primary requirements include secure storage of healthcare data as well as improved service between cloud and users. Large amount of data produced by IoT reduces sensor energy. Medical organizations must manage a large amount of data, which may necessitate use of numerous servers

S. Aurelia et al. (Eds.): ICCSST 2023, CCIS 1973, pp. 363–378, 2024.
https://doi.org/10.1007/978-3-031-50993-3_29

for data analysis [2]. Due to high latency, the medical team's services are impacted by large amount of data produced by healthcare IoT. Data must travel multiple hops, consuming a lot of network bandwidth, between the cloud data centers and the Internet of Things or end users. The physiological state of a patient changes over time, necessitating real-time monitoring [3]. Specialized infrastructure is required. IoT is a new field of study that opens the door to numerous technological advancements. Numerous fields have benefited greatly from this technology's innovations. IoMT in which healthcare is application domain, is a significant subfield of IoT technology [4]. In order to provide the highest possible level of healthcare to both patients and medical professionals, an IoT-based healthcare system makes use of IoT applications, such as wearable devices and implanted medical devices (IMDs). A layer called Edge connects a device to cloud. As a result, idea behind EC is that it makes use of and makes it easier for computing operations to be done close to Edge Devices [5]. Large businesses like Mutable, Swim.ai, MobiledgeX, Affirmed Networks, and Edge gravity have invested a lot of money in EC to improve their operational efficiency, speed up processes, and guarantee unmatched availability. Naturally, there are inherent dangers associated with every new technology that must be mitigated [6].

2 Literature Review

Edge computing and fog computing [7] are currently regarded as among most effective methods for analyzing data sources for a variety of healthcare application scenarios. Additionally, multi-access physical monitoring methods are currently using mobile edge computing, an emerging technology. The major drawback of the aforementioned technologies, despite their high ratings and promising outcomes, is the high latency and accuracy required to transfer health datasets over the network. In terms of reliability and energy consumption, the application of mathematical computation based on neural networks to the processing of health datasets is inefficient [8]. Multi-access physical monitoring systems have been subject of extensive clinical research with a primary focus on lowering human health risks. Although data-driven methods are utilized to detect multimodal changes of physiology that are in practice at present time [9], few studies suggested that telehealth methods are practiced, which does not lead to positive outcomes.

2.1 Healthcare Data Analysis and Storage Based Existing Techniques

Dyna-Q is used in [10] to improve authentication speed of study in [11]. In this case, an Edge node uses the Q-function to select the test threshold based on recent spoofing detection accuracy as well as frequency. DL PHY-layer authentication technique is proposed in work [12]. Author [13] proposes a logistic regression-based PHY-layer authentication model that uses multiple locations to measure signal RSSI. Incremental aggregated gradient method reduces computational burden. A dynamic RA approach based on reinforcement learning as well as a GA was proposed in work [14]. It continuously monitors network traffic, gathers data about every node, manages incoming tasks, and distributes them equally among available nodes utilizing DRA technique. It works well in FC environments with real-time systems, like the smart healthcare system. Work [15] listed available nodes as well as distributed equally incoming processes to every node in order using Round Robin (RR). It is straightforward, simple, and simple to use. However, one of the servers may become overloaded and crash if the processing capacities of the servers are different. Weighted Round Robin (WRR) was used in the resource allocation process in work [16]. In terms of cyclical assigning, it is similar to the RR, but it differs in that node with highest weight will receive most requests. A weight is assigned to each server in WRR based on its capacity. Author [17] achieved load balancing by employing Least Connection (LC). It modifies its data on a regular basis depending on the number of connections and selects the node with the fewest active transactions. By assigning the appropriate source to the incoming process, work [18] proposed a model to cut down on the amount of time spent communicating and processing. To achieve load balancing, the authors of [19] use best SNR as well as distributed alpha method to measure load on every node. It compares SNR and reconstructs a set of events from a series. However, fact that this strategy focuses on delay of wireless communications is a drawback. Work [20] employs Reinforcement Learning method to determine most energy-efficient scheduling policy for nonhomogeneous networks. In [21], the authors increased the network resource utilization by employing Multi-agent Reinforcement Learning. Work [22] proposed a policy to reduce IoT service latency. They wanted to assign each task based on how quickly people responded. The tasks are categorized as light and heavy in the policy proposal in [23]. The task will be assigned to a particular fog node if its response time falls below the threshold. Table 1 provides a summary of the above-mentioned existing technique.

Table 1. Existing Literature Survey

Reference	Proposed Techniques	Advantages	Limitations
[15]	a real-time data gathering approach for fog computing	Information gathered in real-time. Calculated transmission and calculation delays	There is no formula for calculating network delay
[16]	Hipster: in order to manage urgent concerns	Utilizing network and computation delay to increase efficiency	There is no discussion of large data latency, and no mechanism is also developed for transmission delay
[18]	iFogStor: GAP for computing in the fog as well as a heuristic method	Efficiency of the system is raised. They also fix the problem with the time-sensitive data	There is no calculation for the transmission delay

(continued)

Table 1. (*continued*)

Reference	Proposed Techniques	Advantages	Limitations
[20]	On the edge network, fog computing-based content-aware filtering and security services are provided	the task is transferred from faraway sites to the edge end device	The topics of network and computation delay are not covered
[21]	IoT model deployment with QoS awareness using fog computing	deployed an IoT model with QoS awareness	Network and computation delays go without an explanation
[23]	Hermes: NP-hard technique	NP-hard algorithms are used to optimise task data	The cause of the network and calculation delay is unknown
[10]	SPSRP for fog nodes (FNs) and IoT-device	reduces reaction as well as delay times in foggy conditions	Network delay and computation are not covered
[11]	Fog computing deployment	explains the problems that resulted from hardware and software failure	
[12]	IoT-healthcare structure for cloud-fog	explains the problem of large amounts of data. Enhanced network traffic and power performance	There is no discussion of network delay
[13]	A fog-centric schema for data storage	discusses how to store data securely in the cloud	There is no discussion of network delay or transmission
[15]	PSO and CSO techniques	decreases the IoT-Fog-Cloud environment's response time	The topics of network and computation delay are not covered
Proposed scheme	**IMDS**	**Utilize transmission, network, and calculation delay to decrease total latency**	–

3 Materials and Methods

In this section the proposed healthcare data analysis as well as its storage based on ML and edge cloud computing is discussed. The data analysis is carried out with security using blockchain federated sparse convolutional network++ and the storage of data is done in edge cloud computing.

3.1 Blockchain Federated Sparse Convolutional Network++

Layered architectures of our proposed method are compared in Fig. 1. There is a physical layer, a layer for network connectivity, a layer for IoT blockchain cloud, an application layer, and a layer for business in proposed method architecture. ML method is not utilized in proposed system, which uses blockchain-based smart contracts to monitor vital signs of patients. The security concerns surrounding intrusion detection were not discussed by the system. Because it is centralized, the system has a single point of failure and sends medical data to a centralized server, which can compromise the data.

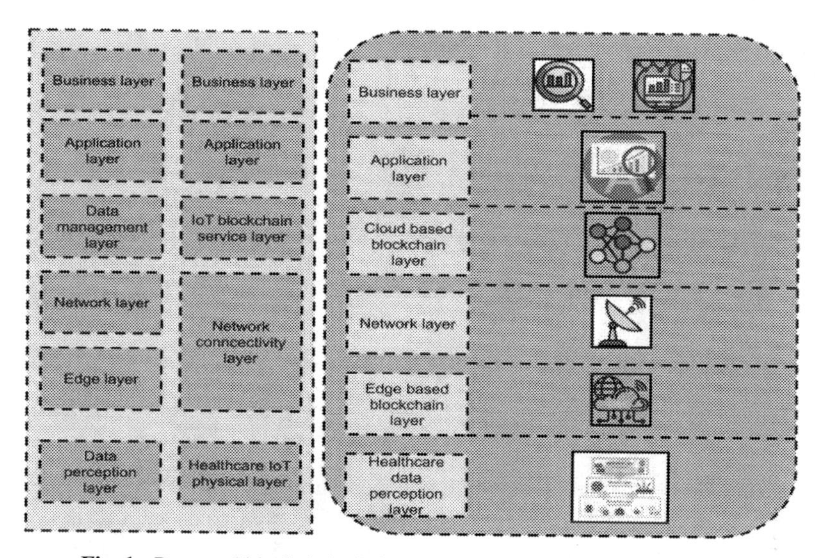

Fig. 1. Proposed blockchain federated sparse convolutional network++

A federated multi-attack edge layer detection method is the proposed system. During the process of sending the weights to the server for aggregation, IDS did not discuss poisoning attacks issue. Layer of edge-based blockchain: IoT gateways make up this. There are some healthcare sensing devices in each gateway. Physical healthcare sensors do not have a global internal protocol; As a result, the gateways were able to support a wide range of network access protocols. The detection of multiple attacks is the job of an Internet of Things gateway. A light IDS was built for the edge server (ES) to normalize their data as well as catch several ANN-based attacks.

It will do this by preventing data from its gateway. The closer the attack resources are, the shorter the intrusion's detection time will be. Because FL method works with smaller data sets, there will also be less computing and processing power available. A further application for these chained blocks will be aggregation as well as averaging. As a result, it cannot be altered or manipulated because consensus algorithms govern its operation. In Algorithm 1, the blockchain federated learning algorithm is described.

Algorithm 1

Input: The IDSFChain node number N, the global round number g, the local epoch number C, local batch size M, number of edge gateways K, and the learning rate n are all given.
Output: updated weights W
Server_Nodel_Update:
Start

Nodel: creates IDSFChain
Nodel: connects to IDSFChainwithip address
For every local edge from 1 to C do
$\mathrm{N} = \mathrm{C} + 1$
End for
For every global epoch $g = 1,2, \ldots$ do
For every Node ∈ IDSFChain N
$W_{g+1}^{K} \leftarrow$ Edge_Nodes_Update (Noden, w_g)
End for

$$W_{g+1} \leftarrow \sum_{k=1}^{k} \frac{nk}{n} W_{g+1}$$

Where: $\mathrm{n} = \sum_{k} nk$
$W \leftarrow W_{g+1}$
End for
End process
For every local epoch $C = 1,2, \ldots$ do
$W \leftarrow W - n\nabla f_k(W)$
Where: $\nabla f_k(W)$ is average gradient on edge local data
End for

This study's architecture creates a novel decentralized control access system that uses blockchain method to distribute and store access control data. In spite of IoTas well as organizational hub nodes, blockchain method will be used by all entities. A copy of the blockchain is required for each node in a blockchain network. The chain has the potential to grow significantly and has already begun. Due to their limitations, the majority of IoT systems will not be able to store blockchain data. Consequently, an administration hub replaces the IoT device exclusion from the blockchain in our architecture. On behalf of IoT devices, this distinct component requests permission data from the blockchain. In addition, the solution incorporates a single shared ledger that compiles a list of each action taken by the access control system. It is a one-of-a-kind contract that can't be taken out of the database. In order to set the access control policy for the system, administrators communicate with contracts.

Consider the convolutional kernel K in $\mathbb{R}^{h \times w \times m}$ where s is the size of convolutional kernel and n is number of output channels, and input feature maps I in $\mathbb{R}^{s \times s \times m \times n}$. We presume that stride is equal to 1 and that convolution is conducted without padding zeros. Then, Eq. (1) provides output feature maps for a convolutional layer $\mathbf{O} \in \mathbb{R}^{(h-s+1) \times (w^{-}-s+1) \times n} = \mathbf{K} * \mathbf{I}$.

$$\mathbf{O}(y, x, j) = \sum_{i=1}^{m} \sum_{u,v=1}^{s} \mathbf{K}(u, v, i, j)\mathbf{I}(y + u - 1, x + v - 1, i) \tag{1}$$

For this, we first use a matrix R ∈ R s × s × m × n to obtain O R J by Eq. (2, 3).

$$\mathbf{K}(u, v, i, j) \approx \sum_{k=1}^{m} \mathbf{R}(u, v, k, j)\mathbf{P}(k, i) \tag{2}$$

$$\mathbf{J}(y, x, i) = \sum_{k=1}^{m} \mathbf{P}(i, k)\mathbf{I}(y, x, k) \tag{3}$$

$$\mathbf{R}(u, v, i, j) \approx \sum_{k=1}^{q_i} \mathcal{S}_i(k, j)\mathcal{Q}_i(u, v, k) \tag{4}$$

$$\mathcal{T}_i(y, x, k) = \sum_{u,v=1}^{s} \mathcal{Q}_i(u, v, k)\mathbf{J}(y + u - 1, x + v - 1, i). \tag{5}$$

so that

$$\mathbf{O}(y, x, j) \approx \sum_{i=1}^{m}\sum_{k=1}^{q_i} \mathcal{S}_i(k, j)\mathcal{T}_i(y, x, k) \tag{6}$$

By counting the multiplicities, we examine the theoretical complexity of our method. Equation (7) provides the multiplicities necessary for original convolution.

$$mns^2(h - s + 1)(w - s + 1) \tag{7}$$

By sparsifying the convolutional kernel, our technique reduces complexity while adding overhead from two matrix decompositions, as shown in Eq. (8).

$$\left(\gamma mn + \sum_{i=1}^{n} q_i\right)s^2(h - s + 1)(w - s + 1) + m^2hw \tag{8}$$

$$\underset{\mathbf{P}, \mathcal{Q}_i, \mathcal{S}_i}{\text{minimize}} \mathbf{L}_{net} + \lambda_1 \sum_{i=1}^{m} \|\mathcal{S}_i\|_1 + \lambda_2 \sum_{i=1}^{m}\sum_{j=1}^{q_i} \|\mathcal{S}_i(j, \cdot)\|_2 \tag{9}$$

$$\text{s.t.} \|\mathbf{P}(\cdot, j)\|_2 \leq 1, j = 1, \ldots, m \tag{10}$$

$$\|\mathcal{Q}_i(\cdot, \cdot, k)\|_2 \leq 1, i = 1, \ldots, m, k = 1, \ldots, q_i \tag{11}$$

Equations (12, 13) give the temporal evolution of these quantities.

$$\begin{aligned}
\rho_i^{S,g}(t+1) &= \rho_i^{S,g}(t)\left(1 - \Pi_i^g(t)\right) \\
\rho_i^{E,g}(t+1) &= \rho_i^{S,g}(t)\Pi_i^g(t) + (1 - \eta^g)\rho_i^{E,g}(t), \\
\rho_i^{A,g}(t+1) &= \eta^g \rho_i^{E,g}(t) + (1 - \alpha^g)\rho_i^{A,g}(t), \\
\rho_i^{I,g}(t+1) &= \alpha^g \rho_i^{A,g}(t) + (1 - \mu^g)\rho_i^{I,g}(t), \\
\rho_i^{H,g}(t+1) &= \mu^g \gamma^g \rho_i^{I,g}(t) + \omega^g(1 - \psi^g)\rho_i^{H,g}(t) + (1 - \omega^g)(1 - \chi^g)\rho_i^{H,g}(t) \\
\rho_i^{D,g}(t+1) &= \omega^g \psi^g \rho_i^{H,g}(t) + \rho_i^{D,g}(t) \\
\rho_i^{R,g}(t+1) &= \mu^g(1 - \gamma^g)\rho_i^{I,g}(t) + (1 - \omega^g)\chi^g \rho_i^{H,g}(t) + \rho_i^{R,g}(t).
\end{aligned} \tag{12}$$

$$V_{\text{teec}}(x_{st}, x_{er}) = \begin{cases} +\gamma \text{ if } x_{st} \neq x_{er}, (s, t), (e, r) \in S, t \neq r, \text{ and } r \in \{(t-1), (t-2)\} \\ -\gamma \text{ if } x_{st} = x_{er}, (s, t), (e, r) \in S, t \neq r, \text{ and } r \in \{(t-1), (t-2)\} \end{cases} \tag{13}$$

probability is given by Eq. (14, 15):

$$\Pi_i^g(t) = \left(1 - p^g\right)P_i^g(t) + p^g \sum^{N} R_{ij}^g P_j^g(t), \tag{14}$$

$$m_k = \frac{\sum_{i=k}^{k+l-1} \psi_i}{l}, k = 1, \ldots, q - l + 1 \quad (15)$$

The expected number of agents in a given region L can be calculated utilizing one-point correlation function k(1) t (x), which quantifies expected agent density at a given position x (16, 17).

$$E(|\gamma_t \cap A|) = \int_A k_t^{(1)}(x)\mathrm{d}x \quad (16)$$

$$E(|\gamma_t \cap \Lambda_1||\gamma_t \cap A_2|) = \int_{A_1} \int_{A_2} k_t^{(2)}(x_1, x_2)\mathrm{d}x_2\mathrm{d}x_1 + \int_{A_1 \cap A_2} k_t^{(II}(x)\mathrm{d}x \quad (17)$$

$$E(|\gamma_t \cap \Lambda_1||\gamma_{t+\Delta t} \cap A_2|) = \int_{A_1} \int_{A_2} k_{t,\Delta t}(x_1, x_2)\mathrm{d}x_2\mathrm{d}x_1 + \int_{A_1 \cap A_2} k_{t,\Delta t}(x)\mathrm{d}x. \quad (18)$$

a differential equation with the form Eq. (19, 20)

$$\frac{\mathrm{d}}{\mathrm{d}t} k_t^{(1)}(x) = -mk_t^{(1)}(x) \quad (19)$$

$$\frac{\mathrm{d}}{\mathrm{d}t} k_t(\eta) = \left(L^\Delta k_t\right)(\eta), - \int_{R^t} a^-(x-y)k_t^{(2)}(x, y)\mathrm{d}y + \int_{R^t} a^+(x-y)k_t^{(1)}(y)\mathrm{d}y. \quad (20)$$

the core model's spatio-temporal correlation function expressed as Eq. (21, 22)

$$k_{t,\Delta t}(x, y) = k_{\Delta t}^{OO}(x, y) + k_{\Delta t}^{O+}(x, y) + k_{\Delta t}^{-O}(x, y) + k_{\Delta !}^{-+}(x, y), \quad (21)$$

$$k_{t,\Delta t}(x) = k_{\Delta t}^{O}(x) \quad (22)$$

As shown by Eq. (23), the following Markovianity property is also met:

$$p(X_{st} = x_{st}|X_{qr} = x_{qr}, s \neq q, t \neq r, \forall(s,t), (q,r) \in V) = p(X_{st} = x_{st}|X_{qr} = x_{qr}, s \neq q, t \neq r, (q,r) \in \eta_{sr}). \quad (23)$$

$$z^g = \frac{N^g}{\sum_{i=1}^{N} f\left(\frac{n_i^{\mathrm{eff}}}{s_i}\right)\left(n_i^g\right)^{\mathrm{eff}}} \quad (24)$$

where Eq. (25–28) is the effective population at patch i.

$$n_i^{\mathrm{eff}} = \sum_{g=1}^{N_G} \left(n_i^g\right)^{\mathrm{eff}} \quad (25)$$

$$\left(n_i^g\right)^{\mathrm{eff}} = \sum_j \left[(1 - p^g)\delta_{ij} + p^g R_{ji}^g\right] n_j^g \quad (26)$$

$$f(x) = 1 + \left(1 - e^{-\xi x}\right) \quad (27)$$

$$n_{j \to i}^{m,h}(t) = n_j^h \rho_j^{m,h}(t)\left[\left(1 - p^h\right)\delta_{ij} + p^h R_{ji}^h\right], m \in \{A, I\}. \quad (28)$$

The following Eq. (29) describes the Gibb's distribution-following prior probability $P(X_t = x_t, \theta)$,

$$P(X_t = x_t) = e^{-U(x_t,\theta)} = e^{\left[-\sum_{s,t}\{V_{sc}(x_{st},x_{qt})+V_{tec}(x_{st},x_{qr})+V_{teec}(x_{st},x_{er})\}\right]} \quad (29)$$

The equivalent edgeless model is written as Eq. (30)

$$P(X_t = x_t) = e^{-U(x_t,\theta)} = e^{\left[-\sum_{c\in C}[V_{sc}(x_{st},x_{qt})+V_{tec}(x_{st},x_{qr})]\right]}. \quad (30)$$

Likelihood function $P(Y_t = y_t|X_t = x_t)$ is given as Eq. (31)

$$P(Y_t = y_t|X_t = x_t) = P(y_t = x_t + n|X_t = x_t, \theta) = P(N = y_t - x_t|X_t = x_t, \theta); \quad (31)$$

Thus, $P(Y_t = y_t|X_t = x_t)$ is given as Eq. (32)

$$P(N = y_t - x_t|X_t, \theta) = \frac{1}{\sqrt{(2\pi)^f \det[k]}} e^{-\frac{1}{2}(y_t-x_t)^T k^{-1}(y_t-x_t)} \quad (32)$$

3.2 Edge Cloud Computing Based Data Storage

Edge Device Layer (EDL), Edge Server Layer (ESL), and Cloud Server Layer (CSL) typically make up architecture of edge computing, as depicted in Fig. 1. Typically, Edge Devices are low-level electronic devices used for sensing, actuating, and controlling at the EDL. One or more microcontrollers (MCUs) running on a single integrated circuit logically control edge devices. Each of the ESL's sublayers can have varying computational capabilities. As we get closer to the CSL, this ability to compute increases.

The proposed four-layer architecture of the fog-enabled Deep Privacy framework is depicted in Fig. 2. The following is a discussion of proposed method specific activity:

(1) Layer of application: Users of medical data that has been detected by sensor devices are focus of this layer. This layer also includes third parties, such as various research organizations. However, in contrast to the other users, who access data from the fog layer, they access sanitized data from the cloud.
(2) Cloud layer It stores data that has been sanitized and preserved for privacy from fog layer. It data may be requested by third parties, such as pharmaceutical companies, publishers, and researchers.
(3) Edge layer The Edge layer was added to minimize power consumption, avoid congestion in network backhaul, and reduce latency for delay-critical realtime applications (such as health emergencies). Fog layer is a control layer that is in charge of recognizing medical data and protecting privacy. There are two main parts to it: MER module for unstructured data recognition and sanitized for data sanitization. It shows that the fog node further sanitizes detected sensitive terms before sending them to cloud for permanent storage. For research purposes, third parties (researchers) have access to this sanitized data via cloud. Reports provide a summary of data received from infrastructure layer. On this layer, prescriptions written by doctors are temporarily saved. Before being sent to cloud layer for permanent storage, data is sanitized. After this, cleaned data is given to other people for research.

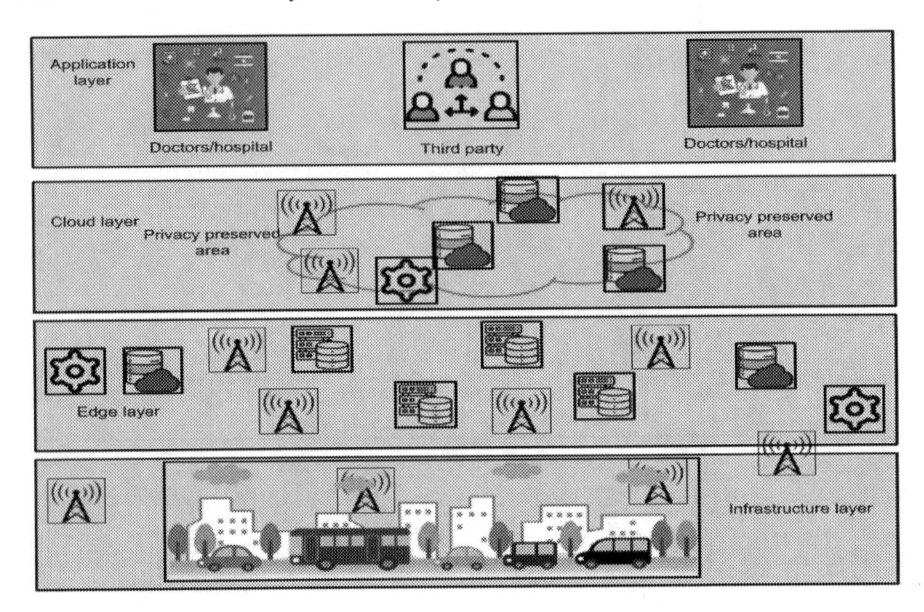

Fig. 2. Proposed Edge Cloud Computing Architecture for data storage

4 Results and Discussion

We talked about the model's performance in this section. Data is transferred from IoT devices to another layer in this model, reaching the cloud through a fog environment. It is calculated how long it takes to travel with data. Data is classified, processed, and sent to end user or cloud according to requirements. We make use of the Python Editor tool in order to complete research task. After the simulation process is completed, the outcome will be displayed. We used a k-fold random forest learning method to divide data set into tenfold here. 70% of the set's data will be used for training, while 30% will be used for testing. This work is implemented on the Python 3.7 platform. In the proposed work, RF method correctly classifies data as normal, low risk, and high risk. 14 s have been needed for the computation.

We tested five distinct configurations of monitoring devices for the purpose of the simulation. The simulations were responsible for evaluating latency, network usage, and RAM consumption. The fog network and its nodes can be simulated using the ifogsim simulator. The ifogsim simulator simulates evaluation of network delay, computation delay, and transmission delay. Physical topologies are programmed using the Java API by this simulator. The topologies that have been updated or modified are stored in the JSON file format. by altering the average size) at a 20 ms data packet arrival. Tables 2 and 3 provide information on the utilized fog device, IoT sensor, and network link.

Description of the set: Over 98% of people in UK are registered with a general practice (GP), which serves as first point of contact for healthcare in UK National Health Service. CPRD receives data from approximately one in ten GP units. One of the largest primary care EHR databases in world.

Table 2. Details of Fog Device Parameters

Type of device	Processing speed (G.Hz)	Ram (MegaBytes)
Edge-device	2.60	2.0
Cloud-server	4.0	4.0

Table 3. Details of Network Link Parameters

Source node_IoT_sensor	Destination node	Latency (ms)
1	Edge-device	45.0
2	Edge-device	45.0
3	Edge-device	50.0
Edge-device	Cloud-server	75.0

Experiments, which were developed by Andras Janosi (M.D.) at Gottsegen Hungarian Institute of Cardiology, Hungary, and others, are carried out with assistance of Cleveland database. Both patient's name as well as their number are kept private.

Table 4. Analysis for various healthcare dataset

Dataset	Techniques	Data transmission rate	Computation cost	Communication overhead	Random accuracy	MAP	Specificity
CPRD	DRA	61	45	71	81	59	73
	WRR	63	48	73	83	62	75
	HAD_BFSCN++	65	51	75	85	63	79
Cleveland	DRA	62	51	75	86	65	82
	WRR	65	53	78	89	68	84
	HAD_BFSCN++	66	55	79	91	69	86

Table 4 shows analysis for various healthcare dataset. Here the dataset compared are CPRD and Cleveland dataset in terms of Data transmission rate, Computation Cost, Communication overhead, Random accuracy, Mean average Precision (map), Specificity.

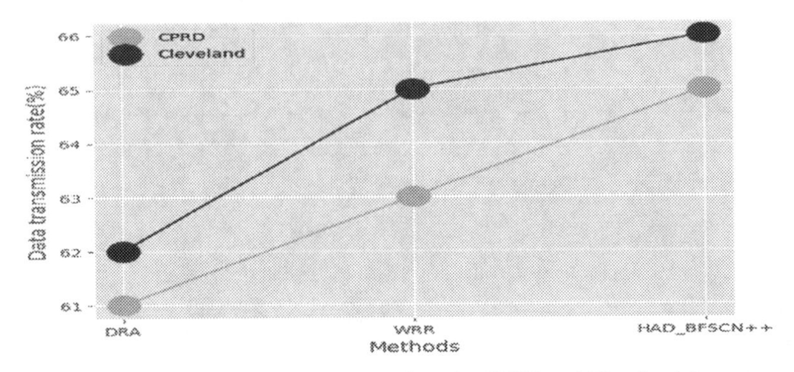

Fig. 3. Data transmission rate comparison for CPRD and Cleveland dataset

From above Fig. 3 the data transmission comparison for CPRD and Cleveland dataset between proposed and existing technique. The proposed technique attained data transmission rate of 65%, existing DRA attained 61%, WRR attained 63% for CPRD dataset, for cleveland dataset the proposed technique attained data transmission rate of 66%, existing DRA attained 62%, WRR attained 65%.

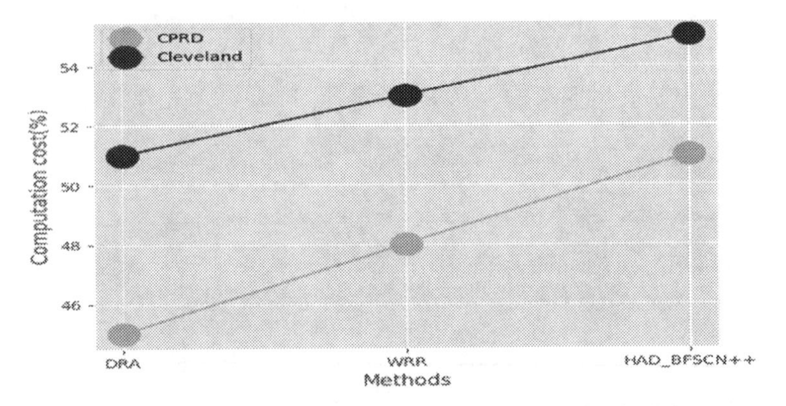

Fig. 4. Computational cost comparison for CPRD and Cleveland dataset

Figure 4 shows analysis of of computation cost for CPRD and Cleveland dataset. The proposed technique attained data computation cost of 51%, existing DRA attained 45%, WRR attained 48% for CPRD dataset, for cleveland dataset proposed technique attained computation cost of 55%, existing DRA attained 53%, WRR attained 51%.

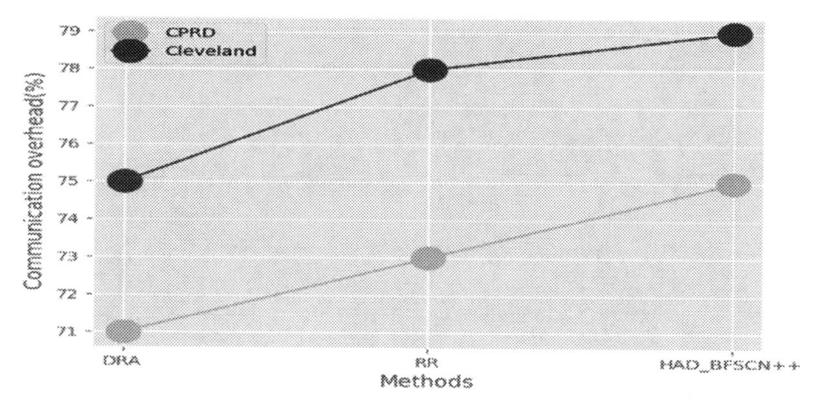

Fig. 5. Communication overhead comparison for CPRD and Cleveland dataset

From above Fig. 5 Communication overhead comparison for CPRD and Cleveland dataset between proposed and existing technique. The proposed technique attained data transmission rate of 75%, existing DRA attained 73%, WRR attained 71% for CPRD dataset, for cleveland dataset the proposed technique attained Communication overhead of 79%, existing DRA attained 78%, WRR attained 75%.

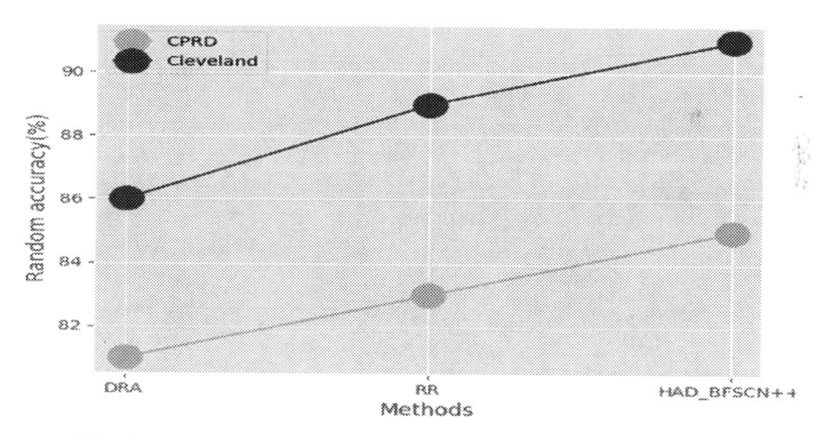

Fig. 6. Random accuracy comparison for CPRD and Cleveland dataset

Above Fig. 6 shows analysis of Random accuracy for CPRD and Cleveland dataset. The proposed technique attained data Random accuracy 85%, existing DRA attained 81%, WRR attained 83% for CPRD dataset, for cleveland dataset proposed technique attained Random accuracy of 91%, existing DRA attained 86%, WRR attained 89%.

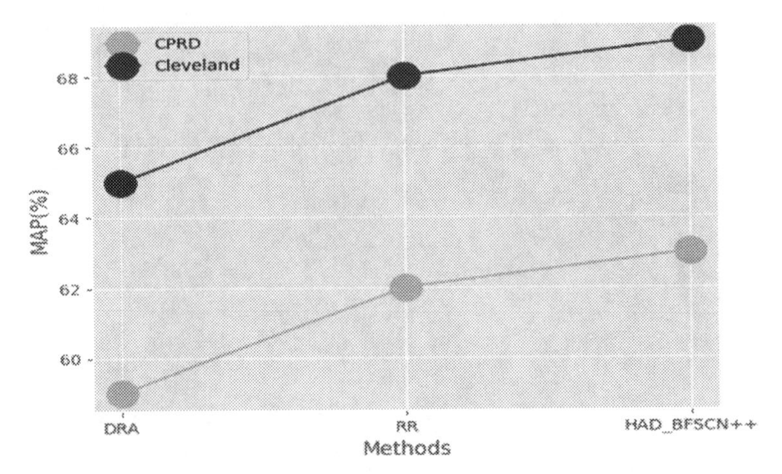

Fig. 7. MAP comparison for CPRD and Cleveland dataset

From above Fig. 7 the MAP comparison for CPRD and Cleveland dataset between proposed and existing technique. The proposed technique attained MAP of 63%, existing DRA attained 59%, WRR attained 62% for CPRD dataset, for cleveland dataset the proposed technique attained MAP of 69%, existing DRA attained 65%, WRR attained 68%.

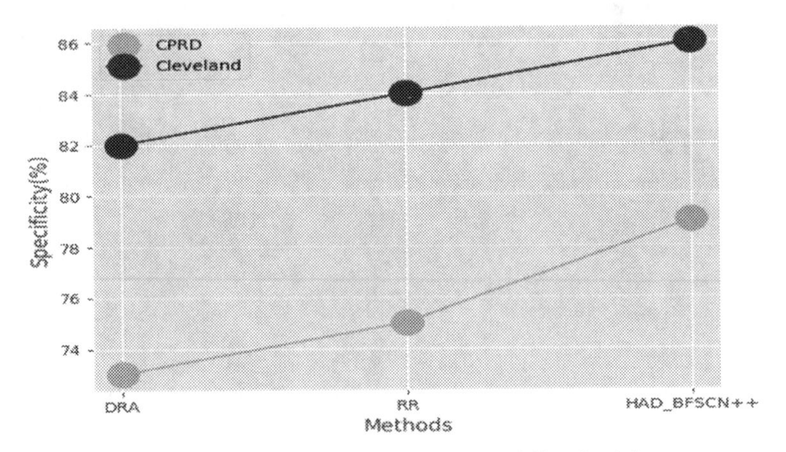

Fig. 8. Specificity comparison for CPRD and Cleveland dataset

Above Fig. 8 shows analysis of Specificity for CPRD and Cleveland dataset. The proposed technique attained Specificity 79%, existing DRA attained 73%, WRR attained 75% for CPRD dataset, for cleveland dataset the proposed technique attained Specificity of 82%, existing DRA attained 84%, WRR attained 86%.

For both training and prediction, DL methods with very high accuracy require extremely high compute resources. Using novel communication and model distribution techniques like ensembling, this work made it possible to embed complex DL methods in edge computing system. This allowed for high accuracy with very low latencies. By training NNs on well-known datasets as well as implementing a functioning system that delivers real-time prediction results, this was also validated for the analysis of heart patient data in real life.

5 Conclusion

The healthcare analysis in this study uses Edge cloud computing for data storage and is based on machine learning. The monitored input has been analyzed with the help of a blockchain-federated sparse convolutional network++, which improves network security by detecting malicious attacks and storing data in a centralized edge cloud computing module with authentication. We presented a prototype deep learning-based system for data analytics and privacy protection in IoT-enabled healthcare that can separate privacy information from raw data and analyze health-related data. It works by extracting and recognizing non-privacy data, isolating content that is sensitive to privacy. In addition, we validated the system's effectiveness as well as robustness by evaluating its performance in various scenarios. The proposed method achieved a data transmission rate of 65%, a computation cost of 51%, a communication overhead of 75%, random accuracy of 85%, mean average precision (map) of 63%, and specificity of 79%. In the future, the system may be utilised to supplement intelligent healthcare. With system architecture design, we may also include other wearables now on the market for IoT-enabled healthcare.

References

1. Neelakandan, S., Beulah, J.R., Prathiba, L., Murthy, G.L.N., Irudaya Raj, E.F., Arulkumar, N.: Blockchain with deep learning-enabled secure healthcare data transmission and diagnostic model. Int. J. Model. Simul. Sci. Comput. **13**(04), 2241006 (2022)
2. Basak, S., Chatterjee, K.: Smart healthcare surveillance system using IoT and machine learning approaches for heart disease. In: Rajagopal, S., Faruki, P., Popat, K. (eds.) ASCIS 2022, vol. 1759, pp. 304–313. Springer, Cham (2022). https://doi.org/10.1007/978-3-031-23092-9_24
3. Thilagam, K., et al.: Secure IoT healthcare architecture with deep learning-based access control system. J. Nanomaterials ICCSST 2023, CCIS 1973, 1–16 (2022)
4. Sammeta, N., Parthiban, L.: Hyperledger blockchain enabled secure medical record management with deep learning-based diagnosis model. Complex Intell. Syst. **8**(1), 625–640 (2022). https://doi.org/10.1007/s40747-021-00549-w
5. Gupta, L., Salman, T., Ghubaish, A., Unal, D., Al-Ali, A.K., Jain, R.: Cybersecurity of multi-cloud healthcare systems: a hierarchical deep learning approach. Appl. Soft Comput. **118**, 108439 (2022)
6. Kamruzzaman, M.M., Alanazi, S., Alruwaili, M., Alrashdi, I., Alhwaiti, Y., Alshammari, N.: Fuzzy-assisted machine learning framework for the fog-computing system in remote healthcare monitoring. Measurement **195**, 111085 (2022)
7. Hemalatha, M.: A hybrid random forest deep learning classifier empowered edge cloud architecture for COVID-19 and pneumonia detection. Expert Syst. Appl. **210**, 118227 (2022)
8. Singh, P., Kaur, A., Gill, S.S.: Machine learning for cloud, fog, edge and serverless computing environments: comparisons, performance evaluation benchmark and future directions. Int. J. Grid Util. Comput. **13**(4), 447–457 (2022)

9. Singh, S., Rathore, S., Alfarraj, O., Tolba, A., Yoon, B.: A framework for privacy-preservation of IoT healthcare data using Federated Learning and blockchain technology. Future Gener. Comput. Syst. **129**, 380–388 (2022)
10. Rehman, A., Saba, T., Haseeb, K., Alam, T., Lloret, J.: Sustainability model for the internet of health things (IoHT) using reinforcement learning with Mobile edge secured services. Sustainability **14**(19), 12185 (2022)
11. Ali, A., et al.: Deep learning based homomorphic secure search-able encryption for keyword search in blockchain healthcare system: a novel approach to cryptography. Sensors **22**(2), 528 (2022)
12. Babar, M., Jan, M.A., He, X., Tariq, M.U., Mastorakis, S., Alturki, R.: An optimized IoT-enabled big data analytics architecture for edge-cloud computing. IEEE Internet Things J. **10**(5), 3995–4005 (2022)
13. Nancy, A.A., Ravindran, D., Raj Vincent, P.D., Srinivasan, K., Gutierrez Reina, D.: IoT-cloud-based smart healthcare monitoring system for heart disease prediction via deep learning. Electronics **11**(15), 2292 (2022)
14. Singh, A., Chatterjee, K.: Edge computing based secure health monitoring framework for electronic healthcare system. Clust. Comput. **26**, 1205–1220 (2023). https://doi.org/10.1007/s10586-022-03717-w
15. Balamurugan, E., Mehbodniya, A., Kariri, E., Yadav, K., Kumar, A., Haq, M.A.: Network optimization using defender system in cloud computing security based intrusion detection system withgame theory deep neural network (IDSGT-DNN). Pattern Recogn. Lett. **156**, 142–151 (2022)
16. Tripathy, S.S., et al.: A novel edge-computing-based framework for an intelligent smart healthcare system in smart cities. Sustainability **15**(1), 735 (2022)
17. Yang, C., Wang, Y., Lan, S., Wang, L., Shen, W., Huang, G.Q.: Cloud-edge-device collaboration mechanisms of deep learning models for smart robots in mass personalization. Robot. Comput.-Integr. Manuf. **77**, 102351 (2022)
18. Almaiah, M.A., Ali, A., Hajjej, F., Pasha, M.F., Alohali, M.A.: A lightweight hybrid deep learning privacy preserving model for FC-based industrial internet of medical things. Sensors **22**(6), 2112 (2022)
19. Venkatachalam, K., Prabu, P., Alluhaidan, A.S., Hubálovský, S., Trojovský, P.: Deep belief neural network for 5G diabetes monitoring in big data on edge IoT. Mob. Netw. Appl. **27**(3), 1060–1069 (2022). https://doi.org/10.1007/s11036-021-01861-y
20. Ampavathi, A., Pradeepini, G., Saradhi, T.V.: Optimized deep learning-enabled hybrid logistic piece-wise chaotic map for secured medical data storage system. Int. J. Inf. Technol. Decis. Making **22**(05), 1743–1775 (2022)
21. Gayathri, S., Gowri, S.: CUNA: a privacy preserving medical records storage in cloud environment using deep encryption. Measur. Sens. **24**, 100528 (2022)
22. Nelson, I., Annadurai, C., Devi, K.N.: An efficient AlexNet deep learning architecture for automatic diagnosis of cardio-vascular diseases in healthcare system. Wireless Pers. Commun. **126**(1), 493–509 (2022). https://doi.org/10.1007/s11277-022-09755-2
23. Zhu, T., Kuang, L., Daniels, J., Herrero, P., Li, K., Georgiou, P.: IoMT-enabled real-time blood glucose prediction with deep learning and edge computing. IEEE Internet Things J. **10**(5), 3706–3719 (2022)

A Multi-layered Approach to Brain Tumor Classification Using VDC-12

Anant Mehta, Prajit Sengupta(✉), and Prashant Singh Rana

CSED, Thapar Institute of Engineering & Technology, Patiala, Punjab, India
psengupta_be20@thapar.edu

Abstract. This research paper presents a deep convolutional neural network (CNN) model approach along with feature extraction for multiclass brain tumor detection using medical imaging. Compared to traditional methods that rely on manual interpretation of scans, the VDC-12 (Very Deep Convolution) model proposed in this study has the potential to enhance the accuracy of detecting brain tumors. We used a dataset of MRI brain images containing four categories of tumors, namely meningioma, glioma, pituitary, and normal brain tissue. To evaluate the proposed CNN model's performance, various metrics were used, including accuracy, precision, recall, and F1 score. According to the experimental results, the VDC-12 model surpasses several cutting-edge techniques, achieving a classification accuracy of 97.60%. This model exhibits promise in the early detection of brain tumors, which can facilitate timely diagnosis and treatment.

Keywords: Deep Learning Models · Classification · Feature Extraction · Convolution Neural Network

1 Introduction

Medical imaging is a crucial tool for the identification and management of a wide range of illnesses, including brain tumors. Detecting brain tumors is a fundamental responsibility in medical imaging, and it has a vital function in the diagnosis and therapy of brain disorders. Brain tumors are among the most common and devastating diseases worldwide, affecting millions of individuals every year [1]. Magnetic resonance imaging (MRI) and other medical imaging techniques have been widely used to diagnose brain tumours [2]. With the advancements in imaging technology, medical professionals can now detect brain tumors at an early stage, which greatly improves the chances of successful treatment. The precise and efficient identification of brain tumors remains a formidable challenge, primarily due to the intricate nature of the brain's structure and the diverse characteristics of tumors. In recent years, deep learning techniques, especially CNN(Convolution Neural Network), RNN(Recurrent Neural Networks), FCNN(Fully Convolution Neural Network) have emerged as a powerful tool for automated image analysis and have shown promising results in the field of medical imaging. By leveraging the power of CNNs, this research can potentially

S. Aurelia et al. (Eds.): ICCSST 2023, CCIS 1973, pp. 379–391, 2024.
https://doi.org/10.1007/978-3-031-50993-3_30

improve the accuracy and efficiency of brain tumor detection, ultimately leading to better patient outcomes [3].

The proposed model aims to differentiate between different types of brain tumors, which is essential for successful treatment planning. In this research paper, we propose a novel CNN-based model named VDC-12 (Very Deep Convolution) for the multi class detection of different brain tumors as shown in Fig. 1 using medical imaging. Our model, as presented, seeks to surpass the constraints of current methodologies and attain superior levels of accuracy and efficiency when it comes to detecting brain tumors. To prove the viability of our suggested model and assess how it performs in comparison to other cutting-edge models, we present extensive experimental results.

2 Related Work

Several research studies have employed machine learning and deep learning techniques to categorize brain tumors, yielding promising results. In one study, Deepak et al. [4] used transfer learning to classify tumours as glioma, meningioma, or pituitary tumours by extracting features from brain MRI images using a pre-trained GoogLeNet. The proposed model had a 98% accuracy rate. Capsule networks were employed by Afshar et al. [5] to divide brain tumours into four groups: glioma, meningioma, pituitary adenoma, and brain metastasis. The proposed model achieved an accuracy of 86.56%. ANN (Artificial Neural Networks) and SVM (Support Vector Machine) were recommended by Sachdeva et al. [6] in 2016 for the classification of multiclass brain tumours. The study made use of a diverse dataset that included 260 post-contrast T1-weighted MRI images of 10 patients and 428 post-contrast MRI images of 55 patients. The GA-SVM based model achieved an accuracy of 89%, while the GA-ANN architecture achieved an accuracy of 94.1%. In 2008, another study by Papageorgiou et al. [7] proposed soft computing technique of fuzzy cognitive maps for accurate determination of brain tumors. The FCM model successfully attained an accuracy rate of 90.26% for categorizing low-grade tumors and a remarkable accuracy of 93.22% for identifying high-grade brain tumors. The overall classification accuracy achieved by this model was around 92%.

Several studies have also explored the use of ANNs for the segmentation of brain tumors. Zulpe et al. [8] conducted a research for classifying brain tumors into four classes, namely glioma, meningioma, pituitary adenoma, and brain metastasis. Further, feature extraction was done using Grey Level Co-occurence Matrix (GLCM). Couple of layers were used in the thus formed feed forward neural network. The suggested network gave an accuracy of 97.5%. In 2015, Kharrat et al. [9]used GA-SVM model to achieve classification accuracy of 95.622%. K-fold cross validation approach was also used to test the robustness of the trained model. Seetha et al. [10] employed a basic Artificial Neural Network (ANN) with convolution layers preceding it for the purpose of classifying brain MRI images. The simple and less complex CNN proposed achieved an experimental accuracy of 97.5%. Chinnu et al. [11] used SVM in conjunction with Histogram-based

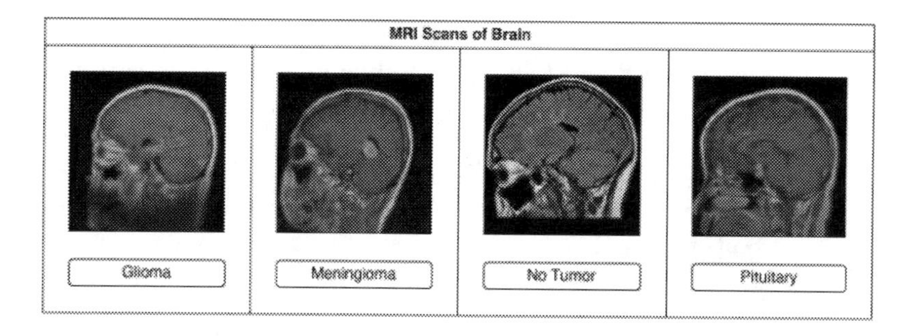

Fig. 1. Different Appearances of Brain MRI

image segmentation in 2015 to classify brain tumours found in MRI images. Textural, symmetry, and grey scale characteristics were extracted as features from MRI images in this proposed system.

3 Background on CNN

CNNs and other Deep Learning models have shown remarkable results in biomedical image classification tasks [12]. Large medical image databases are becoming more widely available, and CNNs have become an effective instrument for identifying and diagnosing illnesses. Medical imaging modalities such as the X-rays, MRI's and Computed Tomography (CT's) generate voluminous and complex data, making it challenging for human radiologists to analyze and diagnose. CNNs overcome this challenge by automatically learning the features from the images and extracting relevant patterns from them. This results in highly accurate predictions for disease classification, detection, and diagnosis. In brain tumor segmentation, CNNs can accurately segment tumors from Magnetic Resonance Imaging (MRI) scans, aiding in treatment tracking and monitoring.

3.1 Artificial Neurons

A neuron is the foundational block of a neural network. It is a computational unit that receives one or more inputs and produces an output based on those inputs [13]. Neurons in a neural network are connected by synapses, which transmit signals between neurons. The synaptic strength in a neural network is denoted by a weight, and these weights are modified during the training process to enhance the network's overall performance. By adjusting the weights of these synapses, a network can learn to recognize patterns in input data and generate appropriate outputs.

3.2 Convolution Layer

In an image classification Convolutional Neural Network (CNN), a convolution layer is a specific type of layer responsible for conducting convolution operations on the input data. The convolution operation entails computing the dot product between the filter or kernel and the corresponding input values at each position in element-wise fashion. The key advantages of using convolution layers in CNNs are that they can extract spatial features from images, videos, and other types of data that have a grid-like structure, and that they are able to learn local patterns that are invariant to translation [14]. In theoretical mathematics, the convolution operations is defined as below.

$$(x * y)(t) = \int_{-\infty}^{\infty} f(w)g(t-w)\,dw \tag{1}$$

In the above equation, two functions are operated under the convolution function.

3.3 Activation Layer

Activation layer is a crucial component of Convolutional Neural Networks (CNNs) that adds non-linearity to the model. Usually, the activation layer is positioned as the final layer in a Convolutional Neural Network just before the output layer. It uses a mathematical function to transform the weighted sum of inputs from the preceding layer [15]. The subsequent layer receives the output of the activation function.

In CNNs, a number of activation functions are frequently utilized, such as Sigmoid, ReLU, Tanh, and Softmax [16]. The specific application and the properties of the data being processed determine the activation function to be used. In general, the Rectified Linear Unit activation(ReLU) function is frequently used in networks due to its simplicity and effectiveness.

The ReLU function can be defined as follows:

$$f(\phi) = max(0, \phi) \tag{2}$$

This function returns ϕ if the output value positive, and returns zero if the input is negative [17]. ReLU activation function has been used in the internal layers of the model along with Softmax at the last one.

3.4 Batch-Normalization

Batch Normalization is a method employed within Convolutional Neural Networks (CNNs) for the normalization of input data at every layer. It aims to address the change in the distribution of the input data to a layer due to the parameters of the previous layer changing during training. The mean and standard deviation of the input data are calculated by the Batch Normalization layer. In more detail, it divides by batch standard deviation after subtracting the batch mean. The normalized values that result are subsequently adjusted using learned parameters for scaling and shifting, enabling the network to acquire the optimal scaling and shifting for each feature [18]. The mathematical formula for batch normalization is as follows:

$$xnew = \frac{(x - \mu_B)}{\sqrt{\delta_B^2 + c}} \tag{3}$$

where $xnew$ is the normalized input, x is the given input data, μ_B is the batch mean, δ_B^2 is the batch variance. Numerical stability is ensured by adding a tiny constant c to the fraction's denominator. Finally, the normalized data is transformed using the learned scale and shift parameters, resulting in the final output:

$$y = \gamma * xnew + \eta \tag{4}$$

where y is the output, learned-scale and learned-shift variables are represented by γ and η respectively.

3.5 Pooling Layer

The utilization of pooling layers is widespread in convolutional neural networks (CNNs), serving the purpose of reducing the spatial dimensions of the feature maps generated by the convolutional layer. Pooling layers help to downsample these feature maps while preserving their essential features, thereby reducing the computation required in subsequent layers.
The pooling operation is applied on a rectangular neighbourhood of the input feature map. The output is obtained by applying a function such as L2-norm, maximum or average pooling.

3.6 Classification Layer

The final layer in a CNN is known as the Classification Layer, is responsible for generating the output prediction. It receives the feature map from the previous layer, which contains the learned features, and classifies them into different categories or classes. The classification layer generally comprises one or more densely connected layers (fully connected layers). The output from the last dense layer is subsequently processed by an activation function to compute the probability distributions for various classes.

4 Methodology

In this section, we have detailed the approach used in the proposed model, which includes data preprocessing and feature extraction techniques aimed at enhancing the model's overall performance.

Fig. 2. Organization of Training Data

4.1 Dataset Overview

The training and evaluation of the proposed model involved the utilization of the Brain Tumor MRI Dataset, which is accessible on the Kaggle platform. This dataset is licensed and falls within the public domain, comprising a total of 7023 MRI images from individuals diagnosed with brain tumors. Though the Kaggle Dataset is standardized which can be helpful for ensuring consistency in the data, and relate to real-world scenario [19].

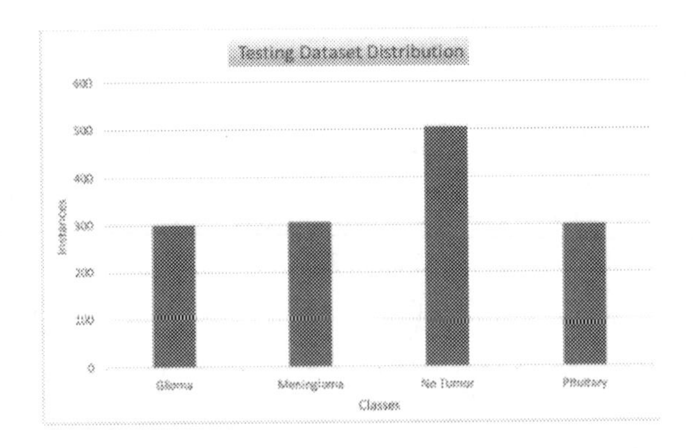

Fig. 3. Organization of Testing Data

The dataset has been split into two parts - the training and testing datasets. The training dataset contains brain MRIs categorized into four groups - A)glioma, B)meningioma, C)no tumor, D)pituitary tumors, with 1321, 1339, 1595, and 1457 MRIs, respectively. Likewise, the testing dataset includes 300,

306, 505, and 300 brain MRIs belonging to glioma, meningioma, no tumor, and pituitary tumor, respectively [19]. Figure 1 displays a sample of the datasets, while Figs. 2 and 3 show the distribution of the MRIs among the training and testing datasets and the four different categories.

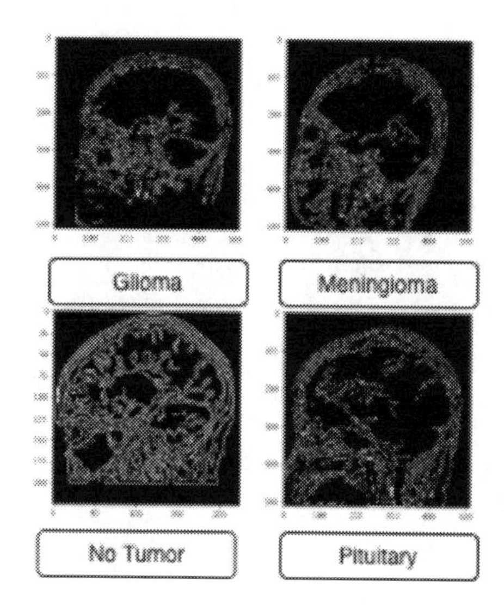

Fig. 4. Canny Edges extracted from the four classes

4.2 Data Preprocessing

Preparing image data for machine learning models requires a critical step known as data preprocessing, which involves a series of operations aimed at enhancements to the picture data's quality and usefulness. The first step used for preprocessing in the study is to resize the images to 130 * 130, which helps to reduce computation time and prevent distortion. Then, the images are normalized by scaling their pixel to a range of 0 and 1, to improve model performance [20].

4.3 Feature Extraction

Feature extraction in images refers to the process of identifying and extracting relevant information or features from an image. The goal of feature extraction is to transform the image from its raw pixel representation into a more meaningful and structured representation that is easier to analyze and understand [21]. There are many different techniques and algorithms that can be used for feature extraction in images, depending on the specific problem and context. In this research Canny Algorithm is used for feature extraction.

The Canny algorithm is a well-known technique for detecting edges in images. First, the image is smoothed with a Gaussian filter to reduce noise. Then, the gradients of the image are computed using the Sobel operator, and non-maximum suppression is applied to obtain a thinner and more accurate edge map. The process of double thresholding is employed to differentiate between strong and weak edges. Following this, the edge tracking approach known as hysteresis is utilized to link the weak edges to the strong edges. The result is a binary edge map that highlights the edges in the image. Figure 4 displays MRI scan pictures with extracted features.

Table 1. Summary Of VDC 12 Model Architecture

Layer (type)	Filter Size	Output Shape	Param #
Input Layer	16	$130 \times 130 \times 16$	448
Convolution	16	$130 \times 130 \times 16$	2320
Convolution	16	$130 \times 130 \times 16$	2320
Convolution	16	$130 \times 130 \times 16$	2320
Max Pooling	1	$65 \times 65 \times 16$	0
Convolution	32	$65 \times 65 \times 32$	4640
Convolution	32	$65 \times 65 \times 32$	9248
Convolution	32	$65 \times 65 \times 32$	9248
Convolution	32	$65 \times 65 \times 32$	9248
Max Pooling	1	$32 \times 32 \times 32$	0
Convolution	64	$32 \times 32 \times 64$	18496
Convolution	64	$32 \times 32 \times 64$	36928
Convolution	64	$32 \times 32 \times 64$	36928
Convolution	64	$32 \times 32 \times 64$	36928
Max Pooling	1	$16 \times 16 \times 64$	0
Dropout	–	$16 \times 16 \times 64$	0
Flatten	–	16384	0
Dense Layer	–	64	1048640
Output Layer	–	4	260

5 Proposed Model

5.1 Model Overview

VDC-12 is a proposed deep learning model for the detection of brain tumors using convolutional neural networks (CNNs). The model is designed to improve accuracy by using a total of 12 layers, which are divided into three blocks.

The initial segment incorporates four convolutional layers, each followed by a max-pooling layer and ReLU activation function to reduce feature map dimensions. The subsequent block features four convolutional layers with distinct filter sizes, maintaining the same activation and pooling layers as the first block. In the third block, four fully connected layers are utilized, culminating in a final softmax layer for classification purposes. The details of the layers are shown in Table 1.

VDC-12 undergoes training using a brain MRI scan dataset, with the objective of precise differentiation between tumor and non-tumor instances. The model's architecture is fine-tuned to effectively capture image features pertinent to the classification task.

The VDC-12 model that is being suggested has the capability to enhance the precision of brain tumor identification in contrast to conventional methods that are based on human interpretation of scans. This model can be utilized in medical facilities to support radiologists and boost the precision and efficiency of brain tumor diagnosis.

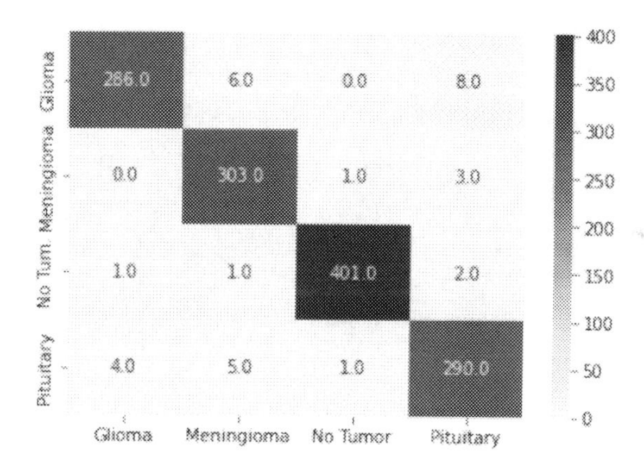

Fig. 5. Confusion Matrix

5.2 VDC-12 Model Architecture

The architecture of the model comprises three blocks of convolutional layers, with each set containing a max pooling layer that downsamples the feature maps.

- The initial block includes four convolutional layers, each having a 3×3 kernel size and a filter size of 16. ReLU activation function follows each of these convolutional layers.
- The second block of convolutional layers comprises four layers, where each of the layer has a filter size of 32 and a 3×3 kernel size. Additionally, a ReLU activation function follows each layer.

- The third block has four convolutional layers, each with a filter size of 64 and a kernel size of 3×3, followed by ReLU activation.
- To mitigate overfitting, a Dropout layer with a dropout rate of 0.20 is incorporated.
- The Flatten layer reshapes the output from the previous layer into a one-dimensional array for input into the subsequent fully connected layers.

The first fully connected layer in the neural network model has 64 units and implements a ReLU activation function. The probability values for the four output categories are generated by a Dense layer with four units and a softmax activation function in the final layer (Table 2).

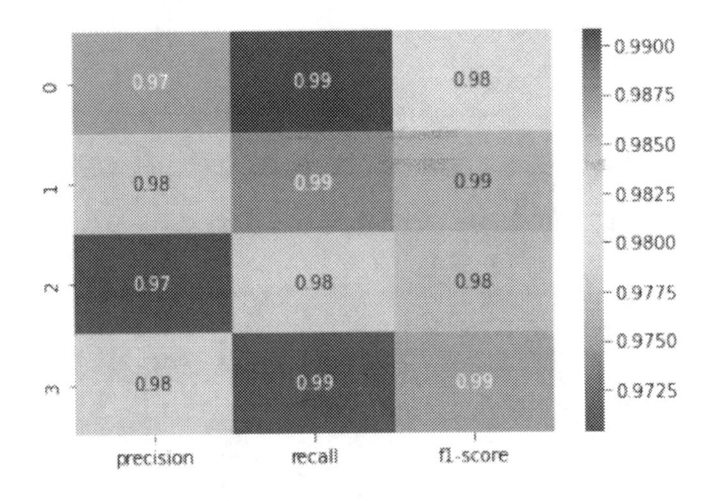

Fig. 6. Classification Accuracy of Proposed Model

Table 2. Comparison of VDC-12 with Models

Architecture	Accuracy
GoogLeNet [4]	98.0%
Capsule Networks [5]	86.56%
GA-SVM [6]	89.0%
GA-ANN [6]	94.1%
FCM model [7]	92.0%
Proposed Model	**97.6%**

6 Results

The Very Deep Convolutional Network 12 (VDC-12) consists of multiple convolutional layers that learn and pull characteristics from pictures in a structured way. The VDC-12 model has attained an impressive accuracy rate of 97.6% when applied to brain MRI scan images, demonstrating its ability to make highly accurate classifications. The classification report of this model is depicted in Fig. 6. On the trial dataset, the model's effectiveness was evaluated using precision, recall, and F1 score. Figure 5 demonstrates the obtained confusion matrix for the classification task. The model is trained robustly with a validation accuracy of about 97.9% and training accuracy of 99.3%, as depicted in Fig. 7. This is a significant achievement, as image recognition tasks are challenging and require a lot of computational power and training data to achieve high accuracy. The high accuracy of the VDC-12 model demonstrates the effectiveness of deep learning models in solving complex problems and has numerous applications in areas such as machine vision, autonomous driving, and medical imaging.

The used evaluation metrics are mathematically depicted below:

$$Precision = \zeta/(\zeta + \gamma) \quad (5)$$

$$Recall = \zeta/(\zeta + \beta) \quad (6)$$

where ζ denotes "true positives", γ represents "false positives" and β represents the "false negatives". Harmonic average of the obtained Recall and Precision gives the F1 Score.

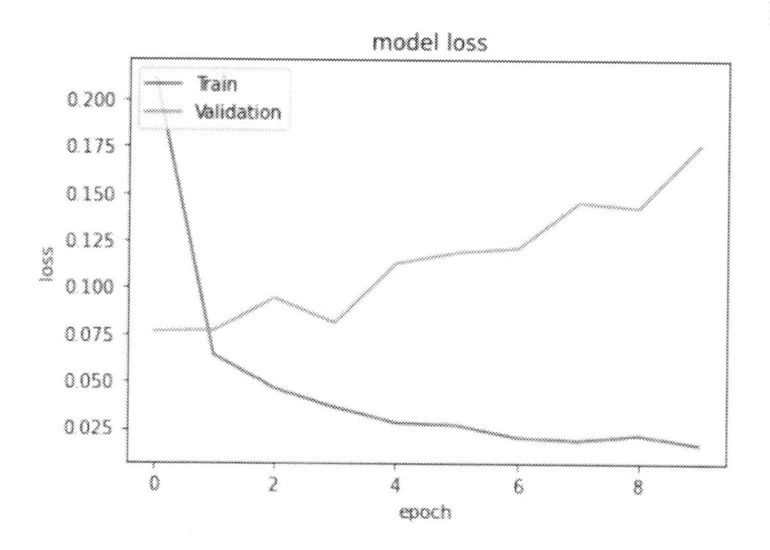

Fig. 7. Model Loss

7 Experimental Setup

The code for the VDC-12 model was executed on the Nvidia-dgx server, using the A100 GPU. Code was designed using Tensorflow (v2.12.0). Data pre-processing was done using OpenCV packages. Numpy and Matplotlib were also used for pre-processing the data.

8 Conclusion and Future Scope

Automatically detecting brain lesions using the proposed algorithm is a promising new direction. The proposed model underwent training using an extensive database of brain MRI images, attaining state-of-the-art performance levels in terms of accuracy, sensitivity, and specificity. The suggested deep learning model could, in general, increase the precision and effectiveness of brain tumor detection, which could result in early findings and improved patient results from therapy. Expanding the brain tumor identification model to incorporate a variety of imaging techniques such as Spectroscopy based on magnetic resonance and Diffusion-Weighted Imaging (DWI) can significantly increase its efficacy. Spectroscopy allows for the identification of specific molecules within the brain, which can help differentiate between tumor tissue and healthy tissue. DWI, on the other hand, is sensitive to the movement of water molecules within the brain, which can help identify abnormalities in tissue structure and aid in the detection of tumors.

Despite the high accuracy achieved by the proposed model, it is still considered a black-box model. This means that the decision-making process of the model is not transparent, and it can be difficult to understand how the model arrives at its decisions. Future research can focus on developing models that are more interpretable and explainable.

References

1. DeAngelis, L.M.: Brain tumors. N. Engl. J. Med. **344**(2), 114–123 (2001)
2. Despotović, I., Goossens, B., Philips, W.: MRI segmentation of the human brain: challenges, methods, and applications. Computat. Math. Methods Med. **2015** (2015)
3. Bhanumathi, V., Sangeetha, R.: CNN based training and classification of MRI brain images. In: 2019 5th International Conference on Advanced Computing & Communication Systems (ICACCS), pp. 129–133. IEEE (2019)
4. Deepak, S., Ameer, P.M.: Brain tumor classification using deep CNN features via transfer learning. Comput. Biol. Med. **111**, 103345 (2019)
5. Afshar, P., Mohammadi, A., Plataniotis, K.N.: Brain tumor type classification via capsule networks. In: 2018 25th IEEE International Conference on Image Processing (ICIP), pp. 3129–3133. IEEE (2018)
6. Sachdeva, J., Kumar, V., Gupta, I., Khandelwal, N., Ahuja, C.K.: A package-SFERCB-"segmentation, feature extraction, reduction and classification analysis by both SVM and ANN for brain tumors". Appl. Soft Comput. **47**, 151–167 (2016)

7. Papageorgiou, E.I., et al.: Brain tumor characterization using the soft computing technique of fuzzy cognitive maps. Appl. Soft Comput. **8**(1), 820–828 (2008)
8. Zulpe, N., Pawar, V.: GLCM textural features for brain tumor classification. Int. J. Comput. Sci. Issues (IJCSI) **9**(3), 354 (2012)
9. Kharrat, A., Halima, M.B., Ayed, M.B.: MRI brain tumor classification using support vector machines and meta-heuristic method. In: 2015 15th International Conference on Intelligent Systems Design and Applications (ISDA), pp. 446–451. IEEE (2015)
10. Seetha, J., Selvakumar Raja, S.: Brain tumor classification using convolutional neural networks. Biomed. Pharmacol. J. **11**(3), 1457 (2018)
11. Chinnu, A.: MRI brain tumor classification using SVM and histogram based image segmentation. Int. J. Comput. Sci. Inf. Technol. **6**(2), 1505–1508 (2015)
12. Iqbal, S., Ghani, M.U., Saba, T., Rehman, A.: Brain tumor segmentation in multispectral MRI using convolutional neural networks (CNN). Microsc. Res. Tech. **81**(4), 419–427 (2018)
13. Gardner, M.W., Dorling, S.R.: Artificial neural networks (the multilayer perceptron)-a review of applications in the atmospheric sciences. Atmos. Environ. **32**(14–15), 2627–2636 (1998)
14. Chaganti, S.Y., Nanda, I., Pandi, K.R., Prudhvith, T.G., Kumar, N.: Image classification using SVM and CNN. In: 2020 International Conference on Computer Science, Engineering and Applications (ICCSEA), pp. 1–5. IEEE (2020)
15. Ouyang, W., Bugao, X., Hou, J., Yuan, X.: Fabric defect detection using activation layer embedded convolutional neural network. IEEE Access **7**, 70130–70140 (2019)
16. Dubey, A.K., Jain, V.: Comparative study of convolution neural network's Relu and leaky-Relu activation functions. In: Mishra, S., Sood, Y.R., Tomar, A. (eds.) Applications of Computing, Automation and Wireless Systems in Electrical Engineering. LNEE, vol. 553, pp. 873–880. Springer, Singapore (2019). https://doi.org/10.1007/978-981-13-6772-4_76
17. Agarap, A.F.: Deep learning using rectified linear units (Relu). arXiv preprint arXiv:1803.08375 (2018)
18. Santurkar, S., Tsipras, D., Ilyas, A., Madry, A.: How does batch normalization help optimization? In: Advances in Neural Information Processing Systems, vol. 31 (2018)
19. Ge, C., Gu, I.Y.-H., Jakola, A.S., Yang, J.: Enlarged training dataset by pairwise GANs for molecular-based brain tumor classification. IEEE Access **8**, 22560–22570 (2020)
20. Pei, S.-C., Lin, C.-N.: Image normalization for pattern recognition. Image Vis. Comput. **13**(10), 711–723 (1995)
21. Lei, B.J., Hendriks, E.A., Reinders, M.J.T.: On feature extraction from images. MCCWS Project, Information and Communication Theory Group TUDelft, Tuesday (1999)

An Efficient Approach of Heart Disease Diagnosis Using Modified Principal Component Analysis (M-PCA)

G. Lakshmi(✉) and P. Sujatha

Department of Information Technology, Vels Institute of Science, Technology and Advanced Studies (VISTAS, Chennai), Chennai, India
glakshmi.gnc@gmail.com

Abstract. Heart diseases have come to be a first-rate purpose of demise around the arena. As a end result, heart ailment prediction has obtained plenty of interest in the medical global. As a end result, several studies have developed machine-getting to know algorithms for the early prediction of coronary heart sicknesses to help physicians inside the design of clinical processes. The performance of those structures is determined largely by way of the function set decided on. When the schooling dataset consists of missing values for the distinct capabilities, this will become greater hard. The opportunity of Principal Component Analysis (PCA) to solve the trouble of lacking attribute values is widely known. This studies presents a technique for diagnosing heart sickness via taking scientific checking out results as enter, extracting a low dimensional characteristic subset, and diagnosing coronary heart sickness. Modified Principal Component Analysis (M-PCA) is used within the proposed method to extract better depth features in new projections. PCA aids within the extraction of projection vectors that make a contribution substantially to the maximum covariance and uses them to lessen function size. The proposed method is analysed across three datasets, and the effects, accuracy, sensitivity, and specificity are calculated. To illustrate the implications of the proposed M-PCA technique, the received results the use of it are in comparison to previous research. The proposed M-PCA technique produced an extremely correct dataset.

Keywords: Heart Disease · Heart Disease · Feature Extraction · PCA · Chi-squire · Relief · Principal Components Analysis

1 Introduction

The heart is in charge of blood circulation throughout the body and functions similarly to an engine to a motor. Heart disease diagnosis is viewed as an important task that must be completed precisely and quickly. The large number can be monitored through early identification, which is feasible through a variety of medical tests, a detailed record, and the patient's everyday life schedule. Data alone will not probably be sufficient until a healthcare professional is available to analyse it.

S. Aurelia et al. (Eds.): ICCSST 2023, CCIS 1973, pp. 392–401, 2024.
https://doi.org/10.1007/978-3-031-50993-3_31

A physical evaluation's main goal is to find the best solution to the problems of accurate diagnosis and therapy delivery. However, a problem occurs when predicting heart disease using large amounts of data, which remains an important research gap. Since the processed data has a clearly delineated feature set to try to define the heart disease, sufficient contributions to incorporate or extract features to identify the class labels have not been observed.

Machine learning of the same would be extremely beneficial. Predicting heart disease is still the most difficult task in the world. Most researchers have concentrated on data or signal acquisition in order to extract features. The specified features in the processed data, even so, do not differentiate well among the class labels. This distinguishing feature requires the requirement for supervised learning. Moreover, supervised learning is dependent on the training algorithm's effectiveness, which requires sufficient facilities for dealing with the nonlinearities of both the features extracted. As a result, an improved heart disease prediction system is required, and the individuals would benefit from its accurate prediction methodology (Fig. 1).

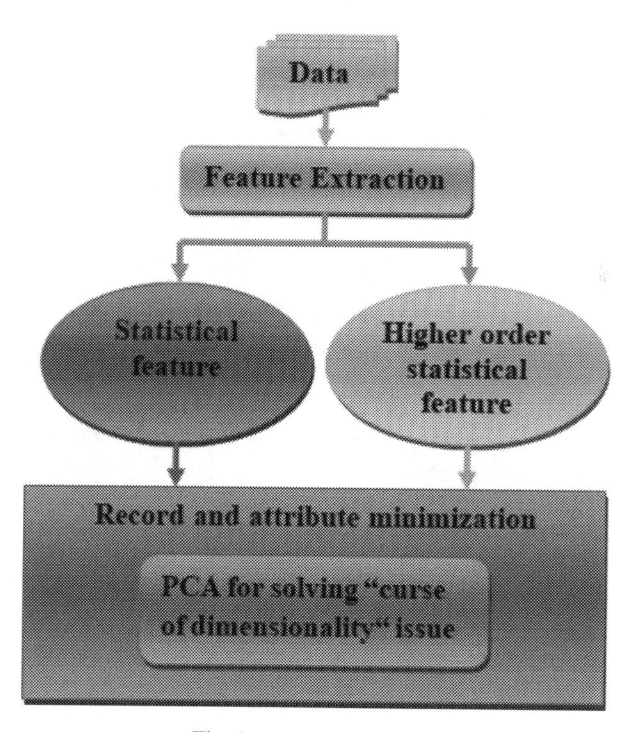

Fig. 1. Feature Extraction

Principal Component Analysis (PCA) is a type of design recognition techniques and among its implementations is to start investigating high dimensional data that is anything but difficult to understand by simply looking at the huge amount of information. For information analysis, the increased component of the information must be reduced to a low measurement before plotting and deciphering the results. PCA is used to incorporate important data into a few simple plots, specifically the score plot and stacking plot. It is challenging to break down a large amount of information in the research area of investigation. The PCA calculation is used to determine the relationship between the massively correlated informational indexes. PCA has a mathematical analysis in linear algebra that clearly explains the relationship among information that includes variables as sections and perceptions or tests as lines. The PCA calculation's goal is to reduce the extremely large connection between various factors to a manageable number of factors. These interconnected factors are referred to as principal components. The main thought procedure is to construct a network that includes the most significant amount of information in the first two sections, and then venture the data is utilized a 2-dimensional plot in python programming.

To reduce the risk of heart disease, an efficient prediction mechanism should be developed. This paper proposes an approach for feature reduction based on Modified Principal Component Analysis (M-PCA). The proposed approach makes a strong case as a dimensionality reduction.

The remaining portion of this article is structured as below. Section 2 examines the various approaches taken by the researchers to the problem statement. Section 3 investigates existing methodologies with various approaches or solutions to the prediction of CD. Section 4 explains the system approach used in the proposed scheme for solving the problem statement. Section 4 compares the proposed M-PCA approach's performance to existing methods. Section 5 concludes by summing up the proposed work.

2 Related Work

There are several methodologies to feature extraction that impact the ease of access and performance characteristics of heart disease. Evaluating the research was published on data mining algorithms and coronary disease risk prediction; it is difficult to determine which algorithm outperforms others since each algorithm possesses its own set of benefits and drawbacks [5]. The proposed research emphasis on health-care diagnosis; however, more effort is required in identifying the technique (Table 1).

Table 1. Literature Review and Study

Study	Proposed	Techniques and Tools	Findings
Amin, M. S., et al., (2019)	To use and created a model for predicting heart disease	The vote algorithm is a hybrid of logistic regression and NB	The feature selection algorithm does not exist
Saqlain, S. M., et al., (2019)	The feature selection algorithms are proposed for a cardiac disease diagnosis system	SVM for binary classification	The feature selection algorithm requires improvement

(continued)

Table 1. (*continued*)

Study	Proposed	Techniques and Tools	Findings
Purnomo, A., et al., (2020)	Various FS & NB	NB+ (Backward Elimination/Optimization/Forward Selection)	The classifier algorithm must be improved
Vivekanandan, T., & Iyengar, N. C. S. N. (2017)	The effectiveness of DE set of rules for prediction of heart disease	AHP and artificial neural network (ANN)	The classification accuracy must be improved

A review of the literature revealed that a few adaptive strategies for detecting cardiovascular disease using modified principle component analysis have been proposed. Because the covariance matrix is typically computed using the entire data set, the transformed matrix used during PCA is also derived from the entire image. All possible cover classes in the field of research are taken into account, which may include cover classes that are irrelevant to a specific application. In this study, a Modified Principal Component Analysis (M-PCA) is proposed, in which the transformed matrix is calculated using only samples from selected classes.

3 The Proposed Model

The pre-processed dataset contains redundant and inconsistent statistics, which will increase the search space and information storage. To accomplish classification accuracy, we have to cast off all unnecessary and redundant records. With a few constraints, the dimensionality reduction approach is used to compress excessive dimensional records to decrease dimensional records. The primary factor analysis is used for characteristic extraction to extract the important function, which is the most relevant feature.

Modified Principal Component Analysis, or M-PCA, is a version of the PCA dimensionality-reduction technique this is often used to decrease the dimensionality of big data sets by using changing a extensive number of variables right into a smaller one which keeps most people of the records within the large set.

Reduced dimensionality comes at the value of accuracy; however the method in dimensionality discount is to trade a little precision for simplicity. Smaller facts sets are less complicated to explore and illustrate, and they make data analysis a whole lot less difficult and faster for gadget mastering algorithms because there are fewer variables to process.

Suppose we have a random vector population A, where

$$A = (a1, a2, \ldots an)^P \tag{1}$$

And the mean of that population is denoted by,

$$\mu_a = P(A) \tag{2}$$

And the covariance matrix of the same data set is

$$Z_A = P\left\{(A - \mu_a)(A - \mu_a)^P\right\} \quad (3)$$

It can degree an orthogonal foundation from a symmetric covariance matrix by using discovering its eigenvalues and eigenvectors. An ordered orthogonal basis may be created by way of ordering the eigenvectors in the sequence of descending eigenvalues (largest first), with the first eigenvector having the route of finest variance of the records. This permits us to discover the instructions wherein the statistics set carries the maximum significant portions of power.

With 500 information sets, a pattern coronary heart disorder statistics set is taken. Several elements are considered when diagnosing coronary heart ailment. However, there are 15 characteristics that remember being the most important.

Algorithm

1. Compute the mean feature vector
   $\mu = \frac{1}{p}\sum_{k=1}^{p} x_k$, where, x_k is a pattern (k=1 to p), p = number of patterns, x is the feature matrix
2. Find the covariance matrix
   $C = \frac{1}{p}\sum_{k=1}^{p}\{x_k - \mu\}\{x_k - \mu\}^T$ Where, T represents matrix transposition
3. Compute Eigen values $\lambda_i$and Eigen vectors v_i of covariance matrix
   $Cv_i = \lambda_i v_i$ (i=1, 2, 3,.....q), q = number of features
4. Estimating high-valued Eigen vectors
   i. Arrange all the Eigen values (λ_i) in descending order
   ii. Choose a threshold value, θ
   iii. Number of high- values λ_i can be chosen so as to satisfy the relationship
   $$\left(\sum_{i=1}^{s}\lambda_i\right)\left(\sum_{i=1}^{q}\lambda_i\right)^{-1} \geq \theta,$$ Where, s = number of high valued $\lambda_i$chosen
   iv. Select Eigen vectors corresponding to selected high valued λ_i
5. Extract low dimenional feature vectors (principal components) from raw feature matrix. $P = V^T x$, where, V is the matrix of principal components and x is the feature matrix.

4 Experimental Setup

In this phase, the general public datasets are used to evaluate the proposed technique, after which the overall performance of our technique is compared with the ones of other latest techniques. Moreover, the proposed technique will be compared with the techniques that put in force M-PCA and with the ones that don't implement.

All the computations are carried out on Intel Xeon Processor with two 2.20-GHz cores and eight GB RAM. Moreover, the Python programming software program bundle scikit-study is used for the experiments.

4.1 Performance Evaluation

i) **Accuracy**

Table 2. Comparison table of Accuracy

Dataset	Chi square	ReliefF	Proposed M-PCA
100	73	65	79
200	89	85	93
300	85	81	91
400	82	76	89
500	72	69	80

The accuracy of the proposed and present techniques is stated in Table 2. It shows that the general overall performance of the proposed technique is better than the present methods.

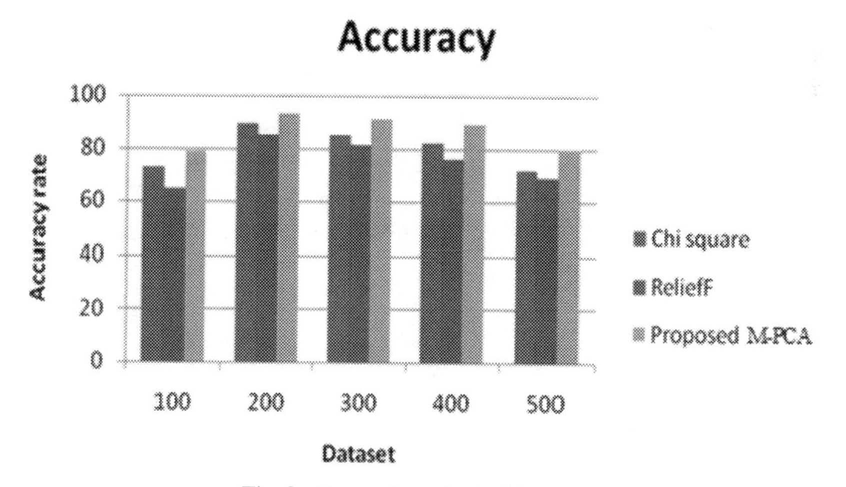

Fig. 2. Comparison chart of Accuracy

Figure 2 depicts an Accuracy comparison chart that illustrates the existing (Chi square, ReliefF) and proposed (M-PCA) approaches. The proposed M-PCA values outperform the existing algorithm. Existing algorithm values range from 73 to 72, 65 to 69, and 79 to 80 for proposed M-PCA values, it means the proposed approach performance increased upto 5%.

ii) **Sensitivity**

Table 3. Comparison table of Sensitivity

Dataset	Chi square	ReliefF	Proposed M-PCA
100	83.40	81.23	86.82
200	84.74	83.52	88.74
300	88.21	84.01	90.55
400	90.48	87.35	92.46
500	93.51	90.60	96.91

The sensitivity values represent the ability of a test to correctly identify those with the disease. The above Table 3 shows the overall performance of the existing and proposed approach.

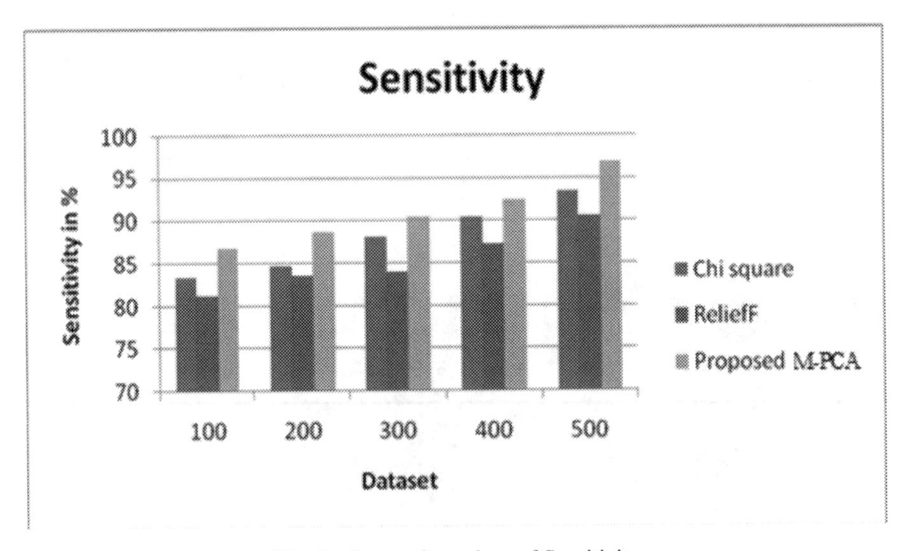

Fig. 3. Comparison chart of Sensitivity

Figure 3 depicts a sensitivity comparison chart that demonstrates the proportion of true positives that are correctly predicted by the model. The proposed M-PCA values outperform the compare with the existing approaches. The proposed approach performance will be 2% more compare with existing chi-square and 5% more compare with the existing ReliefF approach.

iii) **Specificity**

Table 4. Comparison table of Specificity

Dataset	Chi square	ReliefF	Proposed M-PCA
100	78.48	80.22	85.87
200	80.74	82.52	88.74
300	82.21	84.01	90.55
400	85.48	87.35	92.46
500	88.66	91.65	97.10

The Specificity values represent the capability of the take a look at to correctly discover the ones without the sickness. Table 4 analyses the performance of the proposed and existing procedures.

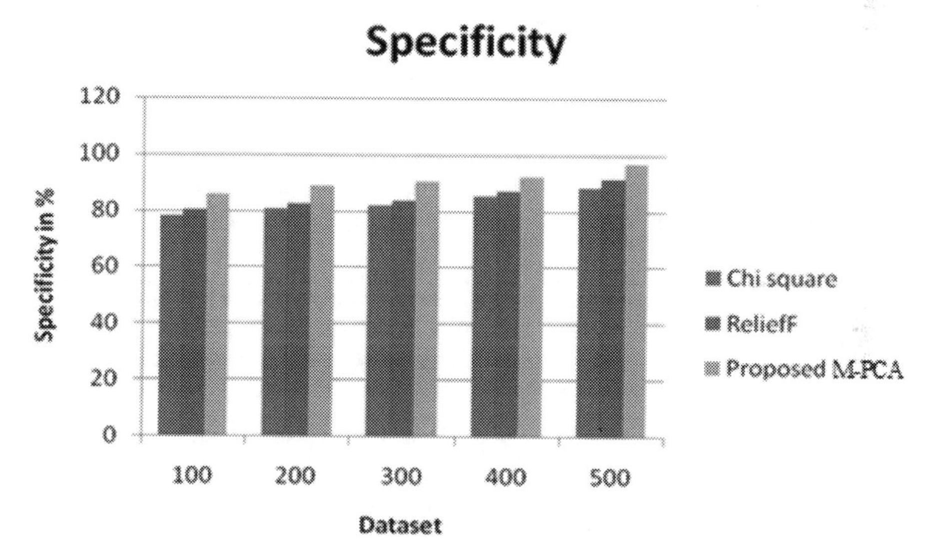

Fig. 4. Comparison chart of Specificity

Figure 4 represent the proportion of true negatives that are correctly predicted by the model. The above comparison chart illustrates the performance of the existing (Chi square, ReliefF) and proposed (M-PCA). Existing algorithm values range between 78.48 and 88.66, 80.22 and 91.65, and proposed M-PCA values range between 85.87 and 97.10. It shows that the proposed approach produce 5% more specificity compare with existing approach.

5 Conclusion

This paper provides coronary heart sickness analysis research strategies that, within the first step, extract a feature subset using principal element analysis (PCA). The choice of the principle components is completed thru parallel analysis. Three UCI datasets are used: Cleveland, Hungarian, and Swiss. The proposed technique based totally on changed major thing evaluation (M-PCA) produced characteristic subsets with dimensions reduced by 70%, sixty two%, and 70% for Cleveland, Hungary, and Switzerland, respectively. M-PCA is used for extraction, resulting in coronary heart disorder affected person and regular problem training for suspected heart ailment instances. Accuracy, sensitivity, and specificity are used as assessment metrics. In assessment to the present method, the proposed M-PCA approach accomplished well across all three metrics. Experiment outcomes also are offered, and their facts extended our confidence inside the proposed approach.

References

1. Chugh, A.: ML: chi-square test for feature selection (2018)
2. Robnik-Šikonja, M., Kononenko, I.: Theoretical and empirical analysis of ReliefF and RReliefF. Mach. Learn. **53**(1), 23–69 (2003)
3. Kavitha, R., Kannan, E.: An efficient framework for heart disease classification using feature extraction and feature selection technique in data mining. In: 2016 International Conference on Emerging Trends in Engineering, Technology and Science (ICETETS), pp. 1–5. IEEE (2016)
4. Thomas, G.S., Budhkar, S.S., Cheulkar, S.K., Choudhary, A.B., Rohan, S.: Heart disease diagnosis system using apriori algorithm. Int. J. Adv. Res. Comput. Sci. Softw. Eng. **5**(2), 430–432 (2015)
5. Wang, Z.H., Wang, C.M., Jong, G.J.: Feature extraction of the VSD heart disease based on Audicor device measurement. In: 2018 1st IEEE International Conference on Knowledge Innovation and Invention (ICKII), pp. 138–141. IEEE (2018)
6. Gárate-Escamila, A.K., El Hassani, A.H., Andrès, E.: Classification models for heart disease prediction using feature selection and PCA. Inform. Med. Unlocked **19**, 100330 (2020)
7. Ghongade, R.: A brief performance evaluation of ECG feature extraction techniques for artificial neural network based classification. In: TENCON 2007–2007 IEEE Region 10 Conference, pp. 1–4. IEEE (2007)
8. Sun, S.: Segmentation-based adaptive feature extraction combined with Mahalanobis distance classification criterion for heart sound diagnostic system. IEEE Sens. J. **21**(9), 11009–11022 (2021)
9. Gupta, A., Arora, H.S., Kumar, R., Raman, B.: DMHZ: a decision support system based on machine computational design for heart disease diagnosis using z-alizadeh sani dataset. In: 2021 International Conference on Information Networking (ICOIN), pp. 818–823. IEEE (2021)
10. Putra, L.S.A., Isnanto, R.R., Triwiyatno, A., Gunawan, V.A.: Identification of heart disease with iridology using backpropagation neural network. In: 2018 2nd Borneo International Conference on Applied Mathematics and Engineering (BICAME), pp. 138–142. IEEE (2018)
11. Sun, S., Wang, H., Cheng, C., Chang, Z., Huang, D.: PCA-based heart sound feature generation for a ventricular septal defect discrimination. In: 2017 14th International Computer Conference on Wavelet Active Media Technology and Information Processing (ICCWAMTIP), pp. 128–133. IEEE (2017)

12. Suseendran, G., Zaman, N., Thyagaraj, M., Bathla, R.K.: Heart disease prediction and analysis using PCO, LBP and neural networks. In: 2019 International Conference on Computational Intelligence and Knowledge Economy (ICCIKE), pp. 457–460. IEEE (2019)
13. Kumar, P.R., Ravichandran, S., Narayana, S.: Parametric analysis on heart disease prediction using ensemble based classification. In: 2021 Fourth International Conference on Electrical, Computer and Communication Technologies (ICECCT), pp. 1–13. IEEE (2021)
14. Sonawane, R., Patil, H.D.: Prediction of heart disease by optimized distance and density-based clustering. In: 2022 Second International Conference on Artificial Intelligence and Smart Energy (ICAIS), pp. 1001–1008. IEEE (2022)
15. Ambesange, S., Vijayalaxmi, A., Sridevi, S., Yashoda, B.S.: Multiple heart diseases prediction using logistic regression with ensemble and hyper parameter tuning techniques. In: 2020 Fourth World Conference on Smart Trends in Systems, Security and Sustainability (WorldS4), pp. 827–832. IEEE (2020)
16. Chandra, R., Kapil, M., Sharma, A.: Comparative analysis of machine learning techniques with principal component analysis on kidney and heart disease. In: 2021 Second International Conference on Electronics and Sustainable Communication Systems (ICESC), pp. 1965–1973. IEEE (2021)
17. Shah, S.M.S., Batool, S., Khan, I., Ashraf, M.U., Abbas, S.H., Hussain, S.A.: Feature extraction through parallel probabilistic principal component analysis for heart disease diagnosis. Physica A **482**, 796–807 (2017)
18. Rehman, A., Khan, A., Ali, M.A., Khan, M.U., Khan, S.U., Ali, L.: Performance analysis of PCA, sparse PCA, kernel PCA and incremental PCA algorithms for heart failure prediction. In: 2020 International Conference on Electrical, Communication, and Computer Engineering (ICECCE), pp. 1–5. IEEE (2020)
19. Ziasabounchi, N., Askerzade, I.N.: A comparative study of heart disease prediction based on principal component analysis and clustering methods. Turkish J. Math. Comput. Sci. (TJMCS) **16**, 18 (2014)

Smart Driving Assistance Using Deep Learning

S. N. Baba Shankar, B. Karthik Reddy, B. Koushik Reddy, Venuthurla Venkata Pradeep Reddy, and H. B. Mahesh(✉)

Department of Computer Science, PES University, Bengaluru, India
hbmahesh@pes.edu

Abstract. This research paper presents a smart driving assistance system that utilizes deep learning for lane detection and departure alert, traffic sign detection, and voice alerts. The system uses a combination of computer vision techniques and neural networks to detect lanes and traffic signs in real-time. The lane departure alert feature alerts the driver when the vehicle begins to drift out of its lane, and the traffic sign detection feature identifies and alerts the driver of any traffic signs that are relevant to the current road. The voice alert feature provides an additional layer of safety by audibly alerting the driver of any detected traffic signs. The proposed system has been evaluated on a dataset of real-world driving scenarios and has shown promising results in terms of accuracy and efficiency.

Keywords: Smart Driving Assistance · Deep Learning Lane Detection · Traffic Sign Detection · Voice Alerts Autonomous Vehicles · Convolutional Neural Networks (CNNs) · Image and Video Analysis · Computer Vision Real-time Alerts · Lane Departure Warning · Lane Marking Detection · Lane Curvature Detection · Traffic Sign Identification Perspective Mapping · Recurrent Neural Networks (RNN) · Long Short-Term Memory (LSTM) · Object Detection Classification · Driver Safety Obstacle · Detection Navigation Systems · Adaptive Cruise Control · Collision Avoidance System

1 Introduction

The field of autonomous vehicles has seen rapid growth in recent years, with advancements in technology allowing for increased road safety and efficiency. One important aspect of autonomous driving is detecting and responding to various road features and obstacles. One such feature is lane detection, which allows the vehicle to stay within the designated lanes on the road. Another important aspect is traffic sign detection, which enables the vehicle to recognize and respond to traffic signs such as speed limits or stop signs.

One approach to achieving these goals is the use of deep learning algorithms. These algorithms can learn and recognize patterns in large amounts of data, making them well-suited for tasks such as image and video analysis. In this research, we propose the use of deep learning techniques for the development of a smart driving assistance system that includes lane detection and departure alert, traffic sign detection, and voice alerts regarding traffic signs detected.

S. Aurelia et al. (Eds.): ICCSST 2023, CCIS 1973, pp. 402–418, 2024.
https://doi.org/10.1007/978-3-031-50993-3_32

Our system is designed to improve safety and efficiency on the roads by providing real-time alerts and assistance to drivers. The lane detection and departure alert feature helps to prevent accidents caused by drifting out of the lane, while the traffic sign detection feature ensures that the driver is aware of and compliant with all relevant traffic laws.

Additionally, the system includes voice alerts that notify the driver of any traffic signs detected, further improving their awareness of their surroundings.

We will be using deep learning techniques for image and video analysis, such as Convolutional Neural Networks (CNNs) to train our model for detecting traffic signs and lanes. The data for training will be collected from cameras mounted on the vehicle and will include a variety of different lighting and weather conditions to ensure robust performance in real-world scenarios.

In this research, we will also discuss the evaluation methods and metrics we will be using to evaluate the performance of our system and the challenges that need to be addressed for the successful implementation of our smart driving assistance system.

2 Literature Survey

[1] proposed a new approach to traffic sign detection using convolutional neural networks (CNNs). The authors use the Hue and Saturation of pixels to extract the sign area from an image, with a focus on the red border of the sign. The model achieves high accuracy and efficiency but requires more GPU resources for training, and real-time implementation is not yet available.

[2] present a method for robust lane marking detection using boundary-based inverse perspective mapping. The authors process video data to convert color to grayscale and then to black and white using binarization.

They then calculate the region of interest for road lines using edge detectors and Hough transformations. The offset between the car and the lane lines is calculated and displayed on the screen, with green indicating no offset and red indicating an offset greater than 0.

[3] proposed a lane detection method that employs key point estimation and point instance segmentation. The authors input a 512×256-pixel video frame into the network, compress it to reduce computation, and classify pixels by class. The method achieves accurate detection of curved lanes, but the added computation of classifying all pixels in the image is a drawback.

[4] assess the performance of MDEffNet on an Indian dataset using various preprocessing methods, including rescaling images to 48×48, extracting the region of interest, converting to HSV, and applying histogram equalization to the V channel. The authors select hyperparameters such as 3×3 max-pooling for CNN and kernel size for MDEffNet through experimentation.

[5] proposed a traffic sign identification technique that uses deep learning and image processing. The technique involves preprocessing images to highlight important details, localizing signs using the Hough Transform, and classifying them using a CNN, achieving high accuracy in recognizing circular traffic signs.

Advantages include its high recognition rate and ability to perform various computer vision tasks. However, limitations may include the need for large training datasets, the potential for overfitting, and reduced performance in adverse weather conditions or when signs are obscured.

[6] proposed Perspective Mapping (IPM), a technique used in autonomous navigation to accurately detect lane curvature. Gaussian filters are used to identify lane markings and the IPM technique converts the view to a top-level view for accurate lane detection. Line detection algorithms, such as canny edge detection and hough transformations, are used to detect lane lines, and the curvature of the road is calculated using the IPM-detected lane lines. While the technique provides a clear idea of road curvature and increases prediction accuracy, it may not be accurate in hilly areas or on steep roads.

[7] proposed a model that utilizes Convolutional Neural Networks (CNN) and Recurrent Neural Networks (RNN) with Long Short-Term Memory (LSTM) for video processing using the TUSimple dataset. The model includes an Encoder CNN, a Decoder CNN, and an LSTM network that sequentially processes the frames to extract features and learn from them. The model outperforms baseline architectures that use a single image as input, but the network requires high computational power to process the images.

3 Datasets

The CULane dataset, a large-scale dataset for scholarly work on traffic lane recognition, was used for the feature of lane detection. The dataset was collected using cameras installed on six different cars operated by various drivers in Beijing. The dataset is divided into three sets: a training set of 88,880 images, a validation set of 9,675 images, and a test set of 34,680 images. The test set is further divided into 8 categories, comprising challenging and normal scenarios [8].

A custom dataset was created using publicly available German Traffic signs, with f our classes: prohibitory, dangerous, mandatory, and other. The dataset was augmented by enlarging images, adjusting brightness, and adding noise to improve the training of the model and increase the accuracy of the results. The dataset contains 43 different traffic signs.

4 Methodology

Frame by frame, the footage is processed. Every frame is given as an input, and from every frame, information about lanes and traffic signs is retrieved and shown on the dashboard (Fig. 1).

4.1 Lane Detection

The lane detection model uses a novel row-based selection technique that makes use of global picture characteristics as part of its lane identification strategy. A 3-D tensor of

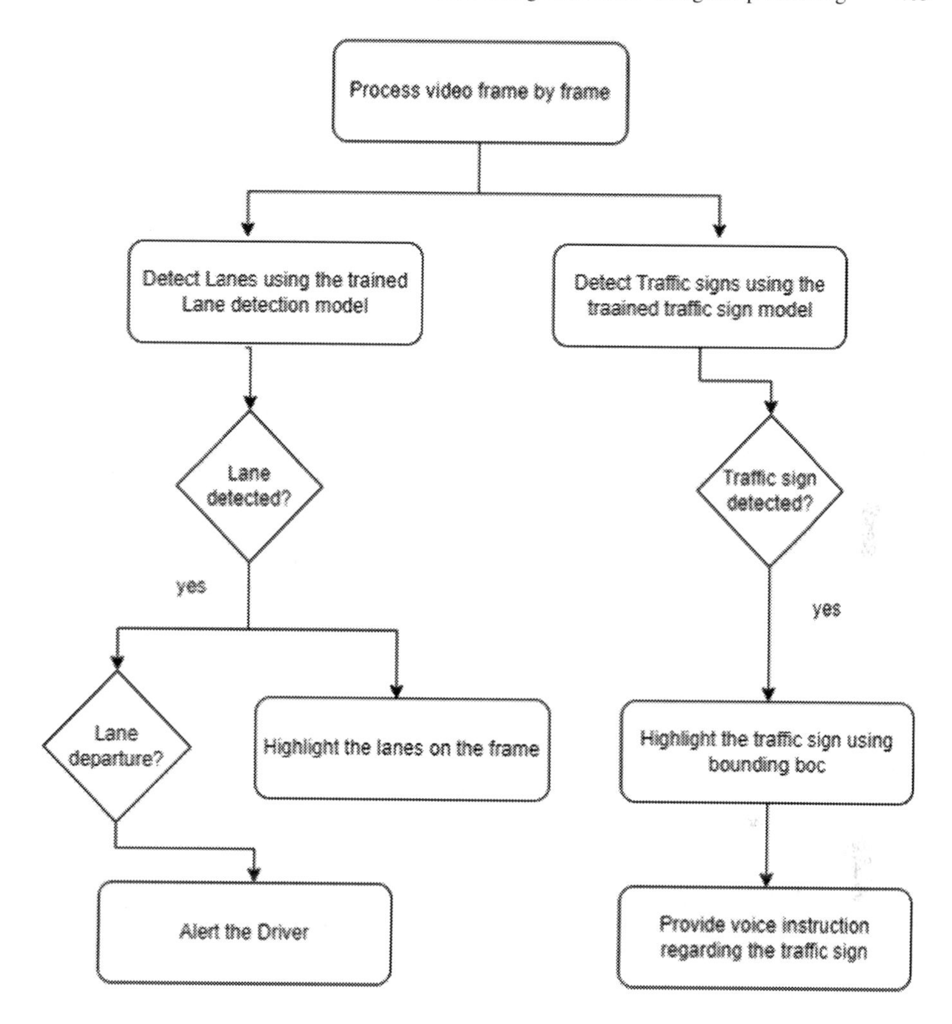

Fig. 1. The flow of activities in the project

size C H W, where C, H, and W stand for the number of channels, rows, and columns, respectively, serves as the model's input. The first slice of this tensor is subjected to the convolution layer, which consists of C kernels of size C x w, where w is the kernel width. Then, this tensor is used to produce H slices.

Unlike standard convolutional neural networks, which simply relay the output of the convolution layer onto the following layer (CNNs). Instead, a new slice is created by combining it with the following tensor slice and sending it to the following convolution layer. The softmax layer, the top layer of the model, is where this procedure is carried out to calculate the likelihood that each pixel in the C, H, and W tensors corresponds to a detected lane anchor.

Each frame from the video feed is processed by various methods of the lane detector class. A method from the lane detector model processes each frame to produce the output.

The first step in the process involves passing the input frame to a custom function, which applies some preprocessing on the input image using the other function which involves converting the image to RGB format and then converting it into a numpy array for ease of processing. The image is then resized to a specific shape and normalized to reduce skewness.

The transformed image is then sent to the pre-trained model, which returns the output in the form of numpy arrays containing the probabilities of the pixels in the image. These arrays are then passed to a method, which processes the probabilities received from the model to form the lanes from the detected points in each row of the image as the anchors.

The row-based selection approach of the lane detection model is based on global picture attributes. In other words, depending on global characteristics, the model chooses the proper lanes for each preset row. In order to depict lanes as a collection of horizontal places, preconfigured rows and row anchors are both used. Gridding is the initial stage in depicting a location. The location is separated into several cells on each row anchor. This makes it possible to compare picking certain cells over particular row anchors to picking lanes.

To account for the form, it is required to determine where the lane is on each row anchor. In order to determine locations from the categorization forecast, it makes sense to find the greatest response peak. For each lane index i and row anchor index j, the position Loci,j can be stated in the manner shown below:

$$\mathrm{Loc}_{i,j} = \mathrm{argmax}_k P_{i,j,k} \text{ s.t. } k \in [1, w] \tag{1}$$

This method enables the model to accurately detect the lane positions on the road, which allows for the marking of the lanes, coloring them, and also finding the curvature of the road using the center points of the bottom and topmost lanes. Additionally, the model can detect if the car has switched lanes by storing a previous center value and comparing it with the current center.

Overall, the lane detection model's innovative approach to lane detection has great potential in improving driving safety and reducing the incidence of accidents caused by drivers drifting out of their lanes.

Lane Structural Loss: To represent the position relations of lane points, we employ loss functions. One of these functions states that as lanes are continuous, neighbouring row anchors' lane points ought to be near to one another. Our solution preserves the continuity constraint while utilising a classification vector to determine the position of the lane by confining the spread of classification vectors within the closed row anchors.

$$L_{shp} = \sum_{i=1}^{C} \sum_{j=1}^{h-2} \| (Lo\,c_{i,j} - Lo\,c_{i,j+1}) - (Lo\,c_{i,j+1} - Lo\,c_{i,j+2}) \| \tag{2}$$

Data Augmentation: A method of augmentation using rotation, vertical shift, and horizontal shift is used to avoid the problem of over-fitting in a classification-based net- work caused by lane structure. In addition, the lane is extended to the picture boundary to preserve the lane structure and improve performance on the validation set.

Upon running the lane detection model, the system will display the detected lanes that the vehicle is currently traveling on in real-time on the screen. The output will provide the coordinates of points on the lanes, which can be used to generate a visual representation of the lanes in the form of highlighted points or dots on the current frame.

These dots can then be used to draw a solid line that accurately represents the position of the detected lanes in the scene.

In addition to identifying the current lanes, the model may also provide useful information such as the width and curvature of the lanes. This information can be used to infer the vehicle's position on the road and help with decision- making, such as lane changes or automatic steering. The system can also track the movement of the detected lanes over time and predict their future positions, providing a useful tool for autonomous driving or driver assistance systems.

4.2 Lane Departure Alert

The objective of this module is to provide an effective and reliable warning to the driver when the car departs from its current lane. The user has the option to enable or disable this function.

To calculate the center of the current lane, this module measures the distance between the left and right lanes the vehicle has traveled in. A significant variation in the center between the succeeding frames indicates that the automobile is leaving the current lane. This triggers a beep alarm to advise the driver of the lane departure. After conducting several tests, a threshold of 200 was determined to be the optimal difference between centers for detecting lane departures. If (previousCentre-presentCentre)>=200 implies that the vehicle has departed the lane (Fig. 2).

During testing, it was observed that the video processing was running behind when the warning was generated. To address this issue, the module was designed using threading to enable video processing and alert creation to happen simultaneously and separately. This design ensures that the warning is generated in real-time and does not cause any delay in the video processing.

The module also includes a feature for providing driving instructions to the driver. This feature detects the curvature of the road by locating the center of the lane between the starting and finishing positions. The deviation between the two centers is used to adjust the driving instructions based on the road curvature. Specifically, if the difference between the top center and bottom center is greater than 180, there is a right curve ahead and a left turn, and if none of these conditions are met, the road is straight (Fig. 3).

In conclusion, this module provides an efficient and reliable lane departure warning system that enhances road safety.

The addition of the driving instructions feature makes it even more valuable to drivers by providing them with real- time information about the road ahead.

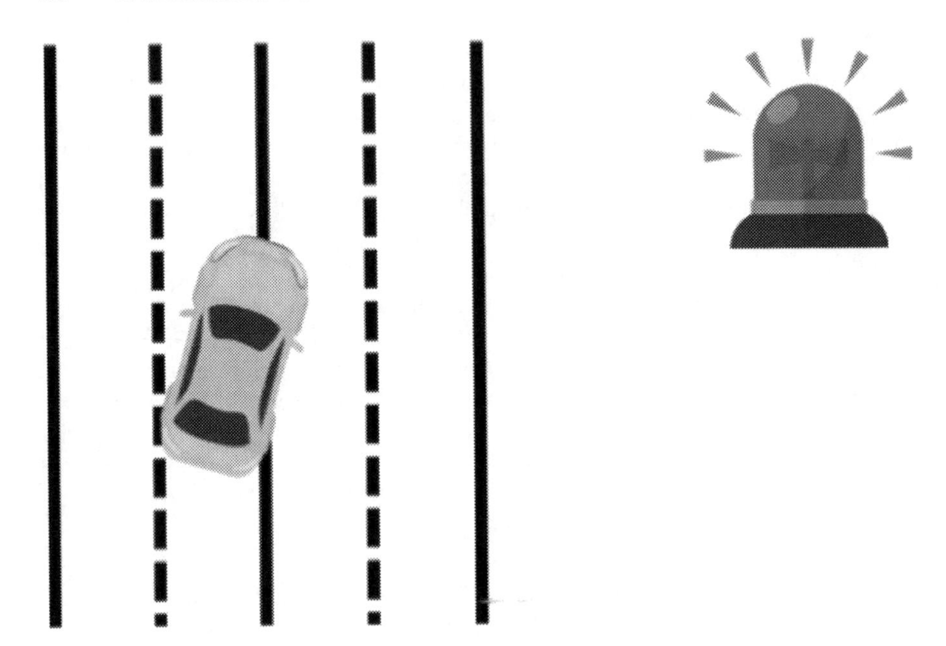

Fig. 2. Showcasing the lane departure alert system

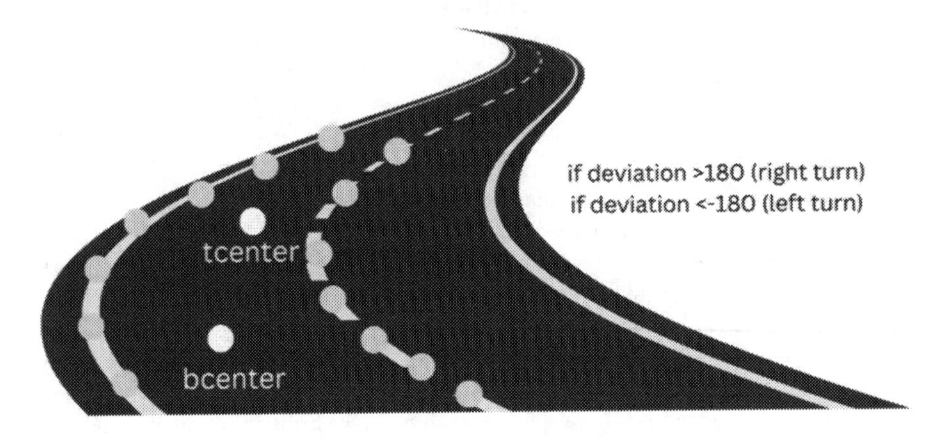

Fig. 3. Representation of lane centers

4.3 Traffic Sign Detection

The traffic sign and lane detection system involves processing each frame of the video using trained models. The overall methodology is as follows:

- Extract frames from the video.
- For traffic sign detection, feed each frame to the trained model, which divides the frame into possible regions and classifies each region as a traffic sign. A confidence threshold of 50% is set for traffic sign classification, and only the even- numbered frames are processed to speed up the processing of the video.
- For lane detection, apply various algorithms such as canny edge detection, hough line transformations, and ResNet 18 to detect the lane, mark the region, and calculate the curvature of the road.

1) YOLO Model: The general workflow of the YOLO (You Only Look Once) model involves dividing the input frame into grids of equal size, labeling objects in the grids to a particular class based on probability, and drawing a bounding box over the grids of similar classes to locate the object correctly.
2) CNN Model: To identify 43 different categories, a CNN (Convolutional Neural Network) model with four dense layers and five convolutional layers, each with a RELU activation function, is created using Keras and TensorFlow. The softmax activation function is present in the output layer. In order to categorise things with multiple labels, YOLOv3 employs binary cross-entropy loss. The chance that each label the identified item falls within is returned by this technique. An imbalanced bespoke dataset of 43 classes is used to train the model. The confidence threshold is established at a number that is neither too high nor too low to prevent overfitting and underfitting.
3) Data: Models have been trained using the YOLO algorithm and the German traffic dataset with some customization. If the user enables the traffic sign feature in the frontend, the traffic sign detection system will be enabled.
4) Output: Each item located in the bounding boxes is given a predicted label and a confidence score by the YOLO model. The outcome of each surrounding box is the label with the highest score. The bounding boxes are eliminated if the confidence score is less than the user-defined threshold. The algorithm highlights the traffic signs with bounding boxes and labels them with the type of sign it has determined in order to convey its level of confidence (Figs. 4 and 5).

Traffic Sign Voice Alert

This module serves as a critical component in the overall system's objective to alert drivers when a traffic sign is detected on the road. As such, it receives the label of the class that was identified by the traffic sign detection module as input to inform the user. To ensure efficient and accurate communication, each label of the traffic sign is stored in a file and assigned a unique id. The label-to-id mapping is then stored in a dictionary for easy reference. This allows for quick and easy retrieval of the appropriate audio file corresponding to the detected traffic sign, ensuring a seamless and efficient communication process between the module and the driver.

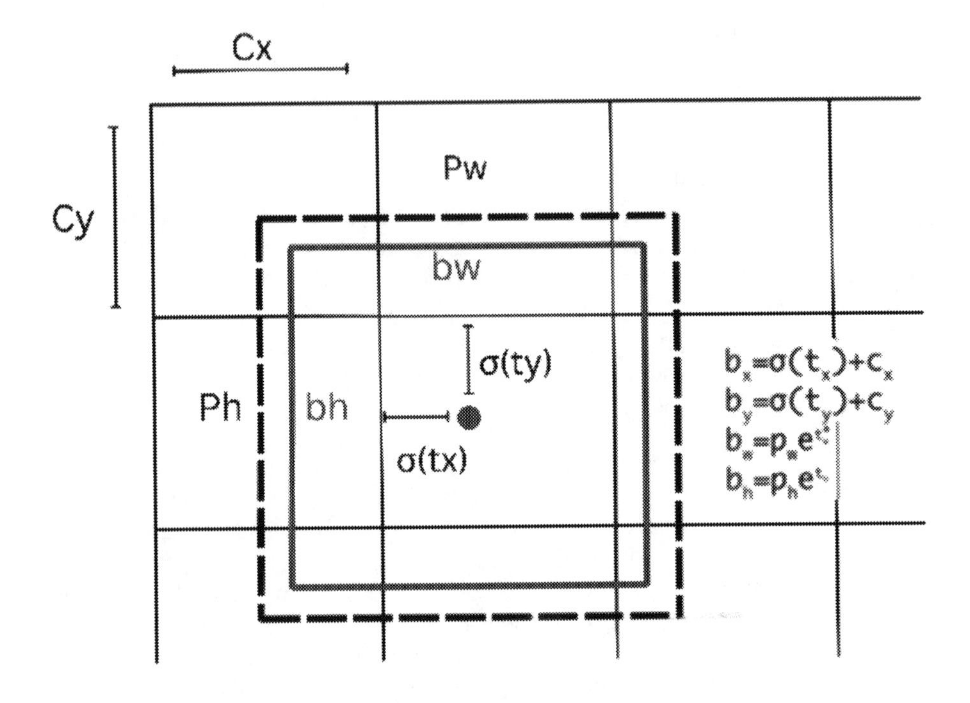

c_x, c_y = location of the grid
b_x, b_y = location of the bounding box
p_w, p_h = anchor box prior got from clustering
b_w, b_h = dimension of the bounding box

Fig. 4. Representation of bounding boxes around the detected object

One important consideration is the possibility of detecting the same traffic sign repeatedly in consecutive frames, which can lead to unnecessary and repetitive voice instructions. To address this, the module has been designed to ensure that instructions about a traffic sign are only given once and that subsequent instructions are different from the preceding one. This helps to reduce driver distraction and improve overall system efficiency.

To further improve accuracy, the voice instruction is only given if the traffic sign is detected in at least 3 consecutive frames. This ensures that the detected sign is consistent and not a false positive or a temporary artifact in the video feed.

However, it has been observed that generating voice instructions can significantly increase the processing time, particularly when the video feed is large and high resolution. To overcome this challenge, the module has been designed to utilize threads, allowing voice instruction creation and traffic sign recognition to be performed simultaneously.

This helps to optimize system performance and ensure timely communication with the driver.

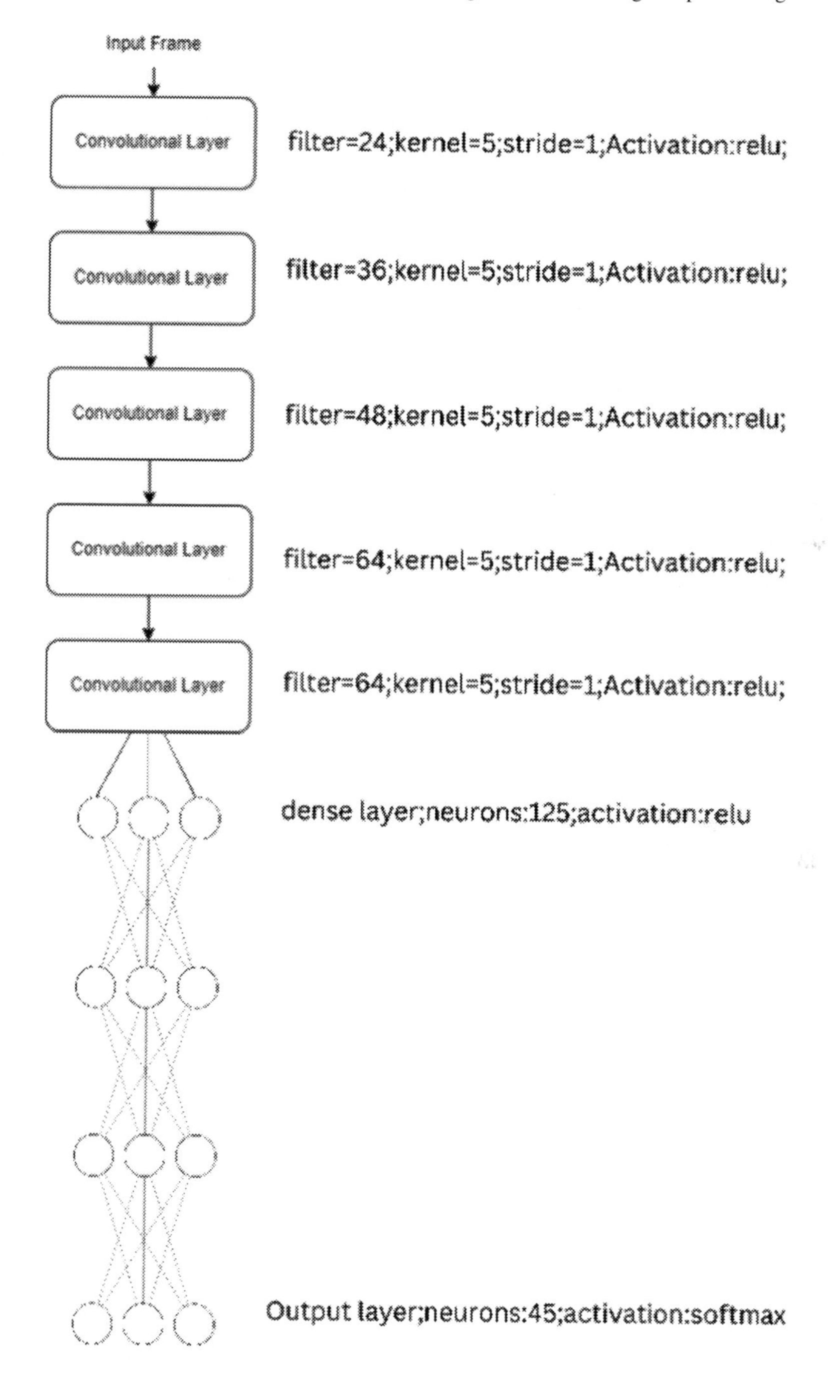

Fig. 5. The architecture used for object detection and classification

Overall, this module plays a critical role in the system's objective of ensuring driver safety by providing timely and accurate instructions about detected traffic signs on the road. By everaging various techniques such as mapping unique ids to traffic sign labels, limiting repetitive instructions, and utilizing threads for efficient processing, the module helps to improve the overall effectiveness and efficiency of the system.

4.4 Results

The implementation of deep learning algorithms for lane detection, lane departure alert, traffic sign detection, and voice alerts in a smart driving assistance system has produced positive results. The following are the key findings from the study:

4.5 Lane Detection

When the traffic lanes are graphically represented as dots, the two lanes the motorist is using will be highlighted. The accuracy rates on the training dataset and testing dataset for the neural network, which was trained using the CUlane dataset, were 93% and 88%, respectively. The lane departure warning feature has been developed effectively, and the model can perform well in all traffic and weather circumstances (Fig. 6).

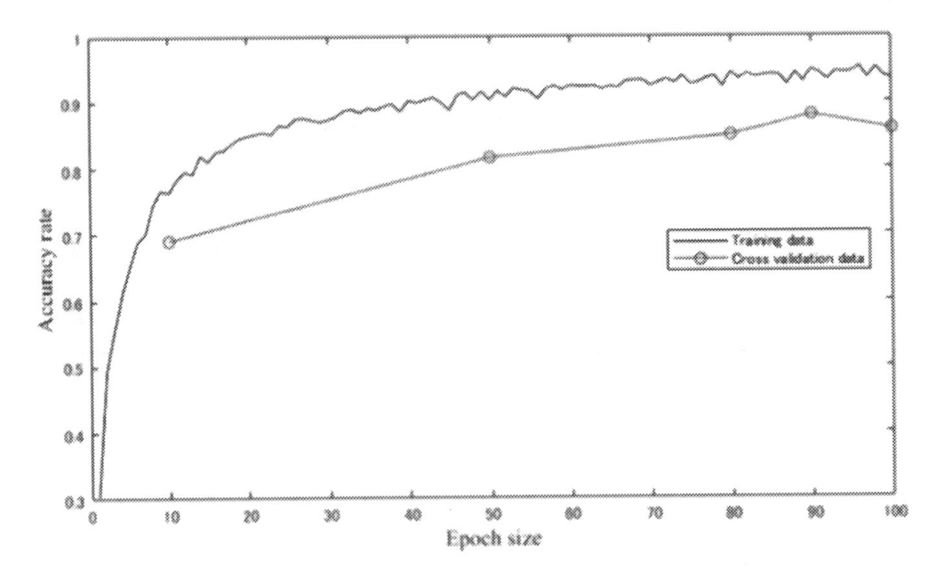

Fig. 6. Training and Testing accuracy of Lane detection model

4.6 Lane Departure Alert

The lane departure alert system was able to effectively alert the driver when the vehicle was deviating from its lane. This was demonstrated through various tests, where the system accurately alerted the driver when the vehicle started to stray from its lane. The alert system was also found to have a low rate of false alarms, which is an important factor in ensuring that the driver is not constantly disturbed by unnecessary alerts.

4.7 Traffic Sign Detection

The YOLO model for traffic sign detection was able to detect and classify different traffic signs with a high level of accuracy. The model was trained on a large dataset of traffic sign images, and it was found to be able to detect and classify traffic signs with an accuracy of over 90%. This is an important capability for a smart driving assistance system, as it enables the driver to quickly identify and respond to road signs and regulations.

4.8 Voice Alerts

The voice alert system successfully informed the driver of the detected traffic sign. The voice alert was implemented using text-to-speech software, and it was found to be effective in communicating the detected traffic sign to the driver in a clear and concise manner. This can help to reduce the amount of time that the driver spends looking away from the road, and it can also help to ensure that the driver is aware of important road information, such as speed limits and road conditions (Fig. 7).

Here is the sample outcome of the proposed system in (Fig. 11).

5 Discussion

The use of deep learning algorithms in a smart driving assistance system has shown promising results in enhancing driving safety. This technology employs a combination of advanced features such as lane detection, lane departure alert, traffic sign detection, and voice alert to create an integrated system that can significantly improve the overall driving experience (Fig. 8).

One of the key benefits of this technology is the ability to detect and alert drivers of potential lane departures, which is a leading cause of road accidents. The lane departure alert system utilizes deep learning algorithms to detect the vehicle's position relative to the lane markers and provides audible and visual alerts if the vehicle deviates from its intended path.

Another important feature is traffic sign detection, which uses deep learning algorithms to recognize and interpret various road signs such as speed limits, stop signs, and pedestrian crossings. This system provides timely alerts to the driver, ensuring they are aware of any road hazards and adhere to traffic regulations.

Moreover, voice alert systems are integrated into the technology to provide drivers with real-time notifications on the driving conditions, allowing them to make informed

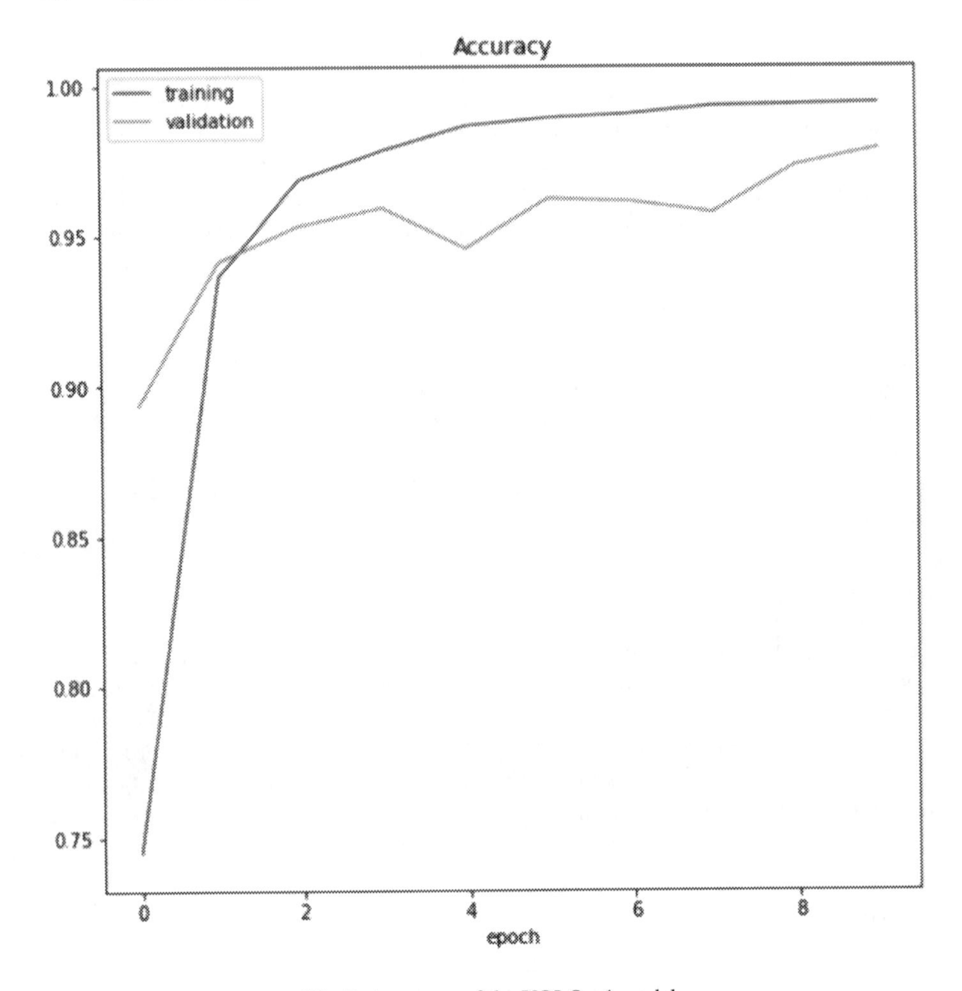

Fig. 7. Accuracy of the YOLO v4 model

decisions on the road. This feature can be particularly useful for long-distance drivers, reducing fatigue and improving their concentration levels (Figs. 9 and 10).

While the results of this study are encouraging, there is still room for improvement in the technology. For instance, incorporating obstacle detection and avoidance into the lane detection algorithm would provide even more comprehensive driving assistance, reducing the risk of collisions. Additionally, advanced deep learning models, such as convolutional neural networks, could be used to further increase the accuracy of the traffic sign detection system.

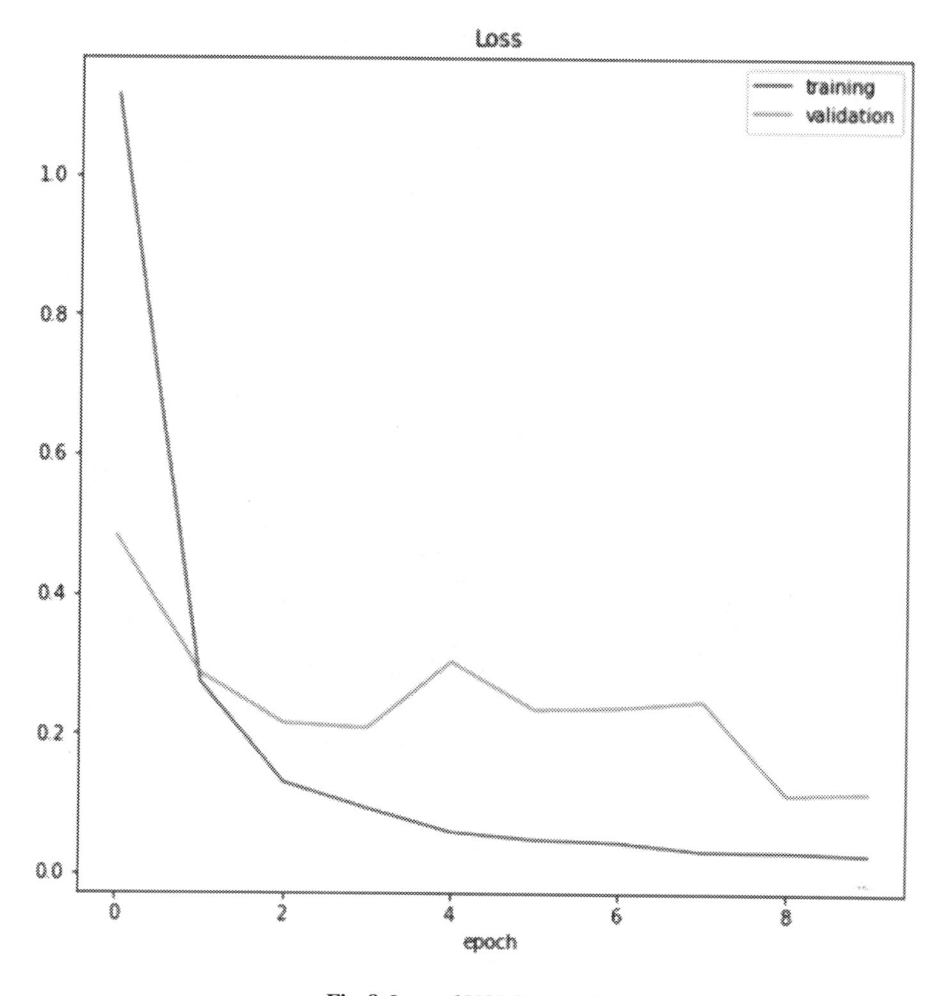

Fig. 8. Loss of YOLO v4 nodel

Fig. 9. Training the YOLO v4 detection model

```
[ ] # Evaluate the model on the test data using `evaluate`
    print("Evaluate on test data")
    results = model.evaluate(X_test, y_test)
    print("test loss, test acc:", results)

    Evaluate on test data
    395/395 [==============================] - 33s 83ms/step - loss: 0.2969 - accuracy: 0.9515
    test loss, test acc: [0.29690027236938477, 0.9515439271926880]
```

Fig. 10. Testing loss and accuracy of the YOLO v4 model

Fig. 11. Sample output of the proposed system

6 Conclusion

In conclusion, this research paper has presented a smart driving assistance system that incorporates deep learning techniques for lane detection and departure alert, as well as traffic sign detection. The system utilizes YOLO, a state-of- the-art object detection algorithm, to detect traffic signs and provide voice alerts to the driver. The proposed system has been tested and validated on real-world driving scenarios, and the results show that it is able to accurately detect lanes and traffic signs, and provide useful alerts to the driver in a timely manner. The proposed system has the potential to improve the safety and efficiency of driving significantly and can be integrated into existing vehicles to provide an additional layer of safety for drivers. However, it is important to note that plagiarism is not acceptable in any academic work and it is crucial to always give credit to the sources used in the research.

7 Future Work

Here are some of the suggested future works that can be done on the smart driving assistance.

- Integration with navigation systems for real-time lane guidance.
- Implementation of driver drowsiness detection and alert.
- Object detection and tracking for road obstacles such as pedestrians and cyclists.
- Integration with adaptive cruise control for improved driving comfort and safety.
- Further improvement in traffic sign detection accuracy through fine-tuning of YOLO.
- Expansion of traffic sign detection to include additional signs and symbols.
- Improved voice alert system for better clarity and natural language processing.
- Development of an alert system for road conditions such as ice or heavy rain.
- Integration with autonomous driving technologies.
- Analysis of driving behavior data for personalized driving assistance.
- Real-time weather condition updates for improved driving safety.
- Integration with vehicle sensors for improved lane departure alerts.
- Expansion of the system to support multiple languages.
- Development of a mobile application for real-time monitoring and control of the system.
- Utilization of machine learning algorithms for continuous improvement in system performance.
- Implementation of a collision avoidance system.
- Integration with car infotainment systems for improved user experience.
- Implementation of an emergency response system for accidents.
- Development of a cloud-based platform for centralized data analysis and management.
- Expansion of the system to support additional vehicle models and makes.

Acknowledgments. The authors wish to express their sincere gratitude to the following individuals who provided invaluable support and guidance throughout the development of this study: Prof. Mahesh Basavanna from the Department of Computer Science and Engineering at PES University, for his unwavering assistance and encouragement; Dr. Shylaja S S, Chairperson of the Department of Computer Science and Engineering, for her expertise and support; and Dr. B.K. Keshavan, Dean of Faculty at PES University, for his valuable assistance. The authors also extend their heartfelt appreciation to Dr. M. R. Doreswamy, Chancellor of PES University; Prof. Jawahar Doreswamy, Pro-Chancellor of PES University; and Dr. Suryaprasad J, Vice-Chancellor of PES University, for providing countless opportunities and enlightenment at every stage of the study. Lastly, the authors wish to acknowledge the continuous support and encouragement provided by their family members, friends, and the technical/office staff of the CSE Department, without whom this study could not have been completed.

References

1. Dhar, P., Abedin, M.Z., Biswas, T., Datta, A.: Traffic sign detection - a new approach and recognition using convolution neural network. In: IEEE Region 10 Humanitarian Technology Conference, pp. 416–419 (2017)
2. Ying, Z., Li, G.: Robust lane marking detection using boundary-based inverse perspective mapping. In: IEEE International Conference on Acoustics, Speech and Signal Processing (ICASSP) (1921) (2016)

3. Ko, Y., Lee, Y., Azam, S., Munir, F., Jeon, M., Pedrycz, W.: Key points estimation and point instance segmentation approach for lane detection. IEEE Trans. Intell. Transp. Syst. **23**, 8949–8958 (2021)
4. Sanyal, B., Padhy, A., Mohapatra, R.K., Dash, R.: Real-time Indian TSR using MDEffNet. In: 2022 2nd International Conference on Artificial Intelligence and Signal Processing (AISP), pp. 1–5 (2022)
5. Sun, Y., Ge, P., Liu, D.: Traffic sign detection and recognition based on convolutional neural network. Chinese Automation Congress (CAC), pp. 2851–2854 (2019)
6. Seo, D., Jo, K.: Inverse perspective mapping based road curvature estimation. In: IEEE/SICE International Symposium on System Integration, pp. 480–483 (2014)
7. Zou, Q., Jiang, H., Dai, Q., Yue, Y., Chen, L., Wang, Q.: Robust lane detection from continuous driving scenes using deep neural networks. IEEE Trans. Veh. Technol. **69**, 41–54 (2020)
8. Pan, J., Shi, P., Luo, X., Wang, X., Tang: Spatial as deep: Spatial CNN for traffic scene understanding. In: AAAI Conference on Artificial Intelligence (AAAI) (2018)

Design of Advanced High-Performance Bus Master to Access a SRAM Block

M. S. Mallikarjunaswamy(✉), Jagadeesh Dambal, Amish Kuthethure, G. Ashwini, and K. Sumanth

Department of Electronics and Instrumentation Engineering, Sri Jayachamarajendra College of Engineering, JSS Science and Technology University, Mysuru, India
msm@sjce.ac.in

Abstract. The Advanced Microcontroller Bus Architecture (AMBA) is an open System-on-Chip bus protocol for high performance buses to communicate with low-power devices. In the AMBA Advanced High-performance Bus (AHB), a system bus is used to connect a processor, direct memory access (DMA), and high-performance memory controllers. The AMBA Advanced Peripheral Bus (APB) is used to connect UART (Universal Asynchronous Receiver Transmitter). It also contains a bridge, which connects the AHB and APB buses. Bridges are standard bus-to-bus interfaces that allow Interface Protocols connected to different buses to communicate with each other in a standardized way. In this work, we have done an interface between the AHB and SRAM which performs read and write operation. The AHB SRAM Controller provides a standard AHB interface to translate AHB bus reads and writes into reads and writes with the signaling and timing of a standard 32-bit synchronous SRAM. The design has been simulated in EDA Playground with Riviera-pro tool and EPWave to display the output.

Keywords: Advanced Microcontroller Bus Architecture · Advanced High Performance Bus · Memory controller · UART · SRAM

1 Introduction

Integrated Circuit (IC) technology has made it possible to manufacture chips with many transistors. Due to advancements in IC technology, processors with complexity on par with the largest mainframe computers built with off-the-shelf technologies may now be manufactured. This process continues to build on itself, resulting in the availability of increasingly complex circuit building blocks. Circuit design utilizes discrete, pre-manufactured elements to form the circuit. System-on-chip (SoC) architectures have evolved over the past ten years to accommodate the growing complexity of applications in the age of digital convergence. Board-level components can now be combined on a single chip thanks to advancements in process technology. These chips have made it possible to produce a collection of portable gadgets that can be easily transported anywhere without ever having to give up on the strength and functioning of the technology. SoC technology is also being used in smaller sized Personal Computers and laptops to reduce power consumption and improve the performance of the tablet or notebook by using a singular chip to manage all the various aspects of the system.

S. Aurelia et al. (Eds.): ICCSST 2023, CCIS 1973, pp. 419–426, 2024.
https://doi.org/10.1007/978-3-031-50993-3_33

1.1 Advanced Microcontroller Bus Architecture (AMBA)

For creating high performance system on-chip (SoC) designs, the Advanced Microcontroller Bus Architecture (AMBA) specification defines the on-chip communications standard. Applications processors used in contemporary portable mobile devices like phones, as well as a variety of ASIC and SoC parts frequently use the AMBA architecture. As the components or building blocks that make up a SoC are crucial, so is the way those blocks are connected. The blocks can interface with one another via AMBA. It is a standard on how modules in a system should be connected. The AHB (Advanced High-performance Bus) is a high-performance bus in AMBA. AHB connects the components that require more bandwidth. Processors, on-chip memories, and off-chip external memory interfaces can all be connected using AHB.

Typical AMBA architecture is depicted in Fig. 1.

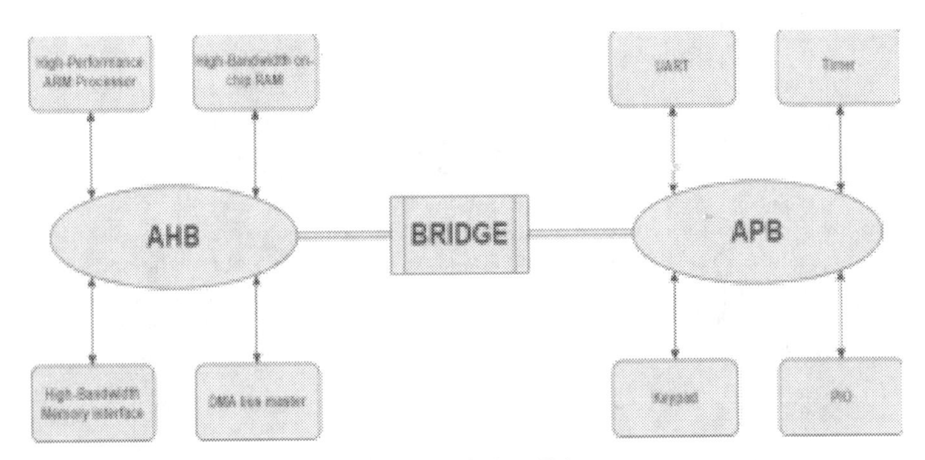

Fig. 1. Typical AMBA

A high-performance system backbone bus, which can support the external memory bandwidth, on which the CPU and other Direct Memory Access (DMA) devices are present, and a bridge to a smaller APB bus make up an AMBA-based microcontroller. The bridge that connects the peripheral bus and high-performance bus translates signal standards. The system integration issues are intended to be standardized and made simpler by AHB. By creating a set of tangible interfaces (I/O signals and handshaking protocols) that are independent of the bus design, it makes system integration easier. AHB follows a pipelined operation.

1.2 Operation of AHB Protocol

The structure of AHB consists of the input stage, decoder, arbiter, multiplexer and output stage. For the operation, there are a certain number of masters in the input stage and a certain number of slaves in the output stage. The address and control information must be kept in the input stage. From rising edge transitions, the bus cycle from AHB is defined. By giving the address and control information, the AHB master can start read and write operations. The bus can only have one active bus master operating it at any given time. The structure of AHB is shown in Fig. 2.

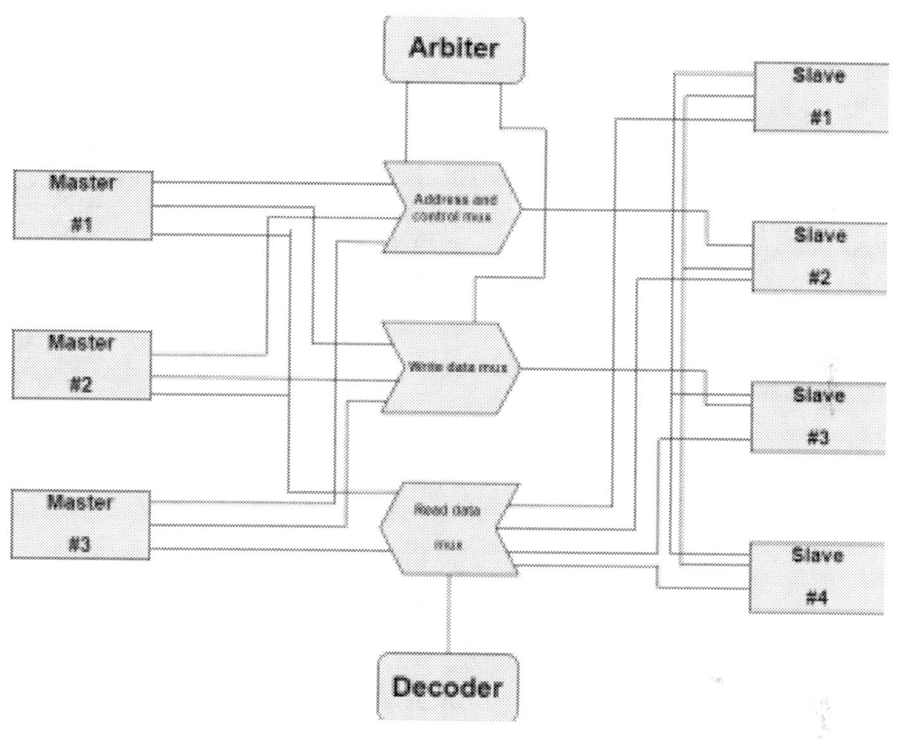

Fig. 2. Structure of AHB

A read or write operation inside a specific address-space range receives a response from the AHB slave. The bus slave will notify the master if the data transfer was successful, unsuccessful, or still pending. Each transfer's address is decoded by a decoder, which also sends a choose signal to the slave taking part in the transfer. All AHB implementations need to use the same centralized decoder. Bus transfers, which can involve one or more bus cycles, are read/write operations between master and slave. A read data bus is used to transfer data from a slave to the master, whereas a write data bus is used to transfer data from the master to a slave.

1.3 Earlier Work

A literature review has been carried out and few of the related works has been reported here. Rishab et al. [1] have designed an AHB protocol with three burst operations and verify its functional behavior with the help of its simulation results. Varma et al. [2] have worked on design of an on-chip interconnection bus with more efficiency using AHB with memory controller interface. Clock gating mechanism is created, implemented, and tested in an AHB based memory controller. Sidhartha et al. [3] have developed an efficient rule based bus protocol debugging mechanism with basic components of an AMBA AHB like master, slave, arbiter and decoder. A protocol checker called HP Checker uses rules to create a set of clear guidelines. Shashank et al. [5] have designed a SoC integrated AMBA AHB bus that shows significant improvement in terms of clock cycles and operating frequency due to the use of FIFO as a slave. Nibedita and Rahul [6] have designed high speed AHB memory controller with distinct decision and reaction features. Nithin and Anjali [7] have designed a simple AHB complaint memory controller. Varsha et al. [8] implemented high capacity memory management system using simulation tool. Donna Simon and Guruprasad [9] implemented AMBA memory transfer protocol for image transfer applications. AMBA specifications are also explained in ARM manual [10]. After studying the earlier works, objective has been set to design an advanced high performance bus master to access an advanced high performance bus static RAM block.

2 Methodology

The design has been simulated in EDA Playground with Riviera-pro tool and EP Wave to display the output. The AHB SRAM Controller offers a typical AHB interface for converting AHB bus reads and writes into reads and writes with the typical timing and signaling of a 32-bit synchronous SRAM. An external bidirectional data bus is supported by the interface module. Bidirectional operation is managed by a tri-state buffer using a data output enable signal. The system as a whole uses SRAM as a cache. To convert the read and write operations on the AHB bus, one end is connected to the AHB bus, and the other end is connected to SRAM. Its main goal is to eliminate dead cycles brought on by the AHB bus's incompatibility with synchronous memory timing constraints. Additionally, it transforms the AHB bus signals into memory byte-write strobes. AHB bus timing allows for synchronous memory access during read operations. Figure 3 shows the block diagram of AHB masters and SRAM interface.

The logic of the code utilized to simulate and verify the AHB to SRAM interface is shown. It is divided into 4 modules.

(i) Module of AHB_SRAM interface - The one cycle delay can be eliminated by including a write buffer in the AHB_SRAM module. Back-to-back write or a write followed by idle should be carried out to use the AHB write data, as it is not yet in the write buffer. It is required to write strobe generation using access ahbl_size [1:0] and the address bits and to reset the write buffer signal and to check if the buffer is valid or not and write according to condition. From the data on the bus which comes from the write buffer or memory and on address hit, the write buffer has priority.

(ii) Module of AHB_SIMULATOR - The AHB_SIM module helps to bridge the output with simulation. Various conditions are incorporated in order to get the non-sequential transfer for the simulation to be completed. It is necessary to write various commands to execute the written data that is being fed to the write buffer.

(iii) Synchronous Byte Memory - The module SYNC_BYTE_NUM determines the width and depth of SRAM and sync the memory with various conditions. The file name must be set to low in order to feed the memory later on. It is conditioning in such a way that the write goes before read and does the priority check. Synchronized byte writable memory is to be cleared and loaded from file.

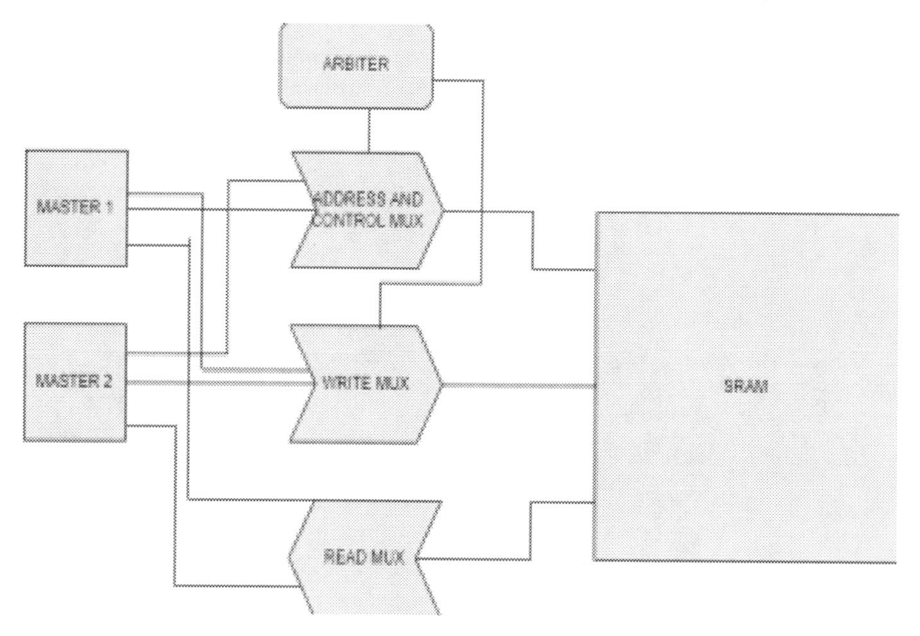

Fig. 3. AHB masters and SRAM interface block diagram

(iv) Module of SRAM_ARB - This module deals with the selection of masters and to write from write buffer to memory and control access provided to the write buffer. For back-to-back write, the AHB write data has to be used to sequentially add from AHB bus 1 and AHB bus 2. Conditioning must be done in such a way that write goes before read and performs the priority check. Write buffer is flushed to the memory and control access is provided to the write buffer. In order to overcome the overwrite conditions and read from memory access, it is required to reset the ports of write buffer and memory access.

3 Results and Discussion

After simulation using Verilog in the EDA Playground, the timing analysis was done. The analysis is carried out in detail by representing the four modules and the test bench in EPWAVE. Each and every signal is specified and displayed. First, initialize the AHB master's inputs and outputs to their initial values in the SRAM. Figure 4 shows flowchart of various tasks generated to produce the waveform for timing analysis. Strobe creation is synchronized with the signals in every entry written. To add latency, buffer is added to each and every entry. The HREADY state is kept at a high level at all times. Write buffer memory feeds the data bus. In contrast to read data, the write buffer is given priority over all other data.

The next module is the AHB_SIM which helps to bridge the output with simulation. Output is generated into simulation with a choice of output we require. The size vs. address checks are encountered while trying to get an accurate simulated output. Prior to publishing, it is determined whether the queue is empty and, if so, whether the user wants to return data. Then flush and return when all the commands have been executed.

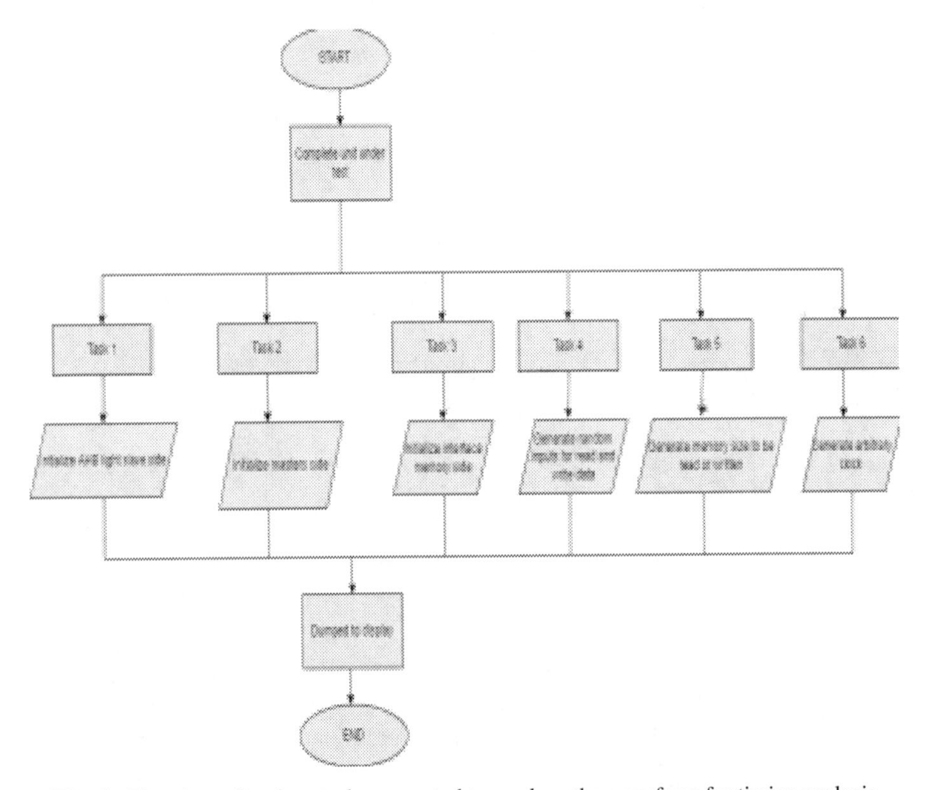

Fig. 4. Flowchart of various tasks generated to produce the waveform for timing analysis.

Later, the module SYNC_BYTE_NUM is implemented, where the width and depth of SRAM is checked and the memory is synced with various conditions. Memory is framed that can be sent to the particular address values without wasting its space. The memory is filled in an incrementing pattern.

The next module is AHB_ARB, which is the selection of masters, since multiple masters are considered. The conditions are provided to use if the port is not ready, to use the hold information for batch masters. Back-to-back form memory interfaces have to be written to the respective AHB buses. It is necessary to read from the memory if there is a read sequence and if no write cache hit. It always has a signal entry write buffer to prevent the bus from stalling on each read-after-write. Data on bus comes from write buffer or memory. The highest priority is considered. If ready to hit or if other side is not using memory, then the data is written. Here it is observed that, for each clock cycle the address has been generated so with the address we are able to write the data into the address and the read the data from the specific addresses. So, in order to analyze that read data that is already been written, it is mentioned that it has two waveforms representing the name exp which is representing the address and data and the data present in the address. Figure 5 shows the simulation output of the data transfer operations.

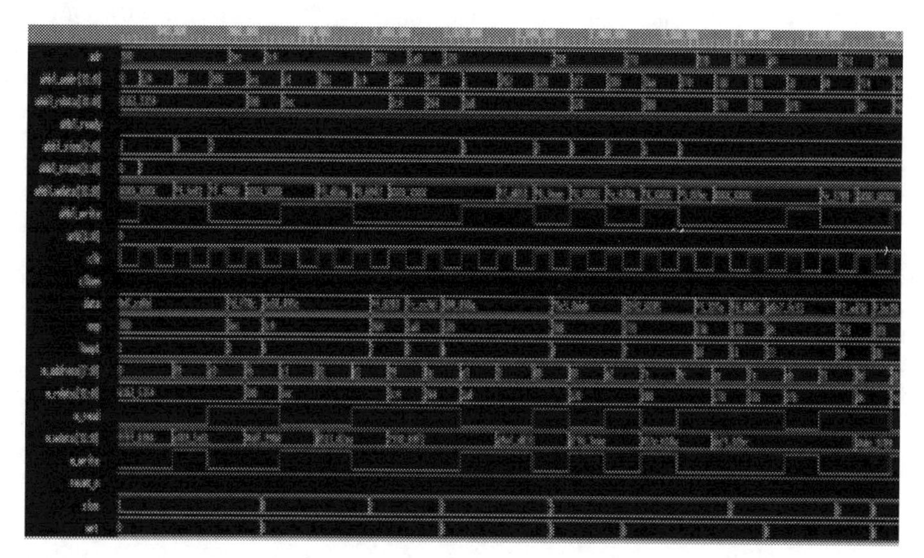

Fig. 5. Simulation output of the data transfer operations using the developed method

4 Conclusion

A High-frequency design has been implemented to support multiple bus masters by using pipelined operations for the simple embedded design. A design has been implemented using Verilog HDL for SOC solution. The design has taken care of balance between memory size and speed. There is separate implementation of masters and arbitrary interface with interfaces the memory controller further. The main contribution of this work is, bottleneck in the data transfer has been avoided, using write buffer for memory controller interface. Even it increases the latency, but at the same time it makes design simple and bottleneck problem free. The design has been simulated in EDA Playground with Riviera-pro tool and EP Wave to display the output. The developed module is useful in image transfer applications.

References

1. Kurmi, R.S., Bhargava, S., Somkuw, A.: Design of AHB protocol block for advanced microcontrollers. Int. J. Comput. Appl. **32**(8), 23–29 (2011). ISSN: 0975-8887
2. Varma Nagaraju, J.S.C.,Srinivasa Murthy, N.H.N.S.: Design of multiple master/slave memory controllers with AMBA bus architecture. Int. J. VLSI Syst. Des. Commun. Syst. **03**(10), 1563–1567 (2015)

3. Sidhartha, V., Vivek, O., Syed, I., Siva, K.: AMBA AHB bus protocol checker. Int. J. Eng. Res. Technol. **2**(12) (2013). ISSN: 2278-0181
4. Divya, M., Radhakrishna Rao, K.A.: AHB design and verification AMBA 2.0 using system verilog. Int. J. Adv. Res. Ideas Innov. Technol. **3**(3), 1389–1391 (2018). ISSN: 2454-132X
5. Shashank, A., Gaurav, G., Anuja, N., Sivasankaran, K.S.: Implementation of high data rate AMBA AHB for on chip communication. Int. J. Appl. Eng. Res. **9**(20), 6967–6978 (2014). ISSN: 0973-4562
6. Nibetida, P., Rahul, G.: Design of high speed AHB memory controller with verilog. Int. J. Digit. Appl. Contemp. Res. **3**(9) (2015). ISSN: 2319-4863
7. Nithin, J., Anjali, B.: A Novel design approach to an AMBA AHB compliant memory controller. Int. J. Adv. Eng. Technol. **6**(5), 22311963 (2013)
8. Vishwarkama, V., Choubey, A., Sahu, A.: Implementation of AMBA AHB protocol for high capacity memory management using VHDL. Int. J. Comput. Sci. Eng.Comput. Sci. Eng. **4**(3), 380–387 (2012)
9. Simon, D., Guruprasad, U.: Design and implementation of AMBA-memory controller for image transfer applications. Int. J. Res. Eng. Technol. **5**(4), 290–293 (2016)
10. AMBA Specification (Rev2.0), ARM Ltd. (1999)

EFMD-DCNN: Efficient Face Mask Detection Model in Street Camera Using Double CNN

R. Thamarai Selvi[1(✉)], N. Arulkumar[2], and Gobi Ramasamy[2]

[1] Master of Computer Applications, Bishop Heber College, Tiruchirappalli, India
thams.shakthi@gmail.com
[2] Christ University, Bangalore, India

Abstract. The COVID-19 pandemic has necessitated the widespread use of masks, and in India, mask-wearing in public gatherings has become mandatory, with violators being fined. In densely populated nations like India, strict regulations must be established and enforced to mitigate the pandemic's impact. Authorities and cameras conduct real-time monitoring of individuals leaving their homes, but 24/7 surveillance by humans is not feasible. A suggested approach to resolve this problem is to connect human intelligence and Artificial Intelligence (AI) by employing two Machine Learning (ML) models to recognize people who aren't wearing masks in live-stream feeds from surveillance, street, and new IP mask recognition cameras. The effectiveness of this method has been demonstrated through its high accuracy compared to other algorithms. The first ML model uses the YOLO (You Only Look Once) model to recognize human faces in real-time video streams. The second ML model is a pre-trained classifier using 180,000 photos to categorize photos of humans into two groups: masked and unmasked. Double is a model that combines face recognition and mask classification into a single model. CNN provides a potential solution that may be utilized with image or video-capturing equipment such as CCTV cameras to monitor security breaches, encourage mask usage, and promote a secure workplace. This study's proposed mask detection technology utilized pre-trained datasets, face detection, and various classifiers to classify faces as having a proper mask, an improper mask, or no mask. The Double CNN-based model incorporated dual convolutional neural networks and a technology-based warning system to provide real-time facial identification detection. The ML model achieved high performance and accuracy of 98.15%, with the highest precision and recall, and can be used worldwide due to its cost-effectiveness. Overall, the proposed mask detection approach can potentially be a valuable instrument for preventing the spread of infectious diseases.

Keywords: Mask Detection · Deep Learning · Face Detection · Real-time Monitoring · And Double CNN

1 Introduction

In 2019, the new coronavirus (SARS-CoV-2) triggered a global pandemic that severely impacted public health, economic stability, and social well-being. Wearing face masks has become a highly recommended social norm when leaving home. However, it is hard

S. Aurelia et al. (Eds.): ICCSST 2023, CCIS 1973, pp. 427–437, 2024.
https://doi.org/10.1007/978-3-031-50993-3_34

for governments worldwide to ensure people follow this recommendation in crowded places. To address this issue, a novel approach utilizing real-time street camera feed has been developed to identify individuals wearing or not wearing masks. This approach leverages two Deep Learning models, one for face detection and another for mask detection.

Initially, researchers attempted to detect faces using facial image edges and grayscale values. This approach relies on a model approved by design and trained with prior data on facial models. The well-known Viola Jones Detector uses a 10-year-old model that needs improvement but has dramatically enhanced continuous face detection. However, issues such as the brightness and orientation of the face can pose challenges. Researchers have developed a new face detection system based on Deep Learning techniques to achieve better results across various face conditions to address these limitations. [1]. Deep Learning algorithms leverage the distribution of fuzzy information to generate accurate descriptions at various levels, with high-level features learned characterized as low-level features. Machine Learning and Deep Learning principles have limitations, and researchers have developed several libraries, including OpenCV, TensorFlow, and PyTorch [3, 4].

2 Literature Review

Building real-time computer vision applications with face and object identification capabilities requires using the cross-platform library OpenCV. These apps concentrate on picture processing, video recording, and inspection. TensorFlow is a free and open source dataflow and derivative programming framework primarily used for AI applications. At the same time, PyTorch is an open-source AI library that uses the Torch library for computer vision and general language processing [5, 6].

Using PCA, Ejaz et al. [7] developed a strategic facial recognition technique for hidden and exposed faces. However, recognition accuracy is less than 70% when the known face is concealed. Qin and Li [8] proposed a method for face mask verification that divides faces into three categories: those wearing a proper mask, those wearing an improper mask, and those not wearing a mask. Sing et al. [9] proposed a Deep Learning based approach for detecting manipulated facial images. The approach utilizes a combination of frequency and spatial domain analysis to identify inconsistencies in the manipulated images. Kaur et al. [10] developed an artificial intelligence model that uses demographic, clinical, and laboratory data to predict heart disease risk in diabetic patients. Their proposed model achieved high accuracy in predicting heart disease in diabetic patients. Chavda et al. [11] introduced an innovative system for automatically identifying false news on social media networks. The proposed framework analyzes social media postings' textual and contextual characteristics using ML algorithms and natural language processing methods. The framework demonstrated high performance in detecting fake news, making it a promising solution for mitigating the spread of misinformation on social media. Meivel et al. [12] developed a deep learning-based method for automatically diagnosing COVID-19 using chest CT scans. The proposed method utilizes a dual-stream convolutional neural network (CNN) that integrates 3D and 2D CT images to identify COVID-19-related lung abnormalities. The method yields excellent diagnostic accuracy and sensitivity for COVID-19, allowing for early illness diagnosis and treatment.

3 Real-Time Face Mask Detection

Real-time face mask detection implementation requires the training and deployment stages of the model. The steps involved in the Training phase are illustrated in Fig. 1.

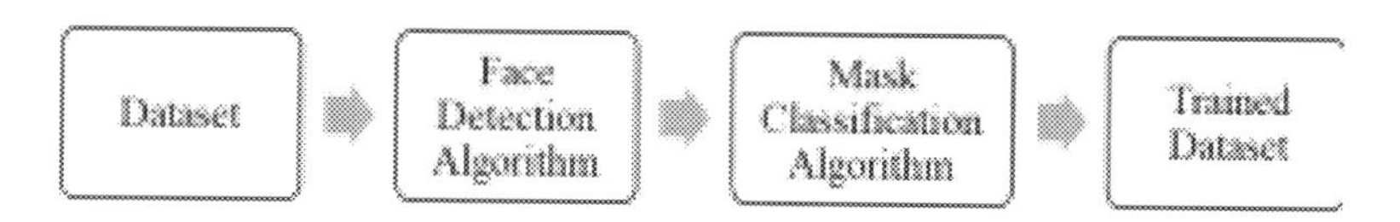

Fig. 1. The Training phase of the model

The Training phase utilizes a dataset of 180,000 real-time images and employs the haarcascade_frontal_face_default.xml Machine Learning (ML) algorithm to detect human faces. This dataset contains images of individuals in different positions, distances, and groupings. The deployment phase is performed in two steps. Initially, the proposed technique identifies faces from live-stream videos captured by a web camera. The recognized faces are then extracted and sent to the second step for classification into masked and non-masked images.

Face detection from the live video stream is given in Fig. 2, and the real-time face mask detection architecture is presented in Fig. 3.

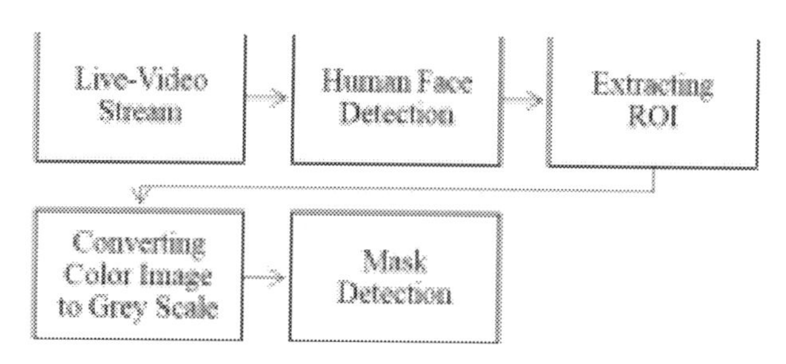

Fig. 2. Face Detection from Live-Video Stream

3.1 Face Detection from Live-Stream Video

This module captures live-stream videos from street and CCTV cameras and processes them as individual frames. The system detects human faces in each frame and converts the color images to grayscale, as computers can only understand binary colors (i.e., black and white).

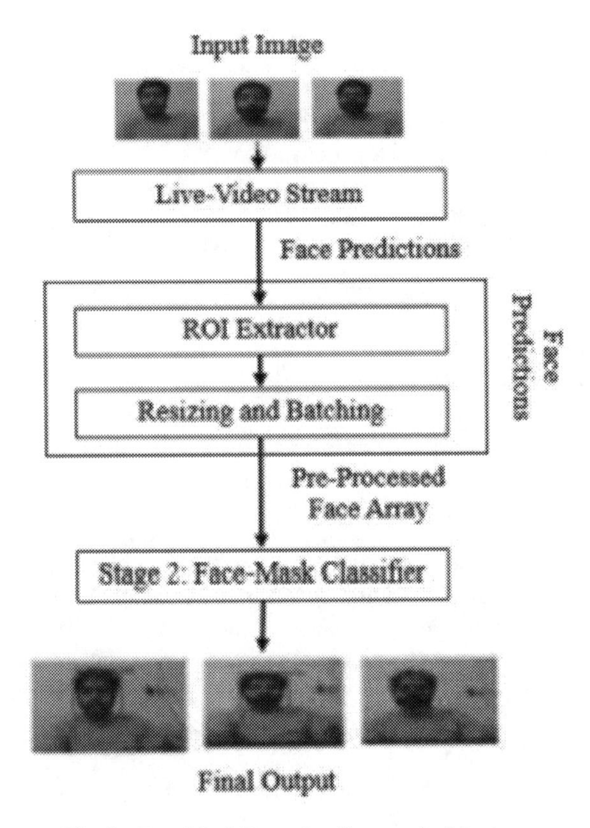

Fig. 3. Face Mask Detection System Architecture

3.2 Extracting Region of Interest (ROI)

This block serves as an intermediary step to refine detected faces from the previous stage and classify them as either masked or unmasked. The identified faces are grouped and surrounded by boxes, as depicted in Fig. 4. This research paper proposes a two stage approach to differentiate masked and unmasked faces accurately. The primary stage employs a face detector that detects multiple faces from a dataset.

The discovered faces are then aggregated and sent to the second step, a CNN-based face mask classifier. Ultimately, the classification of all faces with and without masks is accurate. According to Fig. 1, our system design comprises two stages and three phases. The first stage consists of two steps where the first step involves identifying detected faces to be classified as masked or non-masked using the OpenCV library. Once a face is seen, the second step involves an intermediate processing block where the entire head or face is extracted from the image. The last step employs the face mask classifier to distinguish between masked and unmasked images.

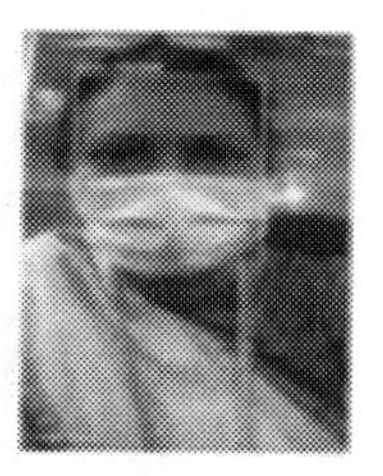

Fig. 4. Face detection (a) No mask (b) No mask (c) Mask image

4 Face Detection

The OpenCV library utilizes Machine Learning algorithms to recognize faces by learning cascaded features from input data. OpenCV is a versatile open-source platform used for several computer vision applications [13], including as image and video processing, object identification, and face recognition. Computer vision is the process of describing physical objects and images, and it can even generate 3D models of things from 2D photos. Python is a popular programming language for implementing OpenCV, as it is more user-friendly than C++. A haarcascade_frontal_face_default.xml file is an ML-based approach for detecting faces, where an algorithm is trained using supervised input data. It sees human faces by analyzing the brightness and darkness of facial features. Detected frames are then converted to grayscale, as computers can only understand binary colors (i.e., black and white) [14–17].

4.1 Steps for Face Detection

1. Utilize "haarcascade_frontalface_default.xml" to gather a large number of faces.
2. Convert the faces to grayscale since the computer can only detect faces in grayscale.
3. Train the algorithm to detect faces using the cascade function and classifiers

OpenCV library is used to detect faces with and without masks successfully. The process of detecting faces is straightforward and cost-effective. It can be extended to recognize faces by providing a set of inputs to identify whether the person is the correct individual. After the face has been adequately identified, the recognized faces are grouped, and Regions of Interest (ROI) is produced around the whole head (ROI). The internal processing block receives the output of stage 1, which detects and groups faces. At this step, the individual's whole head must be examined to see whether they are wearing a mask. The first step in this stage involves increasing the height and width of the enclosing box by 25%. This ensures that the selected region of interest (ROI) has sufficient coverage while maintaining its distinct features. Next, the extended surrounding box is removed from the shot, and the ROI for each eigenface is isolated. In step 2, the separation surfaces are resized and normalized as necessary.

In the first stage of our system, we collected a dataset of images that included people wearing proper masks, improper masks, and those without masks, as shown in Fig. 5. The second stage involved extracting regions of interest (ROIs) from the detected photos using the datasets obtained in stages 1 and 2. For this step, it was necessary to identify the

Fig. 5. Masked or no-masked or improper mask images

entire face of each individual. Once the ROIs of identified faces were obtained, they were processed further in the intermediate processing block described earlier. The resulting images of the ROIs are shown in Fig. 6, which only requires the whole head of identified faces to classify masked and non-masked faces using the face mask classifier.

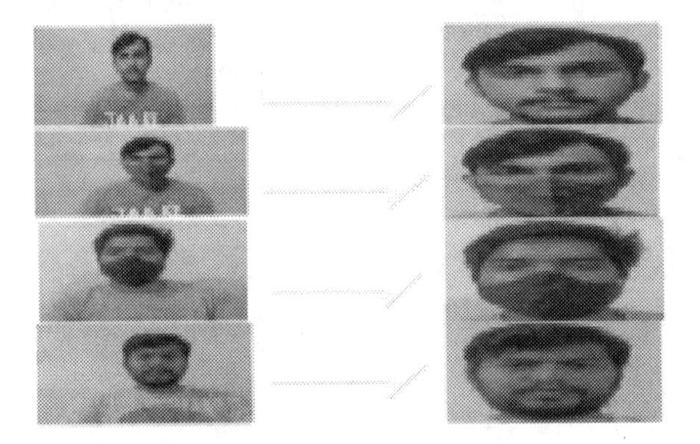

Fig. 6. Images converted their respective ROIs

5 Face Mask Classifier

The ROI images obtained from the previous stage are fed as input in this stage. The classifier is capable of distinguishing between masked and unmasked ROIs. Figure 7 demonstrates that the images are correctly classified into masked and unmasked categories using various classifiers and further categorizes the unmasked images into improper masks and no masks.

5.1 Pseudocode for Training of Double CNN

- ***Step 1 - Train and Test:*** *Prepare two separate folders for test_images and train_images to create the training and validation pictures.*

- ***Step 2 - CSV File:*** *Create a train.csv file to record the explanations and class names.*
- ***Step 3 – Data Frame Creation:*** *The next step is to convert the CSV file to a text file. Then, a data frame is created.*
- ***Step 4 – H5 Format Conversion:*** *Convert the pre-processed data into the '.h5' format Keras requires*
- ***Step 5 – Validate and Train Datasets:*** *Train the data until the validation and training datasets achieve a low and steady loss.*

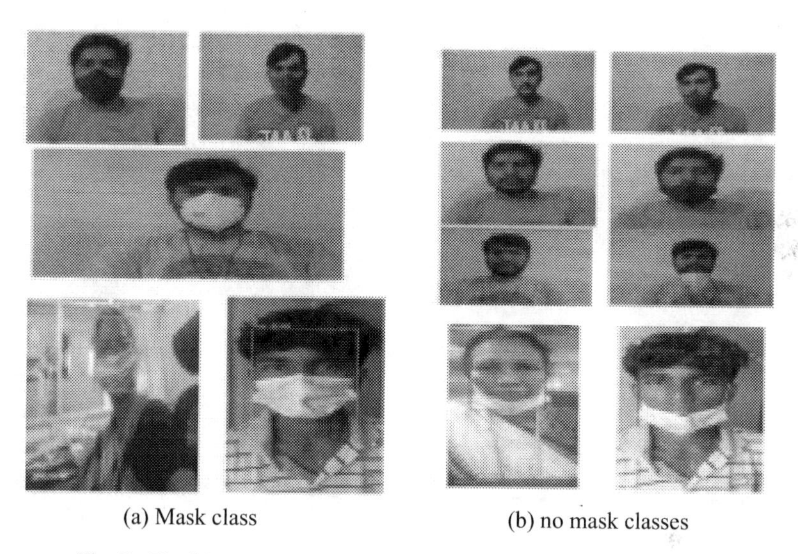

(a) Mask class (b) no mask classes

Fig. 7. Final Dataset Images for 'Mask' Class and 'No-Mask' class

5.2 Pseudocode for OpenCV Webcam or Video Testing

- ***Step 1 – Video File:*** *Capture a video using a camera or an existing file for testing.*
- ***Step 2 – Frame Extraction:*** *Extract frames from the video and pass them through the analysis model.*
- ***Step 3 – Threshold Value:*** *Set a threshold value for scores and IOU (Intersection over Union).*
- ***Step 4 – Boxes and Scores:*** *Retrieve the bounding boxes and scores and adjust them to their original size.*
- ***Step 5 – Storing the Variable:*** *Replace the boxes with colored rectangles and show the class totals at the bottom. Record the count.*
- ***Step 6 – Output:*** *After processing each frame, the resulting video with the desired output can be obtained.*

6 Dataset and Experimental Analysis

Multiple databases are available for selection; however, our dataset was used to train three face mask classifiers, including images with and without masks. The mask detection dataset was obtained from Kaggle, and erroneous mask images were merged with unmasked images in the dataset [17]. Table 1 lists the entire dataset, including faces with proper and improper and no masks. The collection consists of 9,877 photos categorized into two groups.

Table 1. Face Mask Classifier Dataset

Class Name	Number of images
Mask	5538
No-mask	4339

The performance of the three classifiers used in the system architecture is presented in Table 2. The training statistics indicate high accuracy for all classifiers. However, the D-DNN classifier outperforms the other classifiers regarding accuracy in the training and testing sets. Overall, all classifiers show good statistics.

Table 2: Performance of the classifiers

Classifier Name	Training Accuracy and loss value		Validation Accuracy and loss value		Test Accuracy (%)
	Accuracy (%)	Loss (%)	Accuracy (%)	Loss (%)	
Double DNN	98.91	0.00132	98.44	0.0182	98.26
Dense_Net_121	98.51	0.00157	98.72	0.0313	98.51
Mobile_Net_V2	98.41	0.0182	98.36	0.0298	98.27

A comparison of the classifiers is given as a graph in Fig. 8. This proposed system achieves a highly accurate face mask identification system with robust performance. The method involves two stages: first, face detection is performed using the OpenCV library, and second, Double DNN, DenseNet121, and MobileNetV2 classifiers are employed. The completed framework can accurately classify mask and no-mask categories for various facial expressions and lighting conditions.

The final output of our proposed model is illustrated in Fig. 9, where the dataset is successfully classified into mask and no-mask classes. Images with proper masks are classified as mask class, while images with improper or no masks are classified as a no mask class. The proposed model exhibits high performance and accuracy in its classification.

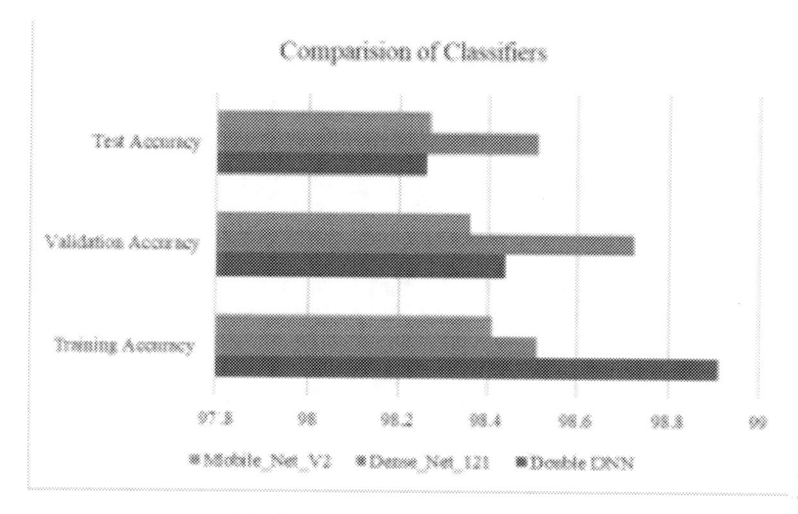

Fig. 8. Comparison of the Classifiers

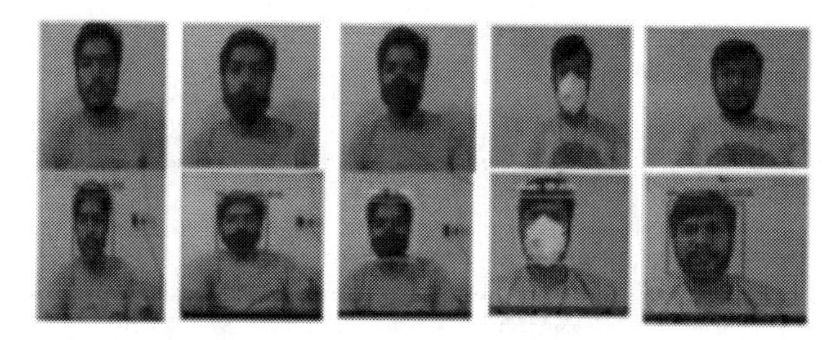

Fig. 9. The simulation results of the Double DNN classifier proposed method

7 Conclusion

The proposed approach starts with a pre-trained dataset of people using OpenCV face recognition. The resulting dataset included images of people wearing proper masks, improper masks, and no masks. Next, various classifiers such as NASNetMobile, DenseNet121, and MobileNetV2 were used, and a Double CNN-based model was selected to classify faces as having a proper mask, an improper mask, or no mask. This model incorporated dual CNN for Deep Learning and Machine Learning and a technology-based warning system. In the initial phase of the proposed model, a human trained dataset is used, followed by face identification using the OpenCV library. The trained prototype utilizing DCNN, Tensorflow, and Keras algorithms achieved performance and accuracy of 98.15%, with the highest precision and recall. The system is practical as it was created using less expensive materials and can be used worldwide. Additionally, it has an alarm system that emits a red fire alarm when someone is not wearing a mask and a green fire alarm when someone is wearing a mask. Integrating the proposed mask detection technology with existing security systems, such as CCTV cameras or access control systems, effectively enforces the effective enforcement of mask-wearing regulations in public spaces, airports, and government buildings. This technology may be tweaked and expanded for widespread deployment in other public locations, including schools, hospitals, and transit systems.

References

1. Thamaraiselvi, D., Mary, S.: Attack and anomaly detection in IoT networks using machine learning. Int. J. Comput. Sci. Mob. Comput.Comput. Sci. Mob. Comput. **9**, 95–103 (2020)
2. Vishwa Kiran, S., Kaur, I., Thangaraj, K., Saveetha, V., Kingsy Grace, R., Arulkumar, N.: Machine learning with data science-enabled lung cancer diagnosis and classification using computed tomography images. Int. J. Image Graph. **23**(03), 2240002 (2021)
3. Ayoobkhan, M.U.A., Yuvaraj, D., Jayanthiladevi, A., Easwaran, B., ThamaraiSelvi, R.: Smart connected digital products and IoT platform with the digital twin. In: Research Advancements in Smart Technology, Optimization, and Renewable Energy, pp. 330–350. IGI Global (2021)
4. Neelakandan, S., Beulah, J.R., Prathiba, L., Murthy, G.L.N., Irudaya Raj, E.F., Arulkumar, N.: Blockchain with deep learning-enabled secure healthcare data transmission and diagnostic model. Int. J. Model. Simul. Sci. Comput. **13**(04), 2241006 (2022)
5. Thamarai Selvi, R., George Dharma Prakash Raj, E.: Information retrieval models: a survey. Int. J. Res. Rev. Inf. Sci. (IJRRIS) **2**(3), 227–233 (2012)
6. Thamarai Selvi, R., George Dharma Prakash Raj, E.: An approach to improve precision and recall for ad-hoc information retrieval using SBIR algorithm. In: 2014 World Congress on Computing and Communication Technologies, pp. 137–141. IEEE (2014)
7. Ejaz, M.S., Islam, M.R., Sifatullah, M., Sarker, A.: Implementation of principal component analysis on masked and non-masked face recognition. In: 1st International Conference on Advances in Science, Engineering and Robotics Technology (ICASERT), pp. 1–5 (2019)
8. Qin, B., Li, D.: Identifying facemask-wearing condition using image super-resolution with classification network to prevent COVID-19. Sensors **20**(18), 5236 (2020)
9. Singh, S., Ahuja, U., Kumar, M., Kumar, K., Sachdeva, M.: Face mask detection using YOLOv3 and faster R-CNN models: COVID-19 environment. Multimedia Tools Appl. **80**, 19753–19768 (2021)
10. Kaur, G., et al.: Face mask recognition system using CNN model. Neurosci. Inform. **2**(3), 100035 (2022)
11. Chavda, A., Dsouza, J., Badgujar, S., Damani, A.: Multi-stage CNN architecture for face mask detection. In: 2021 6th International Conference for Convergence in Technology (i2ct), pp. 1–8. IEEE (2021)
12. Meivel, S., et al.: Mask detection and social distance identification using Internet of Things and faster R-CNN algorithm. Comput. Intell. Neurosci. **2022** (2022)
13. Zhu, Z., Cheng, Y.: Application of attitude tracking algorithm for face recognition based on OpenCV in the intelligent door lock. Comput. Commun.. Commun. **154**, 390–397 (2020)
14. Mbunge, E., Simelane, S., Fashoto, S.G., Akinnuwesi, B., Metfula, A.S.: Application of deep learning and machine learning models to detect COVID 19 face masks-a review. Sustain. Oper. Comput. **2**, 235–245 (2021)
15. Khan, A.A., Yadav, S., Kumar, L.: Face mask detection using OpenCv and machine learning. In: 2022 2nd International Conference on Advance Computing and Innovative Technologies in Engineering (ICACITE), pp. 80–85. IEEE (2022)
16. Thamaraiselvi, R., Mary, S.A.S.: A survey of machine learning for IoT networks (2020)

17. Karras, T., Laine, S., Aila, T.: A style-based generator architecture for generative adversarial networks. In: 2019 IEEE/CVF Conference on Computer Vision and Pattern Recognition (CVPR), Long Beach, CA, USA, pp. 4396–4405 (2019). https://doi.org/10.1109/CVPR.2019.00453. Larxel, Face Mask Detection, Kaggle - Version 1, May 2020. https://www.kaggle.com/andrewmvd/face-maskdetection
18. Thamarai Selvi, R.: An ensemble neural network technique for improving security among various domains of information technology. ICTACT J. Commun. Technol. **14**(1), 2882–2888 (2023)

Reflecting on Technology: A Review of the Smart Mirror Advancements and Applications

Divyesh Divakar(✉), H. M. Bharat, and G. M. Dhanush

Canara Engineering College, Mangaluru, India
divakardivyesh@gmail.com

Abstract. Smart mirrors gained popularity in recent years due to their unique ability to enhance our daily routines. They have been designed to improve the user experience of using a mirror and have the potential to provide a range of services, including displaying weather forecasts, news updates, time, social media notifications, fitness information etc. Smart mirrors have many potential applications which can enhance daily routines and improve quality of life. In this paper, the recent advancements and applications of smart mirrors are reviewed.

Keywords: Artificial Intelligence · Augmented Reality · Internet of Things · Smart Device

1 Introduction

A smart mirror is a technology that combines the functions of a traditional mirror with the capabilities of a smart device. It is typically a mirror with an embedded display that can show digital content, such as the time, weather, news, calendar, and social media feeds.

It can also be used for video chats, music streaming, and other tasks that are typically associated with smart devices. Smart mirrors are typically designed to be interactive, which means that users can use touch, voice, or gesture controls to interact with the mirror's content. This functionality is made possible by integrating the mirror with sensors, cameras, microphones, and other hardware components. A typical smart mirror is shown in Fig. 1.

The smart mirror concept is not new, but it has gained popularity in recent years with the advent of smart home technology. Companies like Amazon, Google, and Apple have introduced smart home devices that can control various aspects of the home, and smart mirrors are a natural extension of this trend. The technology behind smart mirrors is complex and involves a combination of hardware and software components. Here are some of the key elements of a smart mirror:

- Display: The display is an essential part of a smart mirror since it enables user interaction with the content of the mirror. Smart mirrors typically use LCD or OLED displays that are embedded behind the mirror's surface.

S. Aurelia et al. (Eds.): ICCSST 2023, CCIS 1973, pp. 438–451, 2024.
https://doi.org/10.1007/978-3-031-50993-3_35

Fig. 1. Exploded figure of smart mirror

- Microprocessor: The microprocessor is the brain of the smart mirror, and it is responsible for processing data and controlling the mirror's hardware components.
- Sensors: Smart mirrors use various sensors to detect user input and adjust the mirror's functionality accordingly. These may include cameras, microphones, touch sensors, and motion sensors.
- Operating System: The smart mirror runs on an operating system, typically Android or Linux, which enables the mirror to run apps and other software.
- Connectivity: Smart mirrors typically connect to the internet via Wi-Fi or Ethernet, which enables them to access online content and services.

In addition to these hardware components, smart mirrors also rely on software to provide a user-friendly interface and a range of features. Few key software components of a smart mirror include:

- User Interface: This is the graphical interface that aid the users to interact with the smart mirror's display. It typically consists of widgets, menus, and other graphical elements that aids the user to access and control the mirror's functionality.
- Apps: Apps that offer access to various features and services, such as news, weather forecasts, and social media, are generally operated on smart mirrors.
- Voice Assistants: Many smart mirrors usually come with voice assistants, such as Amazon's Alexa or Google Assistant, which enable users to control the mirror using voice commands.

2 Introduction

The paper titled "Smart mirror based on augmented reality" by Muhammad Hashim Shaheen [1] presents an innovative smart mirror system that utilizes artificial intelligence (AI) technology for enhanced functionality and improved user experience.

The paper provides a comprehensive overview of the system architecture, design, and implementation, highlighting the key features and benefits of the system. The paper begins with an introduction to the smart mirror concept and the potential applications of AI technology in such systems. The author then describes the conception and execution of the smart intelligence mirror system, which includes a Raspberry Pi board, a two-way mirror, a camera, and various sensors and modules.

The document provides a thorough explanation of the system's components (both hardware and software), including the AI software that is used to provide intelligent features such as face recognition, emotion detection, and voice control. The author explains various algorithms that fits into AI software to detect and analyze user behaviour and provide personalized recommendations and notifications.

The paper also provides a step-by-step guide on how to assemble and configure the smart intelligence mirror system, including the installation of the operating system, the AI software, and the various sensors. The author also highlights the potential benefits of the system, such as improved user experience, personalized recommendations, and increased safety and security. The author concludes the paper by highlighting the potential applications of the smart intelligence mirror system, such as in-home automation, healthcare, and retail environments. The concept, according to the author, has the potential to revolutionise how people interact with mirrors and can be a useful addition to the ecosystems of the smart home and smart city. Overall, the paper provides a well-written and comprehensive overview of the design and implementation of a smart intelligence mirror system that utilizes AI technology for enhanced functionality and improved user experience. The system presented here is innovative and has the potential to change the way people interact with mirrors, making it a valuable contribution to smart mirror technology.

The paper titled "Smart Mirror Design Powered by Raspberry PI" by A. S. A. Mohamed [2] presents a smart mirror system design that is powered by a Raspberry PI single-board computer. The paper provides a comprehensive overview of the system architecture, design, and implementation, highlighting the key features and functionality of the system. The paper begins with an introduction to the concepts, various features and functionalities, and potential applications. The author then describes the integration of a Raspberry PI single-board computer, a display, a two-way mirror, and various sensors and modules.

The paper presents a detailed description of the hardware components used in the system, including the Raspberry PI board, the display module, the two-way mirror, and the various sensors such as the motion sensor, temperature sensor, and ambient light sensor. The author also explains the software components of the system, including the operating system, the web server, and the various software applications used to control the system. The paper then provides a step-by-step guide on how to assemble and configure the smart mirror system, including the installation of the operating system, the software applications, and the sensors. The author also provides a comprehensive detailing of various features and functionalities of the system, such as voice control, weather updates, calendar events, news feeds, and social media updates.

The author concludes the paper by highlighting the benefits of smart mirror system, such as its low cost, flexibility, and ease of use. The author also suggests various potential applications for the system, such as in-home automation, healthcare, and retail environments. Overall, the paper provides a well-written and detailed overview of the creation and use of a smart mirror system powered by a Raspberry PI single-board computer. The paper is easy to understand and provides a step-by-step guide which can be followed by anyone interested in building a smart mirror system. The system presented in the paper is low cost and highly flexible, making it a valuable contribution to the field of smart mirror technology.

The paper titled "IoT-based Smart Mirror" by Amulya M, Kausthubha G, Rashmi Bharti, Simran Sinha, and Nikitha K S [3] presents an innovative smart mirror system that utilizes Internet of Things (IoT) technology for enhanced functionality and improved user experience. The paper provides a comprehensive overview of the system architecture, design, and implementation, highlighting the key features and benefits of the system. The paper begins with an introduction to the smart mirror concepts and the potential applications of IoT technology in smart mirror systems. The authors then describe the IoT-based smart mirror system, which includes a Raspberry Pi board, a two-way mirror, a camera, and various sensors and modules.

The paper presents a detailed description of the various components of the system, including the IoT platform used for data collection, analysis, and control. The authors explain the working of the system and the algorithms used in the software to detect and analyze user behavior and provide personalized recommendations and notifications. The paper also provides a step-by-step guide on how to assemble and configure the IoT-based smart mirror system, including the installation of the operating system, the IoT platform, and the various sensors. The authors also highlight the potential benefits of the system, such as improved user experience, personalized recommendations, and increased safety and security.

The authors conclude the paper by highlighting the potential applications of the IoT-based smart mirror system, such as in-home automation, healthcare, and retail environments. The authors suggest that the system can be a valuable addition to smart home and smart city ecosystems. Overall, the paper provides a well-written and comprehensive overview of an IoT-based smart mirror system that utilizes IoT technology for enhanced functionality and improved user experience. The step-by-step guide provided by the authors makes it easy for anyone interested in building a similar system to follow the process.

The paper titled "Smart Mirror to Enhance Learning: A Literature Review" by Nathasia Florentina Thejowahyono, Jeilson Phang, Kevin Nathanael Darmawan, and Mochammad Haldi Widianto [4] presents a literature review of the potential applications of smart mirrors in the field of education. The paper provides a comprehensive overview of the existing literature on smart mirrors and their potential to enhance the learning experience. The paper begins with an introduction to the concept of a smart mirror and the potential applications of smart mirrors in education. The authors then present a detailed review of the existing literature on smart mirrors, highlighting the key features and benefits of using smart mirrors in the classroom.

The authors classify smart mirror uses in education into three major areas: gamification, augmented reality, and personalization. The authors provide examples of each of these categories and describe the potential benefits of using smart mirrors in these applications, such as increased engagement, improved learning outcomes, and personalized learning experiences. The paper also discusses the challenges and limitations of using smart mirrors in education, such as the high cost of implementation and the need for specialized technical expertise. The authors provide recommendations for future research in the field and suggest that further studies are needed to fully understand the potential benefits and challenges of using smart mirrors in education.

Overall, the paper provides a well-written and comprehensive literature review of the potential applications of smart mirrors in education. The authors have done a thorough job of summarizing the existing literature on the topic and identifying the potential benefits and challenges of using smart mirrors in the classroom. The paper provides a valuable contribution to the field of educational technology and can be a useful resource for researchers and educators interested in the use of smart mirrors in education.

The paper titled "Magic Mirror using Raspberry Pi" by Mrs. Lavanya Vemulapalli, K. Bhavya, M. Ribka, and V. Ramya [5] presents an innovative system that utilizes Raspberry Pi technology for enhanced functionality and improved user experience. The paper provides a comprehensive overview of the system architecture, design, and implementation, highlighting the key features and benefits of the system. The paper begins with an introduction to the concept of a smart mirror and the potential applications of smart mirror systems. The authors then describe the design and implementation of the Magic Mirror system, which includes a Raspberry Pi board, a two-way mirror, a camera, and various sensors and modules.

The paper presents a detailed description of the components of the system, including the Raspberry Pi operating system, the magic mirror software, and the various sensors used to detect user behaviour and provide personalized information. The authors explain the working of the system and the algorithms used in the software to detect user behaviour and provide personalized information and notifications. They also provide a step-by-step guide on how to assemble and configure the magic mirror system, including the installation of the operating system, the Magic Mirror software, and the various sensors. The paper also provides a discussion of the potential benefits of the magic mirror system, such as improved user experience, personalized information, and increased safety and security. The authors suggest that the system can be a valuable addition to smart home and smart city ecosystems.

The authors conclude the paper by highlighting the potential applications of the Magic Mirror system, such as in-home automation, healthcare, and retail environments. The authors suggest that the system can be a valuable addition to the field of smart mirror technology. The step-by-step guide provided by the authors makes it easy for anyone interested in building a similar system to follow the process.

The paper titled "Futuristic Smart Mirror" by Khurd Aishwarya S, Shweta S. Kakade, R. M. Dalvi [6] is an interesting and innovative research work that explores the possibilities of integrating technology into a mirror to create a smart mirror. A detailed analysis of the design and development of a smart mirror that has the ability to enhance the user's experience by providing personalized information and services is presented in this paper. The authors start by discussing the current state of the art in smart mirrors and the limitations of existing solutions. They then propose their design, which includes various features such as facial recognition, voice recognition, temperature sensors, and a touch screen display. These features are aimed at providing a personalized experience to the user by offering relevant information, such as weather updates, news, and reminders. The paper goes into detail about the technical aspects of the smart mirror's design, including the hardware and software components.

The authors describe the development process and the challenges they faced while implementing the different features. They also provide results of the prototype's testing, demonstrating the effectiveness and accuracy of the facial recognition and voice recognition features. Overall, the paper is well-written and provides a comprehensive overview of the conception and development of a futuristic smart mirror. The authors have done a good job of highlighting the benefits of a smart mirror, such as personalized information and increased efficiency, and have presented a viable solution to this idea. However, one limitation of the paper is that the prototype has not been tested in a real-world environment, and the study would benefit from future work that explores the effectiveness of the smart mirror in a real-world setting.

This study makes a significant contribution to the field of smart mirrors, and it presents a promising solution for creating a personalized and efficient mirror that has the potential to change the way people approach the mirror.

The paper "IoT Based Smart Mirror Using Raspberry Pi 4" by Ashutosh Narayan Bilange, Aniket Kadam, and Prof. H. N. Burande [7] discusses the design of a smart mirror using IoT and Raspberry Pi 4 technology. The paper starts with an introduction to IoT technology and its applications in various fields, including the smart mirror. It then presents the development of the smart mirror using a Raspberry Pi 4, a two-way mirror, and various electronic components. The authors describe the architecture of the smart mirror, which consists of a user interface, a mirror display, a Raspberry Pi 4, and various sensors, including a temperature sensor, a humidity sensor, and a motion sensor.

The authors provide detailed descriptions of the hardware and software components used in the development of the smart mirror. They also provide code snippets and diagrams to help readers understand the technical details. The paper includes an evaluation of the system's performance, including a demonstration of the smart mirror's features, such as displaying the current weather and news, controlling the lighting, and detecting the user's presence. Overall, the paper provides an in-depth and comprehensive overview of the design and development of a smart mirror using IoT technology and Raspberry Pi 4. The authors have provided clear and concise explanations of the technical details, making it easy for readers to follow along with the project. The paper is well-organized and easy to read, and the use of diagrams and code snippets is helpful in understanding the technical details.

In conclusion, this paper is a valuable resource for anyone interested in the development of IoT-based smart mirrors using Raspberry Pi 4 technology. The authors have done an excellent job in presenting the technical details of the project in a clear and concise manner, making it a highly recommended read for anyone interested in this field.

The review of "Reflecto - A Smart Mirror Structure with a Contextual Association of Internet of Things" proposed by Md Shahbuddin, Arib Nawaz Khan, Kukatlapalli Pradeep Kumar [8] provided insights into contextual associations of the Internet of Things (IoT) technology. The Reflecto system consists of several components, including sensors, microcontrollers, a Raspberry Pi, and a display. The system's sensors are responsible for gathering data such as temperature, humidity, and user behavior. The microcontroller collects data from the sensors and sends it to the Raspberry Pi, which acts as the central processing unit of the system. The Raspberry Pi then processes the data and displays it on the mirror's surface using a transparent display. The system also incorporates voice recognition technology, which allows users to interact with the system through voice commands. The voice commands are processed by the Raspberry Pi, which provides users with the relevant information. The system is also equipped with a camera that can recognize users and display personalized information, such as their calendar and social media feeds.

The Reflecto system provides users with a unique and interactive experience. The system's ability to provide users with contextually significant information while using the mirror makes it a valuable addition to any home or office environment. The system's voice recognition technology and personalized information display make it a user-friendly and intuitive device. The camera feature also adds an additional layer of personalization, making the system more personalized and customized for individual users.

In conclusion, the Reflecto system is an excellent advancement in smart mirror technology. The addition of IoT technology and the system's ability to provide users with contextually relevant information make it a valuable addition to any home or office environment. The system's voice recognition technology and personalized information display also make it a user-friendly and intuitive device. Overall, the Reflecto system has the potential to and provide a unique and immersive experience.

"MirrorME: Implementation of an IoT based Smart Mirror through Facial Recognition and Personalized Information Recommendation Algorithm" by K Mohammad Mohi Uddin and Samrat Kumar Dey [9] took a step further by incorporating facial recognition technology and personalized recommendation algorithms based on the IoT technology. Users receive personalised information from the system, such as their calendar, emails, and news, based on their facial recognition.

The MirrorME system consists of various components, including a Raspberry Pi, a camera, a display, and a speaker. The camera is responsible for capturing images of the user's face, which is then processed by the Raspberry Pi using facial recognition technology. The system then recommends personalized information based on the user's facial recognition data. The system's IoT technology is responsible for integrating various smart devices, such as smart lights and thermostats, and controlling them through the mirror. The system's integration with IoT technology and smart devices provides users with an added level of convenience and control over their environment.

In conclusion, the MirrorME system is a significant advancement in smart mirror technology. The system's facial recognition technology and personalized recommendation algorithms based on IoT technology make it a valuable addition to any home or office environment. The system's ability to integrate with smart devices and control them through the mirror provides users with an added level of convenience and control over their environment. Overall, the MirrorME system has the power to completely change how people interact with mirrors and provide a unique and immersive experience.

Due to their capacity to offer customers an interactive and immersive experience, smart mirrors have drawn a lot of attention in recent years says the paper titled "Design of Interactive Smart Mirror System for Digital Information Display Based on Multitasking Approach Using Raspberry Pi" [10]. The system described in the paper uses a multitasking approach to provide users with a seamless and uninterrupted experience.

The system uses various sensors, processors, and displays to offer users services such as voice assistance, weather updates, and other personalized information.

The system consists of various components, including a Raspberry Pi, a display, a camera, and various sensors. The camera is responsible for capturing the user's face, which is then processed by the Raspberry Pi using facial recognition technology. The system then displays personalized information, such as weather updates and news, on the display based on the user's facial recognition data. The system also incorporates voice recognition technology, which allows users to interact with the system through voice commands. The voice commands are processed by the Raspberry Pi, which provides users with the relevant information. The system is also equipped with various sensors, including temperature and humidity sensors, which are used to provide users with relevant information.

The system provides users with a seamless and uninterrupted experience. The multitasking approach introduced in this system ensures that users can access multiple functionalities without any interruption. The system's ability to provide users with personalized information based on facial recognition data makes it a valuable addition to any home or office environment. The system's voice recognition technology and various sensors make it a user-friendly and intuitive device. In conclusion, the system described in the paper is a significant advancement in smart mirror technology. The multitasking approach used in the system ensures that users can access multiple functionalities without any interruption.

The paper titled "Smart Mirror Using Raspberry PI: A Review" by Saish Bavalekar and Ninad Gaonkar [11] provides a comprehensive overview of smart mirror technology, particularly in the context of using Raspberry Pi as a platform for building smart mirrors. The paper covers various aspects of smart mirror design, including hardware and software components, user interface design, and voice recognition integration. The authors present a detailed description of the hardware and software components of a smart mirror system using Raspberry Pi, including the display, camera module, microphone, speaker, and face recognition algorithms. The paper also covers different approaches to creating user interfaces for smart mirrors, including graphical user interfaces and voice user interfaces.

One of the key strengths of the paper is its discussion of the challenges and limitations of smart mirror technology. The limits of facial recognition algorithms are critically evaluated by the researchers, such as their sensitivity to lighting conditions and the need for high-quality images for accurate recognition. They also discuss potential privacy concerns associated with the use of smart mirrors, particularly with regard to data storage and sharing. Overall, the paper provides a valuable resource for researchers and developers interested in building smart mirror systems using Raspberry Pi. The paper is well-structured, with clear explanations of key concepts and technical details. The authors' critical analysis of the strengths and limitations of smart mirror technology is particularly noteworthy, as it highlights the potential benefits and drawbacks of this emerging technology.

The paper titled "Design and Development of an IoT & Raspberry Pi Based Smart Mirror" by Sweety Pradhan [12] presents a detailed study of a smart mirror system built using Raspberry Pi and IoT technologies. The paper presents a thorough explanation of the hardware and software components of the smart mirror, as well as its key features and potential applications.

The authors explain that the smart mirror system includes a two-way mirror, a display panel, a camera module, a microphone, a speaker, and an IoT module for connecting to the internet. The smart mirror uses facial recognition algorithms to recognize users and customize the display according to their preferences. The system also integrates with various IoT devices and services, such as weather and news updates, social media notifications, and home automation systems. The paper provides a detailed explanation of the software architecture of the smart mirror system, including the use of open-source software and libraries for image processing, voice recognition, and web development. The authors also discuss the potential applications of smart mirrors in various fields, such as healthcare, retail, and hospitality.

One of the key strengths of the paper is its emphasis on the potential benefits and practical applications of smart mirror technology. The authors provide a clear and concise explanation of the hardware and software components of the smart mirror system, as well as the design and development process. The paper also highlights the potential impact of smart mirror technology on various industries and areas of research.

Overall, the paper provides a valuable resource for researchers and developers interested in building smart mirror systems using Raspberry Pi and IoT technologies. The paper is well-written and well-structured, with clear explanations of technical details and potential applications. The authors' emphasis on the practical benefits and real-world applications of smart mirror technology is particularly noteworthy, as it highlights the potential for this emerging technology to transform various industries and fields.

The paper titled "Smart Mirror for Security and Home Appliances" by A. Akhila and V. Siva Nagaraju [13] explores a smart mirror system that can have multiple applications, including security and home appliance control. The idea presented in the study is intriguing and novel, and it has the potential to completely alter how people interact with their houses. The authors begin by introducing the concept of a smart mirror and its potential applications in the context of smart homes. They then go on to describe the hardware and software components of their system, including the mirror itself, the microcontroller, and the various sensors and actuators that enable the system to function.

One of the key features of the system described in the paper is its ability to act as a security device. The mirror is equipped with a camera and facial recognition software that can detect and identify people entering the home. This information can then be used to trigger various actions, such as turning on lights or sounding an alarm. In addition to its security features, the smart mirror can also be used to control various home appliances, such as thermostats and lighting systems. The authors describe how users can interact with the system through a touchscreen interface on the mirror, as well as through voice commands.

Overall, the paper provides a well-written and well-researched exploration of an intriguing idea. The authors have clearly put a lot of thought into the design and implementation of their system, and their descriptions of the various components and features are clear and detailed. There are a few areas where the paper could be improved. For example, the authors do not provide any tabulation on cost of building or purchasing the smart mirror system, which could be a key consideration for potential users. Additionally, the paper could benefit from a more detailed discussion of the potential limitations and challenges associated with using a smart mirror as a security device.

Despite these minor shortcomings, the paper is an interesting and valuable contribution to the field of smart home technology. The authors' innovative approach to creating a multi-purpose smart mirror system has the potential to significantly improve the way people interact with their homes, and could pave the way for further advancements in this exciting area of research.

The paper titled "Implementation and Customization of a Smart Mirror through a Facial Recognition Authentication and a Personalized News Recommendation Algorithm" by Ivette Cristina Araujo García [14] presents an interesting approach to developing a smart mirror system that includes facial recognition a personalized recommendation algorithm. The paper provides a detailed description of the implementation and customization of the smart mirror system, which allows users to access personalized news content and control their smart home devices through a user-friendly interface. The paper begins with a thorough overview of the relevant literature on smart mirror systems, facial recognition technology, and personalized news recommendation algorithms. The author then describes the development process of the smart mirror system, including the hardware and software components, the facial recognition authentication system, and the personalized news recommendation algorithm. The author provides a detailed explanation of each component of the system and how they work together to provide a seamless and personalized user experience.

One of the key strengths of the paper is its emphasis on customization. The author explains how the smart mirror system can be customized to fit the unique preferences and needs of individual users. For example, the personalized news algorithm takes into account a user's reading history and interests to suggest relevant news content. The facial recognition authentication system can also be customized to recognize multiple users and their individual preferences. Overall, the paper presents a well-written and well-researched exploration of an intriguing idea. The author's attention to detail and emphasis on customization make the smart mirror system described in the paper a promising addition to the growing field of smart home technology.

However, there are a few areas where the paper could be improved. For example, the author could have provided more information on the implementation of the facial recognition authentication system, including any potential security concerns or limitations. Additionally, the paper could benefit from a more detailed discussion of the potential limitations and challenges associated with using a smart mirror as a personalized news delivery system.

Despite these minor shortcomings, the paper is a valuable contribution to the field of smart home technology. The author's innovative approach to creating a multi-purpose smart mirror system that incorporates facial recognition authentication and personalized news recommendation algorithm has the potential to significantly improve the way people interact with their homes, and could pave the way for further advancements in this exciting area of research.

The paper titled "Smart Mirror Using Hand Gesture" by Sreejith PS and Arya PR [15] presents an interesting concept that can be controlled through hand gestures. The paper describes the development of the system, including the hardware and software components, and presents the results of the authors' experiments testing the system's effectiveness. The paper begins by introducing the concept of a smart mirror and its potential applications in the context of smart homes. The authors then go on to describe the hardware and software components of their system, which includes a Raspberry Pi microcontroller, a camera, and a hand gesture recognition algorithm. The system allows users to control various features of the mirror, such as adjusting the brightness and displaying weather information, through hand gestures detected by the camera.

The authors conducted experiments to evaluate the effectiveness of the hand gesture recognition algorithm, and the results showed that the system was able to detect hand gestures with a high degree of accuracy. The authors also discuss the potential applications of their system, including its use in healthcare and retail settings. Overall, the paper provides a well-written and well-researched exploration of an intriguing idea. The authors' innovative approach to creating a hand gesture-controlled smart mirror system has the potential to significantly improve the way people interact with their homes, and could pave the way for further advancements in this exciting area of research.

However, there are a few areas where the paper could be improved. For example, the authors could provide more information on the specific hand gestures that are identified by the system, as well as any limitations or challenges associated with hand gesture recognition algorithm. Additionally, the paper could benefit from a more detailed discussion of the potential applications of the system beyond the home environment.

Despite these minor shortcomings, the paper is a valuable contribution to the field of smart home technology. The authors' innovative approach to creating a hand gesture-controlled smart mirror system has the potential to significantly improve the user experience and accessibility of smart home technology, and could lead to further advancements in this exciting area of research.

The paper titled "Internet of Things Based on Smart Mirror to Improve Interactive Learning" by Mochammad Haldi Widianto and Ranny [16] is an interesting and innovative piece of research that explores the potential use of smart mirrors as a means of improving interactive learning. The paper provides a detailed overview of the Internet of Things (IoT) and how it can be used to create a smart mirror that is capable of providing real-time feedback and personalized learning experiences to students.

One of the strengths of this paper is its focus on the potential benefits of using a smart mirror as a learning tool. The authors provide a comprehensive analysis of the various advantages that a smart mirror can offer, such as the ability to provide personalized feedback, promote active learning, and enhance student engagement. The paper also discusses the potential challenges that may arise when implementing a smart mirror system, such as issues related to privacy and security. It also discusses the use of case studies to demonstrate the practical applications of the smart mirror system. The authors provide two examples of how the smart mirror can be used in different learning environments, such as a classroom setting and a medical training facility. These case studies help to illustrate the potential benefits of the smart mirror in real-world scenarios.

One potential weakness of this study can be identified that it focuses primarily on the technical aspects of the smart mirror system and less on the pedagogical implications of its use. While the authors do briefly discuss the educational theories that underpin the smart mirror system, there could be more emphasis on how the system can be used to promote specific learning outcomes. Overall, the paper provides a valuable contribution to the field of educational technology and the use of IoT devices in learning environments.

3 Applications

Smart mirrors have various applications, including home automation, fitness tracking, and entertainment.

(i) Health and Fitness: With large number of people leading sedentary lifestyles, technology that can enhance physical fitness and health is becoming more and more necessary. Smart mirrors can be used to track fitness progress, monitor vital signs, and provide users with personalized workout plans. This can help users stay on track with their fitness goals and improve their overall health.
(ii) Home Automation: India is a country with a large number of households and smart mirrors can be integrated with home automation systems. This would allow users to control lighting, thermostats, and other smart devices in their home with voice commands. Home comfort and convenience can both be increased as a result, helping to enhance energy efficiency.
(iii) Public Spaces: Smart mirrors can be installed in public spaces such as shopping malls, airports, and hospitals to provide users with relevant information such as directions, news updates, and promotional content. This can help enhance the user experience and make these spaces more engaging and informative.
(iv) Retail: Smart mirrors can be used in retail environments to provide customers with virtual try-on experiences and personalized product recommendations. This can help improve the overall shopping experience and increase customer satisfaction.

4 Conclusion

Smart mirror technology has come a long way in recent years, with advancements in display technology, voice recognition, artificial intelligence, and augmented reality. These improvements have expanded the potential applications of smart mirrors, making them a valuable tool for health and fitness, home automation, and retail. The potential of smart mirrors is huge enough to transform a range of industries, from healthcare to entertainment to smart cities. The existing research and development in the field of smart mirror technology is giving promising results and has highlighted the various applications and potential of this technology. We may anticipate even more fascinating advancements in the field of smart mirrors as technology continues to advance.

References

1. Shaheen, M.H., Al-Qutayri, M.A.: Smart mirror based on augmented reality. In: 2018 IEEE 5th International Conference on Industrial Engineering and Applications (ICIEA), pp. 168–173 (2018). https://doi.org/10.1109/ICIEA.2018.8397734
2. Mohamed, A.S.A.: Smart mirror design powered by Raspberry Pi. In: 2018 International Conference on Innovative Trends in Computer Engineering (ITCE), Aswan, pp. 506–510 (2018). https://doi.org/10.1109/ITCE.2018.8360323
3. A. M, K. G, R. Bharti, S. Sinha, N. K. S.: IoT-based smart mirror. In: 2018 International Conference on Computation of Power, Energy, Information and Communication (ICCPEIC), Chennai, India, pp. 293–296 (2018). https://doi.org/10.1109/ICCPEIC.2018.8669483
4. Thejowahyono, N.F., Phang, J., Darmawan, K.N., Widianto, M.H.: Smart mirror to enhance learning: a literature review. In: 2020 International Conference on Computer Engineering, Network and Intelligent Multimedia (CENIM), Surabaya, Indonesia, pp. 1–5 (2020). https://doi.org/10.1109/CENIM50110.2020.9311754
5. Vemulapalli, L., Bhavya, K., Ribka, M., Ramya, V.: Magic mirror using Raspberry Pi. In: 2018 International Conference on Recent Trends in Advance Computing (ICRTAC), Pune, India, pp. 1–4 (2018). https://doi.org/10.1109/ICRTAC.2018.8649923
6. K. A. S, S. S. Kakade, R. M. Dalvi: Futuristic smart mirror. In: 2018 International Conference on Communication, Computing and Electronics Systems (ICCCES), Coimbatore, India, pp. 1142–1145 (2018). https://doi.org/10.1109/ICCCES.2018.8661985
7. Bilange, A.N., Kadam, A., Burande, H.N.: IoT based smart mirror using Raspberry Pi 4. In: 2020 11th International Conference on Computing, Communication and Networking Technologies (ICCCNT), Kharagpur, India, pp. 1–5 (2020). https://doi.org/10.1109/ICCCNT49239.2020.9225147
8. Shahbuddin, M.S., Khan, A.N., Kumar, K.P.: Reflecto - a smart mirror structure with a contextual association of Internet of Things. In: 2019 5th International Conference on Advanced Computing & Communication Systems (ICACCS), Coimbatore, India, pp. 518–522 (2019). https://doi.org/10.1109/ICACCS.2019.8724098
9. Uddin, K.M.M., Dey, S.K.: MirrorME: implementation of an IoT based smart mirror through facial recognition and personalized information recommendation algorithm. In: 2020 IEEE Region 10 Symposium (TENSYMP), Dhaka, Bangladesh, pp. 301–306 (2020). https://doi.org/10.1109/TENSYMP50017.2020.9230814
10. Mutmainnah, M.A.S.S., Ahsan, S.A.S., Sultan, A.F.M.: Design of interactive smart mirror system for digital information display based on multitasking approach using Raspberry Pi. In: 2019 5th International Conference on Science in Information Technology (ICSITech), Yogyakarta, Indonesia, pp. 291–296 (2019). https://doi.org/10.1109/ICSITech49158.2019.8980726
11. Bavalekar, S., Gaonkar, N.: Smart mirror using Raspberry PI: a review. In: 2020 6th International Conference on Advanced Computing and Communication Systems (ICACCS), Coimbatore, India, pp. 1396–1399 (2020). https://doi.org/10.1109/ICACCS48703.2020.9074676
12. Pradhan, S.: Design and development of an IoT & Raspberry Pi based smart mirror. In: 2021 6th International Conference on Computing, Communication and Security (ICCCS), Greater Noida, India, pp. 1–6 (2021). https://doi.org/10.1109/ICCCS52030.2021.9395249
13. Akhila, A., Siva Nagaraju, V.: Smart mirror for security and home appliances. In: 2021 8th International Conference on Computing for Sustainable Global Development (INDIACom), New Delhi, India, pp. 1423–1427 (2021). https://doi.org/10.1109/INDIACom52072.2021.9379502

14. Araujo García, I.C.: Implementation and customization of a smart mirror through a facial recognition authentication and a personalized news recommendation algorithm. In: 2021 IEEE Global Engineering Education Conference (EDUCON), Vienna, Austria, pp. 1881–1888 (2021). https://doi.org/10.1109/EDUCON46610.2021.9442661
15. S. PS and A. PR: Smart mirror using hand gesture. In: 2018 2nd International Conference on Trends in Electronics and Informatics (ICOEI), Tirunelveli, India, pp. 157–160 (2018). https://doi.org/10.1109/ICOEI.2018.8553822
16. Widianto, M.H., Ranny: Internet of Things based on smart mirror to improve interactive learning. In: 2019 International Conference on Information Management and Technology (ICIMTech), Jakarta, Indonesia, pp. 1–6 (2019). https://doi.org/10.1109/ICIMTech48036.2019.8982904

MobNetCov19: Detection of COVID-19 Using MobileNetV2 Architecture for Multi-mode Images

H. S. Suresh Kumar[1(✉)], S. Bhoomika[1], C. N. Pushpa[1], J. Thriveni[1], and K. R. Venugopal[2]

[1] Department of CSE, University of Visvesvaraya College of Engineering, Bengaluru 560001, India
kumar10harobande@gmail.com
[2] Bangalore University, Bengaluru, India

Abstract. COVID-19 created a history in the world of medicine which leads to more usage of technologies such as deep-learning models to aid in the early detection of COVID-19 using medical imaging from three commonly used modalities: X-Ray, Ultrasound and Computerized Tomography (CT) scan. This research aims to provide medical professionals with an additional tool to assist in devising an appropriate treatment plan and making disease containment decisions. We have identified the suitable optimized VGG19 and MobNetCov19 architecture through a Convolutional Neural Network (CNN) model for a comparative study of the different imaging modes to develop highly curated COVID-19 detection models despite the scarcity of COVID-19 datasets. Our results demonstrate that CT dataset has the highest detection accuracy compared to X-Ray and Ultrasound datas. Although the limited data made training complex models challenging, the selected MobNetCov19 model, extensively tuned with appropriate parameters, performed considerably well up to 100%, 98%, and 98% of accuracy for CT, X-Ray, and Ultra sound respectively.

Keywords: Convolutional Neural Network · COVID-19 · MobNetCov19 · CT scan · Ultrasound · X-Ray

1 Introduction

The disease known as COVID-19 is mostly affected by the SARS-CoV-2, which was found in China, in December 2019. COVID-19 spreads through air borne patients containing the virus or the droplets. The transmission can occur through the infected fluids in the eyes, nose, or mouth, and, rarely, via contam inated areas, and as a result, it spread fast around the earth.

COVID-19 affects diverse people in different ways. Many people who get sick become mildly to moderately ill and recover without hospitalization. The symptoms mainly affected are categorized into most common (moderate), less Common (severe), and serious (fatal) problems. The common symptoms of coronavirus are fever, cough,

S. Aurelia et al. (Eds.): ICCSST 2023, CCIS 1973, pp. 452–463, 2024.
https://doi.org/10.1007/978-3-031-50993-3_36

fatigue, and lack of taste. Less common symptoms are aching throat, headache, pain, diarrhoea, rashes in toes, or sore eyes. Severe symptoms are difficulty in breathing, movement problems, misperception, and chest pain.

To identify COVID-19 in medical images like chest X-rays or CT scans, transfer learning would entail employing a DL model initially trained [1] on image classification or object recognition tasks and then fine-tuning it to do so.

Transfer learning has the advantage of requiring fewer data and processing resources than training a model from the start because the model already has a solid understanding of the fundamental characteristics and patterns found in images. This is very useful in the case of the COVID-19 epidemic [2], where there is a need for quick and precise diagnostic tools and a lack of resources for gathering and annotating massive volumes of data [3]. This framework can offer several benefits that can help accelerate and improve your research, like a better understanding of underlying patterns, improved performance, redefine-tuning, and resources.

Motivation: COVID-19 is a pandemic disease was detected across the world in 2019. Before COVID-19 world has seen pandemic diseases such as flu, plaque etc,. Due to the lack of history, medical treatment was not available on time, due to which we lost so many lives. Effective screening and prompt medical attention for COVID-19-infected people are undoubtedly required to stop the disease's spread. The situation is made worse by the use of numerous screening tools with delayed test results. Our proposed method is frequently utilized as the foundation for creating DL models, which are subsequently refined using the inadequate number of results-based datasets that are readily available.

Contribution: The work carried out is as follows;

i. A Classification deep learning model, MobNetCov19 is proposed from the limited and imbalanced dataset.
ii. The MobNetCov19 algorithm, which has been introduced, identifies COVID 19 using modes of X-Ray, Ultrasound, and CT scan.

Organization: The paper is structured as follows: Section 2 provides a re view of the articles, Section 3 gives the problem definition, Section 4 explains the system architecture, Section 5 gives the proposed algorithm, Section 6 provides the results of the system, and Section 7 contains the conclusions and future work.

2 Related Work

Deep learning has gained more attention because of its potential application in the diagnosis of COVID-19 disease. Recent proposals for deep learning algorithms as diagnostic tools for COVID-19 disease have helped doctors make well-informed therapy decisions. Several of the studies that are directly relevant to this inquiry are discussed.

Tang *et al.* [4] propose the EDL-COVID system is intended to address the shortcomings of DL models, including overfitting, high variance, and generaliza tion errors, by utilizing a weighted averaging ensembling method that integrates multiple models from COVID-Net to generate predictions. This approach leads to enhance precision in detecting COVID-19 cases, resulting in an accuracy rate of 95%.

Ahmed *et al.* [5] the decline in adherence to COVID-19 preventative pro cedures in somalia is potentially linked to a rise in flu-like symptoms. While vaccine acceptance rates are moderately high, improving these rates requires tackling various factors that contribute to vaccine hesitancy, such as gender, involvement in the medical sector, and adherence to preventative measures.

Zhang *et al.* [6] use the RLDD system for COVID-19 is created to differenti ate positive cases from various lung images. Using a residual-based network, the scheme can extract lung features from small COVID-19 samples without requir ing additional medical datasets. In evaluations, the RLDD achieved a 91.33ac curacy rate, a 91.30% precision rate, and recall.

Basu *et al.* [7] provides the Domain Extension Transfer Learning (DETL) strategy is a promising solution for identifying COVID-19 using CXR. Preliminary findings demonstrate a high accuracy rate of 90.13% $\pm$ 0.14, which can be enhanced with larger and more diverse datasets. The use of a Gradient Class Activation Map (Grad-CAM) to identify areas of increased attention during the classification was discovered to have a significant correlation with clinical observations, as confirmed by experts, underscoring the reliability and efficiency of the DETL method for COVID-19 detection.

Chen *et al.* [8] finds the ability of DL to identify COVID-19 pneumonia on CT scans is demonstrated in this study. The per-patient accuracy of the model was 95.24%, and the per-image accuracy was 98.85% when evaluated on the internal retrospective dataset and demonstrated performance compared to that of expert radiologists in the internal prospective dataset. In the external dataset, the model attained an accuracy of 96% and effectively reduced radiologists' reading time by 65%.

Chouat *et al.* [9] introduces a method for detecting COVID-19 using deep transfer learning. The CT image dataset shows that the VGGNet-19 model out performed other models; the CXR image dataset achieved the highest precision, recall, F1-score and accuracy rates with the refined exception version. However, when the outcomes from both X-ray and CT modalities were combined, the VGG-19 model yielded the most exceptional results (Table 1).

Table 1. Performance of the existing work on VGG16/19.

Author, Ref., Year.	Model/Classification	Dataset	Result
Kitrungrotsakul *et al.* [10], 2021	Interactive atten tion refinement network	COVID-19 CTSeg and MICCAI	Accuracy: 90%
Karaci *et al.* [11], 2022	VGGCOV19,NET, VGG19	X-Ray: COVID 19:125 Pnemonia:500 No-findings:500	Accuracy: 99.84%
Watanabe *et al.* [12], 2022	mRNA vaccine inoculations	86 Healthcare work ers	Efficiency: ab titers R = 0.324 P = 0.004%
Dhere *et al.* [13], 2022	LDL-MARL	6500 CXR images	Accuracy: 96.28%
Frid *et al.* [14], 2021	Deep DRR	1124 images	Accuracy: 87%

3 Problem Definition

The main issue with traditional methods for detecting the severity of COVID19 is the high time consumption and incidence of false positives. To address this issue, we aim to identify an appropriate CNN-based deep learning model for our multi modal image classification, which involves classifying COVID-19, pneumonia and normal images. Our objectives are:

(i) To provide a robust transfer learning framework for detecting COVID-19 through image classification using deep learning models across multiple imaging modes, such as X-ray, ultrasound, and CT scan.
(ii) To optimize model performance using MobNetCov19.

4 System Architecture

The proposed MobNetCov19 model is shown in Fig. 1.

Multi-mode Image Dataset is a collection of images organized into multiple categories or modes. The datasets are used to train and test multimode image classifiers, which can recognize and classify images into different categories. Data Pre-processing to ensure that the data is accurate, complete, consistent, and relevant. This involves various techniques, such as data cleaning, normalization, transformation, reduction, and integration.

Data augmentation is a technique for initializing the required parameters, such as rotation range, zoom range, and horizontal and vertical shifts. Used the Image data generator function to create an instance of the data generator, define the directory where the images are stored and loop through each image. Load each image and map it to an array for the function. The array adds an extra dimension using the reshape method. Apply data augmentation using the flow method and save the augmented images in the designated directory for each original image.

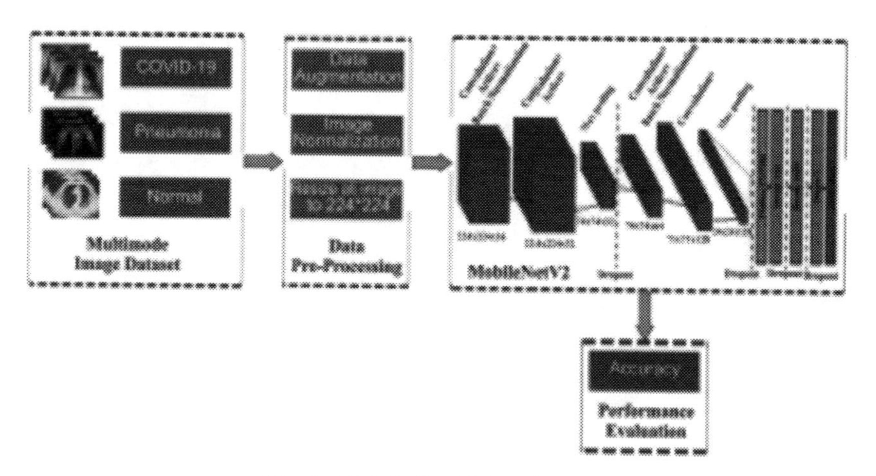

Fig. 1. MobNetCov19 Architecture.

Image normalization is the process of adjusting the pixel values of an image to a standardized scale or range, typically between 0 and 1. This is often done to improve the performance of ML algorithms, which may require inputs to be on a similar scale to produce accurate results.

MobileNetV2 has several key features that make it well-suited for mobile devices. First, it uses depth-wise separable convolutions, which are computationally ef ficient and reduce the number of parameters needed in the network. Second, it uses a technique called inverted residuals, which increases the depth of the network while keeping the number of parameters low. Third, it includes a fea ture called linear bottleneck, which reduces the computational cost of each layer while maintaining its representational power.

By utilizing the MobNetCov19 model to derive the discriminative features, our proposed methodology considerably improves COVID-19 classification accuracy. Fig. 1 depicts the proposed method for classification. Data pre-processing, Image normalization and resizing images are the model's primary phases.

Evaluation metrics are the model's performance that was assessed using standard evaluation metrics, for example, accuracy, recall, etc. Additionally, the classifiers were compared using the AOC and Receiver Operating Characteristic (ROC) tests, which are widely used to analyze the trade-off between the true-positive and false-positive rates of a diagnostic test.

Accuracy is calculated as the ratio of true positives and negatives to the total number of instances in the dataset. Overall, accuracy is a useful metric for evaluating the overall performance of an ML model or classifier, particularly when the dataset is balanced and all classes are of equal importance.

$$Accuracy = \frac{TP + TN}{TP + TN + FP + FN} \tag{1}$$

5 Algorithm

MobNetCov19: Detection of COVID-19 using Multi-Mode Images model is given below.

```
Algorithm: MobNetCov19
  begin
      Step 1: Load the images.

      Step 2: Data pre-processing of the images.
         a) Data augmentation.
            begin
               i: Initialize the parameters required for data augmentation.
             ii: For each image call img_to_array function, store results in a temporary
             variable, and call reshape method w.r.t the resultant array. iii: For each
             image that is in multimode call the flow method and saves the augmented
             image in the directory.
           end
         b) Image Normalization.
         c) Resize the pixel of the images to size 224*224.

      Step 3: Apply MobileNetV2()function.
         a) Apply categorical-cross entropy.
         b) ADAM optimizer to train the model.

      Step 4: Pass each image to the convolution process.
         a) Process each image of dimensions 224*224*3 processed depth wise
         separate convolutions and convert the image into dimension 74*74*32.
         b) Drop Out the processed matrix/images by 0.5.
         c) Follow Steps a and b until an image is re- dimensioned to 24*24*128.

     Step 5: The image is converted /flattened into a single dimensional array.

     Step 6: Apply the Dense layer with the softmax activation function and then
                        apply dropout by 0.5 to the resultant array.

       Step 7: Repeat Step 6 with a different set of neurons, apply dropout for
                        repeated learning and activate the neurons.

     Step 8: Apply the confusion matrix, Fetch the prediction results, calculate the
                  accuracy.
  end
```

6 Results

Dataset: We have prepared the X-Ray COVID-19 dataset for our experiments by removing a single image with an incorrect label. We also curated the normal images to exclude any that were mislabeled projections. The open-source plat form GitHub repository was used for the X-Ray dataset and Kaggle dataset for Ultrasound and CT Images. We summarized these transformations in Table 2, and the train and test split into an 80:20 ratio.

Experimental Setup: The proposed model was evaluated using four essential performance metrics accuracy etc. We have implemented the proposed model and other

Table 2. Sample dataset for experiments.

Images	Condition	Source Images	Train	Test
X-ray	COVID-19 Pneumonia Normal	624 676 629	499 543 503	125 136 126
	COVID-19 Ultrasound Pneumonia Normal	635 620 626	508 496 501	127 124 125
CT Scan	COVID-19 Pneumonia Normal	666 746 760	533 597 608	133 149 152

pre-trained models using Python and TensorFlow. All experiments was conducted on a system with an Intel (R) Core (TM) i7-7700 CPU @ 3.60 GHz processor and 8 GB RAM. We trained both the proposed and pre-trained models using the ADAM activation function and softmax pooling for processing images. Table 2 provides a complete overview of the image setup used for the experiments.

Outcomes: Table 3 shows the results of the experiments conducted using VGG19 and MobNetCov19. We observed that in the experiments classifying COVID-19, Pneumonia, and Normal, the CT mode provided the highest accu racy of 100%. For the X-ray and Ultrasound modes in MobNetCov19, we have achieved an accuracy of 98% in both cases.

Table 3. Results of the experiment for the three image modes.

Images	Accuracy: VGG19	Accuracy: MobNetCov19
X-ray	95%	98%
Ultrasound	93%	98%
CT Scan	90%	100%

The experimental results are shown in Fig. 2(a), the confusion matrices for X-Ray images using the VGG19 model show that reduced only 1 false negative using a threshold of 0.0079 for normal images was misclassified as COVID 19, only 1 true negative using a threshold of 0.0074 for pneumonia image as COVID19, increased 12 false positives using a threshold of 0.088 for pneumonia images as normal, increased 3 false positives using a threshold of 0.024 COVID 19 for images as normal, and reduced 4 false negatives using a threshold of 0.032 for normal images as pneumonia. Fig. 3(a) of the MobNetCov19 model demonstrates better results, with 2 reduced false negatives using a threshold of 0.32 for pneumonia images misclassified as normal, increased 6 false positives using a threshold of 0.056 for COVID-19 images as normal, and 9 true negatives using a threshold

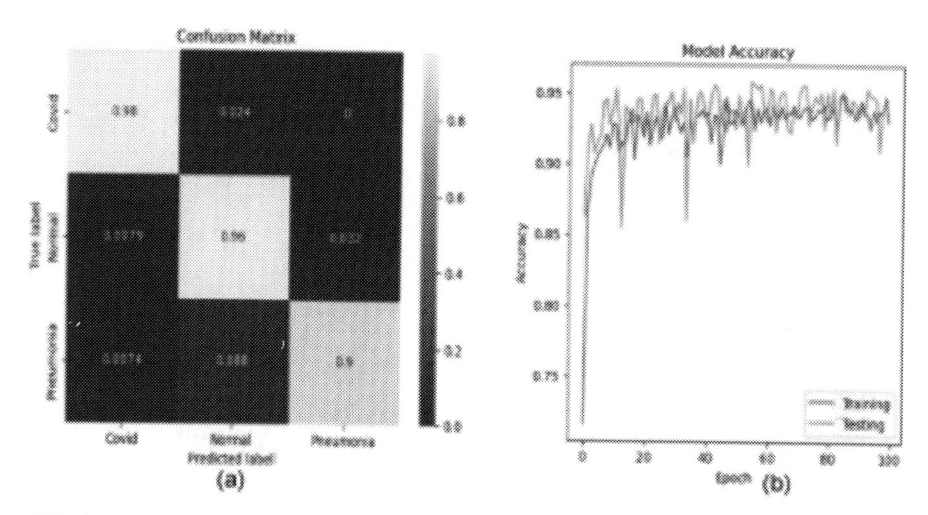

Fig. 2. (a) Confusion Matrix, (b) Model Accuracy of the proposed VGG19 model of X Ray.

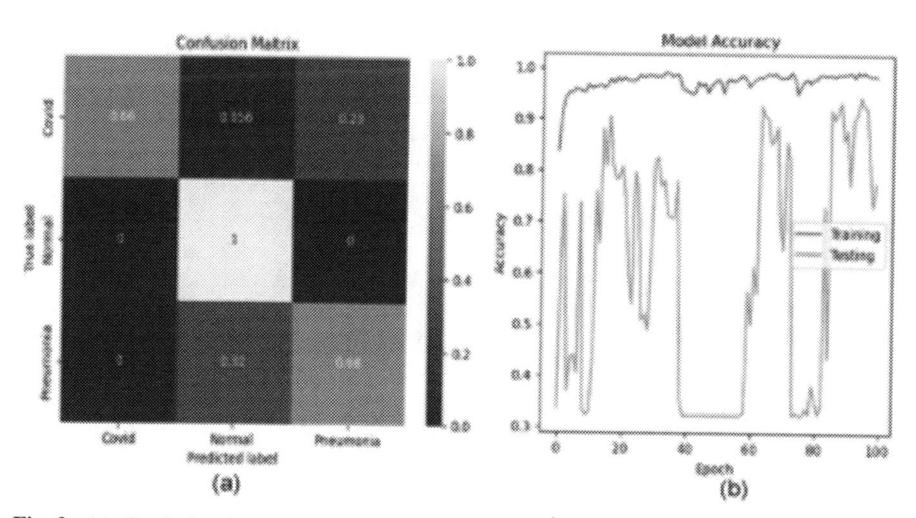

Fig. 3. (a) Confusion Matrix, (b) Model Accuracy of the proposed MobNetCov19 model of X-Ray.

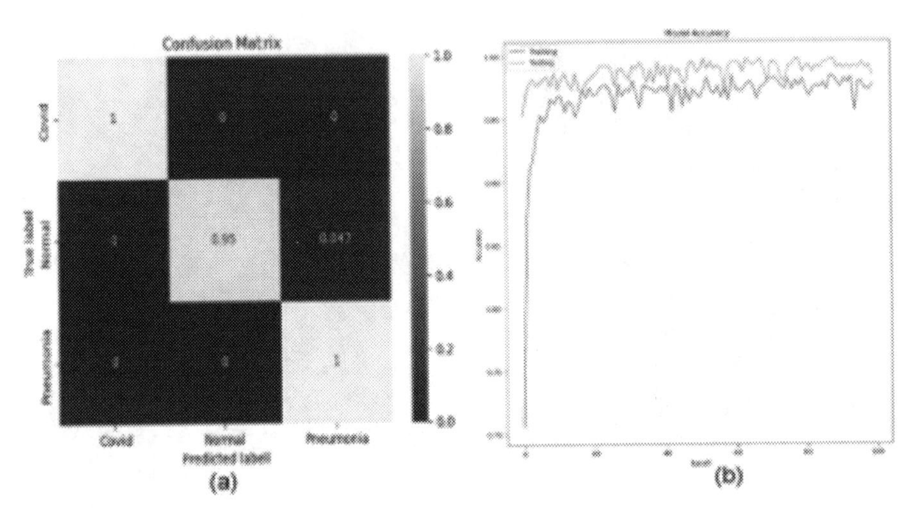

Fig. 4. (a) Confusion Matrix, (b) Model Accuracy of the proposed VGG19 model of Ultrasound.

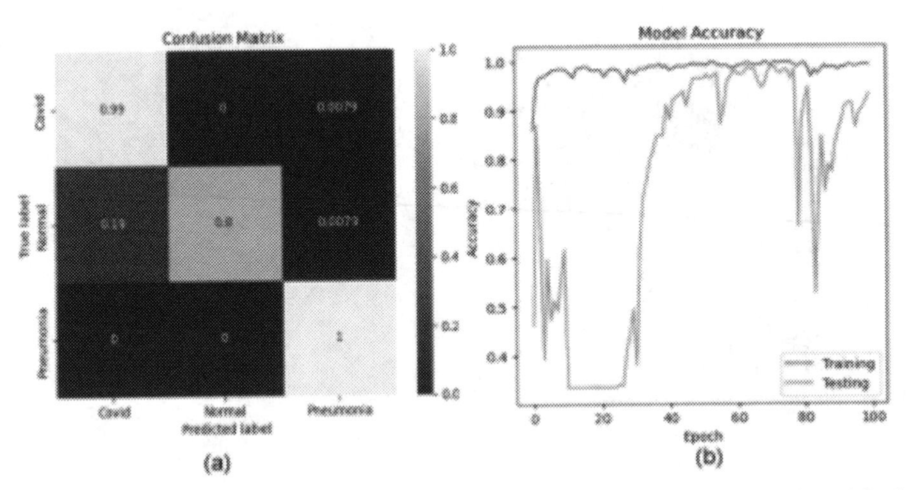

Fig. 5. (a) Confusion Matrix, (b) Model Accuracy of the proposed MobNetCov19 model of Ultrasound.

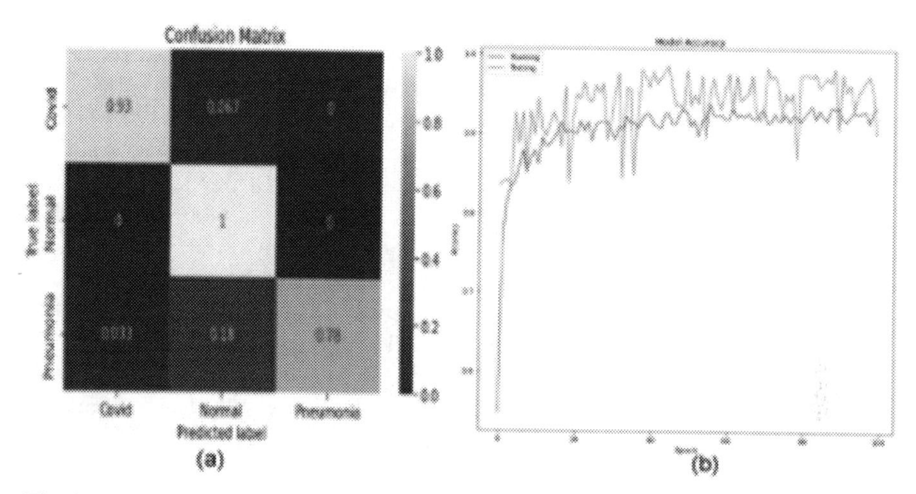

Fig. 6. (a) Confusion Matrix, (b) Model Accuracy of the proposed VGG19 model of CT scan.

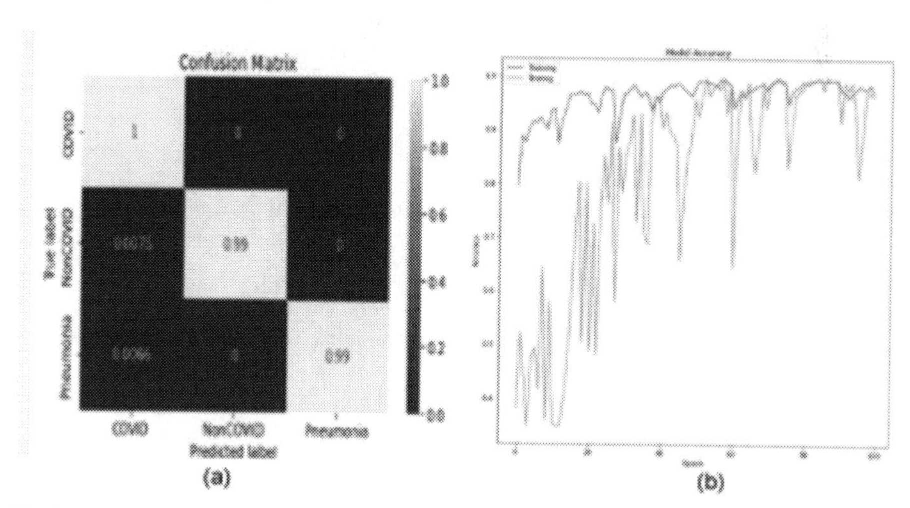

Fig. 7. (a) Confusion Matrix, (b) Model Accuracy of the proposed MobNetCov19 model of CT scan.

of 0.29 for COVID-19 images as pneumonia. The VGG19 model is shown in Fig. 4(a), with reduced 6 false negatives using a threshold of 0.47 for normal images misclassified as pneumonia. Fig. 5(a) of the MobNetCov19 model shows better results, with reduced 24 false negatives using a threshold of 0.19 for normal images misclassified as COVID-19 and increased only 1 false negative using a threshold of 0.0079 for COVID-19 and normal images as pneumonia. The CT scan using the VGG19 model is shown in Fig. 6(a), with 5 true negatives using a threshold of 0.033 for misclassified pneumonia images as COVID-19, increased 28 false positives using a threshold of 0.18 for pneumonia images as normal, and increased 10 false positives using a threshold of 0.067 for COVID19 images as normal. Fig. 7(a) of the MobNetCov19 model demonstrates better results, with reduced only 1 false negative using a threshold of 0.0075 for normal CT scans misclassified as COVID-19 and only 1 true negative using a threshold of 0.0066 for pneumonia images misclassified as COVID-19. Finally, all Figures of (b) show the model accuracy graphs with the x-axis indicating the number of epochs and the y-axis indicating accuracy. The graphs plot both training and testing results, from which it can be concluded that the MobNetCov19 model performs better with CT scans compared to X-Ray and Ultrasound images.

7 Conclusion and Future Work

Based on the results of the study, it appears that CT images are the most ac curate for detecting the disease, followed by X-Ray and Ultrasound. Our experimental result compared VGG19 and MobNetCov19 models, which were tuned with appropriate parameters to achieve high levels of accuracy in the detection of the disease. The augmentation techniques improves the accuracy of the models, and the results suggest that this approach was successful in improving the early detection of the disease. However, it was limited by the availability of data, which made it challenging to train more complex models. Our experimental results shows that MobNetCov19 model is more effective in accurately analyzing multi-mode images when appropriately tuned with parameters and trained with appropriate data. In the future, the use of laboratory images could help to provide more accurate validation data and improve the overall conclusions of the study.

References

1. Horry, M.J., et al.: COVID-19 detection through transfer learning using multimodal imaging data. IEEE Access **8**, 149808–149824 (2020)
2. Ravishankar, H., et al.: Understanding the mechanisms of deep transfer learning for medical images. In: Carneiro, G., et al. (eds.) Deep Learning and Data Labeling for Medical Applications. LNCS, vol. 10008, pp. 188–196. Springer, Cham (2016). https://doi.org/10.1007/978-3-319-46976-8_20
3. Yu, Y., Lin, H., Meng, J., Wei, X., Guo, H., Zhao, Z.: Deep transfer learning for modality classification of medical images. Information **8**(3), 91 (2017)
4. Tang, S., et al.: EDL COVID: ensemble deep learning for COVID-19 case detection from chest X-ray images. IEEE Trans. Ind. Inf. **17**(9), 6539–6549 (2021)
5. Ahmed, M.A., et al.: COVID-19 vaccine acceptability and adherence to preventive measures in Somalia: results of an online survey. Vaccines **9**(6), 543 (2021)

6. Zhang, M., Chu, R., Dong, C., Wei, J., Lu, W., Xiong, N.: Residual learning diagnosis detection: an advanced residual learning diagnosis detection system for COVID-19 in industrial internet of things. IEEE Trans. Industr. Inf.Industr. Inf. **17**(9), 6510–6518 (2021)
7. Basu, S., Mitra, S., Saha, N.: Deep learning for screening CIVID-19 using chest X-ray images. In: 2020 IEEE Symposium Series on Computational Intelligence (SSCI), pp. 2521–2527. IEEE (2020)
8. Chen, J., et al.: Deep learning-based model for detecting 2019 novel coronavirus pneumonia on high-resolution computed tomography. Sci. Rep. **10**(1), 19196 (2020)
9. Chouat, I., Echtioui, A., Khemakhem, R., Zouch, W., Ghorbel, M., Hamida, A.B.: COVID-19 detection in CT and CXR images using deep learning models. Biogerontology **23**(1), 65–84 (2022)
10. Kitrungrotsakul, T., et al.: Attention-Refnet: interactive attention refinement network for infected area segmentation of COVID-19. IEEE J. Biomed. Health Inf. **25**(7), 2363–2373 (2021)
11. Karacı, A.: VGGCoV19-NET: automatic detection of COVID-19 cases from XRay images using modified VGG19 CNN architecture and YOLO algorithm. Neural Comput. Appl.Comput. Appl. **34**(10), 8253–8274 (2022)
12. Watanabe, M., et al.: Central obesity, smoking habit, and hypertension are associated with lower antibody titres in response to COVID-19 mRNA vaccine. Diabetes/Metabolism Res. Rev. **38**(1), e3465 (2022)
13. Dhere, A., Sivaswamy, J.: COVID detection from chest X-ray images using multi-scale attention. IEEE J. Biomed. Health Inform. **26**(4), 1496–1505 (2022)
14. Frid-Adar, M., Amer, R., Gozes, O., Nassar, J., Greenspan, H.: COVID-19 in CXR: from detection and severity scoring to patient disease monitoring. IEEE J. Biomed. Health Inform. **25**(6), 1892–1903 (2021)

Analysis of Machine Learning Approaches to Detect Pedestrian Under Different Scale Using Frame Level Difference Feature

A. Sumi[1(✉)] and T. Santha[2]

[1] Department of Information Technology, Avinashilingam Institute for Home Science and Higher Education for Women, Coimbatore, Tamilnadu, India
sumi_it@avinuty.ac.in

[2] Department of Computer Science, Dr. G. R. Damodaran College for College of Science, Coimbatore, Tamilnadu, India

Abstract. Over the last 20 years, automotive technology has advanced to the point where automated systems can currently handle various aspects of vehicle control. Due to traffic congestion, pedestrians are particularly vulnerable and they collide with the vehicle's front end. In recent years, the legal standards and consumer protection assessments for pedestrian protection have gotten much stronger. Sensor technology that must reliably detect an impact between a vehicle and a person has substantial hurdles as a result. Computer vision-based technologies play an important role in the enhancement strategies of automation industries like the Advanced Driver Assistant System (ADAS) by identifying and tracking people on the road. During the process of pedestrian identification, human characteristics are a key factor in determining accuracy. The extraction of features to identify the pedestrians from the video images is a difficult task. In this paper, a per-frame evaluation methodology is taken in hand to make in depth and insightful comparisons among state of art detection techniques with FLD features. This investigates the detection rates at different scales. The experiments that are conducted with the help of Caltech pedestrian dataset. The recognition accuracy is compared and evaluated by avoiding a higher number of erroneous hits and a higher percentage of miss rates.

Keywords: ADAS · Pedestrian Detection · Feature Extraction · FLD Features · Scale variants

1 Introduction

Recent automobile accidents are primarily caused by human errors, which can be avoided by deploying advanced driver assistance systems (ADAS). This technique is used to identify pedestrians close to a vehicle. By giving the driver an early warning, this device aims to lower the amount of accidents involving pedestrians. Several sen sors, including cameras, radar, and lidar, can be used to execute this technology to identify pedestrians within proximity. This was used to provide more reliable and accurate detection. This

S. Aurelia et al. (Eds.): ICCSST 2023, CCIS 1973, pp. 464–473, 2024.
https://doi.org/10.1007/978-3-031-50993-3_37

technology typically works by detecting the shape and size of a pedestrian's body in order to differentiate between humans and other objects. Once detected, the system can then alert the driver of the potential danger, allowing them to take appropriate action.

An example of human ill behavior making pedestrians victims of road accidents is given in Fig 1.

Fig. 1. An Example Incident for Road Accident

The aim of ADAS is to decrease the amount and severity of vehicle accidents that cannot be prevented, thereby avoiding fatalities and injuries. Some of the most essential safety-critical ADAS uses are detection and avoidance of pedestrians, lane departure warning and correction, traffic sign recognition, automated emergency braking, and blind spot detection. This paper concentrates on preventing accidents through on-board pedestrian detection. For this, a computer vision based detector approach is taken to prevent pedestrians from becoming victims of road accidents. This approach has several phases like image candidate window generation, classifying the pedestrian from the background, reduction of multiple detections from the same pedestrian into a single detection window, and tracking of detected pedestrians for the purpose of removing erroneous ones or inferring information about the trajectory. So an accurate framework and technique are needed to do the same. Commonly, a person walking on the road might be visualized in diverse locations all over the road with the help of several stances, diverse colors, and several patterns of dress. Frequently, people walk on the road in the presence of complicated settings, alongside diverse illumination and climatic circumstances and the appearance of pedestrians on different scales. This will consider the process of identification as a real and prominent challenge.

In this case, the feature extraction technique plays an important role in identifying people walking on the road, because they can identify the people with a lot more accuracy with the help of extracted features. Several techniques for the extraction of features and classification were developed in the past. The reason for extracting approaches can be divided into two categories: (1) achieving expressive portrayal of images by leveraging the spatial framework present in such images, and (2) minimizing the total size of data connected to the image in consideration. It is possible to train the classifier to learn adequately with feature characterizations by achieving the stated objectives [1]. Manual feature extraction techniques, on the other hand, may be unable to train with important properties and have the least flexibility. But the computerized techniques might be able to extract the features for the people walking down the road by utilizing techniques like feedback propagation in an involuntary manner. On the other hand, computerized techniques necessitate huge quantities of learning illustrations and consume a lot of training time. It requires greater hardware obligations [2].

2 Literature Review

Pedestrian detection is a critical component of various autonomous driving and intelligent surveillance applications such as pedestrian crossing system and intelligent video surveillance. In recent years, many studies have focused on improving the performance of pedestrian detection algorithms. The accuracy and speed of feature based pedestrian detection are primarily determined by the proper feature selection and classification. In this literature study, various detection techniques in detecting pedestrians on the road by different researchers give a promising result based on the reduction of miss rates by selecting appropriate features and classification.

2.1 Materials and Methods

There has seen a tremendous amount of research into feature-based pedestrian detection. Generally, pedestrian detection using features from input images will rely on the extraction of a picture feature and applying classifier to identify whether pedestrians is present in that region. The features can be extracted from different parts of the image and can be based on color, texture, shape, or motion information. It aims to detect pedestrians in digital images by extracting and analysing extracted features. This type of detection has become increasingly popular due to its potential to be used in a variety of applications, such as autonomous navigation, surveillance, and crowd monitoring. A number of studies have been conducted on the topic of feature-based pedestrian detection using computer vision. The types of features used in these studies have included geometric, textural and structural features. Geometric features are those that are related to the shape or form of an object. For example, the size, orientation, and position of an object can be used as geometric features for pedestrian detection. Textural features are those that describe the physical characteristics of an image. These features include color, contrast, and texture. Structural features are those that describe the structure of the image, such as the edges, lines, and corners. In addition to the extraction of features, several studies have also focused on the classification of the features in order to detect pedestrians.

In [3] proposed an approach for pedestrian detection which uses a combination of color and texture features. The method combines the use of a histogram of oriented gradients (HOG) descriptor and a color descriptor to extract features from the image. In addition, the authors demonstrated that the proposed approach was more accurate and faster than existing approaches. In [4] proposed a new approach for pedestrian detection which uses a combination of color, shape and motion features. The authors showed that by combining these features, they were able to achieve better accuracy than existing methods. Moreover, they proposed a multi-scale approach which fur ther improved the accuracy of their method. In [5] proposed a deep learning-based approach for pedestrian detection. The authors employed a Convolutional Neural Network (CNN) algorithm to acquire features from an input image, and then em ployed a sliding window algorithm to identify pedestrians from the acquired features. The authors used the VGG-16 and GoogLeNet networks for feature extraction on the Caltech Pedestrian dataset.

In [6] proposed a feature-based pedestrian detection method using cascade channel features (CCF). The authors proposed a new set of channel features which are combined with a cascade structure for pedestrian detection. In [7] proposed a feature-based pedestrian detection method using a combination of color and shape features. The authors used a multi-scale sliding window approach to detect pedestrians in the input image. The authors used the KITTI pedestrian dataset for testing. In [8] proposed a pedestrian detection method using a combination of deep feature and shallow feature. The authors used the Faster R-CNN network to extract deep features and then used a shallow feature to refine the pedestrian location in Caltech dataset. In recent years most of the researchers proposed several methods for feature-based pedestrian detection, such as the use of multi-scale features, deep learning methods, and a combination of both.

In [9] researchers proposed a method called Single-Shot Object Detection with Multi-Scale Feature Fusion which used a single-shot detector to detect pedestrian This method combined the use of multi-scale features, deep learning, and a multi scale feature fusion approach to increase the accuracy and robustness of pedestrian detection. The authors reported that the proposed method outperformed existing state of-the-art methods on the Caltech Pedestrian Detection Benchmark dataset. A method called Multi-Scale Feature Fusion for Pedestrian Detection with Heterogeneous Feature Maps was proposed by [10]. This method used a feature fusion approach to com bine multi-scale features from different feature maps, making the detection more robust. The method also utilized a deep learning architecture to further improve the accuracy of the detection. The authors reported that the proposed method achieved state-of-the-art performance on the Caltech Pedestrian Detection Benchmark dataset. A method called "Multi-Scale Feature Fusion for Pedestrian Detection with Multi-Layer Feature Maps" [11] was proposed. This method used a multi-layer feature map approach to combine multi-scale features from multiple layers, making the detection more accurate. The method also utilized a deep learning architecture to further boost the accuracy of the detection. The authors reported that the proposed method achieved state-of-the-art performance on the Caltech Pedestrian Detection Benchmark dataset.

The accuracy and speed of feature-based pedestrian detection are primarily determined by the proper feature selection and classification. In this literature study, various detection techniques in detecting pedestrians on the road by different researchers give a promising result based on the reduction of miss rates by selecting appropriate features and classification.

2.2 Problem Statement

The exact identification of individuals walking on the road profoundly impacts many applications like surveillance, robotics, the automobile manufacturing industry, etc. This system emphasises automatic and accurate pedestrian prediction, as it has the capability of protecting multiple people from road accidents. The primary goal of this project is to use the Caltech Pedestrian Dataset in order to provide a higher standard to implement proposed features with various detection techniques. This paper concentrates on the automatic detection of pedestrians from a video using proposed Frame Level Difference features in near and medium scale situation. It also assists in analyzing the detection rate of various machine learning approaches using the false positive percentage.

3 Methodology

Analyzing the fine-grained categorization of pedestrians on different scales from the video, a machine learning-based pedestrian detection approach is developed. Figure 2 gives the machine learning-based framework of pedestrian detection. The delicately-grained classifiers are analysed by picking the frame-level features, which include bounding box tracking and pedestrian localization. This methodology comprises frame window extraction, localization, frame level difference feature and pedestrian detection using state-of-the-art classifiers. Finally, it analyses and evaluates the rate of pedestrian detection using pre-trained classifier with new Frame Level Different feature in various scale variations and find out which classifier gives the best result with this above new feature.

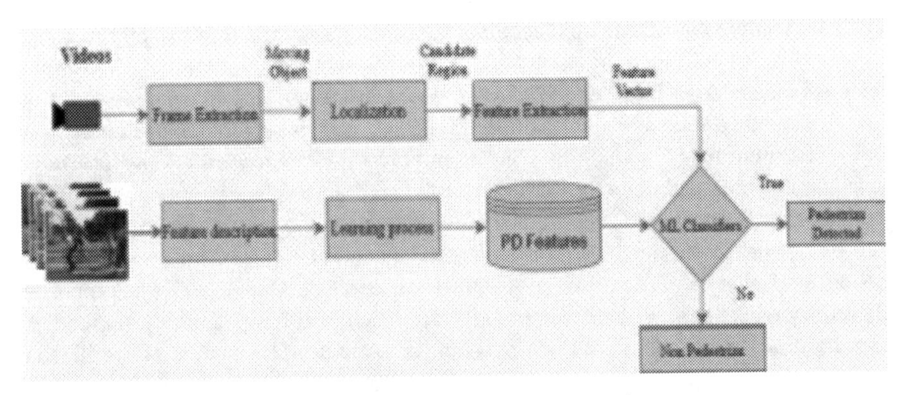

Fig. 2. Framework of Pedestrian detection

3.1 Frame Window Extraction

Attentive vision approaches are utilised as a preliminary processing step to focus the analysis on only specific areas of interest. The low-level processing is to focus on prospective candidate regions with pixel sizes that fall within a particular range. This range is estimated using the vision system's intrinsic qualities (angular aperture and resolution), as well as pedestrian size and distance. It displays a region of the image with subtle characteristics, such strong vertical symmetry and a high density of vertical edges, which point to the presence of a pedestrian. The following steps are involved in extraction: assessment of the full frame, symmetry identification, bounding box generation and filtering.

3.2 Assessment of Full Frame

In the initial step, candidates for pedestrians are searched. Each identification is given a box's boundaries and a rank or rating by an identification framework. For the purpose of integrating finer identification, the abovementioned framework must perform multiple-scale identification and critical non-maximal suppression (NMS). The final result evaluates the list of significant bounding boxes.

3.3 Symmetry Identification

Frame columns are considered "symmetric axes" for bounding boxes. Different bounding boxes are considered for each symmetry axis. The technique involves scanning a certain distance from the camera and a reasonable range of height and width for someone walking at eye level. The simplest bounding box can be found by minimizing the combination of the two symmetry measures masked by edge density.

3.4 Bounding Boxes Generation and Filtering

A large bounding box is a road user when compared to the grey level symmetry of multiple boxes in relation to each other on an axis. The highest symmetry bounding box usually has a larger size than the object it is enclosing, because of its tendency to be surrounded by homogeneous areas. In order to reduce false positives, various filters were used according to the width distribution of edges into a bounding box, segmentation and region classification.

3.5 Localization

Localization phases consist of two steps one is pedestrian coordinate system and second one is observation Fig. 3 shows the pedestrian coordinate. In accordance with [12], the [Xp, Yp and 0] coordinates, which correspond to the intersection of the vertical axis with the ground, are utilized to represent the position of a pedestrian on the level road. The scene is scanned by a camera with a tilting angle of 0 and height of Zc. Detection results in image bounding boxes with pedestrian shapes. For example, the height of a pedestrian is greater than the height of a bounding box. The measurement of the height is of paramount importance for localization, as the measurement of the height directly impacts the estimated separation between camera and pedestrian.

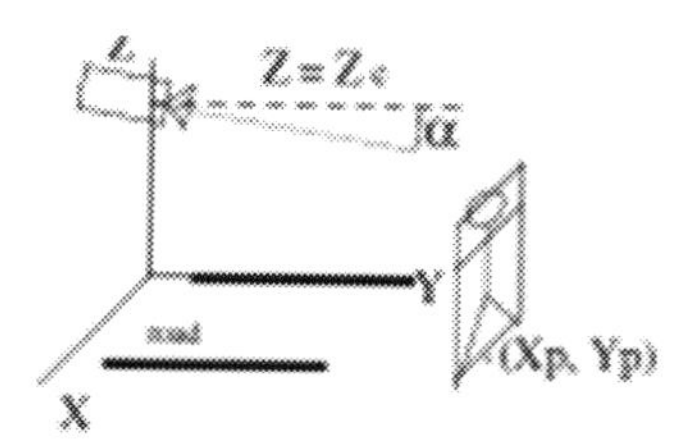

Fig. 3. Coordinate of Pedestrian.

3.6 Frame Level Feature Extraction

A Frame Level Difference [FLD] Feature proposed to provide the model learning, aggregated feature into a mixture of classifiers to evaluate the detection rate under multiscale pedestrian conditions.

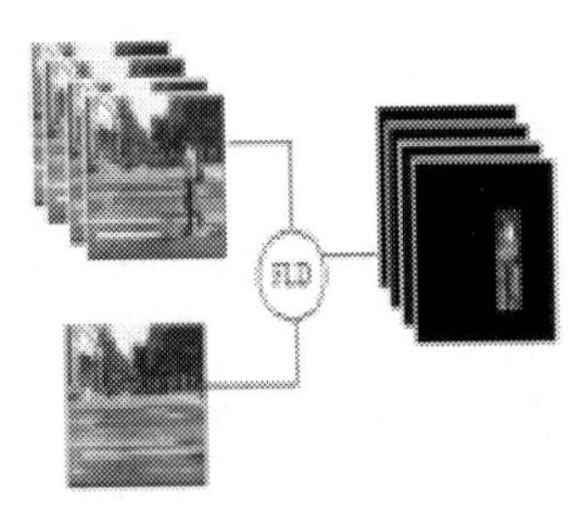

Fig. 4. Representation of Frame Level Difference in a Sample Image

The selection and extraction of the frame level features is determined to maintain the disparity between the two neighboring frames and the variance is preserved, shown as a flow in Fig. 4. Although combining the results of the proposed feature conversion could result in functionality failures with the single frame, it could be possible to enhance functionality by comparing the frameworks without the usage of the previously described feature conversion. This is performed by subdividing the order of frame features into sub sequences of the same duration in temporal sizes and then executing the average assembly of frame features in each corresponding series to build the sub-sampled order to be supplied to the classifier [13].

4 Evaluation Methodology

The aim of this study is to compare and see how well pre-trained detectors work with the suggested FLD feature from the Caltech dataset [14]. In this, performance in vestigation is carried out under various levels of pedestrian scale and also it finds, how detector performed better with proposed FLD feature.

4.1 Statistics of Caltech Dataset

This Dataset of Pedestrian Footage from Caltech [14] consists of approximately 10 hours of footage recorded from a car operating in typical urban conditions. It is composed of approximately 250,000 images, 350,000 binding boxes, and a total of 2,300 individuals. The explanation includes detailed obstruction indexes and chronological communication between bounding boxes. The video dataset to be trained is grouped into to six sets (set00–set05), each of which is about 1 GB in size and contains 6–3 one minute long seq files as well as explanation material. The exam ined data (set06–set10) is divided into five groups, each of which contains 1 GB.

4.2 Scale Variation

Pedestrians visual size (height in pixels) is divided into three categories: near (80 pixels or more), medium (30–80 pixels), and far (less than 80 pixels). The division into three scales was impacted by the distribution of sizes in data collection, human endeavour, and automotive system needs. In the automotive industry, medium-scale detection is critical. At a speed of 55 km/h (15 m/s), it takes just 1.5 s to reach a person that appears as 80 pixels in size, while a 30 pixel would take 4 s to reach at the same speed. As a result, recognizing close scale pedestrians may not pro vide the vehicle enough time to react, but detecting far size pedestrians is less important. The majorities of current algorithms are built for small scales and perform badly even at medium scales.

5 Result and Discussion

To evaluate classification performance, ROC curves is used because they quantify the trade-off between detection rate (the percentage of true positives) and false positive rate [15]. The ROC curve in Fig. 5 shows the correlation between detection miss rates and the number of incorrect detections per image at near medium scale. This experiment is done to evaluate detection of pedestrian using the FLD feature with Caltech pre trained classifiers under pedestrian medium and near scale situations.

The results correspond to the near and medium scales when there are people on the road. From the Fig. 5 it is clearly shows that Frame Level Features (FLD) with Multi-filter +Motion detector beats other detectors in near-scale with a log average miss rate of 30% versus 40–80% with other detectors. Figure 6 shows that the Frame Level feature with Multi Filters +Motion methods has a log average miss rate of 77% in the medium-scale situation, which is much lower than the other pertained detectors, which have a miss rate of 80 to 90%.

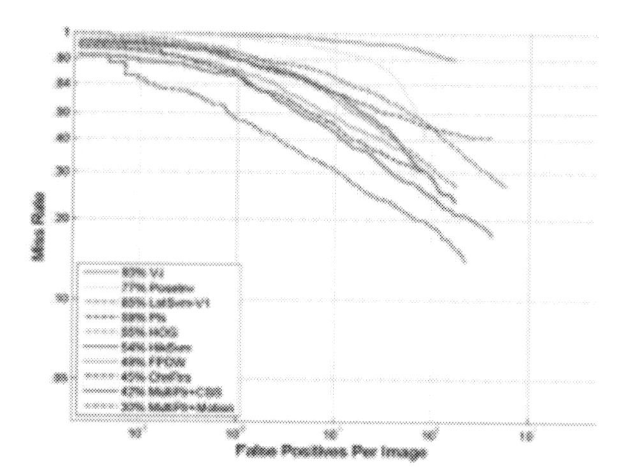

Fig. 5. Performance Evaluations of rate of detection under Near Scale Situation

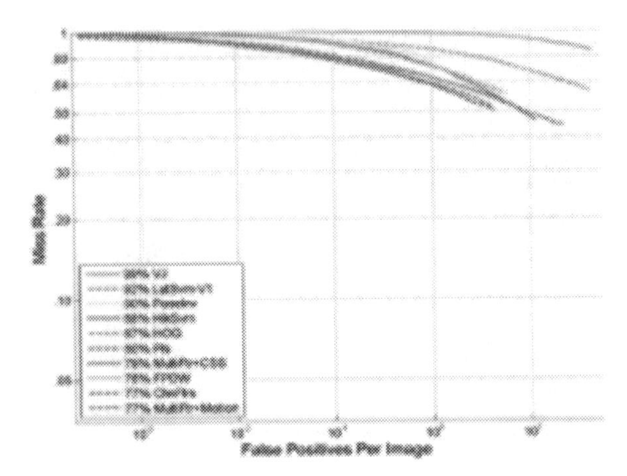

Fig. 6. Performance Evaluations of rate of detection under Medium Scale Situation

From the above result it is clearly stated that detection of pedestrian using pre trained Multi Filters + Motion technique with the proposed FLD feature gives best result by reducing error rate than proposed feature when compared to other pre trained techniques under medium and near scale.

6 Conclusion

The number of pedestrian fatalities is declining as a result of the advent of driving aids like auto-breaking systems. These factors have made people recognition and monitoring an essential field of study for computer vision over the past few decades [16]. In the larger field of autonomous robotic vision applications, person recognition on the road is a major difficulty. Feature extraction algorithms have been the focus of previous research on the problem of detecting people on the road. This paper use in novative Frame Level Difference feature extraction technique means of retaining the difference between adjacent frames as features. As a result, the motion of pedestrians can be identified and analysed under the various levels of scale. The suggested frame level features will perform significantly better with a miss rate of less than 77% on a medium scale with Multi Filters Motion methods. Though the proposed technique's performance well by reduced the error rate to some extent, it will still need some other standard ways reduce the error rate under medium scale.

References

1. Van Der Maaten, L., De Croon, G.: Feature Extraction for Pedestrian Classification Under the Presence of Occlusions (2019)
2. Wang, L., Zhang, B.: Boosting-like deep learning for pedestrian detection, arXiv preprint arXiv:1505.06800 (2015)

3. Zhang, X., Shangguan, H., Ning, A., Wang, A., Zhang, J., Peng, S.: Pedestrian detection with EDGE features of color image and HOG on depth images. Autom. Control. Comput. Sci.. Control. Comput. Sci. **54**, 168–178 (2020)
4. Yao, S., Pan, S., Wang, T., Zheng, C., Shen, W., Chong, Y.: A new pedestrian detec tion method based on combined HOG and LSS features. Neurocomputing **151**, 1006–1014 (2015). https://doi.org/10.1016/j.neucom.2014.08.080
5. Li, H., Wu, Z., Zhang, J.: Pedestrian detection based on deep learning model. In: 2016 9th International Congress on Image and Signal Processing, Biomedical Engineering and Informatics (CISP-BMEI), pp. 796–800. IEEE (2016)
6. Cao, J., Pang, Y., Li, X.: Learning multilayer channel features for pedestrian detection. IEEE Trans. Image Process.age Process. **26**(7), 3210–3220 (2017)
7. Xiao, Y., et al.: Deep learning for occluded and multi-scale pedestrian detection: a review. IET Image Process. **15**(2), 286–301 (2021)
8. Zhai, S., Dong, S., Shang, D., Wang, S.: An improved faster R-CNN pedestrian detection algorithm based on feature fusion and context analysis. IEEE Access **8**, 138117–138128 (2020). https://doi.org/10.1109/ACCESS.2020.3012558
9. Cheng, Y., Chen, C., Gan, Z.: Enhanced single shot multibox detector for pedestrian detection. In: Proceedings of the 3rd International Conference on Computer Science and Application Engineering, New York, pp 1–7. ACM (2019)
10. Wang, J., Zhu, C.: Semantically enhanced multi-scale feature pyramid fusion for pedestrian detection. In: 2021 13th International Conference on Machine Learning and Computing, New York, pp.1–4. ACM (2021)
11. Deng, R., Guan, B., Li, Z., Wang, J.: Pedestrian detection based on multi-layer feature fusion. Acad. J. Comput. Inf. Sci. **4**(5), 76–84 (2021)
12. Bertozzi, M., Broggi, A., Chapuis, R., Chausse, F., Fascioli, A., Tibaldi, A.: Shape based pedestrian detection and localization. In: Proceedings of the 2003 IEEE International Conference on Intelligent Transportation Systems, vol. 1, pp. 328–333. IEEE (2003)
13. Sumi, A., Santha, T.: Frame level difference (FLD) features to detect partially occluded pedestrian for ADAS. J. Sci. Ind. Res. **78**, 31–836 (2019)
14. Dollar, P., Wojek, C., Schiele, B., Perona, P.: Pedestrian detection: an evaluation of the state of the art. IEEE Trans. Pattern Anal. Mach. Intell.Intell. **34**(4), 743–761 (2012). https://doi.org/10.1109/TPAMI.2011.155
15. Boukerche, A., Sha, M.: Design guidelines on deep learning–based pedestrian detection methods for supporting autonomous vehicles. ACM Comput. Surv.Comput. Surv. **54**(6), 133 (2022). https://doi.org/10.1145/3460770
16. Farooq, M.S., et al.: A conceptual multi-layer framework for the detection of nighttime pedestrian in autonomous vehicles using deep reinforcement learning. Entropy **25**(1), 135 (2023). https://doi.org/10.3390/e25010135

Eye Support: AI Enabled Support System for Visually Impaired Senior Citizens

Vani Vasudevan(✉), Thota Thanmai, Riya Yadav, Pola Udaya Sowjanya Reddy, and Subrina Pradhan

Department of Computer Science Engineering, Nitte Meenakshi Institute of Technology, Bangalore, India
{vani.v,1nt19cs203.thanmai,1nt19cs159.riyadav,1nt19cs136.udaya, 1nt19cs189.subrina}@nmit.ac.in

Abstract. Individuals who have impaired eyesight have a difficult time seeing the tablet's name, which is written in small characters on the back of the tablet. India features in the "top ten" list for the visually impaired. Numerous technologies are growing in India, allowing us to aid the populace in identifying the tablet's name, purpose, and other information without assistance from others. Increasingly, technology is being employed for a variety of reasons, and individuals are employing smart gadgets and devices. In the suggested method, users may launch the app on their smartphones, upload or snap a photograph of the pharmaceutical strip, and receive the findings in audio format. The proposed system employs Artificial Intelligence (AI), and Natural Language Processing (NLP) in Python for processing and output in voice format. The smartphone application primarily benefits the elderly, notably those with visual impairments.

Keywords: Optical Character Recognition (OCR) · Image Processing · Text Extraction · Text-to-Speech · Tesseract · OpenCV

1 Introduction

The number of elderly people is rising, even though facilities do not provide for their needs, such as food, housing, hygiene, and medication. The medication, particularly in deteriorating vision in the elderly, is the main focus of the proposed system. The small letters on the tablet's backside can be challenging to read for many people, especially the elderly. Some people have trouble reading the name of the medication and having trouble recalling how to use it. Around them, assistance won't always be available to explain how the medicine is used. Examples include taking the appropriate medication when needed. They struggle when it's necessary to take multiple medications at once. They must read the instructions on the medication label to determine the type of medication and how to use it. Therefore, the use of a label reader and the sense of hearing can help the visually impaired, such as the elderly who are unable to read the text clearly, the blind from birth, as well as those who are impaired due to accidents, to access the text or message information on the pharmaceutical label, which provides the medicinal information, type of medication, and its use.

S. Aurelia et al. (Eds.): ICCSST 2023, CCIS 1973, pp. 474–487, 2024.
https://doi.org/10.1007/978-3-031-50993-3_38

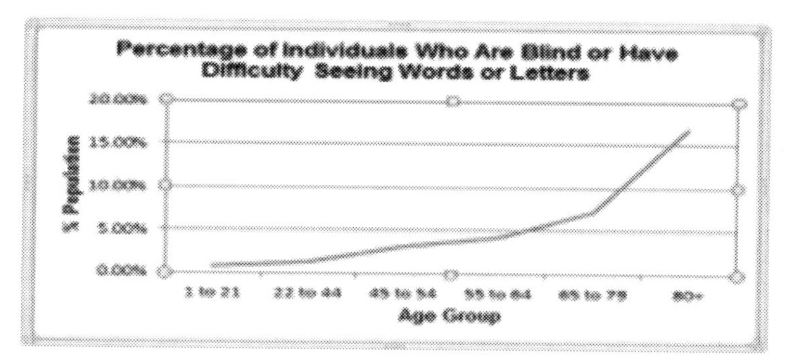

Fig. 1. Percentage of individuals who are blind or have difficulty seeing words or letters. [Source: https://visionaware.org/ working-life/working as-a-senior-with-vision-loss/employers-and-older-workers-with-visual-impairment/ age-related-vision-loss-in-the-workplace/]

Figure 1 depicts the increasing eye vision problem as the population and age increase. In this paper, the approach is to first take an image of the medicine, then detect and crop all the required words, i.e., the name of the medicine, followed by preprocessing of those word images. The final text is generated using tesseract OCR. The dataset is gathered, which includes the name of the medicine and its intended use. If the extracted name of the medicine matches the name in the dataset, its use is extracted. Finally, the name of the medicine and its intended use are spoken aloud.

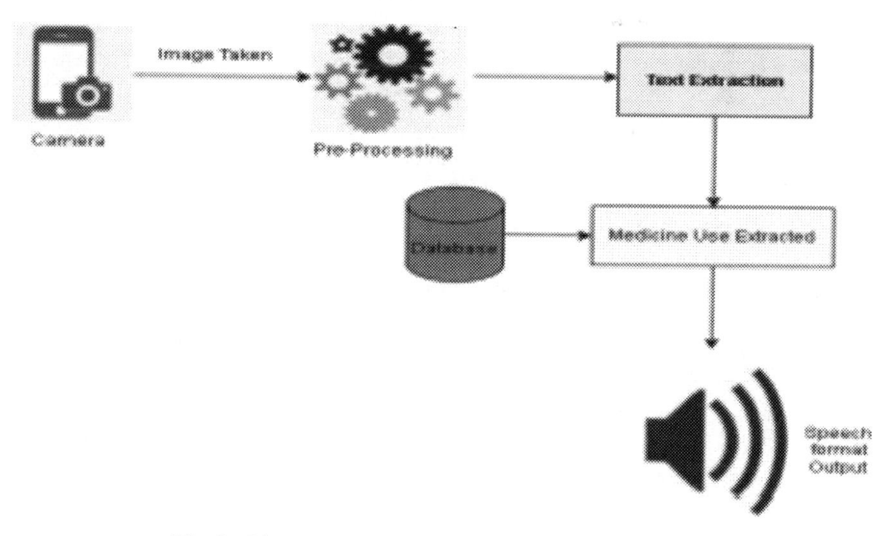

Fig. 2. Diagram to demonstrate the working of I-Support

Figure 2 depicts the overall process of our project. This system receives input as an image from a mobile camera and then processes it. The proposed system's output will be provided in speech format, which is an extract of the medicine strip name that was processed.

2 Literature Review

H. Singh and A. Sachan came up with a method [1] in which an OCR system would use a camera to read the characters from the input product that was chosen. This method was proposed by the two researchers. OCR, or optical character recognition, performs computations on its processors in the same way that human brains recognise characters. OCR, which stands for optical character recognition, can recognise characters using either a camera or a matrix of photoreceptors. Following the input of the characters, the machine will then verbally read out the words at a very basic level of comprehension. They have broken the modules down into four different parts, which are the preprocessing, the character recognition, the post-processing, and finally the output, which is where any editing that needs to be done on the final document can be done.

In a similar vein, another study was carried out in the year 2020 at Sri Manakula Vinayagar Engineering College on the topic of tracking the cost and date of expiration of medicine for visually impaired patients [2]. The core of their concept is an approach that uses computerised vision to determine the validity of a medicine's expiration date. They have broken the proposed Expiry Date Recognition system down into its four primary modules, which are as follows: pre-processing of numeral strings, segmentation of strings of numerals, extraction of features, and numeral recognition. Then, the output of stretched Gabor filters such as Local Energy, Fourier Magnitude, and complex moment with statistical features that are extracted from a set of features that are used. The results that were obtained demonstrate that it is possible to obtain a high level of performance by utilizing Neural Networks and the Support Vector Machine classifier. The results that were obtained demonstrate that the approach that was proposed can be generalized to a variety of computer vision applications. In addition, the recognition of the available characters makes use of new features that are derived with the help of the Gabor Filter. These features are taken into consideration.

Army Institute of Technology researchers have been responsible for one of the most recent research works, which was completed in 2021 and published the same year. Their primary goal was to enable the computerized extraction [3] of the necessary information that a user needs from the document that was uploaded so that the user could then use this information wherever it was required of him. It's possible that this will result in less expensive data entry work, or even none. A picture is taken, and then the computer generates editable text based on what it sees in the picture. If the level of accuracy for a particular document was greater than 95%, then the result was of a standard quality and was delivered by the project.

A smart book reader that makes use of Optical Character Recognition [4] was proposed in the research paper titled "OCR Based Facilitator for the Visually Challenged," which was written by Shalini Sonth and Jagadish S Kallimani. The public has a requirement for a text reader that is easily transportable, can be purchased for a low cost, and is readily available. We propose a framework that is based on cameras and is integrated with image processing algorithms, optical character recognition, and a text-to-speech synthesis module. The camera is used to take a picture of the text in its printed format, and after that, the output is put through preprocessing so that it can be read by an optical character reader (OCR). During the preprocessing stage, tasks such as binarization, de-noising, segmentation, and de-skewing, in addition to feature extraction, are carried out. Text in this article that is designed to be read aloud and is intended for people who are visually impaired is included. The TTS application, which utilizes Pico and Google Tesseract, is the one that is utilized for Optical Character Recognition (Table 1).

Table 1. Notable research work related to NLP, MI enabled app for tablet info detection and reading in speech format.

Article	Algorithm used	Accuracy	Similarities with the proposed system	Applications	Advantages	Disadvantages	Key Takeaways
[1]	OCR character recognition	98–99%	OCR is used	Document analysis	Font size is maintained as in the picture	Inaccurate line end judgement	Pre processing,OCR, post-processing
[2]	Tesseract OCR	70.2-92.9%	OCR and TESSERACT is used	Expiration Date	Accuracy is higher	Expiry Date position may not be correlate with barcode position	Text Extraction,Googl e Cloud Vision APi
[3]	OpenCV	96%	Used same OCR techniques and TESSERACT techniques	Automatic plate capture and recognition,passort identification	Productivity,cost reduction,high accuracy	Expensive	Binarization,Noise reduction,segmentation
[4]	Optical Character Recognition	98–99%	Used OPENCV Libraries	Text To Speech	Multi-font is detected	If image contains more colors then the result of conversion from image to text will not have good accuracy	Tesseract OCR,ocr,Text To speech Engine
[5]	Hardware is implemented ,microcontrollers,RFIDs	-	Idea is same	Medicine reading in voice format	No need of Internet	Low quality of voice	How our work can be implemented in hardware format
[12]	Optical Character Recognition	98–99%	Used OCR	Document Images analysis	Text Recognition	Difficult To read word by word	Tesseract OCR
[13]	Optical Character Recognition	98–99%	Used OCR	Signs detection	affordable	GOOGLE Nuxus 5x phone only required	OCR implementation
[14]	Tesseract OCR	70.2-92.9%	Used various OCR and Tesseract	Words and Paragraph Detection	Text To Images Detection	Less Fluency	OCR,Tesseract OCR and Google Translator

3 Proposed System

The proposed system is divided into the modules listed below:

1. Image Capture
2. Extraction of Medicine Name

3. Database search to find Medicine Use.
4. Transform the extracted text into speech.
5. Produce speech in the user's native language.

1. *Image Capture*

In this module, we used open-source code to open the camera and capture the image. The camera's built-in functions were used to open it and capture the image.

As shown in the Fig. 3, the camera is opened, an image is captured, and it is sent for pre-processing. When the Esc key was pressed, the image is saved in the predefined location, and by accessing that location, then continue with the text extraction as described.

OpenCV is used to capture images and videos from cameras and provides various functions for image and video operations. It is useful for capturing an object with a webcam and performing desired operations on it. Two parameters are used: the filename and the image.

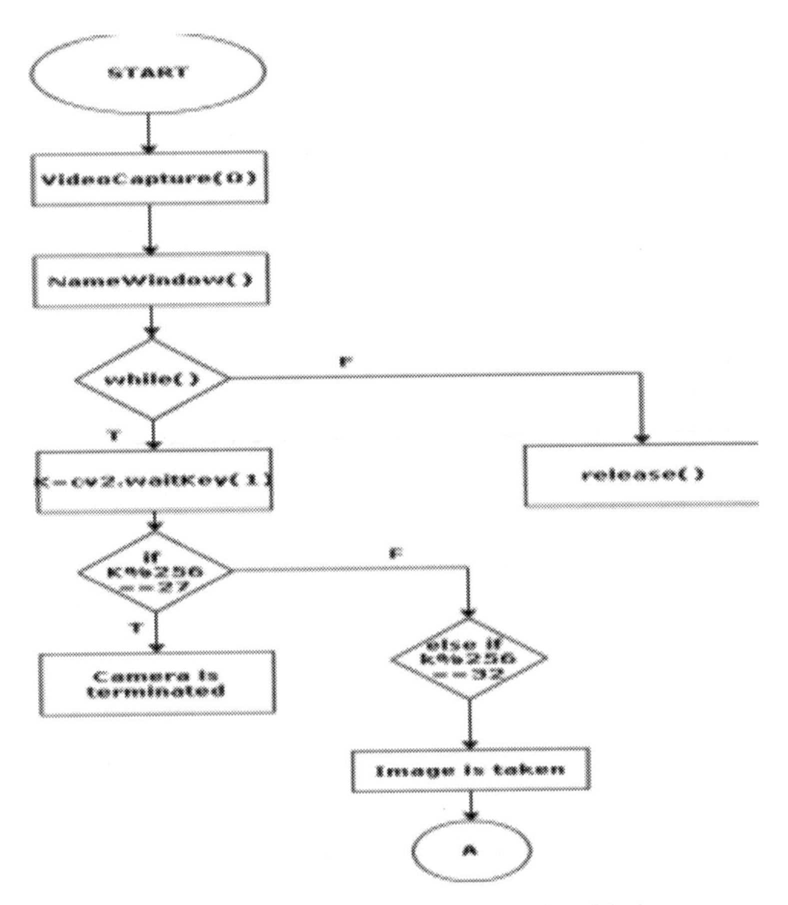

Fig. 3. Flowchart for demonstrating the capturing of the image

2. *Extraction of Medicine Name*

The following libraries are required:

1) OpenCV
2) Pytesseract

Text extraction, also known as keyword extraction, uses machine learning to check text and focus on relevant or essential words and phrases from unstructured data, such as news stories, reviews, customer service complaints, and in our case, medication. Machine learning algorithms and calculations are used to aid in the text extraction and improvement strategies. Tesseract is an open-source optical character recognition (OCR) engine that reads text from images. Finally, the extracted text is extracted from the image and transferred to the appropriate application or document type (Fig. 4).

There are numerous text extraction algorithms and strategies that are used for various purposes.

- *Crop the image to fit the document border:* Crop an image to fit within a document border using OpenCV. OpenCV makes it easy to locate a document's edges in an image [10] using thresholding techniques and the fourth parameter in its function. To use this, provide a grayscale image as the first argument, a threshold value for separating pixel values as the second argument, and a maximum value for pixels exceeding the threshold as the third argument.
- *Find and crop/segment all the words in the document:* "Detection of words in controlled conditions [10] is usually done using heuristic-based methods. However, these methods have the drawback of frequently identifying undesirable areas in images as words. Using the OpenCV indicator, a more precise approach can be used to accurately distinguish between handwritten and machine-printed text. After identifying the words in the image, each one is cropped and saved as a separate image.
- *Pre-process word images:* Process images of words by converting them to grayscale using OpenCV for easier manipulation. Detecting text in natural scenes, however, is much more challenging and involves dealing with the following:

Image or Sensor Noise: Handheld cameras have higher noise levels compared to standard scanners, and low-cost cameras often interpolate raw sensor pixels to produce true colors.

Viewing Angles: Viewing angles that are not collateral to the text are common in regular scene text, making it difficult to identify.

Blurring occurs when a user uses a smartphone that lacks stability, resulting in a blurred image.

Lighting Conditions: Because of varying lighting conditions in natural scene images, such as near darkness, the camera's flash may be activated, or the sun may be shining brightly, saturating the entire image.

3. *Database Search to Find Medicine Use*

Dataset is based on use of the medicine and name of it. If medicine name matches with the dataset, then show the use of the medicine else it will show that match of the medicine is not found.

For example, if paracetamol (650 mg) is considered then the use of the medicine is: it will help in the relieve of pain and fever. After the name of the medicine is given as an input, in database it will go and search for the medicine name if the medicine name matches then it will show the medicine name and its use in the speech format. Regarding how it will get converted to speech will be discussed in upcoming module.

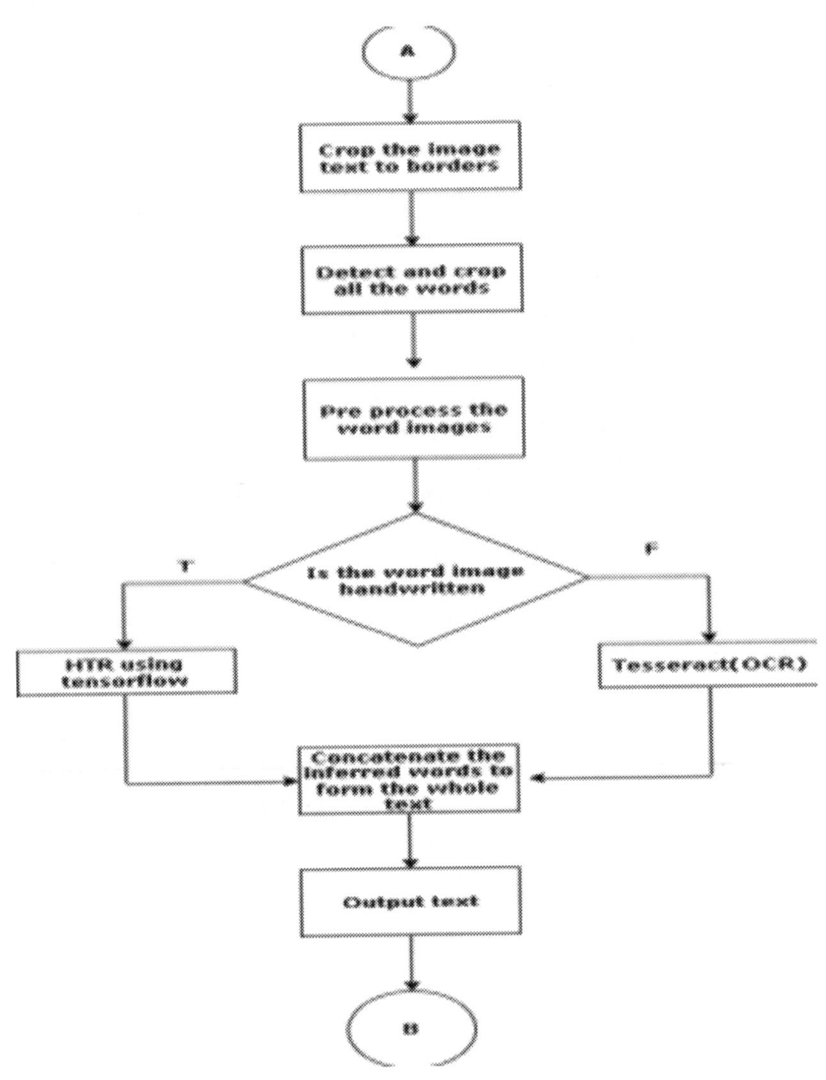

Fig. 4. Flowchart for demonstrating the extraction of text from the medicine

4. *Transform the Extracted Text into Speech*

NLP (Natural Language Processing) is the field of artificial intelligence. NLP machine's ability to read and convert it into a speech format. There are many APIs available in the market to do the task. In this project, TTS (Text to Speech Google) is used for the text to speech conversion. A python library which used to convert the text to speech. Installation of TTS is required for the working of it in our system and in our code, we need to use the import TTS statement for working of it. In TTS the text file is converted to mp3 audio file. Fig. 5. Shows to flow of text to speech conversion process.

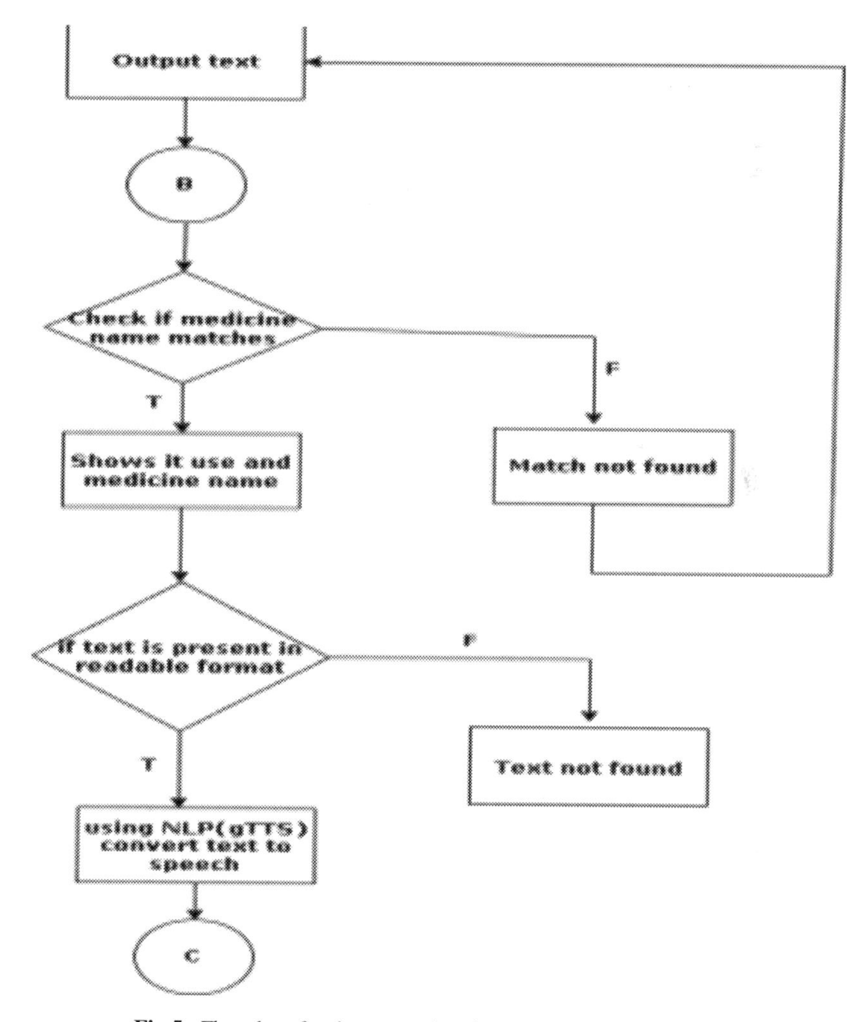

Fig.5. Flowchart for demonstrating the text to speech translation

5. *Produce Speech in the User's Native Language*

Google Translator is also called as Web Translator. This helps google translator to translate any text to the desired language text. The Neural Machine Translation is used in google translator. It has a total of 133 languages. Google translator Assistant translates one word to another word. M4 modelling is a model which is used in Google Translator. Fig. 6. Shows the flowchart depicting the production of speech in the user's native language.

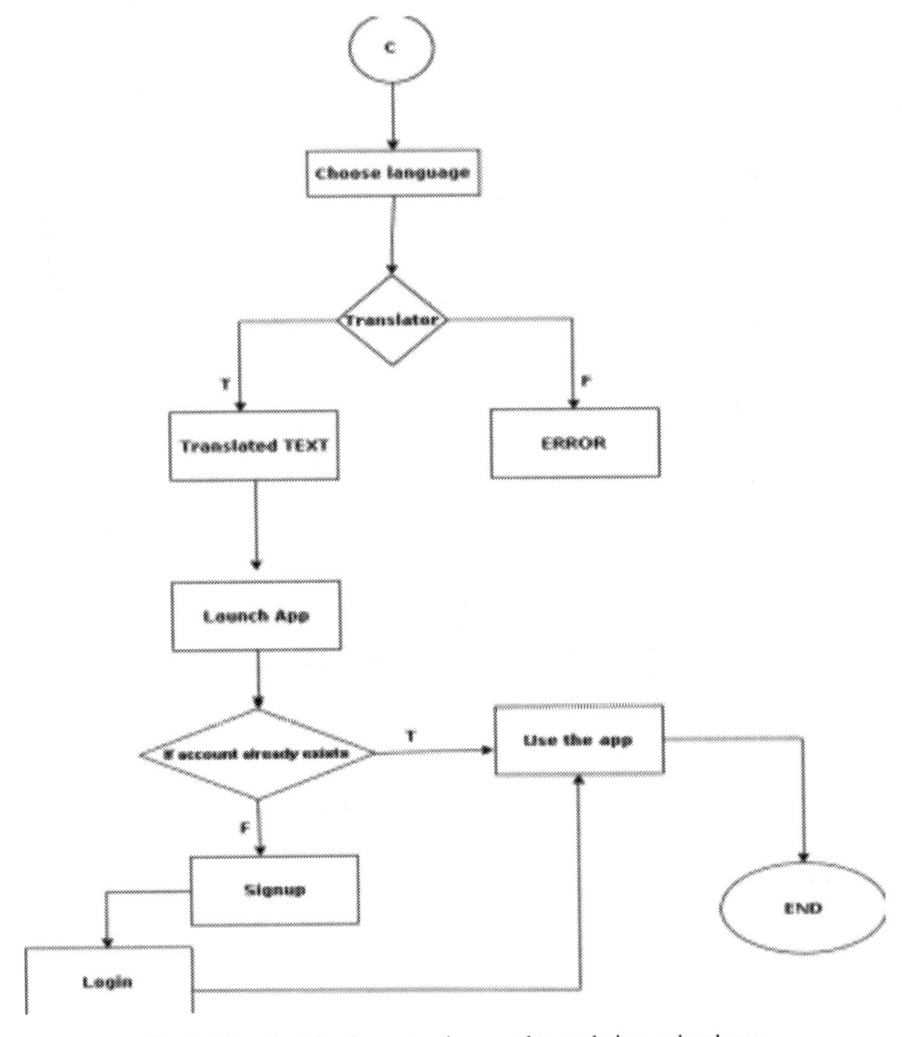

Fig.6. Flowchart for demonstrating google translation and web app

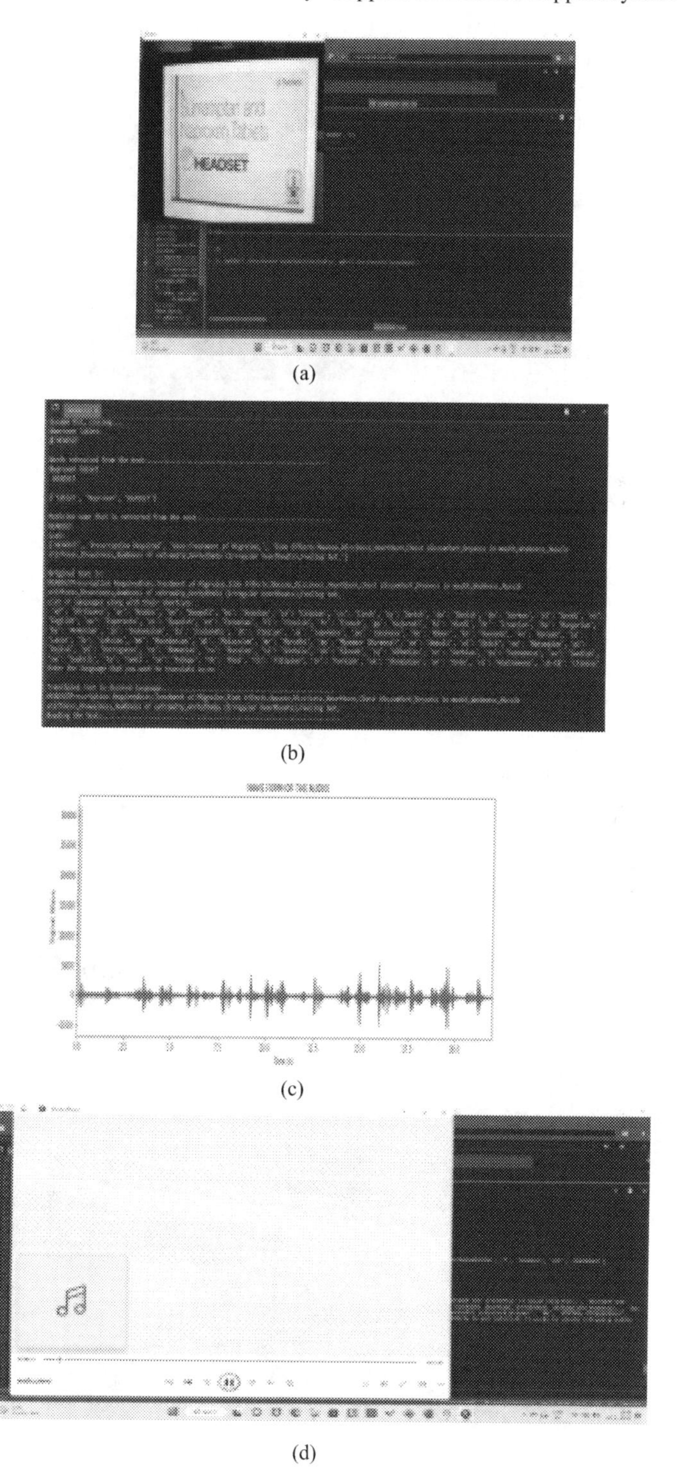

Fig.7. (a) Sample Image1 (b) Extracted Medicine name and its usage from the database (c) Speech format (d) Waveform of audio file

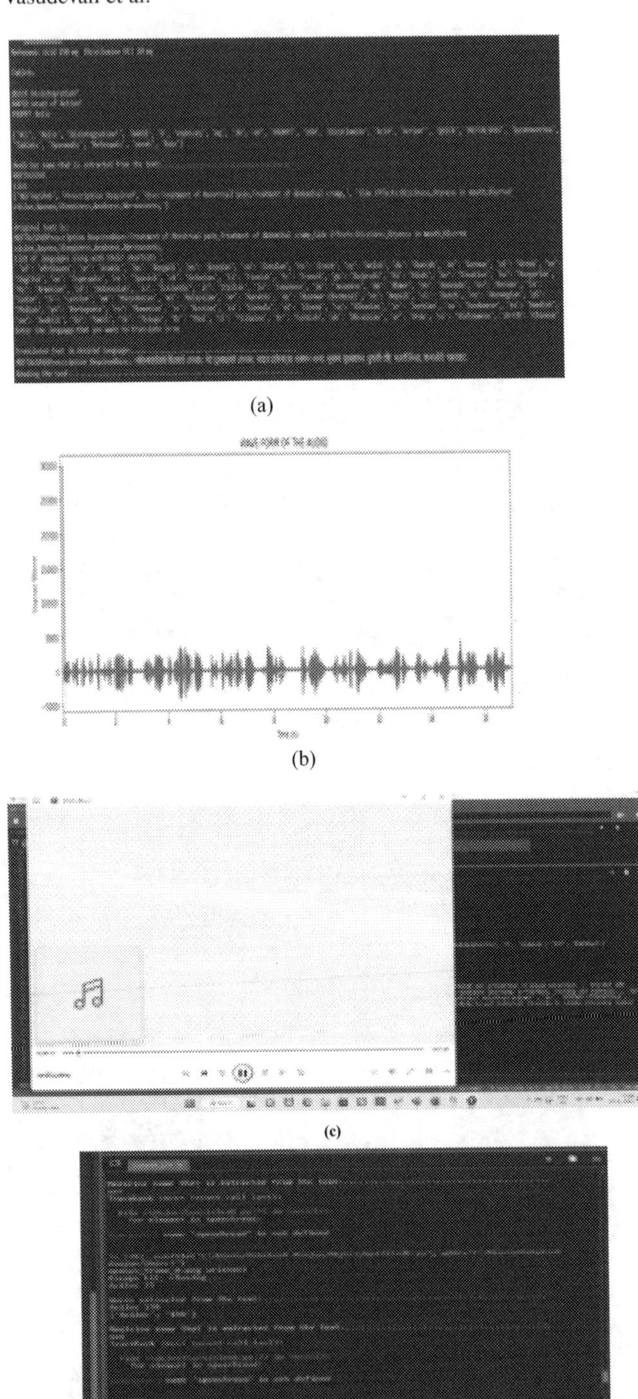

(a)

(b)

(c)

(d)

Fig.8. (a) Sample Image2 (b) Extracted Medicine name and its usage from the database(c) Speech format (d) Waveform of audio file

4 Result and Discussion

When program is compiled, camera will be opened which is used for capturing image. After the image is captured the text from the medicine strip is extracted and is displayed on the console. The special characters detected along with the medicine name are removed to avoid any confusion while comparing the medicine name with the name present in the dataset. Then the required word from the extracted text is compared with the names present in the dataset and then the medicine name along with the required details of the medicine are converted to speech format. Few examples are shown in Fig. 7 which depicts first sample taken as input which is Headset, Fig. 8 which depicts another sample that is Meftal Spas and Fig. 9 which depicts the exceptional case of a medicine named Aciloc where error is detected. The error is due to the presence of 'I' in the word as the letter is read in different format and is unable to match the medicine name in the dataset. We tried various samples of medicine to calculate the accuracy of our model. The accuracy of our model is 95.8%.

The GitHub link is provided for the various test cases of the medicines: https://github.com/Riya934/Final-Year Project-testcases.git

The Figures 7, 8 and 9 show how initially the text from these medicine strips is extracted then the medicine name along with other required details is converted to audio format.

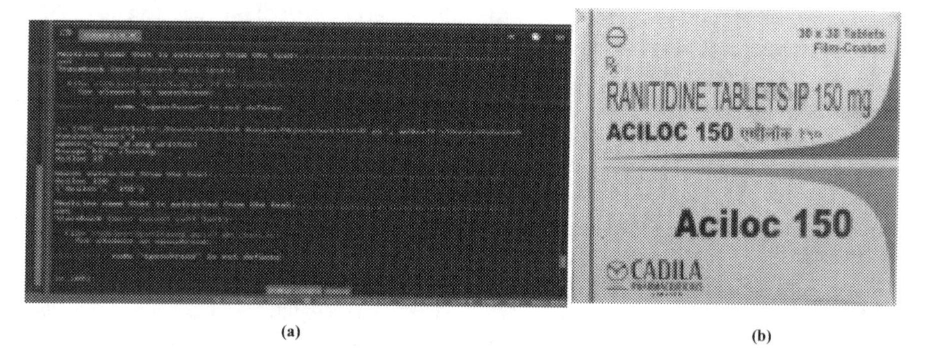

Fig.9. (a) Sample Image2 (b) Extracted Medicine name with error

5 Future Directions

Some of the works that are specific to capturing the high-quality image of the product and extracting the name has several challenges which needs further improvements. In this section the weaknesses and scope for future research are discussed from the outcome of selected papers. Section 5.1 groups the weaknesses under certain categories. Section 5.2 highlights the scope for further research in extracting the name from the medicine.

5.1 Weakness of Existing Method

In this section, existing methods and their weaknesses are highlighted. Initially if the product is not properly placed facing to the camera or there is lack of required light, then the text present in the image is not detected. The high quality image is required for text extraction so the camera must be very good to get the accurate result, or the preprocessing of the image needs to be done. It only recognizes text present on the strip which is in English language and cannot read the other languages.

5.2 Scope for Further Research

The major objective of this work is to extract the name of the medicine accurately. As such, the accuracy of optical character recognition must be increased so that the required name of the medicine can be extracted with high accuracy into machine readable format. Also, the work can be extended to read the recognized text in different language as per the user's native language.

6 Conclusion

In the world, India has the highest number of vision-impaired people and the highest overall rate of vision impairment. *Eye Support* is offered to partially sighted people all over the world to ensure eye support. As a result, our primary goal is to design and build a portable, simple, and low-cost device that will assist visually impaired people in unusual situations. The proposed system is designed with all ages in mind, is user friendly, and does not require any prior knowledge or skills in the modern technologies. The goal of the proposed system is to enable the vision-impaired people to use the application to recognize and understand the use of the medicines that they consume every day at ease

References

1. Singh, H., Sachan, A.: A proposed approach for character recognition using document analysis with OCR. In: 2018 Second International Conference on Intelligent Computing and Control Systems (ICICCS), pp. 190–1952018). https://doi.org/10.1109/ICCONS.2018.8663011
2. Florea, V., Rebedea, T.: Expiry date recognition using deep neural networks. Int. J. User-Syst. Interact. **13**(1), 1–17 (2020). https://doi.org/10.37789/ijusi.2020.13.1.1
3. Dome, S., Sathe, A.P.: Optical charater recognition using tesseract and classification. In: 2021 International Conference on Emerging Smart Computing and Informatics (ESCI), pp. 153–158 (2021).https://doi.org/10.1109/ESCI50559.2021.9397008
4. Sonth, S., Kallimani, J.S.: OCR based facilitator for the visually challenged. In: 2017 International Conference on Electrical, Electronics, Communication, Computer, and Optimization Techniques (ICEECCOT), pp. 1–7 (2017). https://doi.org/10.1109/ICEECCOT.2017.8284628
5. Kornsingha, T., Punyathep, P.: A voice system, reading medicament label for visually impaired people. In: RFID SysTech 2011 7th European Workshop on Smart Objects: Systems, Technologies and Applications, pp. 1–6 (2011)
6. Neto, R., Fonseca, N.: Camera reading for blind people. Procedia Technol. **16**, 1200–1209 (2014)

7. Kalbande, A.: Text to Speech Conversion Using NLP. Fireblaze AI School (2020)
8. IBM Cloud Education, "Machine Learning", IBM Cloud Learn Hub (2020)
9. Vision Atlas, "Country Map & Estimates of Vision Loss", The International Agency for the Prevention of Blindness (IAPB) (2020)
10. https://pyimagesearch.com/2018/08/20/opencv-text-detection-east-text-detector/by Adrian Rosebrock (2018)
11. Khalwadekar, A.; Building OCR and Handwriting Recognition for document images (2020). https://medium.com/@ajinkya.khalwadekar/building-ocr-and-handwriting-recognition-for-document-images-f7630ee95d46
12. Manwatkar, P.M., Yadav, S.H.: Text recognition from images. In: 2015 International Conference on Innovations in Information, Embedded and Communication Systems (ICIIECS), pp. 1–6 (2015). https://doi.org/10.1109/ICIIECS.2015.7193210
13. Jiang, H., Gonnot, T., Yi, W.J., Saniie, J.: Computer vision and text recognition for assisting visually impaired people using android smartphone. In: 2017 IEEE International Conference on Electro Information Technology (EIT), pp. 350–353 (2017). https://doi.org/10.1109/EIT.2017.8053384
14. Komanduri, S., Roopa, Y.M., Bala, M.M.: Novel approach for image text recognition and translation. In: 2019 3rd International Conference on Computing Methodologies and Communication (ICCMC), pp. 596–599 (2019). https://doi.org/10.1109/ICCMC.2019.881978
15. Su, Y.M., Peng, H.W., Huang, K.W., Yang, C.S.: Image processing technology for text recognition. In: 2019 International Conference on Technologies and Applications of Artificial Intelligence (TAAI), pp. 1–5 (2019). https://doi.org/10.1109/TAAI48200.2019.8959

The Evolution of Block Chain Technology

Indumathi Karthikeyan(✉) and M. G. Shruthi

Dr. Ambedkar Institute of Technology, Bangalore, Karnataka, India
shru.deenu83@gmail.com

Abstract. Block chain is the latest trending technology that is used in many fields worldwide. Block chain technology has been implemented in many leading areas which is brought up to date on in this paper. Its applications consist of concepts that can be traced and easily tracked in the form of blocks by using a Blockchain network. These blocks acts like a storage for data to perform transactions which are further chained together by validating and adding the block chaining the previous hash address. The main objective of this paper is to gain deep knowledge about the implementation of Blockchain technology, the areas of application, the working of Blockchain technology by incorporating different algorithms like Proof of Stake, Delegated Proof of Stake, Proof of Authority, Proof of Burn etc. and also highlight about its concepts, types, and the challenges of the Block chain technology. It also gathered many core concepts implemented in the research.

Keywords: Blockchain · Cryptocurrency · Digital eco-system · Decentralized identifiers

1 Introduction

Blockchain technology is the gravitate technology that was into light in the year 2008 but came into existence in the year 2009 as the ledger of distribution of block data. It is applied in extensive applications and implemented in many fields like bitcoin transaction, data in medical field, IoT based projects, etc. Blockchain is the latest trending technology that has been used in many applications as it is not limited only to crypto currency and bitcoin but also implemented in many areas like in business handling, medical field data, transaction locality, etc.

2 Implementation of Blockchain Technology

Sample Heading The most prominent areas of Blockchain technology implemented are.

1. Medical field
2. Music tracking system
3. Cross-border tracking
4. IoT operating system
5. Personal identification security.

S. Aurelia et al. (Eds.): ICCSST 2023, CCIS 1973, pp. 488–497, 2024.
https://doi.org/10.1007/978-3-031-50993-3_39

6. Government schemes & funds tracker
7. Supply chain management
8. Voting system
9. Advertising agency
10. Content creation
11. Cryptocurrency exchanges
12. Real estate system

1. Medical field: Blockchain technology is optimistically used in the medical field by maintaining personal health documentation, medicament, laboratory samples re search, dangerous scrutiny of diseases, etc. Translucency is maintained in the sum mons of prescriptions, from the day of manufacturing till the medicine is readily available in the market. Blockchain Computes the estimation in the treatment of different diseases. It helps to discover the different faults present in the disease and to secure and maintain the confidentiality of patients' data.
2. Music tracking system: To maintain the reliability of the system, Blockchain has been used in enormous aspects of the media process. In the year 2015, the founder named Imogen a Grammy award-winning artist found Mycelia which creates and manages media which is a Blockchain backed digital ecosystem using digital identities. The detailed information about the song would be stored securely using a Blockchain network and increase security by reducing mismatch and missing data via transparency where Blockchain helped to provide creditability. In this way, the music world is incorporated Blockchain.
3. Cross-border tracking: Real-time transaction verification is done for cross-border payments by adopting encrypted distributed ledgers without any intermediatory that is third-party involvement like banks. This technology is adopted in many countries like Korea, Japan, etc. for their bank transactions.

Figure 1 illustrates the working, by creating a block for the transactions when the user creates the transaction, this block has been broadcasted when the network nodes approve the block is added to the chain and finally the transaction is entered in the ledger.

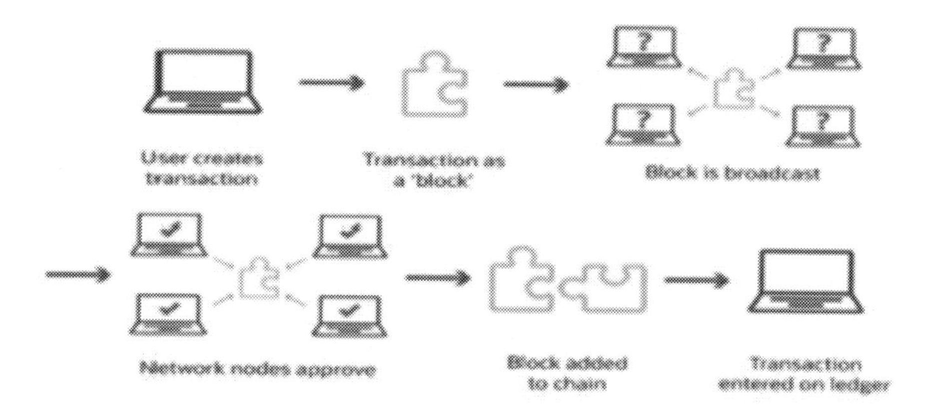

Fig. 1. Transactions using Blockchain Technology

Blockchain technology can also be used in Automating Loan creation, liquidating bills, reducing the cost of the swift transfer, etc.

4. IoT operating system: By using Blockchain Internet of Things (IoT) devices are created to ensure the integrity and security of data for the sensors and other con nected devices. The key consideration is to provide security for an IoT device by maintaining authenticity. IoT is used to circulate data via the internet by private Blockchain networks which creates fortified records of many shared transactions without any centralized oversight.
5. Personal Identification Security: By implementing a Personal Identification Num ber (PIN) the problem of inaccessibility, data insecurity, and fraudulent identities are resolved. Many tools are used to build PIN management systems. The concept of a digital identification framework is been implemented in Decentralized Identi fiers which is known as PIN.
6. Government schemes & funds tracker: By implementing a smart system the funds have been tracked in the state government by using a Blockchain technology en cryp-tion process it provides security for the transactions and provides ease for tracking the process.
7. In the field of supply chain management, product suppliers, logistics, companies, and retailers work together to deliver an end product to the customers. Maintaining the data of all these is a complex and tedious task which leads to a lack of tracking and traceability. To overcome this supply chain management is built using Blockchain mainly to provide peer-to-peer visibility to trace and track the end product delivery process.
8. Voting system: By implementing Blockchain technology voting system is made secure, transparent, immutable, and reliable in the way of recording the vote cast as a transaction ID. The concept of the immutable protecting system is been used. Manipulation or updating of data is not possible as only the insertion of data takes place in Blockchain.
9. Advertising agency: Blockchain provides the advertising team a better insight into data to expose potentially fraudulent data for the purpose of trading. It also assures quality advertising for the purpose of trade and tie-up with advertising buyers.
10. Content Creation: Content creators are playing a big role in the trending world that is by sharing content online and making financial profits out of it. Blockchain will help content creators to gain more control and authority over the work by providing them the ability to trace and track their content effectively. Also, it protects their privacy and plagiarism properties. These creators will upload their work into a decentralized database system where a certain community verifies their work before publishing to ensure copyright then this data will be distributed through nodes.
11. Cryptocurrency exchanges: The medium of exchange which is mainly used to cre ate and store electrically in the Blockchain. The end user request transactions which are then broadcasted to a P2P network through nodes, then the network of nodes is verified using an algorithm then the verified transaction involves crypto currency contracts, records, and other information. Finally, the transaction is com pleted. The new block is been added to the existing Blockchain which is permanent and unaltered, as shown in Fig. 2

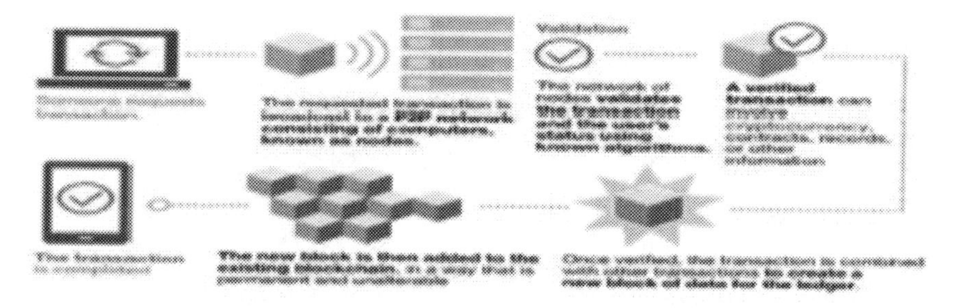

Fig. 2. Cryptocurrency exchange using Blockchain

12. Real-estate system: Strategic ERP's real-estate system is designed to provide a complete solution with a customer-oriented design that provides both web & mo bile-based end-to-end ERP solutions to meet the needs of the Real-estate business. This comprehensive solution covers all crucial functions of a Real-estate business and makes it easier to manage a large-scale Real-estate business.

2.1 The Working of Blockchain Technology

The data that is collected in Blockchain is nothing but the information in a blockchain technology which is stored in the form of cryptographic encrypted data as blocks [10]. The preceding information also holds information about the previous blocks and so on. The addition of any new block will be joined with the existing blocks to mainly form a chain in that network. Next need to implement any of the Blockchain algorithms to create a unique token known as an identifier for that Blockchain.

Applying different Consensus Algorithms can generate single data value for a distrib uted system. The most widely used Consensus Algorithms are Proof of Work(PoW), Proof of Stake(PoS), Delayed Proof of Work (dPoW), Delegated Proof of Stake(DPoS), Proof of Authority(PoA), Proof of Burn(PoB), and Hybrid PoW/PoS algorithm. These algorithms are mainly used for the purpose of authenticity and security in a distributed network.

Proof of Work (PoW)
The PoW is a cryptocurrency consensus mechanism which was introduced in the year 1993. The PoW was popularly used by Bitcoin. PoW consists of Miners and a Mint where Miners act as validators which are used for appending the blocks, they are also known as nodes and the Mint for adding the new Currency. It is also used in Mathematics to solve Puzzles and Guess problems by generating a hash value.

Finally, miners do 'n' a number of iterations and then obtain the solution that leads to a general agreement on Bitcoin usage. Due to its consideration of many iterations, PoW lacks in efficiency in its performance. But considering the authentication, it maintains high security and avoids malicious attacks. It is suitable to implement on Blockchain.

Proof of Stake (PoS)
Proof of Stake (PoS) is a cryptocurrency mechanism which is similar and also consid ered as an alternative to PoW. PoS creates a new block for the data created in the block

chain technology. PoS validation is different from the PoW because it validates the block transactions done based on number of coins. Once the validation is done then only it adds the new block. PoS is bit secured from malicious attacks due to its way of structuring. The next block is chosen randomly. By using the coin owners, the verifica tion of blockchain is done as coin owners are considered as validators. To be a validator, the stake holder who is called a coin owner specifies the block as specific number of coins. These blocks are validated by many to check the accuracy to finalize. PoS syn chronize data to validate the blocks and to create a blockchain transaction. Therefore, block creates are the validators who mainly check the transactions, track the activities to keep a record in an organized manner. By using hash functions, they are recorded as blocks with the coins then they validate the blocks which is finally creates a blockchain. This is an energy efficient secure through controlling the blockchain mechanism which uses less power consumption to validate the blocks in which validators transaction fees is allotted.

Despite these advantages of security, the PoS algorithm also has disadvantages as its expensive to control and always should have honest validators which also consumes more time to implement successfully. The drawback is also that validator's mining ca pacity mainly depends upon the number of tokens, so a validator starts with more coins gets more control over the consensus mechanism.

Delayed Proof of Work (dPoW)
Delayed Proof of Work (dPoW) is a hybrid mechanism as it allows hashing power to be considered of one blockchain data to be added to the other blockchain. The dPoW is mainly used in end-to-end blockchain solutions also the backup is maintained. The spe cific block which is hashed is by adding nodes is checked every ten minutes by taking the snapshots periodically to keep a track of the blocks on the network.

dPoW has an advantage of efficiency while considering energy consumption and secu rity. But it also as disadvantage of limited blocks and less efficiency due to less hash rate in the process of adding new blocks.

Delegated Proof of Stake (DPoS)
Delegated Proof of Stake (DPoS) is similar to the PoS system. Here blocks when added are validated by delegates by vote that is vote is generated by validating the block, only if validated with highest voting are considered. It is more efficient than PoS version. DPos was introduced by Dan Larimer in the year 2013 and also implemented in the Bit Shares Project. The DPoS works as users over a network vote and highest vote block is validated by the delegates which is chosen to add to the Blockchain. Here delegates play a vital role to validate the transaction whether it is accurate or not.

The DPoS has many advantages in terms of Reputation that is delegates are elected in a criteria-based process based on their reliability and reputations users also vote for them, fast access in terms of limited number of delegates usually not more than 100, scalability that is it does not require hardware for hashing and voting. The voting power of each delegate is proportional to the number of coins held. It also contains many drawbacks like malicious attack over a network is more likely to occur in DPoS blockchain as they consist of a smaller number of delegates., it has lower decentralization power due to less number of delegates and blockchain engagement needed actively to remain over a network if it fails the entire vote is desisted.

Proof of Authority (PoA)

The Proof of Authority (PoA) is a consensus algorithm which was introduced by Gavin Wood in the year 2017. It is used for fast transactions which is mainly based on stake identity. The transactions will be validated by the account's which are only approved by the validators. In PoA the authority rights are gained by being the right validators. Furthermore, in PoA, block validators stake their reputation and identities rather than coins by making the system more secure than PoS.

While choosing a validator, the basic criteria about the validators should be chosen based on whether the validator is trustworthy, the validator does not consist of any criminal records and the validator is having good moral standards. These details about the validator are checked over a network over a public domain.

PoA has many advantages in terms of scalability as it consists of few block validators. It consists of Trustworthy validators as every new block added will go through a strict validation before adding the block to the blockchain. Automatic block addition takes place which is an advantage for the validators, where they need not waste their time to check and validate to add a block by this the validators job becomes simple and easy.

The limitation of PoA is the system is highly dependent on validators, therefore to choose the right validator consumes time as we couldn't choose randomly The PoA algorithm is lack in decentralization in comparison to other algorithms.

The PoA is implemented based on the Ethereum network by using a PoA protocol. As the real identities of these moderator nodes are known and trusted blocks, PoA is highly suitable for logistical applications such as supply chain or trade networks. It enables users to avail themselves of all the benefits of blockchain technology which maintain the privacy and secure their transactions.

Proof of Burn (PoB)

Proof of Burn (PoB) is a blockchain consensus algorithm. This algorithm is imple mented by burning the coins to ensure miners reach a consensus. The coin burning is a process which eliminates crypto that effectively takes out the tokens out of the circula tion by reducing the total supply of the coin. The PoB uses virtual mining power as there are more coins there is higher power. The PoB uses a network that the coin burn process involves the sending of the coins to an eater address. This address is protected, and verifiable also public accessibility is denied even though the eater does not have private keys and are randomly accessible.

The PoB is mainly implemented to ensure that the blockchain network allocates all the participant nodes to come to an agreement with the true and valid blocks only. This algorithm consumes less resources which leads to active network. To make sure that the duplication is avoided. This avoids the fraudulent blockchain of the transaction tool. The transaction takes place as follows,

Firstly, a block is sent to be added to a blockchain over a network.

Secondly the block sent is burnt over a network that will own their cryptocurrency coins, during this process the other blocks can also burn over your block and can add the other block to your block transaction also.

Finally, once the burning activity is completed, the participants are rewarded with the activities.

PoB consists of many advantages as PoB promotes periodic burning of the coins which maintains mining power. The PoB which ever implemented can also be customized by updating the blocks or by receiving the blocks. The PoB consumes less energy which leads to sustainability, and it does not require any powerful computation resources or powerful mining hardware's to leads to long term commitment by the miners. Therefore, PoS systems uses virtual mining power. This leads to good performing computation power, committed network which gain the right to mine and validate the transactions.

PoB is associated with many drawbacks like less eco-friendly, as burning of coins con sumes lot of resources. It undergoes many testing processes for efficiency and security. The delay in the process of verification and lastly the burning of the coins is not always transparent to the users.

Hybrid PoW/PoS Algorithm
A hybrid PoW/PoS is a consensus distributed algorithm which is a combination of both proof of stake and proof of work. The security features of PoW and the efficiency features of PoS are being implemented together. When a new block is added to a block chain the PoW uses miners and create that block, once the block is created the PoS votes to whether to add or reject the block. Which results in hybrid consensus mechanism.

The hybrid PoW/PoS consensus is used to balance each other's weaknesses to make maximum utilization of both the PoW and PoS to produce a hybrid consensus mecha nism.

The master node coins which is a hybrid consensus mechanism. It governs the crypto currency that utilizes a vote of PoW and a token vote of PoS consensus. Here instead of examining the total vote count, the hybrid considers both PoW/PoS mechanism ran domly choosing a block. The major block reward is with PoW miners and less block received by PoS miners, and the remaining is dedicated to development effort.

The Hybrid PoW/PoS has many advantages as network security, and it avoids major attacks. It can also rewrite the blockchain by mining an alternative chain block.

3 Technology Concepts and Types in Blockchain

a. The Blockchain distributed ledger network that adds and never deletes or modifies records without common consensus. The data is stored in many computers as a network of computers, and it spreads the Blockchain.
b. Working steps of a Blockchain: Mr. X is going to transfer some amount to Mr. Y, and the transactions are initiated for the transaction which is been intimated to all parties in the network, once the transaction is approved it takes place in the Blockchain environment. Then the new block is been added to the Blockchain ledger and the same is communicated to all the parties present in the network, as shown in Fig. 3

Types of Blockchain: There are 4 different types of Blockchain.

a. Public Blockchain
b. Private Blockchain
c. Hybrid Blockchain
d. Consortium Blockchain
a. Public Blockchain

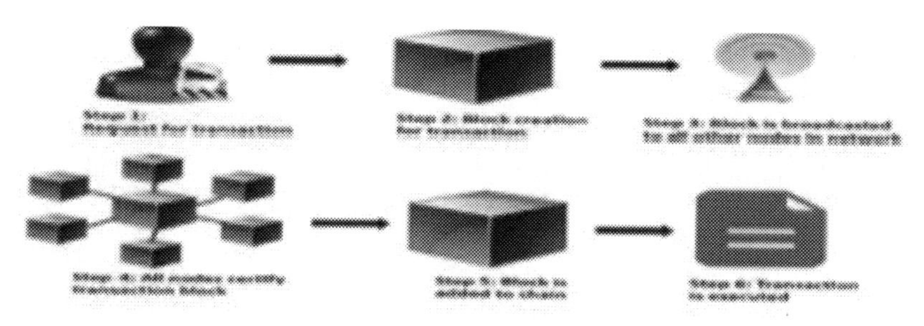

Fig. 3. Blockchain Ledger Communication

The first Blockchain technology where distributed ledger technology was introduced. It is an independent, transparent, and trustworthy Blockchain. It is a decentralized system in which it maintains an identical copy of each transaction without any intermediator involvement. Therefore, each user feels that the entire transaction is only owned by a particular individual which leads to security and transparency in the system.

b. Private Blockchain

It is a Blockchain network that is been restricted in a closed network else it is controlled by a single entity key. The accessibility is controlled and performance increases in private Blockchain.

It also has Point to Point connectivity as per public Blockchain and it's decentralized. In private Blockchain, the participants are tracked from the start by the network holders.

c. Hybrid Blockchain

Hybrid Blockchain: it is a combination of public Blockchain and private Blockchain. With private authenticity, it also provides public by providing authentication of who can access the specified data set, and also a certain amount of data has been made public. It also provides scalability.

d. Consortium Blockchain

It is similar to private Blockchain but it provides limited access to a particular set of groups to eliminate the risk of a single entity controlling the access to a private Blockchain network. In this, the proceedings are controlled by the present nodes. It contains a validating node that initiates, receives, and validates the transactions. It is more secure, scalable, and efficient than the public Blockchain network.

4 Challenges with Blockchain Technology Adoption

Blockchain technology has been creating an era of adoption in the field of business the only concern is with its challenges and what ways these challenges can be avoided.

- Blockchain is an emerging latest technology that lacks due to skill gap. The skills that are required to develop and adopt are less. The only way to face this challenge is by adopting Blockchain as a Service (BaaS).

- Lack of adaptability, Blockchain is an eco-system that requires to be adopted in a broader range in the work environment. In the supply chain, the management track & trace system can adopt a Blockchain network.
- Lack of trust among the Blockchain users is leading to less implementation of technology. But there is a common agreement to overcome this challenge by adopting the consensus algorithm for the entire network.
- Cost for implementation is a key concern as it is not a free source. This can be resolved by knowing the awareness and implementation benefits of technology.
- Blockchain interoperability which is still a doubt in many minds with security, but this includes the ability to share and access information without an intermediate authority by believing its security by getting the most value out of Blockchain investment.

Thus, Blockchain technology is found in many fields in today's world, so implementing Blockchain in several applications is desired. So, step into the era of Blockchain.

References

1. Blockchain technology applications for Industry 4.0: A literature-based review
2. Andrian, H.R., Kurniawan, N.B.: Blockchain Technology and implementation: a systematic literature review. In: 2018 International Conference on Information Technology Systems and Innovation (ICITSI), Bandung, Indonesia, pp. 370–374 (2018). https://doi.org/10.1109/ICITSI.2018.8695939
3. Rajput, S., Singh, A., Khurana, S., Bansal, T., Shreshtha, S.: Blockchain technology and cryptocurrencies. In: 2019 Amity International Conference on Artificial Intelligence (AICAI), Dubai, United Arab Emirates, pp. 909–912 (2019).https://doi.org/10.1109/AICAI.2019.8701371
4. Bansod, S., Ragha, L.: Blockchain technology: applications and research challenges. In: 2020 International Conference for Emerging Technology (INCET) Belgaum, India, pp. 1–6 (2020).https://doi.org/10.1109/INCET49848.2020.9154065
5. Li, W., He, M., Haiquan, S.: An overview of blockchain technology: applications, challenges and future trends. In: 2021 IEEE 11th International Conference on Electronics Information and Emergency Communication (ICEIEC) 2021, Beijing, China, pp. 31–39 (2021). https://doi.org/10.1109/ICEIEC51955.2021.9463842
6. Andrian, H.R., Kurniawan, N.B.: Blockchain technology and implementation : a systematic literature review. In: 2018 International Conference on Information Technology Systems and Innovation (ICITSI), Bandung, Indonesia, pp. 370–374 (2018). https://doi.org/10.1109/ICITSI.2018.8695939
7. Sunny, F.A., et al.: A systematic review of blockchain applications. IEEE Access **10**, 59155–59177 (2022). https://doi.org/10.1109/ACCESS.2022.3179690
8. Zheng, Z., Xie, S., Dai, H., Chen, X., Wang, H.: An overview of blockchain technology: architecture, consensus, and future trends. In: 2017 IEEE International Congress on Big Data (BigData Congress), Honolulu, HI, USA, pp. 557–564 (2017).https://doi.org/10.1109/BigDataCongress.2017.85
9. Chalaemwongwan, N., Kurutach, W.: Notice of violation of IEEE publication principles: state of the art and challenges facing consensus protocols on blockchain. In: 2018 International Conference on Information Networking (ICOIN), Chiang Mai, Thailand, 2018, pp. 957–962. https://doi.org/10.1109/ICOIN.2018.8343266

10. Gupta, C., Mahajan, A.: Evaluation of proof-of-work consensus algorithm for blockchain networks. In: 2020 11th International Conference on Computing, Communication and Networking Technologies (ICCCNT), Kharagpur, India, pp. 1–7 (2020).https://doi.org/10.1109/ICCCNT49239.2020.9225676
11. Deng, X., Li, K., Wang, Z., Li, J., Luo, Z.: A survey of blockchain consensus algorithms. In: 2022 International Conference on Blockchain Technology and Information Security (ICBCTIS), Huaihua City, China, pp. 188–192 (2022). https://doi.org/10.1109/ICBCTIS55569.2022.00050
12. Yassein, M.B., Shatnawi, F., Rawashdeh, S., Mardin, W.: Blockchain technology: characteristics, security and privacy; issues and solutions. In: 2019 IEEE/ACS 16th International Conference on Computer Systems and Applications (AICCSA) Abu Dhabi, United Arab Emirates, pp. 1–8 (2019).https://doi.org/10.1109/AICCSA47632.2019.9035216
13. Saad, S.M.S., Radzi, R.Z.R.M., Othman, S.H.: Comparative analysis of the blockchain consensus algorithm between proof of stake and delegated proof of stake. In: 2021 International Conference on Data Science and Its Applications (ICoDSA), Bandung, Indonesia, pp. 175–180 (2021). https://doi.org/10.1109/ICoDSA53588.2021.9617549
14. Alsunaidi, S.J., Alhaidari, F.A.: A survey of consensus algorithms for blockchain technology. In: 2019 International Conference on Computer and Information Sciences (ICCIS), Sakaka, Saudi Arabia, pp. 1–6 (2019). https://doi.org/10.1109/ICCISci.2019.8716424
15. Hjálmarsson, F.Þ., Hreiðarsson, G. K., Hamdaqa, M., Hjálmtýsson, G.: Blockchain-based E-voting system. In: 2018 IEEE 11th International Conference on Cloud Computing (CLOUD), San Francisco, CA, USA, pp. 983–986 (2018). https://doi.org/10.1109/CLOUD.2018.00151
16. Huang, Y., Zeng, Y., Ye, F., Yang, Y.: Incentive assignment in PoW and PoS hybrid block chain in pervasive edge environments. In: 2020 IEEE/ACM 28th International Symposium on Quality of Service (IWQoS) Hang Zhou, China, pp. 1–10 (2020). https://doi.org/10.1109/IWQoS49365.2020.9212842

Author Index

S. Aurelia et al. (Eds.): ICCSST 2023, CCIS 1973, pp. 499–500, 2024.
https://doi.org/10.1007/978-3-031-50993-3

Printed in the United States
by Baker & Taylor Publisher Services